HOLISTIC DARWINISM

HOLISTIC
DARWINISM

Synergy, Cybernetics, and the Bioeconomics of Evolution

Peter A. Corning

The University of Chicago Press · *Chicago & London*

Peter Corning is director of the Institute for the Study of Complex Systems. He is the author of *The Synergism Hypothesis* and *Nature's Magic*, among other books.

The University of Chicago Press, Chicago 60637
The University of Chicago Press, Ltd., London
© 2005 by The University of Chicago
All rights reserved. Published 2005
Printed in the United States of America

14 13 12 11 10 09 08 07 06 05   1 2 3 4 5

ISBN: 0-226-11613-1 (cloth)
ISBN: 0-226-11616-6 (paper)

Library of Congress Cataloging-in-Publication Data

Corning, Peter A., 1935–
    Holistic Darwinism : synergy, cybernetics, and bioeconomics of evolution /
Peter A. Corning.
        p.   cm.
    Includes bibliographical references (p.   ) and index.
    ISBN 0-226-11613-1 (cloth : alk. paper) ISBN 0-226-11616-6 (pbk. : alk. paper)
    1. Social evolution. 2. Evolutionary economics. 3. Sociobiology. I. Title.

JC336.C848 2005
303.4–dc22   2005011956

⊗ The paper used in this publication meets the minimum requirements of the American
National Standard for Information Sciences—Permanence of Paper for Printed Library
Materials, ANSI Z39.48-1992.

For Susan, proof of synergy

The whole is over and above its parts,
and not just the sum of them all.
—Aristotle, *Metaphysics*

Contents

The New Evolutionary Paradigm

A major paradigm shift is currently underway in evolutionary theory. Neo-Darwinism—the reductionist, mechanistic, gene-centered approach to evolution epitomized by the selfish gene metaphor of Richard Dawkins—has come under assault from various quarters. These attacks have included the following:

- **A growing appreciation for the fact that evolution is a multilevel process, from genes to ecosystems;** coevolution has come to be recognized as a many-faceted phenomenon.
- **A revitalization of group selection theory,** which was banned (too hastily) from evolutionary biology more than thirty years ago.
- **An increasing recognition that symbiosis is an important phenomenon in nature and that symbiogenesis is a major source of innovation in evolution.**
- **A broad array of new, more advanced game theory models,** which support the growing evidence that cooperation is commonplace in nature and not a rare exception.
- **New research and theoretical work that stresses the role of developmental dynamics, "phenotypic plasticity," and organism-environment interactions in evolutionary continuity and change;** an inextricable relationship between nature and nurture are the rule, rather than the exception.
- **A flood of publications on the role of behavior, social learning, and cultural transmission as pacemakers of evolutionary change,** a development that is especially relevant in relation to the evolution of humankind.
- **New insights into the nature of the genome,** and increasing respect for the fact that the genome is neither a "bean bag" (in biologist Ernst Mayr's caricature)

1

nor a gladiatorial arena for competing genes but a complex, interdependent, cooperating system.

- **The emergence of hierarchy theory,** which stresses that the natural world is structured and influenced by hierarchies of various kinds.
- **The rise of systems biology,** a new field that emphasizes the systemic properties of living organisms; one scientist, writing in the journal *Science,* called it "whole-istic biology."
- **The claims advanced by various theorists for the role of autocatalysis, self-organization, network dynamics, and even "laws" of evolution** (though I remain guarded about them).

The crux of the debate, however, has to do with the evolution of complexity. An individualistic, gene-centered theory seems insufficient to account for the evolution of more complex, multileveled biological systems over time. It is increasingly evident that the selfish gene metaphor is inadequate. A more appropriate metaphor is the cooperative gene (see Corning 1996a; Ridley 2001). Thus, a major challenge for evolutionary theory is to develop a better understanding of cooperation and complexity in the natural world. Many theorists are now looking beyond the individualistic, mutation-competition-selection paradigm.

Accordingly, the term *Holistic Darwinism* is not an oxymoron. Nor does it refer to some metaphysical abstraction. It is a candidate name for the new paradigm that is emerging as an alternative to Neo-Darwinism. It also involves a theory about the role of "wholes" in evolution.

First and foremost, Holistic Darwinism views evolution as a dynamic, multilevel process in which there is both "upward causation" (from the genes to the phenotype and higher levels of organization) and "downward causation" (phenotypic influences on differential survival and reproduction), and even "horizontal causation" (between organisms). In this paradigm, the emergence of higher-level "individuals" (in Michael Ghiselin's characterization) are not epiphenomena; they *act* as wholes and exert causal influences as distinct evolutionary units. They may constrain, control, and even differentially determine the fate of various parts (genes). To borrow a metaphor from the biologist Egbert Leigh (1977), selfish genes are subject to the "parliament of the genes." I call it the "selfish genome" model, but the principle applies equally well to selfish groups, selfish colonies, selfish packs, selfish bands, and even selfish nation-states, as we shall see (see also Corning 1997a).

A second major feature of Holistic Darwinism is that it serves as an umbrella for a broad theory of cooperation and complexity in nature first proposed in *The Synergism Hypothesis* (Corning 1983). This theory has recently been gaining support among biologists and social scientists. Both the historical background and more recent developments relating to this theory are detailed in chapter 1.

A critique of Neo-Darwinism and the case for Holistic Darwinism are presented in chapter 2, along with a brief introduction to the Synergism Hypothesis. The hypothesis, in brief, is that synergy—a vaguely familiar term to many of us—has been a wellspring of creativity in the natural world and has played a key role in the evolution of cooperation and complexity at all levels of living systems. This theory asserts that synergy is more than a class of interesting and ubiquitous effects in nature. It has also been a major *causal agency* in evolution; it represents a unifying explanation for complexity at all levels of living systems.

Moreover, this theory is fully consistent with Darwin's theory. It involves only a different perspective on the evolutionary process. In contrast with gene-centered theories, or postulates of self-organization and emergent "laws" of complexity, the Synergism Hypothesis represents, in essence, an economic (or, more precisely, bioeconomic) theory of complexity. It is postulated that the functional payoffs produced by various kinds of synergy have been the drivers of this important evolutionary trend. After a brief survey of the many different kinds of synergy in the natural world (including examples drawn from a variety of disciplines), the case for the Synergism Hypothesis is presented in detail in chapter 3, along with some ways of testing the theory.

A third major feature of Holistic Darwinism is that it fully acknowledges the "teleonomy" (purposiveness) of living systems and incorporates this important aspect of the natural world into the causal dynamics of the evolutionary process itself. This pertains especially to behavior, which has often served as a "pacemaker" of evolutionary change (in Ernst Mayr's famous characterization). Sometimes this is referred to as the "Baldwin effect" (see chapter 2), but credit for the idea that a change in an organism's "habits" can influence the course of evolution should properly be given to Jean Baptiste de Lamarck (even though he guessed wrong about the "mechanism" involved). The role of teleonomy and self-determination in evolution, as well as the relationship between synergy and self-organization, are discussed in chapter 4.

A fourth feature of Holistic Darwinism is that it also encompasses the phenomena associated with emergence—the recently rediscovered nineteenth century term for the idea (traceable back to Aristotle) that wholes may have distinct properties that transcend their parts. The re-emergence of emergence is reviewed in some detail in chapter 5, along with a discussion of how emergence relates to synergy and the Synergism Hypothesis.

Another important aspect of Holistic Darwinism is that it also applies to human evolution and to the evolution of human cultures and their political systems. In fact, synergy played a key causal role in the evolution of humankind, and so did cybernetic (political) processes—decision making, social communications, social control, and feedback. Nor are such processes unique to our species. Analogous cybernetic processes are found in slime molds, leaf-cutter ants, naked mole rats, killer whales, and our closest primate relatives, among others; they are a fundamental feature of social life. The intimate relationship between synergy, cybernetics, and the bioeconomics of sociality is discussed in chapter 6. Devolution, which provides a major opportunity for testing this theory, is defined and explored in chapter 7.

Finally, Holistic Darwinism embraces the recently revitalized "superorganism" concept, which has regained respectability in evolutionary biology after a long, ideologically tainted period in the wilderness. Actually, the so-called organismic analogy has ancient roots. It was first articulated by the classical Greek theorists as a way of characterizing human societies, and it has been utilized by many other political theorists over the past two thousand years. However, the term *superorganism* itself was coined by the nineteenth century polymath Herbert Spencer, who focused especially on the division of labor and the problem of functional integration in complex systems. The history of this concept is briefly reviewed in chapter 8, along with a survey of the many examples found in nature. It is also stressed that cybernetic processes are essential concomitants of superorganisms of all kinds. More important, this broad evolutionary perspective also applies to the ongoing political evolution of human societies, including the prospects for global governance.

Part 2 of this volume comprises three chapters on the subject of bioeconomics—a key element of the paradigm shift identified with Holistic Darwinism. Over the years, much has been made of the relationship between "the economy of nature" (a term of art that Darwin borrowed from Linnaeus) and human economies. More recently, there has also been much cross-fertilization between biology and economics, especially in areas such as behavioral ecology, evolutionary game theory, evolutionary economics, ecological economics, and the like. However, the connection between biology and economics goes beyond analogies, and beyond methodologies. There is also a deeper-level homology, which the new interdiscipline of bioeconomics is helping to illuminate. In a nutshell, the fundamental linkage between biology and economics derives from the fact that humans share with all other living species the fundamental problems of survival and reproduction. This bedrock challenge is multifaceted, ongoing, and inescapable; it can never be permanently solved. Indeed, whether we are aware of it or not, the overwhelming majority of our activities as a species are devoted to various aspects of the survival problem (either directly or indirectly). A human society represents, quintessentially, a "collective survival enterprise." This important metatheoretical issue is addressed in chapter 9 in the course of a review essay on how the new interdiscipline of evolutionary economics should be defined and developed.

Chapter 10, "Bioeconomics as a Subversive Science," elaborates on this theme and mounts a frontal challenge to the basic premises of traditional (neoclassical) economics. In effect, bioeconomics redefines the nature and purpose of a society, and an economy. The "ground-zero premise" (so to speak) of the life sciences is that survival and reproduction represent the basic problem for all living organisms, and this bedrock challenge applies also to human societies. It is the "paradigmatic problem" for all economies, and economic performance must ultimately be judged in these terms, not in relation to economic growth or gross domestic product or other conventional economic measures. Indeed, even traditional income and standard of living measures may be woefully inadequate. Chapter 11, an expansion on an essay that originally appeared in the *Journal of Bioeconomics* (Corning 2000), follows through on this theme. It applies the concept of biological adaptation specifically to human societies and develops a detailed framework of fourteen basic needs that define the parameters of the survival/reproduction problem for humankind. There is also a brief introduction to the "Survival Indicators" program, which represents an effort to deploy an array of concrete measuring rods, or gauges, for adaptation similar to the economic and social indicators that are already widely used by social scientists and policy makers.

Part 3 then addresses the theoretical foundations of evolutionary theory in general—and Holistic Darwinism in particular—at a much deeper level. Two major areas of modern physics, namely thermodynamics and information theory, have made significant incursions into evolutionary theory over the past twenty years or so, and some major claims have been made on behalf of both the so-called entropy law and Claude Shannon's statistical information theory. In chapter 12, based on a paper coauthored by the late Stephen Jay Kline (emeritus professor of engineering at Stanford University and a leading expert on thermodynamics), the pretensions of these theoretical schools are sharply criticized. Major alternatives to these formulations are then developed in chapter 13 (on thermoeconomics) and chapter 14 (on control information). These new formulations are entirely compatible with Darwin's theory and with the teleonomic, cybernetic approach to the role of information in evolution that is described in part 1. The term *thermoeconomics* refers to the use of economic criteria to understand the role of energy in evolution, and *control information* describes a new, cybernetic (functional) kind of information that is measured in terms of the energy that can be controlled in a given context. Examples are provided to illustrate each concept.

Finally, part 4 includes some writings that examine the long-standing and vexed debate over evolutionary ethics—an inescapable aspect of any paradigm shift in evolutionary theory. The history of this debate is briefly summarized in a review essay (chapter 15), and it is argued that evolutionary ethics is indeed an idea whose time has come. Once the narrow, constricted, individualist caricature associated with Neo-Darwinism is replaced by the more balanced, ecumenical, economically oriented paradigm of Holistic Darwinism (not to mention a more balanced view of human nature and the role of cooperation in human evolution), the main theoretical impediment to a robust evolutionary ethics is removed. In chapter 16, this perspective is applied specifically to a critique of a recent volume on the sociobiology of democracy. The authors of that work hold a dour view of democracy's prospects, but I disagree with their analysis.

The ethical implications of Holistic Darwinism are more fully articulated in a final essay entitled "Fair Shares" (chapter 17), which seems especially relevant in light of recent economic trends. The two major twentieth century political ideologies are critiqued and the durable concept of fairness—now a "hot" research area in the social sciences—is advanced as a more balanced alternative. The important advantage of this middle-ground alternative is that it is grounded in an evolutionary and biological framework; it has a strong empirical basis. (This is reflected in the subtitle, "a biological approach to social justice.") Chapter 17 could also have been subtitled "beyond John Rawls," for it does not rely on an artificial philosophical construct; it grounds (and justifies) the concept of fairness in the biological sciences, and it implies certain specific principles for how to apply this criterion in human societies. It reaches the conclusion that we cannot avoid making ethical choices and that many of these have significant consequences for our ultimate survival and reproductive success. As the biologist Garrett Hardin (1972, p. 360) pointed out:

We cannot *predict* history but we can *make* it; and we can *make* evolution. More: we cannot avoid making evolution. Every reform deliberately instituted in the structure of society changes both history and the selective forces that affect evolution—though evolution may be the farthest thing from our minds as reformers. We are not free to avoid producing evolution; we are only free to close our eyes to what we are doing.

One final note: Many of the chapters in this volume previously appeared in peer-reviewed journals during the past few years, though they have been edited and updated in various ways. Others served as the basis for presentations at professional meetings, including the International Society for the Systems Sciences, the New England Complex Systems Institute, the Human Behavior and Evolution Society, the International Society for Human Ethology, the International Society for Endocytobiology, and the Association for Politics and the Life Sciences. The relevant citations and acknowledgments are included at the end of the book. However, it should be stressed that this is not simply a disparate collection of writings on various subjects. Each chapter forms an essential part—a building block—for the theoretical structure that I have called Holistic Darwinism; each chapter is a part of a synergistic whole that, it is hoped, will help to advance the emerging new, post–Neo-Darwinian evolutionary paradigm. Three of the major aspects of this new paradigm, to reiterate, are synergy, cybernetics, and bioeconomics. More important, in this paradigm, selfish genes are the servants, not the masters. To borrow a punch line from a later chapter, many "engines" have been proposed to account for the evolution of complexity, but the engine is nothing without the car. It is time to focus on the car.

PART I

Synergy and Evolution: From the Origins of Life
to Global Governance

— ℰℑ —

There is a tide in the affairs of men, which taken at the flood leads on to fortune; omitted, all the voyage of their life is bound in shallows and in miseries.
—Shakespeare, *Julius Caesar*

SUMMARY: Synergy—as we shall see—is an energizer, a creative wellspring underlying the evolution of complexity, in nature and human societies alike. It constitutes the theoretical core of the new paradigm that I call Holistic Darwinism. Although the title of this chapter ends with a question mark, in fact the time may have come to replace the question mark with an exclamation point.

1

Synergy: Another Idea Whose Time Has Come?

Catching the Flood Tide

Shakespeare's famous metaphor has been borrowed by many modern authors, perhaps because it captures an eternal truth. In the 1930s the historian Arthur Schlesinger (senior) used this image in a widely acclaimed article called "Tides of American Politics" (1939). In the 1960s, the French historian Jacques Pirenne wrote a magisterial volume that was translated and published in English as *The Tides of History* (1962). Political scientist Karl Deutsch also used the metaphor in the title of his classic text *Tides among Nations* (1979).

More recently, a search of the Internet bookseller amazon.com produced a total of 274 current titles that include the word "tides." There are books on corporate tides, the tides of power, tides of migration, tides of change, the tides of reform, China against the tides, NATO and the tides of discontent, the tides of war, the tides of love, and political tides in the Arab world, as well as, of course, many volumes related to ocean tides.

Our everyday lives are also subject to such tidal influences, especially in the business world, in the arts, and in politics. This year's fad is often next year's remainder or closeout sale item. This year's titanic blockbuster movie will be available for rental next year for a pittance. And this year's hot political issue may be ignored by the media next year, even though the underlying problem still exists.

Although we like to think that science is free from such extraneous influences, of course this is not so. Thomas Kuhn, in his celebrated volume *The Structure of Scientific Revolutions* (1962), argued that science is very much influenced by the tidal effects associated with different paradigms. Ideas and theories that fit within or support the currently dominant framework of basic

11

assumptions and theories in a given discipline are more likely to be favorably received. On the other hand, conflicting work, especially if it challenges the dominant paradigm, is often ignored or rejected. Kuhn's specific scenario for scientific revolutions has been much debated. Nevertheless, there seems to be widespread agreement that Kuhn's core idea is valid, even if the dynamics may be somewhat different from his original formulation.

A classic case in point is biologist Barbara McClintock's work on the so-called jumping genes—genetic rearrangements during ontogeny via what are now called transposons (or transposable elements) that can produce variations in the phenotype of an organism (such as the different color patterns in maize). This phenomenon, painstakingly documented by McClintock over twenty years, remained in the shadows until late in her life. The reason was that it contradicted the then-reigning central dogma of molecular biology—namely, that the genome is expressed during ontogenesis in a linear, deterministic fashion (DNA to RNA to proteins). Now, of course, it is recognized that ontogeny is a much more complex process and that a variety of nonlinear, feedback-dependent influences may affect the outcome (see E. F. Keller 1983).

In a similar fashion, the dominant paradigm in the social sciences for the better part of the past century utilized as its core premise the assumption that human behavior and cultural processes are determined (caused) by the socio-cultural environment, and that biological influences are largely irrelevant. According to the widely quoted dictum of Emile Durkheim, one of the founding fathers of sociology, "Every time that a social phenomenon is directly explained by a psychological phenomenon, we may be sure that the explanation is false" (1938, p. 104). Among the many consequences of this dogmatism was a wall of prejudice against any purported facts that conflicted with socio-economic and cultural explanations. Accordingly, Edward O. Wilson's paradigm-shattering textbook, *Sociobiology: The New Synthesis* (1975), was greeted by many mainstream social scientists with great hostility. This is not surprising; Wilson threatened their core assumptions and challenged the hegemony of their explanatory apparatus. (The term *sociobiology* was actually coined by the pioneer biopsychologist John Paul Scott, but Wilson made it famous.)

Now it seems that another, somewhat less contentious tide change is underway, one that is affecting both evolutionary biology and the social sciences. It is a shift that, hopefully, will result in a more balanced, multileveled, interactional perspective on the evolutionary process generally and the ongoing evolution of the human species in particular. Over much of the past twenty-five years, evolutionary theory has been dominated by the "selfish gene" (or Neo-Darwinian) paradigm, so named after biologist Richard Dawkins's famous 1976 book by that title. The selfish gene metaphor epitomizes a reductionist perspective in which atomistic individual competition is viewed as the predominant, if not exclusive, shaping force in evolution. In this view, cooperative phenomena are not only very limited in scope but are reducible to gene self-interest; higher-level cooperative relationships are even considered by some theorists to be epiphenomena that are not causally important in their own right.

Given this predisposition among many evolutionary theorists of the 1980s, a new theory about the role of synergy in evolution—about cooperative effects of various kinds as a causal mechanism in the evolution of complexity—was, in retrospect, launched on a strongly unfavorable tide. The theory was developed in a book-length monograph called *The Synergism Hypothesis: A Theory of Progressive Evolution* (Corning 1983), and it was largely ignored at the time that it was published. Not only did this theory challenge the dominant Neo-Darwinian paradigm, shifting the focus from competition to cooperation (or, better said, to competition via cooperation), but it directed attention away from genes and stressed the functional dynamics of living systems at various levels of organization—that is, the functional effects produced by the phenotypes. As a corollary, this theory also proposed to shift the explanatory focus to the economics of survival and reproduction.

Paradoxically, at the time this theory was first proposed, the concept of synergy was already widely used in biochemistry, physiology, pharmacology, and related disciplines. (A search of a biological database for the year 1988 using the keyword "synerg" identified 613 references, of which 95 percent were related to these hard sciences.) However, in evolutionary theory and the behavioral sciences the concept of synergy was largely ignored during those years—aside from a few eccentric uses by the anthropologist Ruth Benedict, the engineer-inventor Buckminster Fuller, and a handful of others. Of course, the term *synergy* is often used—and misused—in the business world, most notably in relation to corporate mergers and the like.

A Tide Change in Evolutionary Theory

Today there is every indication that the tide has turned. One early sign was the adoption of the synergy concept by biologist John Maynard Smith (1982a, 1983, 1989), who developed a synergistic selection model to characterize the interdependent functional effects that may arise from altruistic cooperation. (Maynard Smith later broadened the concept to accord with a strictly functional interpretation, whether altruistic or not.) The work of political scientist Robert Axelrod and biologist William Hamilton (1981; also see Axelrod 1984) on the evolution of cooperation, which relies on the game theory methodology pioneered by Maynard Smith, was also important.

Another significant contribution was made by biologist Leo Buss in his 1987 book on the evolution of higher levels of organization, which invoked the concept of synergy, albeit in a narrow sense and without much elaboration. The biologically oriented psychologist David Smillie (1993) has also utilized the concept of synergy in relation to social interactions in nature. Biologist David Sloan Wilson and various colleagues have also played a role with their dogged efforts over the past twenty-five years to put the concept of group selection on a new footing (D. S. Wilson 1975, 1980; Wilson and Sober 1994; Wilson and Dugatkin 1997; Sober and Wilson 1998). Although Wilson's paradigm remains gene-centered, he stresses the role of what he calls a "shared fate" among individual cooperators, which implies a functional interdependency.

Especially important, though, is the work of biologist Lynn Margulis on the role of "symbiogenesis" in evolution (particularly in relation to the origin of eukaryotic cells). Now recognized as a major theoretical contribution, this concept has focused attention on an area in which synergistic functional effects have played a key role (see Margulis 1981, 1993; Margulis and Fester 1991; Margulis and Sagan 1995). Indeed, the relatively new discipline of endocytobiology—inspired in part by Margulis's work but centered in Europe—is concerned especially with investigating symbiotic and synergistic phenomena of various kinds at the cellular level.

Perhaps the most significant sign that a favorable tide now exists for the synergy concept is the publication of two books coauthored by John Maynard Smith and Eörs Szathmáry on the evolution of complexity, *The Major Transitions in Evolution* (1995) and *The Origins of Life* (1999), which feature the role of synergy at various levels of biological organization. Maynard Smith came to recognize the universal importance of functional synergy (personal communication), as did Ernst Mayr (personal communication). Nowadays, articles about synergy in evolution are routinely accepted for publication, whereas fifteen years ago they were routinely rejected.

Complexity is also recognized by many theorists these days to be a distinct emergent phenomenon that requires higher-level explanations. In fact, there is a rapidly growing literature in complexity theory—much of it powered by the mathematics of nonlinear dynamical systems theory—that is richly synergistic in character; it is primarily concerned with collective properties and collective effects. To be sure, much (but not all) of the work in complexity theory involves a radically different view of the evolutionary process from the functional, selectionist paradigm within which the Synergism Hypothesis fits. For instance, the biophysicist Stuart Kauffman's work (e.g., 1993, 1995, 2000) is directed toward trying to identify overarching laws of biological order. His metatheoretical premise is that much of the order found in nature is self-organized—"order for free" as he puts it. (Ultimately, I believe that both self-organizing influences and synergistic functional influences will be recognized as important mechanisms in the evolution of complex systems. For more on this issue, see chapter 4.)

Even the concept of progressive evolution—lately denigrated as an outmoded idea (see especially Nitecki 1988; S. J. Gould 1996)—has also been resuscitated. For instance, John Stewart (1997) proposes that progressive evolution, meaning the trend toward the emergence of higher levels of organization, has been catalyzed and sustained by the functional advantages of cooperation and the ability of managers to control cheaters and free riders. Stewart boldly projects this process forward with a futuristic vision of government on a "planetary scale." A similar vision can be found in Robert Wright's *Non Zero* (2000). There is also much work in biology these days on emergence and the evolution of higher-level individuals (e.g., Michod 1997, 1999; Frank 2003; also see Ghiselin 1997). Again, we will explore these matters further in chapters 2 to 5.

So, the question is, will the rising tide lead on to fortune for the concept of synergy? A firm prediction would be risky, of course, but there do seem to be a number of favorable indications. One is the case for it made by Maynard Smith and Szathmáry in their two volumes on major transitions theory. Several of my recent publications also seek to advance the concept (see especially Corning 1995, 1996a, 1998, 2003).

There is also a recognition, only now emerging, that synergistic functional effects are a fundamental aspect of virtually every scientific discipline (see chapter 3). The reason why the universality of this functional principle has not been widely appreciated in the past is that synergy has traveled under many different aliases: emergent effects, cooperativity, symbiosis, a division of labor (or, more precisely, a combination of labor), epistasis, threshold effects, phase transitions, coevolution, heterosis, dynamical attractors, holistic effects, mutualism, complementarity—even interactions and cooperation.

Finally, there are currently several convergent theoretical developments that focus in various ways on synergistic phenomena, even though they may not employ the term *synergy* explicitly. These developments include, among others, (1) *network theory and network dynamics* (see, for example, Barabási 2002; Buchanan 2002; Strohman 2002; Fewell 2003; Strogatz 2003); (2) *niche construction theory* (Laland et al. 2000; Odling-Smee et al. 2003); (3) *emergence theory* (J. Goldstein 2002; S. Johnson 2001; Morowitz 2002); (4) *evolutionary developmental systems theory* or evo-devo (Rollo 1995; Oyama 2000; Pigliucci 2001; W. Arthur 2002; West-Eberhard 2003); (5) *systems biology* (Kitano 2001, 2002; Chong and Ray 2002; Csete and Doyle 2002); and (6) *gene-culture co-evolution theory* (Cavalli-Sforza and Feldman 1981; Boyd and Richerson 1985; Durham 1991; Thompson 1994; Weingart et al. 1997; P. R. Ehrlich 2000; Hammerstein 2003; Richerson and Boyd 2004).

Of course, it is one thing to recognize synergy as a ubiquitous phenomenon. It is another thing to assign to it a major causal/explanatory role in various domains, particularly biological evolution, human evolution, and the evolution of complex societies. This is what the Synergism Hypothesis encompasses, and the case for this theory, along with an argument for using synergy as a unifying concept for cooperative effects of all kinds in various scientific disciplines, will be presented in chapters 2 and 3.

— ☙ —

Often the most important contribution a scientist can make is to discover a new way of seeing old theories or facts.

—Richard Dawkins

The power and majesty of nature in all its aspects is lost on one who contemplates it merely in the detail of its parts and not as a whole.

—Pliny the Elder

SUMMARY: "Holistic Darwinism" is a candidate name for a post–Neo-Darwinian evolutionary paradigm. When two functionally linked genes are selected together, or when two symbionts (say a ruminant and its gut bacteria) are jointly favored, or when a group of communally nesting female wasps reproduce in greater abundance, the unit of differential survival and reproduction (in functional terms) is the whole—the combined (synergistic) effects produced by the cooperating parts. Holistic Darwinism is not a different theory; it involves a different perspective on the evolutionary process. To borrow Richard Dawkins's image, it is an alternative way of viewing the theoretical Necker cube. Holistic Darwinism is distinctive in that it is concerned especially with the bioeconomics—the functional costs and benefits—of cooperative phenomena of all kinds. It does not contradict the Neo-Darwinian assumption of gene self-interest but highlights the paradoxical interdependence of genes and their "vessels." Indeed, it is argued that the units of replication (genes, genomes, gene pools) and their genetic relationships are less important as determinants of cooperative phenomena than are the functional properties and survival consequences of cooperation, as the data on such interactions clearly suggest. (Maynard Smith has termed it "synergistic selection.") Many hypotheses have been advanced to explain the evolution of complexity—an undisputed historical trend if not a "law" as some theorists have claimed. Holistic Darwinism focuses on the causal role of functional synergy.

2

Holistic Darwinism: Synergistic Selection and the Evolutionary Process

Introduction: The Perils of Group Selection

The emotionally charged group selection debate in biology—which will celebrate an unofficial thirty-ninth anniversary in 2005—provides a classic example of a controversy based largely on a misconception. To Darwin and many of his contemporaries, group selection was a perfectly respectable concept. Indeed, it was Darwin who first proposed, in 1871 in *The Descent of Man*, the then-unexceptional idea that differential group selection may have played an important role in human evolution, along with what he called family selection (now known as inclusive fitness or kin selection theory) and individual reciprocities (now variously called mutualism and reciprocal altruism). Darwin's tripartite explanation of human evolution was quite subtle, but his view of the role played by group selection is illuminated in this brief passage: "All that we know about savages, or may infer from their traditions and old monuments, the history of which is quite forgotten by the present inhabitants, show that from the remotest times successful tribes have supplanted other tribes" (1874 p. 147). Herbert Spencer, one of the outstanding theorists of the nineteenth century, expressed a similar view in *The Principles of Sociology* (1897), and many of the pioneer anthropologists of that period also seem to have agreed.

In the first half of the twentieth century, the founding fathers of modern genetics and population biology, notably including Haldane, Wright, Fisher, Morgan, and Dobzhansky (plus some non-geneticists such as Huxley, Mayr, and Simpson) redefined evolutionary theory in quantitative genetic terms. However, the so-called modern synthesis was also deemed to be compatible with group selection of various kinds. For instance, Sewall Wright (1968–78) at the University of Chicago coined the term *interdemic selection*—i.e., selection between discrete breeding groups, or demes—and developed what he called a shifting balance model, which he believed was of the utmost importance in producing evolutionary changes. Ernst Mayr, likewise, spoke of evolutionary change as a population-level phenomenon, meaning that populations and species are the ultimate units of evolutionary change, not individuals. Mayr also developed what he called the founder principle, which envisions small, reproductively isolated groups as a significant source of evolutionary innovation (Mayr 1963, 1976). More recently, the paleontologist Niles Eldredge (1995) and the paleonologist-popularizer Stephen Jay Gould (2002) have championed a higher-level species selection paradigm. Meanwhile, various students of animal behavior, such as William Morton Wheeler and Warder C. Allee, stressed the cooperative aspect of animal behavior and social life. Wheeler (1927) also promoted the idea of emergent evolution, and he borrowed from Spencer the idea that a socially organized group can be likened to a superorganism (Wheeler 1928; see chapter 8).

However, a theoretical punctuated equilibrium occurred in 1962. In his subsequently much-maligned book *Animal Dispersion in Relation to Social Behaviour,* Vero C. Wynne-Edwards (1962) made himself a stalking horse, in Edward O. Wilson's characterization, by propounding a seriously over-stated version of the group selection hypothesis. Wynne-Edwards asserted that group-living animals regularly display behaviors that involve the curtailment of their own personal fitness for the good of the group (for example, through conventional controls on personal reproduction that serve to limit population densities). "The great benefit of sociality," he claimed in a companion article in *Nature* (1963), "arises from its capacity to override the advantage of individual members in the interest of the survival of the group as a whole." Some of Wynne-Edwards's critics, playing loose with the facts, accused him of a Pollyanna-like naivete that violated Darwinian theory, but in fact he clearly stated that altruistic, group-serving behaviors could arise only if natural selection were to operate between social groups "as evolutionary units." Notwithstanding, Wynne-Edwards became a pariah in evolutionary biology and has been routinely chastised for his heresy ever since—rather like the treatment accorded to Lamarck.

Although the assault on group selection theory began with William D. Hamilton's now-classic papers on "The Genetical Evolution of Social Behavior" (1964a, 1964b), it was fully elaborated in George C. Williams's New Testament, *Adaptation and Natural Selection* (1966). Williams's near-legendary book was in many respects a therapeutic cold bath that served to purge evolutionary theory of some sloppy thinking. However, Williams also took an extreme position, from which he has since retreated, to the effect that selection at any level higher than that of an individual is essentially "impotent" and is "not an appreciable factor in evolution" (1966, p. 8; cf. Williams 1992).

Edward O. Wilson was more moderate by comparison in his discipline-defining volume, *Sociobiology* (1975), but he also (inadvertently) propagated a conceptual muddle that caused much confusion and inadvertent mischief in evolutionary theory.[1] Wilson launched his massive synthesis with the startling assertion that altruism is "the central theoretical problem of sociobiology" (p. 3). The implication, which guided much subsequent work in this new interdiscipline, was that social life is founded on altruism. Therefore, cooperative behaviors are inherently a theoretical problem that can be overcome only under extraordinary circumstances—such as via group selection, kin selection, and maybe Robert Trivers's (1971) "reciprocal altruism." In opposition to Wynne-Edwards, Wilson considered "pure" group selection—that is, among non-kin—to be highly improbable, a rare occurrence confined to humans and perhaps a few other species. His detailed, chapter-length discussion of group selection included a review both of the available evidence and of various formal models, but his conclusion was preordained by the assumption that "pure" group selection necessarily implied genetic altruism (E. O. Wilson 1975, pp. 106–29).

Another broadside against group selection theory came in 1976 in the form of Richard Dawkins's ideologically tinged popularization with the cunningly anthropomorphic title *The Selfish Gene.* "I think 'nature red in tooth and claw' sums up our modern understanding of natural selection admirably," Dawkins wrote with evident relish (1989, p. 2). Not surprisingly, *The Selfish Gene* became a controversial best seller. In retrospect, the selfish gene metaphor has proven to be a powerful heuristic tool. It has led to many new insights about the interactions within and among various functional units in nature and to much productive research. On the other hand, it also introduced a simplistic and seriously distorting perspective into evolutionary theory.

The short-term consequence of this rancorous theoretical debate was a wholesale rejection of the concept of group selection. Nevertheless, as noted earlier, for the past twenty-five years or so David Sloan Wilson (lately with the collaboration of Elliott Sober and with parallel efforts from a growing number of other workers) has been attempting to resurrect group selection on a new foundation. What Wilson calls "trait group selection" (D. S. Wilson

1975, 1980; Wilson and Sober 1989, 1994; Sober and Wilson 1998) refers to a model in which there may be linkages (a "shared fate" in Wilson's term) between two or more individuals (genotypes) in a randomly breeding population, such that the linkage between the two becomes a unit of differential survival and reproduction. Initially, Wilson assumed that one of the two was an altruist, for he was then intent on accounting for the evolution of altruism without recourse to kin selection. As noted earlier, John Maynard Smith developed a similar model, which he dubbed "synergistic selection." (See also Matessi and Jayakar 1976; Wade 1977, 1985; and the discussion in Dugatkin et al. 1992.)

The current revival of group selection theory may perhaps be attributed, in considerable measure, to the growing recognition that it can also entail win-win processes. Cooperating groups might provide mutual advantages for their members, so that the net benefits to all participants outweigh the costs. In other words, cooperation is not equivalent to altruism and does not by definition require sacrifices, or genes for altruism. I refer to it as "egoistic cooperation," to distinguish it from altruism. This, in essence, is what game theory models of cooperation (such as the classic prisoner's dilemma game) tacitly postulate (see Maynard Smith 1982b, 1984, 1989; Axelrod and Hamilton 1981), which is why game theory formulations are largely indifferent to the degree of relatedness, if any, between the cooperators. And game theory models of cooperation (along with experimental research on the subject) have been growing exponentially in number over the past decade or so (see especially Sigmund 1993; Binmore 1994a, 1994b; Gintis 2000a; Stephens et al. 2002; Sigmund et al. 2002; Bowles et al. 2003). (More on this below.)

Moreover, game theory provides a window into a vastly larger galaxy of cooperative phenomena that, I submit, reduces the group selection controversy to a tempest in a teapot. This alternative formulation was originally developed in *The Synergism Hypothesis: A Theory of Progressive Evolution* (Corning 1983) and will be summarized in chapter 3. It was also developed independently by Maynard Smith and Szathmáry (1995, 1999), and it is supported by an accumulating body of research findings across many different specialized disciplines, from molecular biology and microbiology to behavioral ecology, primatology, and sociobiology—not to mention the social sciences. This alternative paradigm might be characterized as Holistic Darwinism. (See also Dugatkin and Reeve 1994, and Dugatkin and Mesterton-Gibbons 1996, on indirect "by-product mutualism" in evolution; and D. S. Wilson and Dugatkin 1997, on the role of "assortative interactions," or behavioral selection, as a mechanism of group selection.)

Holistic Darwinism Defined

Holistic Darwinism—to repeat—is not an oxymoron. The term was coined as a way of highlighting the paradox that selfish genes are, without exception, selected in the context of their functional consequences (if any) for various wholes. Holistic Darwinism is strictly Darwinian in its underlying assumptions about natural selection and the evolutionary process. It has no fundamental quarrel with the theoretical premise of gene selfishness. Rather, it involves a different perspective on the causal dynamics of evolution. In his preface to the second edition of *The Selfish Gene*, Dawkins uses the metaphor of a Necker cube—a two-dimensional drawing of a three-dimensional object that can be perceived in different ways—to characterize the intent behind his inspired metaphor: "My point was that there are two ways of looking at natural selection, the gene's angle and that of the individual. . . . It is a different way of seeing, not a different theory" (1989, pp. x–xi).

Actually, there are more than two ways of looking at natural selection, and Holistic Darwinism focuses not on genes, or individuals, or even groups as units of selection but on the functional relationships among the units at various levels of biological organization, from genomes to ecosystems, and on their consequences for differential survival and reproduction. It involves refocusing the Necker cube on the interactions between genes, between cells, between organisms, and between organisms and their environments. Perforce, Holistic Darwinism is also about the role of synergy—the combined effects produced by phenomena that cooperate (operate together)—as a major cause of evolutionary continuity and change.

A word is in order here about my use of the word *holism*. To some the term may seem problematic—perhaps idealistic or vaguely metaphysical. To the contrary, my intent is to characterize a theoretical shift of focus to the combined (synergistic) effects produced by various combinations of parts (wholes) at various levels of biological organization. In contrast with such early holists as Jan Smuts (1926), who evidently coined the term *holism* and who discerned an inherent "driving force" in nature toward the emergence of wholes, Holistic Darwinism represents a functional, Darwinian approach. Living wholes are contingent products of evolution, and of natural selection.

It should also be stressed that the term *cooperation* will be used here in a strictly functional sense; it refers to functional interactions. In this conceptualization, cooperation may or may not also be considered selfish or altruistic, mutualistic or parasitic, positive or negative. Such attributes involve additional, post hoc judgments about the consequences of a cooperative relationship with respect to some separately specified goal or value. (Of course, in Darwinian theory the operative value is survival and reproductive success.) By the same token, a cooperative relationship may or may not be voluntary. Slavery, in nature and in human societies alike, involves a form of involuntary cooperation, and so (presumably) does the host's role in a parasitic relationship.

Accordingly, a key point about cooperation as a functional concept is that it is found at every level of living systems. Beginning with the very origins of life, it is a common denominator in all of the various formal hypotheses about the earliest steps in the evolutionary process (reviewed in Corning 1996a). All share the common assumption that cooperative interactions among various component parts played a central role in catalyzing living systems.

Similarly, at the level of the genome, it goes without saying that genes do not act alone, even when major single-gene effects are involved. Indeed, the human genome sequencing project has established, among many other things, that there are in fact 1,195 distinct genes associated with the human heart, 2,164 with white blood cells, and 3,195 with the human brain (see Little 1995). The functional (morphogenetic) implications behind those numbers are awesome to contemplate. As Richard Dawkins himself so eloquently put it in a later book, *The Blind Watchmaker*:

> In a sense, the whole process of embryonic development can be looked upon as a cooperative venture, jointly run by thousands of genes together. Embryos are put together by all the working genes in the developing organism, in collaboration with one another. . . . We have a picture of teams of genes all evolving toward cooperative solutions to problems. . . . It is the "team" that evolves. (1987, pp. 170, 171)

The origin of chromosomes, likewise, may have involved a cooperative/symbiotic process (see Maynard Smith and Szathmáry 1993). Sexual reproduction, one of the major outstanding puzzles in evolutionary theory, is also a cooperative phenomenon, as the term is used here. Although there is still great uncertainty about the precise nature of the benefits, it is assumed that sexual reproduction is, by and large, a mutually beneficial joint venture.

As one moves upward in "the great chain of being" (to borrow a durable anachronism), one finds further variations on the theme of functional cooperation. Once upon a time bacteria were considered to be mostly loners, but no longer. It is now recognized that large-scale, sophisticated cooperative efforts—complete with a division of labor—are commonplace among bacteria and can be traced back at least to the origin of the so-called stromatolites (rocky mineral deposits) that, it is believed, were first constructed by bacterial colonies some 3.5 billion years ago (J. A. Shapiro 1988; J. A. Shapiro and Dworkin 1997; Margulis 1993; Bloom 1997). Shapiro suggests that bacterial colonies can be likened to multicellular organisms.

Eukaryotic cells can also be characterized as cooperative ventures—obligate federations that may have originated as symbiotic unions (parasitic, predatory, or perhaps mutualistic) between ancient prokaryote hosts and that have now become the cytoplasmic organelles, particularly the mitochondria, the chloroplasts and, possibly, eukaryotic undulipodia (cilia) and certain internal structures that may have evolved from structurally similar spirochete ancestors (Margulis 1993). The phenomenon of symbiosis, by definition a category of cooperative relationships in nature, provides yet another example. Not only has the darker side of symbiosis—parasitism—gained new prominence over the past decade or so but more benign commensalistic and mutualistic forms of symbiosis are also more widely appreciated (see below).

Sociobiology is also, by definition, concerned with cooperative relationships among conspecifics, interactions that can provide a variety of adaptive consequences for the participants. As shown by the many field studies and laboratory experiments that were inspired by inclusive fitness theory and game theory, the social interactions that occur in nature among members of the same species may be perturbed by "free riders," defectors, exploiters, conspecific parasites, and so on, yet the fact remains that within-species cooperative behaviors are fairly common and encompass a broad array of survival-related functions, including (1) hunting and foraging collaboratively, which may serve to increase capture efficiency, the size of the prey that can be pursued, or the likelihood of finding food patches; (2) joint detection and avoidance of, and defense against, predators, using behaviors that range from mobbing and other kinds of coordinated attacks to flocking, herding, communal nesting, and synchronized reproduction; (3) shared protection of jointly acquired food caches, notably among many insects and some birds; (4) cooperative movement and migration, including the use of formations that increase aerodynamic or hydrodynamic efficiency, reduce individual energy costs, and/or facilitate navigation; (5) cooperation in reproduction, which can include joint nest building, joint feeding, and joint protection of the young; and (6) shared environmental conditioning.

Neo-Darwinian theory—as purified by the selfish gene perspective—attributes evolutionary change to competition among the replicators—the ultimate units of information transfer in evolution. In the classical Neo-Darwinian model, cooperation plays a decidedly subsidiary role. But if we shift our perspective and view evolution as an ecological and economic process—a survival enterprise in which living systems and their replicators are embedded—then differential reproductive success may be viewed as the result of a complex interplay of competitive and cooperative interactions (along with a variety of other factors), both within and among functionally interdependent units of ecological interaction. Our focus shifts to the activities of the "vehicles" (in Richard Dawkins's terminology) or the "interactors" (in the terminology of David Hull 1980)—and, more important, to the bioeconomic consequences of their functional interactions. (For a classic paper on this subject, see Paine 1966.)

It has been a cardinal assumption of Neo-Darwinism that cooperation in nature is a phenomenon that is at odds with the basic principle of gene competition, and that extraordinary conditions are required to overcome the inherent selective bias against the evolution of cooperation. This assumption is what accounts for the importance attached to inclusive fitness theory (or "kin selection," in Maynard Smith's usage) and to game theory. However, a functional/bioeconomic perspective on the evolutionary process challenges that point of view. Not only is cooperation (broadly defined) fairly common in nature, but synergistic effects (the functional consequences of cooperation), it is argued, have played an important causal role in evolution, especially in relation to the evolution of complexity. To put it baldly, functional synergy explains the evolution of cooperation in nature, not the other way around. In other words, functional groups (in the sense of functionally integrated teams of cooperators of various kinds) have been important units of evolutionary change at all levels of biological organization; functional group selection is thus a ubiquitous aspect of the evolutionary process.[2] This is obviously a highly contentious assertion. In the next section, I will briefly summarize the evidence.

Consider the Evidence

If cooperation in nature is not largely dependent on inclusive fitness, we would expect to find a significant degree of decoupling in the natural world between genetic relatedness and cooperation, and, in fact, there are at least four sources of evidence for this proposition. First, there is the entire domain of symbioses. Here we can observe a wide range of cooperative relationships that can only be accounted for in bioeconomic, cost-benefit terms. Kinship is largely irrelevant. Indeed, many types of symbioses—such as the estimated 20,000 species of lichen partnerships involving approximately 300 different genera of fungi and algae, or the rhizobia and similar bacteria that form root nodules with some 17,500 species in 600 genera of plants—reflect a plethora of independent inventions. In other words, many different species may discover and utilize the same functionally advantageous cooperative relationships. As Maynard Smith (1989) has noted, extreme nonspecificity is the rule among mutualists, whereas parasitism is highly specific. The case for symbiogenesis as a significant factor in evolution was documented by participants at a 1989 conference on the subject and in a subsequent volume edited by Margulis and Fester (1991). (Symbiogenesis will be discussed in more detail in chapters 3 and 4.) The following is some of the extensive evidence that was presented at the conference:

- Mutualistic or commensalistic associations (not to mention parasitism) exist in all five kingdoms of organisms, as defined by Whittaker and modified by Margulis and Schwartz (1982). Most extant species may, in fact, be either a product of or currently involved in (or both) endo- or ecto-symbioses. Elsewhere, Bermudes and Margulis (1987) documented that twenty-seven of seventy-five phyla in the four eukaryotic kingdoms (or 37 percent) exhibit symbiotic relationships.

- Over 90 percent of all modern land plants establish mutually beneficial associations with the mycorrhizal fungi that are ubiquitous in fertile soils (Lewis 1991), and Silurian and Devonian plant fossils have been found to contain structures closely resembling the symbiotic vesicles produced by modern mycorrhiza (D. C. Smith and Douglas 1987).

- Land plants may have arisen through a merger between fungal and algal genomes, as sort of inside-out lichens. In any case, it is evident that modern land plants represent a joint venture between fungi and green algae (Pirozynski and Malloch 1975; Atsatt 1988, 1991).

- Approximately one third of all known fungi are involved in mutualistic symbioses (e.g., lichens), many of which have conferred on their partnerships the ability to colonize environments that would not otherwise have been accessible to each one alone (Kendrick 1991).

- Virtually all species of ruminants, including some two thousand termites, ten thousand wood-boring beetles, and two hundred Artiodactyla (deer, camels, antelope, etc.), are dependent upon endoparasitic bacteria, protoctists, or fungi for the breakdown of plant cellulose into usable cellulases (P. W. Price 1991).

- Within the teeming communities of organisms that have recently been discovered in proximity to various sea floor hydrothermal vents, there are a number of symbiotic partnerships between chemoautotrophic (sulfur-oxidizing) bacteria and various invertebrates, which rely on the bacteria for their carbon and energy requirements (Vetter 1991).

•   Most bacterial cells congregate and reproduce in large, mixed colonies with many endosymbionts (virus-like plasmids and prophages) and ectosymbionts (metabolically complementary bacterial strains). These congregations call into question the classical notion of a species, in the sense of competitive exclusion and reproductive isolation (Sonea 1991; also J. A. Shapiro 1988; J. A. Shapiro and Dworkin 1997).

A second body of supporting evidence can be found in the various game theoretic models of cooperation between *unrelated* individuals, along with the substantial research literature that these models have inspired. (These will be discussed further below.) Third, there is the entire category of outbreeding reproduction, a class of cooperative behaviors that, by definition, falls outside of the inclusive fitness model. Finally, over the past decade or so there have been many field and laboratory studies of cooperation among conspecifics that are inconsistent with inclusive fitness theory and suggest that the particular behaviors in question are more satisfactorily explained in bioeconomic terms, although cooperation remains more likely to occur in closely related, or at least familiar, animals.

A detailed summary of this discordant evidence (including twenty-eight recent field and laboratory studies and seven reviews of the older literature) can be found in Corning (1996a); see also the careful analysis by Goodnight and Stevens (1997). One particularly well-documented illustration is the food-sharing behavior among vampire bats (*Desmodus rotundus*), which clearly demonstrates the power of functional/bioeconomic factors to transcend the influence of genetic relatedness in shaping cooperative behaviors (G. S. Wilkinson 1984, 1988, 1990). If gene competition were of overriding importance, the sharing of blood among vampire bats (their exclusive food source) would be confined to close relatives. The reason is that blood sharing in this species has very high fitness value; an individual bat that fails to feed for two nights in a row will die. In field studies as well as in controlled observations in captive groups over a ten-year period, Wilkinson found that blood sharing both between relatives (matrilines) and nonrelatives was extensive. Both relatedness and prior association proved to be important facilitators. Moreover, quantitative cost-benefit analyses showed that the cost to donors was relatively low (in effect, they were sharing their surpluses), whereas the fitness benefits to recipients was relatively high. When this was combined with the fact that the donors' generosity was

usually reciprocated later (i.e., reciprocal altruism *sensu* Trivers 1971), there
was a significant increase in the mutualists' joint fitness. Wilkinson concludes, "Reciprocity is likely to be more beneficial than kin selection—provided that cheaters can be detected and excluded from the system" (1990,
p. 82). (For a more recent example of non-kin cooperation, in red-winged
blackbirds, see Olendorf and Getty 2004.)

Two themes stand out in the many other examples that are described in
Corning (1996a, 2003): (1) the importance of bioeconomic cost benefit
considerations in cooperative relationships and (2) the presence of synergy—combined functional effects (payoffs) that are jointly produced and
provide benefits to the cooperators that are greater than would otherwise be
possible. As Maynard Smith and Szathmáry put it in *The Major Transitions
in Evolution* (1995), if an individual can produce two offspring on its own
but by cooperating in a group consisting of n individuals can produce $3n$
offspring, it pays to cooperate. (An application of this perspective to avian
species can be found in Emlen 1996.)

Game Theory Revisited

Game theory models of cooperation, viewed in the proper light, are also
consistent with Holistic Darwinism. Game theory suggests that the evolution of cooperative behaviors depends on an appropriate set of strategic circumstances. Although the focus has always been on the behavioral context
and the strategies of the players, if one looks closely at the various game theory formalizations they tacitly depend on an interaction between the behavior of the players and the structure of the payoff matrix. And if one looks
closely at the payoff matrices in some of the classic formulations, such as tit-
for-tat, the cooperative strategies in turn depend on synergy. In Axelrod and
Hamilton's (1981) model, mutual defection yielded one point each, asymmetrical cooperation (parasitism?) yielded five points for the defector and
none for the cooperator, and mutual cooperation yielded a total of six
points, evenly divided. Furthermore, defectors would be penalized in subsequent rounds (it was conceived as an iterated game) so that mutual cooperation becomes an increasingly rewarding option over time. In effect, this
amounts to a quantification of synergy; the implicit economic benefits of
the game are critically important.

But what about the problem of cheating or defection (the prisoner's
dilemma)? Maynard Smith and Szathmáry (1995) have proposed a response
in terms of game theory, as illustrated in the two diagrams in figure 1. (I have
taken the liberty of revising the payoff values that were utilized by Maynard

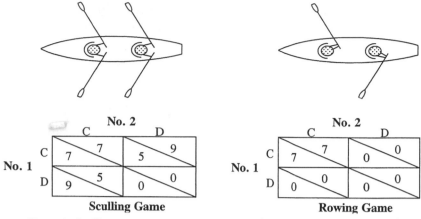

Figure 1. Sculling game versus rowing game. The matrices show the payoff values for each oarsman if he cooperates (C) or defects (D).

Smith and Szathmáry to accord with a more explicit assumption about the object of the game, namely, that the oarsmen are both seeking to cross a river.) The left-hand diagram involves a sculling model, in which two oarsmen each have a pair of oars and row in tandem. In this situation, it is easy for one oarsman to slack off and let the other one do the heavy work. This corresponds to the classical two-person game. However, in a two-person rowing model, as depicted in the right-hand diagram, each oarsman has only one opposing oar. Now their relationship to the performance of the boat is interdependent. If one oarsman slacks off, the boat will go in circles. In this case, mutual cooperation becomes an evolutionarily stable strategy and defection is totally unrewarding; in the absence of teamwork, the boat will not reach its goal.

Maynard Smith and Szathmáry conclude that the rowing model is a better representation of how cooperation evolves in nature: "The intellectual fascination of the Prisoner's Dilemma game may have led us to overestimate its evolutionary importance" (1995, p. 261). Indeed, as Peck (1993, p. 195) observed, "The position of [stable] equilibria (and hence the frequency of cooperators) depends on the size of the various payoffs that define the Prisoner's Dilemma game." (See also Dugatkin et al. 1992; Brembs 1996.)

An Evolutionary Theory of Government

If many forms of cooperation are functionally interdependent and thus self-policing, many more are not. The problems of cheating, defection, and free riders—phenomena that the selfish gene metaphor has helped to illuminate—are real. But, in retrospect, the problem may have loomed much larger in theory than it does in fact; our models may have been too pessimistic about the constraints on errant behavior in cooperative relationships. In effect, the games may have been unintentionally rigged. Consider some of the common assumptions in classical two-person games: The games are always voluntary and democratic; each player is free to choose his or her own preferred strategy, and the opposing player has no means available for coercing choices or compliance. Also, the players are not allowed to communicate with one another in an effort to reduce the uncertainties in the interactions. Furthermore, defectors are usually rewarded handsomely for cheating while the cooperators are denied the power to prevent defectors from enjoying the rewards, much less to punish them for defection. Such grade inflation for defection biases the game in favor of cheating. Worse yet, in iterative games the players are forced to continue playing; they cannot exclude or ostracize a defector. They can only retaliate by themselves defecting and hoping thereby to penalize the other player (see also Binmore 2004).

A tacit rebuttal to this formulation was incorporated into a new kind of prisoner's dilemma model developed by Nowak and Sigmund (1993) called Pavlov, which the authors suggested can outperform tit-for-tat. They called their strategy "win-stay, lose-shift," and the significance of this innovation is that, in contrast with an iterated game in which the players must continue playing regardless of the outcome, in Pavlov they have the choice of leaving the game if they don't like the results. In other words, a player may also have the power to exercise some control over the behavior of a defector by denying to that player future access to the game and its potential benefits. Punishments as well as rewards may be utilized as a means of keeping the game honest and, more important, as a means of restricting the game over time to mutual cooperators.

In addition to such suggestive formalizations, there is increasing evidence that a policing function does in fact exist in nature. Indeed, there may even be altruistic punishment. (Among the outpouring of publications on this subject, see especially Boyd and Richerson 1992; Clutton-Brock and Parker 1995; Frank 1995, 1996; Michod 1996; Fehr and Gächter 2000a, 2000b, 2002; Gintis 2000b; Falk et al. 2001; Henrich and Boyd 2001; Bowles and Gintis 2002; Boyd et al. 2003; Gintis et al. 2003; Binmore 2004.) As Clutton-Brock and Parker point out in the summary of their review article on the subject, "In social animals, retaliatory aggression is common. Individuals often punish other group members that infringe their interests, and punishments can cause subordinates to desist from behaviour likely to reduce the fitness of dominant animals. Punishing strategies are used to establish and maintain dominance relationships, to discourage parasites and cheats, to discipline offspring or prospective sexual partners and to maintain cooperative behaviour" (1995, p. 209). Evidence of a policing function has also been documented in, among others, social insects (Ratnieks and Visscher 1989), naked mole rats (Sherman et al. 1991), primates (de Waal 1996), and, needless to say, *Homo sapiens.*

From a functional (synergy) perspective, if cooperation offers sufficient benefits it may be in the interest of some individuals to invest in coercing the cooperation of others. Inclusive fitness provides one possible explanation for punishment as a successful strategy in social groups. Another explanation might be the sort of individual fitness trade-offs referred to above. Group selection may also provide a mechanism. The enforcement of cooperation might have significant fitness-enhancing value for groups that are in competition with other groups, or other species. Maynard Smith's (1982a, 1983, 1989) "synergistic selection" model is relevant here. The model suggests that, if cooperative interactions among two or more individuals—related or unrelated—produce selectively advantageous synergistic effects for all parties (on average), the cooperating players may become a unit of selection. A synergistic functional group might be favored in competition with other groups or with ecological competitors from other species. Or a cooperating group may gain an advantage in comparison with the statistical probability of its survival and reproduction in the absence of cooperation. More broadly, synergistic selection can be defined in terms of gene combinations that enable or induce synergistic functional effects at various levels of biological organization. (See Michod 1996 for a model related to the multicellular level.) The evolution of "government" will be considered more fully in chapter 6.

Synergistic Selection

The concept of functional group selection, or synergistic selection, can be illustrated by returning to Maynard Smith and Szathmáry's sculling and rowing models, as described previously (and shown in figure 1). What if the objective of the game were changed? Rather than wanting to merely cross a river (say), now the two oarsmen in each boat share the objective of winning a race against the other boat. Now it has become a functional group selection game (see figure 2). In this situation, if either oarsman were to defect, their team might lose the race; only all-out cooperation would provide

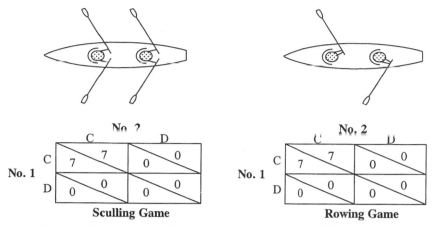

Figure 2. A group competition or group selection game. The matrices show payoff values for each oarsman, under the assumption that the goal is winning a race against the other boat.

rewards for either player. (Note that the two payoff matrices are now identical.) Now the sculling and the rowing games are functionally equivalent in the sense that the performance of either boat depends upon both of its oarsmen; each set has become a functional group; there is "synergistic selection." Furthermore, it is irrelevant whether or not the oarsmen are related to one another.

The following are a few specific examples of synergistic selection:

- In insects, Page and Robinson (1991) conducted an analysis of their own and other researchers' data on the division of labor in honeybees, including a number of computer simulations, and concluded that natural selection operated on colony-level parameters. Oldroyd and others (Oldroyd et al. 1992a, 1992b) also studied the genetics of honeybee colonies and concluded that colony performance was also influenced by the interactions among subfamilies, a colony-level parameter. Fewell and Winston (1992) conducted a study that examined the relationship between pollen storage levels in honeybee colonies (a group-level parameter) and individual forager efforts; not only was the correlation strong, but the researchers detected evidence of a homeostatic "set point." And Guzmán-Novoa and others (1994) reported on a study that focused on the relationship between colony-level natural selection and the level of effort associated with various components of the division of labor in honeybee colonies (see also Calderone and Page 1992).
- An older study by Hoogland and Sherman (1976) examined in detail the influence of six possible disadvantages and three potential advantages of

colonial nesting in 54 colonies of the bank swallow (*Riparia*), ranging in size from 2 to 451 members. Hoogland and Sherman concluded that the disadvantages were not very burdensome and, more important, that the maintenance of coloniality was most strongly associated with group-level defensive measures, which differentially benefited the larger colonies. Although potential predators were not more frequent visitors to large groups, they were detected much more quickly and were mobbed by greater numbers of defenders; predators were also subject to more vocal commotion, and, bottom line, larger colonies were more effective overall in deterring predators.

- Scheel and Packer (1991), in a study of female African lions, found that the average degree of relatedness among the animals had no bearing on their propensity to engage in group hunting. The key variable was the potential for synergy; successful hunting of larger prey required group hunting. In a separate study by Packer and others (1990), it was concluded that the dynamics of female lion grouping were also strongly influenced by the need to defend their cubs (often a group-level function) and to compete against neighboring prides. In both situations, larger groups had an advantage.

- Maynard Smith illustrated his 1982 article on synergistic selection with, among others, the following examples: groups of fifteen to twenty female orb-web spiders (*Metabus gravidus*) that cooperate in building a joint web to span a stream where prey are abundant, tropical wasps (*Metapolybia aztecoides*) that establish joint nests, and coalitions of male lions (*Panthera leo*) that cooperate in taking over and holding a pride.

- Finally, a recent study of spotted hyenas (*Crocuta crocuta*) by Russel Van Horn and his associates (2004) showed that, contrary to kin selection theory, individual matrilines commonly aggregate into larger "clans" of unrelated groups when confronted with dangerous competitors—including even matrilines that are closely related. It should also be noted that D. S. Wilson and Sober (1994), in an in-depth target article on the subject, provide a compendium of over two hundred references on group selection, of which thirty-five are identified as field or laboratory research efforts. See also the in-depth study of group selection in social bees in Moritz and Southwick (1992).

Downward Causation

Closely related to the notion of functional group selection, or synergistic selection, as an evolutionary phenomenon is the concept of "downward causation." The term was actually coined by psychobiologist Roger Sperry

(1969, 1991) in connection with the functional organization and operation of the human brain—that is, cybernetic control processes. (It may be that psychologist Donald Campbell [1974] developed the concept independently.) Sperry was fond of using as an illustration the metaphor of a wheel rolling downhill—its rim, all of its spokes, indeed all of its atoms, are compelled to go along for the ride.

We will use the term here in a slightly different sense. Downward causation in this context refers to the selective influences that have shaped the evolution of cooperative phenomena generally and complexity in particular. Why do selfish genes cooperate in ways that produce teamwork, which, in turn, leads to interdependency? What compels them to subordinate their interests to the interests of the whole? To be specific, how did morphological castes and a division of labor evolve in army ants? How do reproductive controls evolve in mutualistic symbioses where, as Margulis (1993) points out, there must of necessity be reproductive synchronization if the relationship is to remain stable? (See also the discussion of sociogenesis in E. O. Wilson 1985; also Buss 1987; D. C. Smith 1992; and the examples cited in Leigh 1991.) Equally important, how can the potential for cheating among selfish genes (or selfish individuals) be constrained?

Downward causation in an evolutionary context refers to the fact that the functional (synergistic) properties of the whole become a selective screen—a significant influence on the differential survival and reproduction of the parts. Sometimes individual parts are disadvantaged (e.g., nonreproductive workers), and kin selection may help us to understand how such sacrifices for the common good may occur. But, as the evidence previously cited indicates, kinship is not a *sine qua non*. The whole may also be sustained by fitness trade-offs; that is, the costs may be offset by commensurate benefits. For instance, an animal that is at risk from predators may suffer a reduction in its relative reproductive fitness in a social group setting, but it may also enjoy greatly enhanced odds of survival and absolute fitness. (This may help to explain why defeated contenders for breeding privileges sometimes stay on in the group and may even serve as helpers.) To quote Dawkins again, "In natural selection, genes are always selected for their capacity to flourish in the environment in which they find themselves. . . . But from each gene's point of view, perhaps the most important part of its environment *is all the other genes that it encounters* [emphasis in original]. . . . Doing well in such environments will turn out to be equivalent to 'collaborating' with these other genes" (1987, pp. 170–71).

In some cases, the whole may represent an unalloyed benefit for the parts with little or no cost. Many cases of mutualistic symbioses seem to fit into this category. For instance, Margulis (1993) is adamant about the cooperativeness, promiscuity, and evolutionary significance of bacterial colonies. (See also the parallel argument of J. A. Shapiro 1988.) Thus, an isolated bacterium would be cut off from access to the extensive gene swapping and the collective environmental intelligence (information) that commonly exist in bacterial colonies, not to mention the advantages of a division of labor and various collaborative efforts. Social mammals may also exhibit many of these higher-level properties. Some of the most compelling recent field research has illuminated the surprisingly sophisticated social organization, mutualism, and even "culture" in whales and dolphins (see especially Würsig 1988, 1989; Mann et al. 2000; Gygax 2002; Yurk et al. 2002; Whitehead and Rendell 2004). Conversely, the power of a social group to isolate or ostracize a free rider can be a significant deterrent and an agency of negative (downward) selection.

In any case, the synergies that result from cooperation may selectively reinforce cooperative behavior (to use the terminology of behaviorist psychology), and this may in turn differentially favor the evolution of relevant morphological and psychological characteristics over time. Thus, army ant submajors have acquired anatomical specializations that facilitate their role as porters, and humans have evolved psychological predispositions that help us to orchestrate (and even enjoy) our participation in group activities.

In sum, the relevant factors for explaining cooperative phenomena in nature (and in human societies) may include genetic relatedness, but kinship is neither necessary nor sufficient. The key lies in functional synergy and its bioeconomic consequences for differential survival and reproduction in a specific context; functional synergy is the frequently unappreciated common denominator in various models of cooperative behavior in sociobiology.

From Evo-Devo to the Baldwin Effect

Four convergent developments in theory and research, mentioned earlier, should also be highlighted briefly. One is the work on what has been variously called developmental systems theory (Oyama 2000), phenotypic plasticity theory (Rollo 1995; Pigliucci 2001; West-Eberhard 2003), and, simply, evolutionary development theory, or evo-devo (various authors but see especially S. J. Gould 2002). In effect, this movement represents an effort to meld the traditional, gene-centered evolutionary theory with the expanding body of evidence that developmental processes—which involve an inextricable interaction between an organism and its environment—can also be an important source of evolutionary novelties.

Some adherents claim this is the path to a new evolutionary synthesis, much like the so-called "modern synthesis" of the 1930s, which melded Mendelian genetics and Darwinian theory. However, this claim falls short because it excludes another important development in evolutionary theory—a renewed focus on the role of behavior, and "mind," as an innovative agency in evolution. The roots of this idea can be traced back to Jean Baptiste de Lamarck's 1809 *Zoological Philosophy* (1963). Lamarck argued that, in the evolutionary process, changes in an animal's "habits" often come first and that morphological changes may then follow. At the turn of the twentieth century a movement arose that attempted to Darwinize Lamarck's insight, postulating that behavioral changes could reshape the context and the criteria for natural selection. This phenomenon—renamed the Baldwin effect by paleontologist George Gaylord Simpson (1953) after one of the leading advocates of what was originally known as organic selection theory, psychologist James Mark Baldwin—has gained an increasing number of adherents over the years (see, among others, Roe and Simpson 1958; E. Mayr 1960; Corning 1983, 2003; P. P. G. Bateson 1988; Plotkin 1988; Avital and Jablonka 1994, 2000; Deacon 1997; Weber and Depew 2003). I prefer to call this mechanism neo-Lamarckian selection, in honor of the theorist who first recognized its importance. But, under any name, behavioral innovations have often been the "pacemakers" of evolutionary change—in Ernst Mayr's (1960) characterization.

Another convergent trend in evolutionary theory involves the important work by Odling-Smee and his colleagues on "niche construction theory" (Laland et al. 2001; Odling-Smee et al. 2003). These models, and the supporting evidence, show clearly that living organisms at all levels are not passive recipients of environmental conditions but actively shape their environments—and even the entire biosphere—to suit their needs.

Finally, there is a rapidly expanding body of work—most relevant to human evolution but not exclusively so—that is referred to as gene-culture coevolution theory. Theoretical work on this goes back to the 1980s (see especially Cavalli-Sforza and Feldman 1981; Corning 1983; Boyd and Richerson 1985; Durham 1991). However, the last few years have seen a rapid increase in empirical work that is supportive of this paradigm, along with further theoretical refinements. This has been stimulated in part by a growing recognition that culture (broadly defined as the social transmission of adaptive behavioral information) also exists in other species—from songbirds to cetaceans and, especially, our primate relatives. (For more extensive discussions, see Weingart et al. 1997; Dunbar et al. 1999; Corning 2003; Hammerstein 2003; Richerson and Boyd 2004.) Indeed, a case can be made for the view that, in some species (most notably humans), it might be more accurate to call it culture-gene coevolution theory. We will have more to say on this subject in the next section and in chapter 6.

The Selfish Genome

All of this rampant holism begs a question, however. Do wholes have goals that transcend the goals of the parts? Can wholes come to exercise a degree of autonomous control as wholes? In other words, can we postulate a "selfish genome?" The Neo-Darwinian response, it appears, is a somewhat equivocal "no." Richard Dawkins (1989) became famous for the assertion that organisms are merely "robot vehicles" that have been blindly programmed to serve the interests of the genes, yet (as noted earlier) he also allowed that genes can be selected for their ability to serve the interests of the gene team. And George Williams (1966), although acknowledging the wholeness and unity of organisms, characterized many of the claims regarding superorganisms as figments of a "romantic imagination" (p. 220). In truth, some of these superorganismic claims were inflated, but Williams's view of this issue was perhaps a bit too jaundiced: "A wolf can live on elk only when it attacks its prey in the company of other wolves with similar dietary tendencies. I am not aware, however, of any evidence of functional organization of wolf packs" (pp. 217–18).

In contrast, Holistic Darwinism postulates that wholes at various levels of biological organization may evolve mechanisms that permit partially autonomous control over the parts and their actions. Some insight into how superordinate controls can evolve in nature is provided in Egbert Leigh's various discussions (as noted earlier) of how groups might act to contain or override individual advantages for the good of the group—what he calls the "parliament of the genes" (Leigh 1971, 1977, 1983, 1991; see also Michod 1996, 1997, 1999; Frank 2003; Rainey and Rainey 2003). Leigh's argument, in essence, is that if the potential payoffs (synergies) for each of the participants in a cooperative relationship are high enough, this could also provide an incentive for the imposition of "government" in the "public interest." Leigh even draws on Adam Smith's reasoning, not from his 1776 *The Wealth of Nations* but from his earlier and less well-known *The Theory of Moral Sentiments* (1759). Although it has not been widely appreciated until quite recently, Smith argued for the necessity of a system of laws and appropriate means of enforcement in human societies to resist the dangers of unfettered self-interest. (1976, pp. 86, 88–9, 340–41)—a theme that has gained many adherents of late. Government is, in fact, an essential functional element in any socially organized group.

A key to understanding the evolution of government at various levels of biological organization may lie in what could be called the "paradox of dependency." Although cooperative interactions may produce individual fitness-enhancing synergies, a trade-off may be that the more valuable the benefits the more likely it is that the parts will become dependent upon the whole. As the benefits of cooperation increase, so may the costs of not cooperating. Wholes may then become obligatory survival units, one consequence of which may be that a decrement in the performance of the whole might result in the demise of the parts. An example can be found in a long-term study by Jeon (1972, 1983). A strain of *Amoeba proteus* was initially infected with bacterial parasites that were resistant to the hosts' digestive enzymes. After two hundred generations, or eighteen months, a mutualistic relationship had become established, and after ten years the symbionts had developed complete interdependence. (Jeon [1992] has also illuminated some of the biochemistry of these changes.) It should also be noted that Margulis (1993) makes a similar argument with respect to the integration of symbiotic organelles in the ancestral eukaryotic cells. An obvious implication is that the incentives (both proximate and ultimate) for imposing government over the parts are likely to increase in relation to the degree of interdependency among the parts, and the advantages of operating as a superorganism (more on this in chapters 6 and 8).[3]

In fact, in what may appear to be an utter contradiction of classical Neo-Darwinism, it may often be the case that it is in the interest of a gene, or an individual, to promote the well-being of an interdependent other, simply because functional interdependence means just that; it's "one for all, and all for one," to borrow a legendary slogan. Consider this hypothetical example: If one of the two oarsmen in the rowing game (depicted in figure 1) should suffer from thirst and dehydration in the summer heat (he forgot his water bottle), his partner might decide to share his water supply, in the interest of reaching their joint goal. Or, to cite an example from nature, consider the exquisitely complex energy-production services that the mitochondria provide for eukaryotic cells, in their own direct self-interest. Or, for that matter, consider the innumerable situations in human societies where our well-being depends, unequivocally, upon the performance of others—airline pilots, railroad engineers, surgeons, and the other motorists that we encounter on the highways, to name a few. How do we explain these cooperative relationships? They have nothing to do with altruism, kin selection, reciprocal altruism, or even (strictly speaking) tit-for-tat mutualism. They are sustained by pure self-interest.

One final point related to the concepts of downward causation, government, and the selfish genome. At the most basic level of biological organization (the genome itself), there is mounting evidence that the genes do not inhabit a "bean bag" (to reiterate Ernst Mayr's felicitous caricature), and that morphogenesis is not a mindless process. Rather, it is an organized, cybernetic process that entails the extensive use of superordinate feedback controls (the very essence of a teleonomic system). In other words, selfish genes are only citizens on good behavior in the selfish genome, and the outlaws, tax evaders, and parasites among them do not have a license to pursue their antisocial interests ad libitum.

Evolution as a Multilevel Process

There is one other aspect to Holistic Darwinism that should be mentioned briefly. It relates to the traditional distinctions between parts and wholes, individuals and groups, even the concepts of "self-interest" and the "public interest" in political theory. During the past decade or so, there has been a growing appreciation of the fact that evolution is a multilevel, hierarchical (some prefer the term *holarchical*) process, just as survival (and reproduction) is a multifaceted problem (see especially Koestler 1967; Corning 1983; Brandon and Burian 1984; Eldredge and Salthe 1984; Salthe 1985; Eldredge 1985, 1995; Buss 1987; Grene 1987; D. S. Wilson and Sober 1994; Maynard Smith and Szathmáry 1995; Michod 1996; Sober and Wilson 1998; S. J. Gould 2002). In essence, there is a recognition that natural selection operates at various levels of biological organization—from genes to ecosystems—often simultaneously. Indeed, Eldredge and Salthe (1984; also Eldredge 1995) have shown that there are different kinds of hierarchies in the natural world and that these may be in conflict with one another. One implication of this more complex view of evolution is that both competition and cooperation may coexist at different levels of organization, or in relation to different aspects of the survival enterprise. There may be a delicately balanced interplay between these supposedly polar relationships. The following examples illustrate this interplay.

- Eusocial insect species can generally occupy a broader spectrum of habitats and are often able to dominate and even exclude potential competitors among solitary and primitively social species, as noted earlier (see Hölldobler and Wilson 1990). Nevertheless, eusocial insect societies are not the harmonious communities that we once supposed. Among other things, there may be intense competition for breeding rights among potential queens and there is evidence of nepotism among the patrilines in polyandrous species.

- A number of ant species establish pleometrotic colonies; multiple foundresses cooperate in initial nest construction and brood production. In at least one case, the desert seed-harvester ant *Messor pergandei*, a study by Rissing and Pollack (1991) has shown that pleometrotic colonies are able to prevail in direct ecological competition with single-foundress colonies; multiple-foundress colonies are able to produce a larger raiding force more quickly, and this apparently provides a decisive competitive advantage (group selection). However, other studies of these colonies suggest that one trade-off may be internal competition among co-foundresses and their offspring—all very suggestive of human societies.

- Members of African lion prides cooperate and compete with one another in a variety of ways: Females typically hunt large prey in groups, share food, and even share in guarding cubs and defending the pride. As Packer and Ruttan (1988) observe, there is evidence of synergy. For instance, a group of females can more effectively defend a kill against scavengers, including other lion groups. Likewise, a group of males can successfully defend access to a group of females, whereas single males cannot. However, there is also much intra-coalition competition among the males for mating privileges.

- One of the more dramatic examples of the interplay between competition and cooperation concerns the northern elephant seals (*Mirounga angustirostris*). Males of this species, which can weigh up to 4,500 kg, are legendary for their prolonged and bloody battles for dominance and mating privileges when they come ashore to breed in the winter and early spring. However, the males will only fight when estrous females have formed "harems" of fifty or more. And when the fighting is over, the alpha males commonly form coalitions with a half-dozen or more beta males, who will defend the perimeter of the harem against other marauding males (in return for which the beta males get limited mating privileges for themselves). Elephant seals generally feed at sea alone, and at great depths, but whenever they are ashore they congregate peacefully in tightly packed rookeries that facilitate defense and heat sharing (a critically important function in these animals). Males collaborate in this way during their summer molting season; nonbreeding males also aggregate into loser groups during the breeding season; females huddle closely together to share heat and defend their pups during the breeding season; and pods of weanling pups huddle for warmth and mutual self-defense before setting off on their initial feeding expeditions (Le Boeuf 1985; Le Boeuf and Laws 1994).

The Evolution Of Humankind

Human evolution may provide a singular illustration of the synergistic, functional group selection hypothesis, and of Holistic Darwinism. In effect, the principles that were elucidated in the previous section can also be observed in the evolution of the human species, and in our cultural evolution as well. For various reasons, the evolution of humankind has often been portrayed as a process that is sui generis. Of course, this overlooks the fact that all of evolution can be said to be sui generis, given its historical and situation-specific causal dynamics. As Darwin himself put it in *The Descent of Man* (1874), any evolutionary innovation depends upon many "concurrent favorable developments" that are always "tentative" (p. 150). Nevertheless, the evolution of humankind is undeniably one of the more remarkable episodes in evolutionary history. (This matter will be discussed briefly here and considered further in chapters 6 and 17.)

A number of suggestive and thoughtfully argued theories of human evolution have been advanced over the years. These theories were reviewed and critiqued in depth in Corning (1983), and a synthetic explanation was offered there that, in effect, combined Darwin's tripartite selection theory of human evolution—that is, family (kin) selection, mutualism (including reciprocity), and group selection—with the concept of functional synergism. As Darwin pointed out—and this point is crucial—the three modes of selection need not be opposed to one another; they can be complementary and mutually reinforcing. In addition, the Synergism Hypothesis posits, in essence, that it was the bioeconomic payoffs (the synergies) associated with various forms of social cooperation that produced—in combination—the ultimate directional trend over a period of several million years, from the earliest bipedal hominids to modern *Homo sapiens*. That is, the synergies produced by various collaborative behavioral innovations provided proximate rewards or reinforcements (as the behaviorists would say) that were substantial enough to create a behavioral "pacemaker" (*sensu* Ernst Mayr 1960) for the progressive evolution over time of our distinctive wardrobe of biological characteristics. In other words, we invented ourselves (in effect) in response to various ecological pressures and opportunities—a paradigm that may be more widely applicable to evolutionary change than is generally appreciated (see Corning 1996a, 2003). Here I can only summarize the argument.[4]

The traditional approach to explaining human evolution has been to propose a "prime mover" theory, which is typically portrayed as the "engine" that has powered the course of human evolution. Darwin, in *The Descent of*

Man (1874), singled out the role of toolmaking. E. O. Wilson (1975) stressed our primate "preadaptations" and speculated about the possible role of an unspecified "autocatalysis." Bipedalism, which we now know preceded the development of the "big brain," is currently viewed by many theorists as the "breakthrough" development (e.g., Donald Johanson, Richard Leakey, Timothy White, etc.). Major climate changes during the Miocene and Pliocene have also been suggested as important precipitating factors (e.g., Yves Coppens, Elizabeth Vrba). Then there are the various competing group theories: group hunting (Dart, Washburn, Ardrey, Thompson, Stanford and others); group scavenging (Potts, Blumenshine, Shipman, etc.); female gathering (Zihlman and Tanner); the nuclear family and male provisioning (Lovejoy); collective defense against predation (Kruuk, Kortlandt, Alexander, etc.); and the ever-popular group conflict hypothesis, which traces back to Darwin and Spencer and which has been championed in the past century by Dart, Keith, Ardrey, Lorenz, Bigelow, Otterbein, and Alexander, among others. In the latter stages of human evolution, climate change, population growth, food surpluses, the adoption of fire, language development, increased intelligence, and warfare have also been singled out as prime movers by various theorists at various times.

Holistic Darwinism suggests the contrarian view that all of these factors were important but that none was sufficient. To repeat, the engine is nothing without the car. The answer lies in the unique combination of factors that produced, over time, many compatible and mutually supportive cooperative effects (functional synergies). Indeed, objections can be mounted against every one of the factors cited above, taken individually. For example, bipedalism is not unique to humans; birds are bipedal, and kangaroo forelimbs have atrophied rather than becoming instruments for the skilled manipulation of tools. In fact, hominid bipedalism existed for some millions of years before the "big brain" emerged. Toolmaking is also an insufficient explanation; we now know that many species make and use tools, especially our closest relatives the chimpanzees. Also, crude stone tools were used by our hominid ancestors for perhaps a million years before the more refined and standardized Acheulean toolkit appeared (Leakey 1994; Tattersall 1995). Group hunting and scavenging are also inadequate explanations, for the same reasons; our ancestors were hardly unique in this regard. The primeval gathering scenario and the nuclear family scenario are appealing, though difficult to support empirically; yet they are in any case insufficient because they overlook other factors—namely, the often serious threat from potential predators and the premium associated with meat getting (via scavenging, hunting, or both) in the more open and diversified environments in which some of the later hominid developments most likely occurred. Even the conflict hypothesis—which Alexander (1979) asserts is both necessary and sufficient—begs the question: why are there no nation-states composed of chimpanzees, which we now know can be quite warlike? The very absurdity of that idea highlights the fact that there had to be many other factors that worked together to propel the process. Indeed, the extensive hominid migrations out of Africa over time suggest that conflict avoidance might well have been a common adaptive strategy.

In a major critique of cultural evolution theories, anthropologist Elman Service (1971) came to this emphatic conclusion: "down with prime movers!" (p. 25). The same can be said, equally emphatically, of the larger process of human evolution. Prime mover arguments invariably take for granted some, or all, of the other requisites for survival and reproduction. Very often they reflect a kind of ecological naivete; they discount the many life-and-death challenges associated with living (and evolving) in a demanding and changeable environment over a period of several million years. But if no one factor alone can provide a sufficient explanation for the evolution of humankind, then what is sufficient? The answer is that all of the important human traits were necessary and none were sufficient. In effect, there was a mutually reinforcing synergy among the key innovations—combined effects that would not otherwise have been possible.

"Man Makes Himself"

However, the dominant theme of human evolution may have been the expansion of various modes of social cooperation (including cooperative modes of competition), which have been rewarded with commensurate bioeconomic benefits. To reiterate, competition and cooperation are not mutually exclusive explanations for human evolution; each played an important role in shaping our evolution. Nevertheless, the thesis here is that increasingly potent (and selectively advantageous) forms of social cooperation may have given our ancestors their competitive edge.

As Edward O. Wilson (1975, 1985) has noted, a multifaceted group-living ecological strategy is a relatively rare occurrence in nature. We rightly admire the complex social organization of honeybees, naked mole rats, army ants, killer whales, and a small number of other highly social species, including some of our close primate relatives. The synergies that have made such collective survival strategies rewarding for various social species are increasingly well documented. We are among that select company, and it has been the key to our evolutionary success. A human society can be characterized as a "collective survival enterprise." We meet our basic survival needs through elaborate networks of social cooperation.

We do not know, and likely never will know, the full story of our evolution as a species, although we are gradually adding more details to the outline and making better-informed guesses. However, there is reason to believe that behavioral changes in the direction of greater social cooperation for specific functional purposes were the "pacemakers" that precipitated supportive morphological changes. In a very real sense, as anthropologist V. Gordon Childe (1951) put it in the title of his famous 1936 book on the rise of civilization, the human species may have "invented" itself. The real key to human evolution, accordingly, was not any single prime mover but the entire suite of cooperative behavioral, cultural, and morphological inventions—a synergy of synergies.

An oft-used (and important) illustration of this dynamic is the adoption by evolving hominids of the controlled use of fire (or, more broadly, various exogenous forms of energy). This is a uniquely human cultural invention and is still a major factor in our ongoing evolution—a point that various energy-oriented theorists have thoroughly documented (e.g., L. A. White 1943, 1949; Cottrell 1953; H. T. Odum 1971; Adams 1975, 1988). The earliest strong evidence for the use of fire by our hominid ancestors is identified with the Middle Pleistocene, perhaps 200 to 400,000 years ago. However, some theorists argue on plausibility grounds, albeit with more fragmentary evidence, for a much earlier date. (See especially the cautious review by James [1989] and the offsetting commentary by Lewis in the same volume.) The controlled use of fire by hominids (in effect, a cooperative animal-tool symbiosis) had enormous long-term benefits. Over the course of time, fire was most likely used as an effective means of defense against predators; it was a source of warmth that facilitated migration into colder climates; it might well have served as an insect repellant and as a means for obtaining honey from bee hives (as a bee suppressant); it probably became a weapon for driving and capturing prey animals; it was a means for shaping and hardening tools; it could be used for conditioning the environment (as in slash and burn horticulture); and, not least, it enabled our ancestors to add to their diets many foods that would otherwise have been toxic, indigestible, or infectious if eaten raw (Leopold and Ardrey 1972; Stahl 1984). See also the case that is developed in Wrangham et al. (1999) for what they call the "cooking hypothesis" (also Wrangham 2001).

In any event, fire represented the functional equivalent of a major morphological development. With the acquisition of fire, our ancestors were able to greatly expand their niche, which in turn changed the selective forces to which our ancestors were subject. Furthermore, fire most likely became another focal point of social cooperation. Fire-keeping was a collective good that required a division of labor—for gathering firewood, fire tending, fire transport, and, eventually, fire-making. In other words, this primordial hominid technology, like most human technologies, was at once a source of bioeconomic benefits and a generator of social cooperation and social organization.

How can this synergistic theory of human evolution be tested? One way is to try a few thought experiments: Take away fire (along with other energy sources in modern societies); we are utterly dependent upon exogenous forms of energy. Or, take away language, or bipedalism, or tools and technology. In short, there is no major adaptive modality that we could do without; they are all necessary parts of an interdependent, synergistic system.[5]

Conclusion: A Post-Neo-Darwinian Paradigm?

I believe that Holistic Darwinism can plausibly be viewed as a candidate for a post–Neo-Darwinian theoretical paradigm. It refocuses evolutionary theory on the "vessels" and their functional properties as the vanguard of evolutionary change. In fact, that is where natural selection as a causal dynamic actually occurs; to use an older turn of phrase, it is the phenotypes that are "tested" in the environment. Holistic Darwinism shifts our focus from the anthropomorphic purposes of selfish genes in theoretical isolation to the evolved, emergent purposiveness of the living systems as wholes, and to the functional interactions and relationships (adaptations in specific environments) that result in differential survival and reproduction. It also stresses the "synergistic selection" of various combinations.

Equally important, Holistic Darwinism de-emphasizes (without denying) the role of genetic mutations, recombinations, transpositions, etc., as sources of creativity in evolution and emphasizes purposeful innovations which may occur at the behavioral, cognitive, even social levels (inclusive of symbiogenesis). In this model, proximate "neo-Lamarckian selections" by wholes (i.e., adapting organisms and, in some cases, adapting groups) assume a much more important role in evolutionary change than is acknowledged in Neo-Darwinism.

Finally, and perhaps most contentious of all, I maintain that Holistic Darwinism also encompasses human evolution and, indeed, the ongoing biocultural evolution of our species. (Again, see the further discussion in chapters 6 and 17.) It is a seamless theoretical framework that does not require any additional causal principle or "mechanism" to account for humankind.[6] We must recognize human culture, human economies, even human political systems for what they truly are—an augmentation of adaptive modalities that can be found in rudimentary form in many other species. (In this paradigm, the activities of humankind virtually everywhere on earth are viewed as an integral part of the evolutionary process, not something that is separate from it.) But, having recognized that, we must also acknowledge our uniqueness as a species. Complex human societies are as different from those of honeybees or naked mole-rats or even chimpanzees as complex multicellular organisms are from single-celled protists. The cumulative, synergistic effects of many differences in degree have produced a difference in kind—and a dynamic of rapid change at the behavioral/cultural level that is obviously unique. This is an evolutionary development that Holistic Darwinism can fully comprehend. It is not even conceivable in a theoretical world that barely recognizes the existence of wholes.

I believe that it is time to refocus the Necker cube on the problem of explaining the evolution of complex systems in a way that is fully consistent with Darwin's vision. In the long run, I believe that the Darwinian, functional explanation of complexity will prevail over various orthogenetic theories of self-organization—in reality a teleological black box that begs the "how" question (see chapter 4), or theories that postulate a random "drunkard's walk" (S. J. Gould 1996). Gould's argument is especially surprising, coming as it does from such a sophisticated and articulate student of evolution. It is a formulation which tacitly ignores the functional costs and benefits related to the evolution of biological complexity; complexity is not a free lunch but a cumulation of adaptive innovations over the eons. I can only second the conclusion of George Williams in the peroration of his famous book (1966, p. 273): "It may not, in any absolute or permanent sense, represent the truth, but I am convinced that it is the light and the way."

— ℰℐℐ —

Generalizations derived from a juxtaposition of facts are not fruitful unless some conceptual, theoretical scheme guided the generalizations and, incidentally, the selection of facts. . . .

—Anatol Rapoport

Experiments unguided by an appropriate theoretical framework usually amount to little more than 'watching the pot boil.' . . . We need experiments to inform theory, but without theory all is lost.

—John H. Holland

The purpose of scientific theory is to unite apparently disparate observations into a coherent set of generalizations with predictive power.

—Lynn Margulis

SUMMARY: Synergy—here defined broadly as the combined (interdependent) effects produced by two or more parts, elements, or individuals—is a ubiquitous phenomenon in nature and human societies alike. Although it plays a prominent part in most, if not all, of the scientific disciplines, its importance is not widely appreciated because it travels under many different aliases. (A number of examples are provided to illustrate.) At the very least, the term *synergy* could be utilized as a pan-disciplinary *lingua franca* for the functional effects produced by cooperative phenomena of various kinds; a terminological shift would underscore the fact that the differently named phenomena studied by various disciplines are in fact variations on a common theme in the natural world. More important, synergistic effects of various kinds have also played a major causal role in the evolutionary process; in particular, synergistic effects have provided the underlying functional basis for the evolution of complex systems, in nature and human societies alike. The so-called Synergism Hypothesis is here briefly described, and the accumulating empirical support for this theory is summarized. Some methods for testing the theory are also discussed.

3

The Synergism Hypothesis: On the Concept of Synergy and Its Role in the Evolution of Complex Systems

Introduction

It is one of the paradoxes of our age that as the tools of scientific research have grown ever more powerful—from positron emission tomography to electron microscopy, nuclear magnetic resonance, and massively parallel computers—the phenomena that we are able to investigate (and their causal dynamics) seem to grow ever more complex. The relentless reductionism of particle physics, polymer chemistry, molecular biology, and neurobiology, among other disciplines, has not (so far) revealed the decisive mechanisms or underlying laws of the phenomenal world. Instead, the "microcosmos" (to borrow Lynn Margulis's term) displays profound complexity, interactionism, interrelatedness and, not least, historical specificity.

It has been suggested that our era should be called the age of complexity. While this sobriquet (or epithet, depending on your values) may be appropriate, complexity is certainly not a newly discovered aspect of the natural world (see chapter 4). The debate over the relative importance of wholes and parts (or holism and reductionism) can be traced at least to Periclean Athens and, especially, to the writings of Aristotle. Although scholars these days have a propensity for forgetting their forebears, over the course of the past century there have been successive waves of wholistic and reductionist theorizing—a sort of transgenerational dialectic—in which many of our most distinguished scientists have played a part. After reaching an apogee of sorts with the imposing theoretical edifice of the nineteenth century polymath Herbert Spencer (1892, 1897), wholistic theorizing was all but banished at the turn of the past century by the supporters of Darwin's theory, and of Weismannism and mutation theory.

However, in the 1920s wholism (especially the concept of "emergent evolution") reappeared, thanks to the writings of C. Lloyd Morgan (1923), Jan Smuts (1926), William Morton Wheeler (1927), and others. (The concept of emergence is discussed further in chapter 5.) Following another hiatus in the 1940s, wholism, recast as systems theory, was revived again in the 1950s with the emergence of the systems sciences. See especially the work of Ludwig von Bertalanffy (1950, 1956, 1968); Kenneth Boulding (1956, 1977); H. R. Ashby (1958); Anatol Rapoport (1968); Arthur Koestler and John R. Smythies (1969); Ervin Laszlo (1972); and James Grier Miller (1995). Nowadays, systems theory—which is partial to cybernetics and feedback models—seems to have been temporarily eclipsed by complexity theory—which is partial to chaos models and hypotheses of self-organization. However, the two disciplines are really close kin, and the recent emergence of systems biology is a very positive sign (see Kitano 2001, 2002; Chong and Ray 2002; Csete and Doyle 2002).

What sets the present era apart is that the scientific enterprise seems to be in the process of bridging the theoretical chasm between wholism and reductionism; as noted above, there is evidence of a growing appreciation of the inextricable relationships between (and within) wholes and parts, and between various levels of organization—relationships that necessitate multileveled, multidisciplinary, interactional analyses (see Corning 1983; Kline 1995; Polanyi 1968; Anderson 1972; Ghiselin 1981, 1997; Eldredge 1985, 1995; Buss 1987; Grene 1987; Maynard Smith and Szathmáry 1995, 1999; J. G. Miller 1995; D. S. Wilson 1997a, 1997b; Sober and Wilson 1998). Witness Francis Crick (1994), a Nobel laureate (for the double helix) and a reformed arch-reductionist who now embraces the phenomenon of emergence in his recent book on the nature of consciousness (discussed further below). Indeed, the very terms *mechanism* and *law* seem increasingly to be naive formulations in light of the enormously complex, dynamic processes that we can observe (and model) in ever more sophisticated ways. Consider just a few examples: quantum nonlocality and quantum entanglement in physics; the highly conserved homeobox domain, consisting of some sixty amino acids, which plays a key role in morphogenesis; the awesome functional organization of the human immune system, which includes at least nine different subsystems; the elaborate cortical substrate of human vision, which involves many millions of neurons and at least twenty distinct visual areas in the brain; the intricate relationships and multileveled feedback processes associated with even a relatively simple ecosystem; and the daunting interconnections between world population growth, technology, economic activity, and vested political interests and rivalries, on the one hand, and the problems of environmental pollution, habitat destruction, and resource depletion.

There have been many efforts in recent years to gain greater theoretical control over this overwhelming complexity. Best known, perhaps, are the nonlinear dynamical system models that are capable of exploiting the computing power of supercomputers (see Casti 1979, 1997; Yates et al. 1987; Kauffman 1993, 1995; Mosekilde and Mosekilde 1991; Mittenthal and Baskin 1992, Holland 1992, 1995, 1998, Petersson 1996, also, see the Proceedings of the Santa Fe Institute Studies in the Sciences of Complexity). This has proven to be a fertile and productive enterprise, and we can at present barely glimpse its ultimate potential. For instance, computer scientist John H. Holland has undertaken an ambitious attempt to model the evolution, aggregate behavior (emergence), and anticipation (purposiveness and cybernetic feedback processes) of what he characterizes as "complex adaptive systems." (See also Chauvet 1993.)

Another important development is the emergence of agent-based modeling—simulations in which intentional agents are allowed to interact with one another and with the social structures that arise from these interactions (Axelrod 2001; Luna and Perrone 2001).

On the Concept of Synergy

Here I will describe a complementary approach. It involves, in effect, a conceptual revisioning of the phenomenal world—a paradigm shift—that directs our attention to an underlying causal principle that is concerned with structural and functional relationships of various kinds and with the concrete consequences or effects that they produce. Albert Einstein many years ago observed that "we should make things as simple as possible, but not simpler." Theoretical simplifications, or generalizations, may serve to identify key features, common properties, or important relationships among various phenomena. Equally important, a concept that encompasses a broad range of phenomena may also serve as the anchor for a theoretical framework that, in turn, may catalyze specific hypotheses, predictions, or tests.

One example is the concept of natural selection. Evolutionists often speak metaphorically about natural selection (as did Darwin himself) as if it were an active selecting agency, or a mechanism. But in fact natural selection is an umbrella category that refers to whatever functionally significant factors (as distinct from, say, stochastic or teleological influences) are responsible in a given context for causing the differential survival and reproduction of genes, genic "interaction systems" (in Sewall Wright's terminology), genomes, groups, populations, and species. Genes are the units that are selected, but it is the *functional consequences* of the genes that (by and large) determine their ultimate fate. The classical population genetics definition of natural selection as a change in gene frequencies in a population is—as Wimsatt (1980) has pointed out—inadequate because it focuses on the informational and bookkeeping aspect of a larger, iterative functional process.

Accordingly, as a theory of evolutionary change, natural selection makes no global predictions about the overall course of evolution or the future of any given species, in contrast with various orthogenetic or law-like theories of evolution. Nevertheless, the concept leads to many situation-specific explanations, predictions, and postdictions about the properties of various organisms, about the relationships among species (and between any species and its environment), and about the causes of various directional changes over time.

Another example of an umbrella term is the concept of *hierarchies.* The basic principle was well understood by Aristotle, and by nineteenth century taxonomists and evolutionists, but the term itself apparently traces to the outset of the twentieth century (reviewed in Grene 1987). Today the term is used in a variety of ways, with each usage having its own theoretical connotations (see the discussions in P. A. Weiss 1971; Pattee 1973; and the references for multileveled organization cited previously). Thus, the postulate of a taxonomic hierarchy, which entails a classification of various species into more inclusive groupings (genera, families, orders, etc.), also implies that a given species has certain characteristics and evolutionary relationships in common with (or different from) other species, both extant and extinct. The physiologists, in contrast, associate the term *hierarchy* with organelles, cells, tissues, organs, and so on, a scheme which implies a nested set of functional parts-wholes relationships. Likewise, to political scientists a hierarchy refers to structured relationships of power, rule, or authority—to different levels of cybernetic (political) control. And when biologists Niles Eldredge and Stanley Salthe (1984) drew a distinction between genealogical and ecological hierarchies in nature, they were also implicitly making certain claims about the causal dynamics of the evolutionary process (see also Ghiselin 1981, 1997; Eldredge 1985, 1995; Salthe 1985).

Synergy (from the Greek word *synergos*) is another such umbrella term. Although it is often overlooked, underrated, or misunderstood (or called by a different name), synergy is a ubiquitous and fundamentally important aspect of the natural world. (For an in-depth discussion, see Corning 1983; also 1996a and chapter 2.) Synergy, broadly defined, refers to combined or cooperative effects—literally, the effects produced by things that operate together (parts, elements, or individuals). The term is frequently associated with the slogan "the whole is greater than the sum of its parts" (which traces back to Aristotle in the *Metaphysics*) or "2 + 2 = 5", but, as we shall see, this is actually a caricature, a narrow and perhaps even misleading definition of a multifaceted concept. I prefer to say that the effects produced by wholes are *different* from what the parts can produce separately.

There are innumerable illustrations of synergy. (Some were cited above.) One pointedly *nonquantitative* example has to do with pattern recognition, or what is referred to in psychology as gestalt phenomena (reviewed in Rock and Palmer 1990). The two-letter combinations PT, TP, RT, and TR, mean nothing to most of us (except perhaps to old salts of the World War II era, who may remember the PT boats). But adding a vowel—either an "o" or an "a"—to each of these consonant pairs, gives the results shown in table 1. The three-letter combinations in the table are now meaningful (at least to English-literate readers), although, interestingly enough, the combinations shown in column four are utilized only as acronyms, perhaps because they are more difficult to pronounce. (Some of these acronyms are "please turn over," "Parent-Teacher Association," and "Rapid Transit Authority." No doubt there are others as well.)

Unless you happen to be a fan of crossword puzzles, you may not recognize in the table the more obscure words *tor* (rocky promontory) and *ort* (food scrap). This illustrates an important point about the nature of synergistic phenomena. Synergy refers specifically to the structural or functional effects that are produced by various combinations of things. In this case, it refers to the effects that the words produce in the reader's mind; it is not something that is inherent in the patterns themselves. Thus if a word evokes no mental image, it is at best an example of latent synergy.

There are obviously many different kinds of cooperative or synergistic effects. Some arise from linear or additive phenomena. I call them "synergies of scale." Larger size, frequently the result of an aggregation of similar units, may provide a collective advantage. For instance, colonies of the

Table 1: Pattern recognition

POT	TOP	OPT	PTO
PAT	TAP	APT	PTA
ROT	TOR	ORT	RTO
RAT	TAR	ART	RTA

predatory myxobacterium *Myxococcus xanthus* are jointly able to engulf much larger prey than any one or a few could do and, more important, are able collectively to secrete digestive enzymes in concentrations that would otherwise be dissipated in the surrounding medium (Bonner 1988; J. A. Shapiro 1988).

A variant of this type of synergy involves frequency- or density-dependent phenomena. Brood parasitism in birds is a case in point. The effectiveness of this freeloading reproductive strategy depends upon, among other things, the availability of nest sites, the number of eggs laid by the hosts, and the number of eggs laid by their parasites (Read and Harvey 1993). Density-dependent effects are also involved in the well-known correlations between bacterial colony size and a bacterium's ability to cause infections or resist drugs (e.g., *Staphylococcus aureus*).

So-called emergent phenomena are a particularly important class of synergistic effects. (I restrict the term *emergence* to the subset of synergistic effects in which new physical wholes are synthesized—see chapter 5.) Thus, stainless steel is an alloy of steel (itself an alloy) together with nickel and chromium, a combination that exhibits rust- and tarnish-resistance and increased durability. Duralumin, which is a compound of aluminum, copper, manganese, and magnesium, combines the light weight of aluminum and the strength of steel. And the so-called superalloys—comprised of nickel, cobalt, and other elements—are favored for jet engines and spacecraft because they can resist very high temperatures, high pressures, and oxidation.

The division of labor (or, very often, various combinations of labor), a phenomenon appreciated by Plato and further articulated by Adam Smith and the classical economists, represents another important category of synergy. To illustrate, one important component of the reproductive machinery in living systems involves a division of labor and coordinated efforts of three different kinds of RNA—messenger RNA, transfer RNA, and ribosomal RNA. Darnell et al. (1990, p. 88) argue in their textbook on molecular and cell biology that "the development of three distinct functions of RNA was probably the molecular key to the origin of life."

Among the many other examples of a division of labor found in nature, some of the most remarkable appear in very primitive life-forms: bacterial colonies, eukaryotic protists, cellular slime molds, and others. One case in point involves *Anabaena*, a genus of cyanobacteria that engages in both nitrogen fixation and photosynthesis, a dual capability that gives it a significant functional advantage. However, these two functional processes happen to be chemically incompatible; the oxygen produced by photosynthesis can inactivate the nitrogenase required for nitrogen fixation. *Anabaena* has solved this problem by complexifying. When nitrogen is abundantly available in the environment, the cells are uniform in character. However, when ambient nitrogen levels are low, specialized heterocysts are developed that lack chlorophyll but are able to synthesize nitrogenase. The heterocysts are then connected to the primary photosynthesizing cells by filaments. Thus, a compartmentalization and specialization of functions exists that benefits the whole (J. A. Shapiro 1988).

Synergy is also found in a variety of mutually enhancing or augmenting functional effects in nature. Hemoglobin, a tetrameric protein whose four monomers cooperatively bind oxygen, is one well-known example. (Indeed, there is an area of biochemistry called cooperativity theory, which is focused on the study of the many kinds of synergistic phenomena that occur at the biochemical level; see Hill [1985].) Another example at the microlevel concerns the observed error rate in normal cellular DNA replication, which is remarkably low (about 10^{-10} to 10^{-8} per base pair) compared with the theoretical potential, given the rate of ambient sources of decay, damage, and copying errors of about 10^{-2} per base pair. The explanation for this discrepancy is that it is the combined result of a complex set of mechanisms that work together, including proofreading by DNA polymerases, methylation-instructed mismatch correction, enzymatic systems that repair or bypass potentially lethal or mutagenic DNA damage, processes that neutralize or detoxify mutagenic molecules, the regulation of nucleotide precursor pools, and, of course, the redundancy achieved by double-stranded genetic material (Haynes 1991; Ridley 2001).

There is also in nature a broad category of what might be called bioeconomic efficiencies that are derived from cooperative behaviors of various kinds, such as joint environmental conditioning, cost sharing, risk sharing, information sharing, and so on. Thus, emperor penguins (*Aptenodytes forsteri*) are able to buffer themselves against the intense Antarctic cold by huddling together during the winter months in dense, heat-sharing colonies numbering in the tens of thousands. Experiments have shown that, by so doing, the penguins are able to reduce their individual energy expenditures by 20 to 50 percent (Le Maho 1977). Similarly, honeybees, through joint heat production or fanning activities as the need arises, are able to maintain the core temperature of their hives within a narrow range (J. L. Gould and C. J. Gould 1995). Also, Partridge (1982) and his colleagues have shown that fish schools, which can involve the active coordination of behaviors, may be highly adaptive for individual members. For instance, the evasive maneuvers utilized by dwarf herring against predatory barracudas dramatically reduce the joint risk of being eaten.

Information sharing (wittingly or otherwise) may also be highly synergistic. Social insects and communally nesting birds frequently share information about food sources. Many animals engage in alarm calling, which may alert nearby members of their own and other species. And many flocking birds and herd animals share in the tasks—and energy costs—associated with scanning for potential predators (King 1955; E. O. Wilson 1975; Caraco and Wolf 1975; de Groot 1980).

Many social animals also benefit from various forms of swarm intelligence, or collective decision making, from insects (Franks 1989; Seeley and Levien 1987; Seeley 1995) to humans (Boehm 1996, 1997, 1999; also Sober and Wilson 1998).

Another preliminary point about the concept of synergy is that it is value-neutral. Over the years various writers have equated the term with mutualism or even altruism. However, this is not correct. Synergy refers to combined effects of all kinds. These effects may be considered eufunctional (positive synergy), dysfunctional (negative synergy), or neutral, depending on the context. For instance, the mutation-enhancing effects of gamma rays and metallic salts in combination might be viewed positively by a geneticist who wanted to enhance mutation rates in a laboratory study. By the same token, the synergies achieved by pack-hunting social animals (in terms of, say, capture efficiency or the size of the prey) may be viewed positively from the point of view of the predators but rather negatively from the point of view of their victims.

A Pan-Disciplinary Phenomenon?

As already noted, synergistic phenomena can be found in many different disciplines. In some disciplines, the term *synergy* is used extensively. (A literature search of a database for the biological sciences for the five-year period from 1991 through 1995 produced over three thousand "synerg" references, concentrated for the most part in such "hard sciences" as endocrinology, neurochemistry, and pharmacology.) Yet in other biological disciplines, paradoxically, the term *synergy* is used sparingly, if at all. Consider sociobiology, for instance, which is primarily concerned with cooperative behaviors among conspecifics in nature. A highly organized social species, such as honeybees, army ants, or naked mole rats, exhibits many different forms of social synergy—from joint thermoregulation to information sharing, cooperative foraging, mutual grooming, mutual defense, and cooperative nest building. Nevertheless, these behaviors are typically characterized as kin selection, or reciprocal altruism, a division of labor, emergence, mutualism, or simply cooperation. Likewise, as noted above, the various formal models of cooperation in sociobiology also implicitly depend on synergy. Yet the dependence of these models on synergy is seldom acknowledged.

It is not possible to provide here a detailed, discipline-by-discipline analysis of the many different kinds of synergy and the terminology that is used to describe it. As an alternative, a selected sample of the scientific disciplines is presented in table 2, along with representative examples and some

Table 2: Synergy in selected scientific disciplines

Scientific discipline	Representative example(s)	Associated terminology
Quantum physics	Quantum coherence	holism, ordering
Physics	Chaotic phenomena	emergence, interactions, attractors, order
	Self-organized criticality	interactions, holism
	Phase transitions	cooperative effects, symmetry breaking,
Thermodynamics	Dissipative structures	order/disorder, low entropy, negative entropy, order, emergence
Biophysics	Hypercycles	cooperation, interactions, coordination, order, emergence
Chemistry	Molecular macrostructures	symmetry, collective stability, order
Biochemistry	Supramolecules	interaction, functional integration, coordination
Molecular biology	DNA	complementarity, epistasis, heterosis
Developmental biology	Homeotic genes	organization, coordination, cooperation
Neurobiology	Neuronal transmission	cooperativity, threshold effects, emergence
Ecology	Coevolution	interactions, mutualism, parasitism
Behavioral biology	Symbiosis	mutualism, cooperation
	Sociobiology	mutualism, reciprocal altruism, emergence, cooperation, division of labor
Anthropology	Cultural evolution	symbiosis, cooperation, division of labor

associated terminology. The following sections then briefly describe each example, with special reference to its synergistic aspects.

Quantum Coherence

Sometimes referred to as Bose-Einstein condensation, quantum coherence involves situations in which a large number of particles participate collectively in a quantum state; the wave function that describes the set of particles is unentangled with its environment and behaves as if it described a single particle (Frölich 1970, 1975). Large-scale versions of quantum coherence can be observed in superconductivity and superfluidity. In any case, collective quantum effects are synergistic in nature.

Chaotic Phenomena

Chaos involves unpredictable but deterministic interactions among phenomena that may result in coherent, collective states of order or disorder that are synergistic. Laminar flow and turbulence in fluid dynamics are examples. The so-called strange attractors or dynamical attractors that are highlighted in chaos models represent states whose stability properties arise from the interactions among the variables (Crutchfield et al. 1986; Ditto and Pecora 1993).

Self-Organized Criticality

A process in which large composite systems that are involved in dynamic interactions may self-evolve to a critical state in which any further change can result in a global transformation (Bak and Chen 1991). Avalanches are an example, and it has been discovered that such phenomena are the result of the global properties of the whole.

Phase Transitions

Physicist Herman Haken (1973, 1977, 1983) and various colleagues have spent more than twenty years developing a science of cooperative phenomena called synergetics, the objective of which is the elucidation of a set of general principles of cooperativity and cooperative effects. Phase transitions, which involve a collective change of state (e.g., laser beams, water turning into ice crystals or to steam at various critical temperatures or pressures, or the loss of magnetism in iron crystals at 774°C.) have been among the many important areas of synergetic analysis.

Dissipative Structures

Ilya Prigogine (1978) and his coworkers have built upon Erwin Schrödinger's idea that there is a class of systems in nature that can defy the second law of thermodynamics by being open, or energy-processing, in nature. They feed on throughputs of energy to sustain order, or negative entropy, and can thereby remain in a sustained state of thermodynamic disequilibrium. Dissipative structures, according to Prigogine, may also be self-organizing. They may arise spontaneously and then spontaneously evolve toward greater complexity. This may occur when an open system is driven so far from an equilibrium condition that nonlinear discontinuities, or threshold instabilities, may occur that transform the system in the direction of greater complexity and more structural stability. In any event, the behavior of dissipative structures and the transitions between system states are holistic phenomena; they involve cooperative effects. Indeed, an often-overlooked point about thermodynamic processes is that those with negative entropy, tending toward more highly ordered energy states, also entail synergy—they involve a concentration of energy such that they are capable, collectively, of doing work.

Hypercycles

A major hypothesis about prebiotic evolution, the hypercycle model envisions a cyclical, mutually reinforcing set of catalytic processes among interrelated RNA-like precursors. As Manfred Eigen and Peter Schuster (1977, 1979) note, the buildup of functionally proficient translation machinery in evolution required the integration of several different replicative components. This integration, they maintain, could only be achieved by a "mechanism" such as a hypercycle, which they characterize as a "cooperative" process. Stuart Kauffman (1993), on the other hand, has proposed an alternative model based on the idea of a collective phase transition involving connected sequences of biochemical transformations. He likens the process to one in which a set of pegs distributed on a floor are gradually connected by strings until, in due course, they reach a critical point where they combine to form a net—a synergistic effect. Indeed, Kauffman characterizes the origins of life as an "emergent property" of complex systems of polymer catalysts; life, he says, has an "innate holism."

Molecular Macrostructures

The macrolevel material world exhibits many kinds of synergy. Indeed, Mendeleev's periodic table is a monument to the extraordinarily diverse ways in which the basic building blocks of nature can be combined to produce emergent new phenomena with a great variety of synergistic physical properties and effects. Likewise, the multifarious chemical compounds that are found in nature (or synthesized in our laboratories and factories) are products of the covalent, ionic, and coordinate bonds that glue atoms together. Some are so commonplace that we take them for granted: water, table salt, ammonia. Others are more exotic. Buckminster fullerenes (C_{60} and several variants) provide a particularly apt example, because these recently synthesized molecules of pure carbon are named for the well-known engineer who invented geodesic domes (and incidentally promoted the concept of synergy). Nicknamed the "buckyball," C_{60} was given this moniker because it derives its extraordinary stability from its physical resemblance to a geodesic dome, or a soccer ball. Most important, it is the gestalt—the geometry of the whole—that gives the buckyball its distinctive collective properties (Curl and Smalley 1988).

Supramolecular Chemistry

The chemistry of molecular assemblies and intermolecular interactions is a fast-developing, interdisciplinary enterprise. It is focused on the processes by which substrates bind to enzymes, how drugs find their targets, how coordinated actions occur among molecular assemblies, how signals propagate between cells, and so on. Among the remarkable features of this research domain is the fact that information plays a key role, both in the processes of polymolecular self-assembly and in the intermolecular interactions, with results that are systemic in nature (Lehn 1993).

DNA

In the age of biotechnology and recombinant DNA, the renowned three-letter acronym for deoxyribonucleic acid is a household word. Most schoolchildren now learn about the double helix and the four nucleotide letters that make up the genetic code. What is often glossed over is the fact that the properties of DNA are highly cooperative in nature. There is, first of all, the double-stranded, antiparallel backbone in which phosphate groups alternate with molecules of a sugar, deoxyribose, to form covalently linked chains—a structure that hangs together because its atoms share pairs of electrons. Also, the four nucleotide bases—each a complex synthesis of carbon, nitrogen, hydrogen, and (except for adenine) oxygen—can only perform their vital informational function because of their very specific complementarities: adenine pairs only with thymine, and guanine only with cytosine. It is the order in which the bases are arrayed in various three-letter "codons" (similar to the words in table 1) that gives DNA its capacity for constructing amino acids. Furthermore, the functional capabilities of DNA depend on the highly coordinated transcriptional role played by three distinct forms of RNA (a division of labor), as noted above. Finally, the construction of a living organism requires a complex, multilayered fabrication process. An estimated 2.5 billion base pairs are required to define the 30,000 plus genes in the human genome. (Even a simple virus like the much-studied SV40 in monkeys has 5,243 base pairs and five genes.) The genes function cooperatively to construct the twenty different amino acids, which in turn are used to build several thousand different proteins. All of these cooperative processes must operate flawlessly; if there is even a single alteration in the gene sequence that codes for, say, hemoglobin, the result may be sickle-cell anemia, or worse.

The Homeobox

It has long been appreciated that genes generally do not act alone in producing the phenotypes of the next generation. Many years ago, the geneticist Theodosius Dobzhansky (1937) demonstrated that, even in *Drosophila* (fruit flies), factors on all of the chromosomes (*Drosophila* have four) may contribute to as simple a trait as the size of the testes, and many experiments since then have confirmed the cooperative nature of the genome's functional organization. A particularly dramatic example is the homeobox domain, which was discovered in the 1980s (Gehring 1985; Maynard Smith and Szathmáry 1995). The homeobox is a distinctive DNA segment, consisting of some sixty amino acids, that is found in all of the so-called homeotic genes and, remarkably, has been conserved over many millions of years of evolution in organisms ranging from fruit flies to humans. (There are about ten homeobox segments in *Drosophila* and at least forty, arranged in four complexes on different chromosomes, in humans.) The homeotic genes determine the basic body plan of a given organism and serve as key regulators of embryonic development; they establish the body's architecture and tell the developing embryo what kinds of structures to make in each location. This is, in effect, a microscopic example of a combination of labor that is oriented to the production of a combined, emergent result (De Robertis et al. 1990). More recently, human genome mapping projects have greatly expanded our appreciation for the synergies that are involved in morphogenesis (Goodfellow 1995; Little 1995).

Neuronal Transmission

The human brain and nervous system are characterized by a complex and as yet only partially understood division (and synthesis) of labor among numerous functionally specialized areas that are interconnected by an estimated one hundred billion neurons. Many aspects of the brain's *modus operandi* (particularly the "binding" process by which the activities of the various parts are integrated into the whole that constitutes our conscious experience) still elude us. However, what we do know affirms that the workings of the brain are synergistic. We know that the neurons are in constant communication with one another via a network of staggering complexity. An individual neuron may have anywhere from five hundred to twenty thousand synaptic connections with other neurons, and at any given moment millions of neurons are firing in concert in a highly cooperative process. Even the transmission of a signal (an electrochemical impulse) by a single neuron turns out to be synergistic. In brief, the neuronal firing process involves what Francis Crick (1994) characterizes as a "complex dynamic sum" of both excitatory and inhibitory inputs from all of the other neurons with which the neuron is in contact via the synapses (or junctions) between its own dendrites and the axonal endings (or knobs) of neighboring neurons. The way in which signals bridge the synaptic cleft, the gap between neuronal junctions, is also highly cooperative. Even the mechanics of the transmission process within each neuron are, in Crick's words, a "chemical miracle." It is not at all like electricity flowing through a wire; it is an intricate process of chemical (ionic) balance shifts in electrical potentials, a dynamic that is facilitated by an elaborate system of molecular gates and metabolic pumps. As Crick concludes, "A neuron, then, is tantalizingly simple. . . . It is only when we try to figure out exactly how it responds . . . that we are overwhelmed by the inherent complexity of its behavior. . . . All this shows, if nothing else, that we cannot just consider one neuron at a time. It is the combined effect of many neurons that we have to consider" (1994, pp. 103–4).

Coevolution

A term coined by biologists Paul Ehrlich and Peter Raven (1964) and subsequently developed by a number of other biologists (Thompson 1982; Futuyma and Slatkin 1983; Nitecki 1983), *coevolution* refers to the aspect of the evolutionary process that is driven by the interactions among species. Some are mutually beneficial; some are commensalistic (with benefits to one or more species without apparent detriment to others); and some are competitive, parasitic, or predatory. Broadly defined, coevolution could encompass a major part of the total evolutionary process. But the consensus seems to be that the term should only be applied to situations in which one species becomes a selection pressure for another species in such a way that it precipitates stepwise directional changes over a number of generations in two (or more) species via reciprocal causation—a sort of arms race or, conversely, a process of progressive accommodation and mutual facilitation. The paradigm-defining example, provided by Ehrlich and Raven, is butterflies and the plants on which they feed. Over the course of time, many plants have evolved chemical compounds that are apparently without physiological functions (alkaloids, quinones, glycosides, flavonoids, etc.) that seem to be repugnant to otherwise predatory butterflies. The butterflies, in turn, appear to have diversified their diets (and their habitats) and to have evolved appropriate new digestive and concealment adaptations. In more extreme cases, coevolution may resemble what biologist Leigh Van Valen characterized as a Red Queen's race (from Lewis Carroll's *Through the Looking Glass*), in which interacting species must run as fast as they can just to stay in place (see also Vermeij 1987). But, in any case, coevolutionary processes are relational and synergistic.

Symbiosis

The term *symbiosis* is generally used by biologists to connote the living together of "dissimilar" organisms—sometimes for their mutual benefit and sometimes not. The classic example of a mutualistic symbiosis is lichen, a generic label for the roughly twenty thousand different species of partnerships between some three hundred genera of fungi and various species of cyanobacteria and green algae. Although many lichen partners can apparently live independently, in combination they enjoy significant functional advantages (synergies). (Empirical support can be found in a recent quantitative analysis by Raven 1992.) Fungi have gripping and water-retention capabilities that can be especially advantageous in a relatively harsh or barren environment, and the cyanobacteria or green algae bring photosynthesizing capabilities to the partnership. The symbionts also commonly combine forces to produce a thallus. Some lichens even reproduce together (asexually) via symbiotic diaspores. Although symbiosis is often equated with mutualism, it also includes many examples of parasitism—relationships that may or may not be deleterious (negative synergy) for one of the partners, including many cases in which the functional consequences vary with the circumstances. For example, the so-called VAM (vesicular-arbuscular mycorrhizal) fungi that are models of mutualism with many species of plants do in fact enhance plant growth in low-phosphorus soils, but in high-phosphorus soils or in low-light conditions (when photosynthetic activity is reduced), they may become parasitic and reduce plant growth (Bethlenfalvay et al. 1983; Daft and El-Giahmi 1978). Ten years ago, symbiosis was still considered by many biologists to be a minor theme in evolution. However, a number of subsequent developments have elevated symbiosis to a place at the head table. Not only is the endosymbiotic origin of eukaryotic cells—a major turning point in evolution—now widely accepted, but there is a recognition that mutualistic and commensalistic associations (not to mention parasitism) are widespread throughout the living world. For instance, there is growing evidence of mutualism between many species of plants (Margulis 1993; Hunter and Aarssen 1988). Perhaps most remarkable, we are discovering a vast new domain of mutualistic and parasitic interactions at the microlevel, among bacteria, viruses, and plasmids (Sonea and Panisset 1983; Weinberg 1985; Margulis 1993; Margulis and Sagan 1995).

Sociobiology

As defined by Edward O. Wilson (1975), sociobiology is concerned with behavioral relationships among members of the same species, ranging from parent-offspring interactions to elephant seal harems to the tightly integrated division of labor in a number of social insect species. One of the most impressive examples of the latter is *Eciton burchelli,* a species of army ants found in Central and South America. *E. burchelli* form highly organized colonies of about five hundred thousand members, with four morphologically distinct castes (in addition to the queen) that divide up the responsibilities for colony defense, foraging, transport, nest making, and care of the brood. The so-called submajors (or porters), for instance, team up to carry sometimes very large prey that, even if split up into pieces, would be more than the individual ants could carry alone. *E. burchelli's* highly coordinated foraging system is legendary. In a single day, a raiding party of up to two hundred thousand individuals, marching in a dense phalanx, may cover as much as two hundred meters and reap some thirty thousand prey items, many of which must be transported back to the nest. (Army ants are actually top carnivores; so far as is known, no species preys on them, so formidable are their combined defensive capabilities.) One of the most remarkable features of *E. burchelli's* adaptive strategy is the fact that the workers form nests out of their own interlinked bodies and are able to maintain an internal nest temperature within ±1°C. Moreover, during their fifteen-day nomadic periods, the colony moves its nest daily (to provide sufficient food for its growing brood), a process that involves a highly complex, coordinated maneuver. As ecologist Nigel Franks notes, army ant colonies also display flexible, problem-solving behaviors as an emergent property of the colony; their actions are not centrally controlled (Franks 1989; Franks et al. 1991).

Cultural Evolution

The Igorot provide an example of how even a primitive human economy may depend on an intricate network of ecological, technological, social, and political components. When they were studied in the 1970s by anthropologist Charles B. Drucker (1978), the Igorot occupied a mountainous area of Luzon, in the Philippine Islands, where for centuries they had practiced irrigated rice cultivation within an awe-inspiring system of earthwork terraces, dams, and canals that were laboriously carved with simple tools out of the precipitous mountainsides. One key to the Igorot's subsistence mode was the remarkable, sustained fecundity achieved by the constant replenishment of soil nutrients, especially nitrogen. This depended on the presence in the rice ponds of nitrogen-fixing cyanobacteria that maintained a symbiotic relationship with the rice plants. Thus, over a period of several centuries, the Igorot were able to grow almost enough staple food on a single hectare to feed a family of five. Yet this is only part of the story. The Igorot's cultural adaptation also depended on a cooperative set of social arrangements. Whereas the ancestral Igorot had lived in small family groups and practiced a form of shifting, small-scale (slash and burn) plant cultivation, the successful adoption of the rice terrace mode of production required the coalescence of an integrated social organization. Sustained cooperative efforts became necessary, first to build the rice-terrace system and then to utilize, maintain, and expand it over time, for without constant weeding and repairs the system would soon deteriorate. In effect, the productivity of each individual family plot depended on the productivity of the whole. Accordingly, the Igorot had to invent a set of social and political structures and processes (and cultural norms) to coordinate in a disciplined manner the activities of the many previously isolated family groups. (A more recent study of the Balinese water temple system and its relationship to ecological and subsistence patterns, utilizing a dynamical systems model too complex to detail here, suggests the possibility that cultural practices with emergent, collective effects may also arise from self-organizing processes. See Lansing and Kremer [1993].)

Some Implications

As the sampler provided in the previous section suggests, a broad definition of synergy as cooperative or combined effects of all kinds casts a very wide net over the phenomenal world. Can such a broad definition be useful? I believe the answer is emphatically yes. Cooperative interactions are everywhere in nature, to be sure, but the particular focus of a synergy orientation is the subset of all imaginable interactions that have combined functional effects (positive or negative) for those aspects of the material world that we wish to understand more fully. Synergy shifts our theoretical focus from mechanisms, objects, or discrete bounded entities to the *relationships* among things, and, more important, to the functional effects that these relationships produce. Synergistic causation is configurational; synergistic effects are always codetermined.

A point that was made earlier should also be stressed at this juncture. Synergy is real. Its effects are measurable, or quantifiable: for example, economies of scale, increased efficiencies, reduced costs, higher yields, lower mortality rates, and a larger number of viable offspring. More subtle measuring rods for synergistic effects include enhanced stability properties, greater stress tolerance, increased fidelity in reproduction, the melding of functional complementarities to achieve new properties, and so forth. A frequently invoked example of nutritional synergy can be used to illustrate: One-half cup of beans provides the nutritional equivalent (in terms of usable protein) of two ounces of steak, and three cups of whole wheat flour provide the equivalent of five ounces of steak. Eaten at separate times, the two food items contribute the equivalent of seven ounces of steak. But because of the complementarity of their amino acids, if the two substances are consumed together they provide the equivalent of nine ounces of steak, or 33 percent more useable protein. Here is a case where, literally, the whole is greater than the sum of its parts.

Another illustration can be derived from rowing, a sport that seems to be popular with biologists. In his near-legendary book *The Selfish Gene*, Richard Dawkins (1989) concedes that genes are not really free and independent agents: "They collaborate and interact in inextricably complex ways. . . . Building a leg is a multi-gene cooperative enterprise" (p. 37). To underscore the point, Dawkins employs a rowing metaphor: "One oarsman on his own cannot win the Oxford and Cambridge boat race. He needs eight colleagues. . . . Rowing the boat is a cooperative venture" (p. 38). Furthermore, Dawkins notes, "One of the qualities of a good oarsman is teamwork, the ability to fit in and cooperate with the rest of a crew" (p. 39). (The obvious inconsistencies between these statements by Dawkins and his ruling selfish gene metaphor are treated at length in Corning 1996a, 1997.) Recall also that Maynard Smith and Szathmáry (1995) used as a metaphor specifically for synergy the image of two men in a rowboat, each with one oar. If only one oarsman is rowing, the boat will go in circles.

A quantitative example can be added to these nautical metaphors: A world-class varsity eight (plus coxswain) can cover two thousand meters over the water in about 5.5 minutes. However, a single sculler can at best row the same distance in about 7 minutes. The difference is a synergistic effect, and if rowing faster were a matter of survival (and it may very well have been at various times in our history as a species), the cooperators would be the fittest.

Toward a Multilevel Paradigm in the Sciences

The emergence of a multilevel selection paradigm in evolutionary biology was noted in chapter 2. Here we will elaborate on this important development. The idea that living systems have a multilayered structure, and dynamics, is certainly not new. In the first quarter of the twentieth century this idea was espoused by, among others, Lester Frank Ward, Conwy Lloyd Morgan, Samuel Alexander, Roy Wood Sellers, and others. In the 1960s the idea was championed by biologist Paul Weiss (1969, 1971), who characterized living systems as hierarchical structures in which, as he put it, the causal dynamics are "stratified" (see also Pattee 1973). But perhaps the most influential statement of this idea was an article in *Science* by the chemist Michael Polanyi (1968) entitled "Life's Irreducible Structure." Polanyi argued that the natural world consists of a hierarchy of "levels" that can be identified empirically in relation to distinct "boundary conditions" that impose more or less inclusive constraints on the laws of nature. Each level works under principles that are irreducible to the principles governing lower levels. Thus, the "laws" governing the properties of DNA are not reducible to the laws of physics and chemistry. Nor are the principles governing morphogenesis reducible to those that govern nucleic acids. Equally important, the principles that control higher levels may serve to restrict, order, and "harness" lower levels. To use one of Polanyi's examples, the grammatical rules that govern the structure of various human languages utilize but also subsume the principles of phonetics. Accordingly, any hierarchically organized phenomenon may embody several distinct sets of level-specific principles.

In another classic article in *Science* a few years later, entitled "More Is Different," physicist Philip Anderson (1972) reinforced and expanded on Polanyi's argument. "The reductionist hypothesis does not by any means imply a 'constructionist' one: The ability to reduce everything to simple fundamental laws does not imply the ability to start from those laws and reconstruct the universe. . . . The constructionist hypothesis breaks down when confronted with the twin difficulties of scale and complexity. . . . At each level of complexity entirely new properties appear. . . . Psychology is not applied biology, nor is biology applied chemistry" (p. 393). Anderson used the now-familiar term *broken symmetry* to characterize such qualitative shifts. Examples cited by Anderson included complex organic molecules, superconductivity, and crystal lattices. "We can see how the whole becomes not merely more but very different from the sum of its parts," Anderson concluded. (Needless to say, these arguments have been reiterated many times since the early 1970s.)

Although these landmark publications helped to legitimize the development of such umbrella disciplines as systems theory, theoretical ecology, hierarchy theory, chaos theory, dynamical systems theory, and complexity theory, neither one squarely addressed the central questions: What precisely are the relationships between levels (and between disciplines)? What are the causal factors in the phenomenal world that are responsible for producing the phase transitions and dynamical attractors that are simulated in our mathematical models? In other words, what are the functional relationships between parts and wholes? The frequent use by scientists of such descriptive terms as *emergence, interdependency, interactions, positive cooperativity, codetermination,* and even *synergy* represent at least a tacit acknowledgment that the various levels of organization in nature are connected to one another. What is less frequently acknowledged—and sometimes even denied—is the fact that various levels may be interdependent; wholes and parts may interact, and coevolve, in complex ways. Indeed, the very concept of hierarchical levels may sometimes become an obstacle to understanding. (For more extended discussions of these issues, see P. A. Weiss 1971; Pattee 1972; Wimsatt 1974; Kline 1995; see also chapter 5).

One implication of a synergy perspective is that it points us to a paradigm that explicitly focuses on both wholes and parts, and especially the interactions that occur among the parts, between parts and wholes, and between wholes at various levels of interaction and causation. The synergy perspective might be called a science of relationships, as distinct from a science of mechanisms or laws. Thus, the phenomenon of consciousness may well be a synergistic product of a vastly complex set of interactions within the machinery of the brain and between the brain and its environment. Moreover, synergistic wholes are also units of causation in the phenomenal world—and in evolutionary change (as discussed in chapter 2).

Of course, the relationship between parts and wholes is often more complicated than that. Parts may constrain the wholes in various ways. The mobility of a particular animal species, for instance, is strictly limited by the capabilities of its locomotive machinery, which in turn has been shaped by a complex nexus of influences ranging from the principles of physics and thermodynamics to the particular evolutionary history of its lineage. An organism's parts may also establish functional priorities for the whole (e.g., its pressing survival and reproductive needs). Conversely, higher levels in the biological hierarchy may set priorities for lower levels. Thus, social insects, we now appreciate, respond in complex ways to the cues and signals that arise in their social environments—at the "superorganism" level of organization (in Herbert Spencer's original formulation) and even beyond. (Indeed, many of these higher-level cues may be combined properties of the whole.) Thus, even though the proximate mechanisms of behavioral control in social insects may be distributed, the organizational and functional principles are nevertheless superordinate and holistic.

To reiterate, then, there is both upward and downward causation in nature, and very often a synthesis of the two. Moreover, wholes of various kinds may become interdependent units of selection and evolution, just as, conversely, the evolution of various parts may be shaped by the functional requisites of the whole. An example of the latter are the army ant submajors described above; their large size and long legs are morphological adaptations that reflect their role in the army ants' legendary division of labor. In like manner, symbionts (say mitochondria and their primitive protist hosts) may coevolve adaptations that serve the functional needs of the partnership as an emergent unit of selection (an obvious example is the synchronization of reproductive efforts in symbionts).

Multiple Levels and Multiple Disciplines

Accordingly, this perspective implies a multileveled, interactional research paradigm, one that gives equal weight to both reductionist and holistic perspectives and invites both intra- and interlevel analyses and explanatory models. (A number of philosophers of science have discoursed on this issue in recent years. In particular, see Wimsatt 1974 and Bechtel 1986.)

Water, perhaps the most studied of all substances, provides an illustration from the physical sciences. Despite our considerable knowledge of the remarkable substance that covers 70 percent of the earth's surface and comprises about 65 percent of our bodies by weight, there are still properties and aspects of its behavior that we do not understand. The basic atomic properties of water have been understood for almost two centuries, thanks to John Dalton. We also know a great deal about the chemistry, statics, dynamics, and thermodynamics of water, which is subject to numerous macrolevel physical principles (as Polanyi pointed out). We understand, for instance, how the constituent atoms of hydrogen and oxygen are linked together by covalent bonds. We know that quantum theory is required to explain some of the remarkable energetic properties of water. Additional principles of chemistry are needed to account for the state changes that produce water from its constituent gases and, under appropriate conditions, the changes that can reverse the process. Still other principles are required to account for the macroscopic properties of water as a liquid medium: its compressibility, surface tension, cohesion, adhesion, and capillarity. Thermodynamic principles are needed to understand the dynamics of temperature changes in water. Static principles relating to density and specific gravity must be invoked to account for the buoyancy of rowboats and varsity eights. Hydraulics is needed to understand how water reacts to a force exerted upon it. Dynamics and Newton's laws are relevant for understanding the tidal action of water in large bodies, and hydrodynamics is required to explain the behavior of water flowing through a pipe, or in a river bed. Here Bernoulli's principle also becomes relevant.

And yet, despite all of this knowledge, we still do not know exactly how water molecules "network" with other water molecules—a key to understanding how water can be so fluid and yet have such an anomalously large capacity for absorbing heat and holding other substances in suspension (Amato 1992). Significantly, progress in studying various kinds of intermolecular interactions in water is being made via interdisciplinary efforts. In a published report on molecular clusters in water, Colson and Dunning (1994) conclude, "This work also illustrates the synergism that has developed between experimental and theoretical studies in modern chemical physics."

By the same token, at the most inclusive geophysical level, the problem of understanding the role of water in world climate patterns presents a formidable research challenge that has necessitated multileveled, multidisciplinary modeling efforts. Larry Goldberg (1994), a philosopher of science, has studied this research domain intensively. The complexity of the problem arises from the interdependence of various component subsystems—the atmosphere (troposphere and stratosphere), the oceans and other large water bodies, the so-called cryosphere (continental ice sheets), the lithosphere (the earth's crust and upper mantle), and the biosphere (the activities of the earth's biota). Each of these subsystems, which cut across the subject matter of at least half a dozen different disciplines, presents a complex set of modeling problems in its own right. Yet they also interact in different ways depending on the particular spatial location and timeframe. Consider, for example, the impacts on the oceans from fluctuations in solar output over various scales: days (r mode oscillations), years (quasi biennial oscillations), tens of years (eleven- and twenty-two-year solar magnetic cycles), and hundreds of years (Maunder-minimum type cycles), not to mention such regularities as the time of day and seasonal cycles, and variables such as cloud cover and cloud density. Thus, as Goldberg notes, at any given location, date, and time of day, the level of solar radiation being absorbed by the oceans depends upon an extraordinarily complex set of interacting (synergistic) causal factors. These synergies demand multidisciplinary analyses.

Similar multidisciplinary challenges confront the life sciences. A particularly striking example at the micro level is a recent study reported by Wang and his coworkers (1993) showing that the movement, shape, and polarity of individual cells in a multicellular cluster depends on close cooperation among proteins outside the cell in the extracellular matrix (ECM), proteins that are found on the surface of the cell (cell adhesion molecules, or integrins), and proteins inside the cell (the cytoskeleton). Not only are there close interconnections between these system components but they are organized according to the so-called tensegrity (tensional integrity) architecture that underlies Buckminster Fuller's geodesic domes. Furthermore, this parts-whole-environment interaction has been illuminated by the melding of two separate disciplines—cytomechanics and detailed biochemistry (Heidemann 1993).

The same sort of challenge applies to the macrobiotic level, where there have been various efforts by theoretical ecologists in recent years to address parts-wholes interactions and the complex feedbacks that exist among various ecosystem levels. Particularly notable is the hierarchical network approach developed by ecologist Claudia Pahl-Wostl (1993, 1995), which utilizes information theory and patterns of interlevel feedback in an effort to capture the spatiotemporal dynamics of an ecosystem. Pahl-Wostl (1993, p. 31) concludes, "These order parameters arise from the interactions among the components of the systems through processes of self-organization. Along this line of reasoning the dichotomy between top-down and bottom-up control converges to a mutual and inseparable dependence on both factors. Neither a purely reductionist approach nor a merely holistic perspective is sufficient to encompass the intrinsic nature of the system's behavior."

A further implication is that the phenomenon of synergy is more than simply the end point or outcome of the processes that drive the phenomenal world. Synergy is also an important source of causation in the ongoing evolutionary process. Indeed, a synergy focus directs our attention to one of the major wellsprings of creativity in evolution. The novelist and polymath Arthur Koestler (Koestler and Smythies 1969) observed that "true novelty occurs when things are put together for the first time that had been separate." A number of examples were cited above: the emergent properties of chemical compounds; the mitochondria that provide eukaryotic cells with specialized metabolic capabilities; the functional complementarity of the lichen partnerships; the exotic compounds that comprise super alloys and composite materials.

The Synergism Hypothesis

In *The Synergism Hypothesis* (Corning 1983), it was proposed that synergistic phenomena of various kinds have played a key causal role in the evolution of cooperation generally and the evolution of complex systems in particular; it was argued that a common functional principle has been associated with the various steps in this important directional trend. The reasoning behind this hypothesis can be briefly summarized.

First, it is necessary to return to the problem of defining natural selection—a much-debated subject and an issue that may seem tiresome to those who are already familiar with the debate. Despite the volume of material on the subject, misunderstandings persist. As previously mentioned, evolutionists often speak metaphorically about natural selection. Thus, George Gaylord Simpson (1967, p. 219) asserted that "[t]he mechanism of adaptation is natural selection. . . . [It] usually operates in favor of maintained or increased adaptation to a given way of life." Similarly, Ernst Mayr (1976, p. 365) informed us that "[n]atural selection does its best to favor the production of programs guaranteeing behavior that increases fitness." In his discipline-defining volume *Sociobiology* (1975, p. 67), E. O. Wilson assured us that "natural selection is the agent that molds virtually all of the characters of species." More recently, E. O. Wilson (1987) provided a more ecologically oriented definition of natural selection as "all the events that cause differential survival and reproduction." Nor does it clarify matters when Dawkins (1989, p. v) characterizes living organisms as "robot vehicles blindly programmed to preserve the selfish molecules known as genes," which implies that genes are the locus of evolutionary causation (see also Endler 1992).

The problem is that natural selection is not a mechanism. Natural selection does not *do* anything; nothing is ever actively selected (although sexual selection and artificial selection are special cases). Nor can the sources of causation be localized either within an organism or externally in its environment. In fact, the term *natural selection* identifies an aspect of an ongoing dynamic process. It is an umbrella concept that refers to whatever functionally significant factors are responsible in a given context for causing differential survival and reproduction. Properly conceptualized, these factors are always interactional and relational; they are defined by both the organisms and their environments.

This crucially important point can be illustrated with a textbook example of evolutionary change—industrial melanism. Until the Industrial Revolution, a "cryptic" (light-colored) species of the peppered moth (*Biston betularia*) predominated in the English countryside over a darker "melanic" form (*carbonaria*). The wing coloration of *B. betularia* provided camouflage from avian predators as the moths rested on the trunks of lichen-encrusted trees, an advantage that was not shared by the darker form. But as soot blackened the tree trunks in areas near growing industrial cities, in due course the relative frequency of the two forms was reversed; the birds began to prey more heavily on the now more visible cryptic strain (Kettlewell 1973).

The question is, where in this example was natural selection located? The short answer is that natural selection encompasses the entire *configuration* of factors that combined to influence differential survival and reproduction. In this case, an alteration in the relationship between the coloration of the trees and the wing pigmentation of the moths, as a consequence of industrial pollution, was an important proximate factor. But this factor was important only because of the inflexible resting behavior of the moths and the feeding habits and perceptual abilities of the birds. Had the moths been prey only to insect-eating bats that use "sonar" rather than a visual detection system to catch insects on the wing, the change in background coloration would not have been significant. Nor would it have been significant had there not been genetically based patterns of wing coloration in the two forms that were available for selection. (Later studies concerning the additional influence of air pollution can be left out of the discussion for our purpose.)

Accordingly, one cannot properly speak of mechanisms or fix on a particular selection pressure in explaining the causes of evolutionary change via natural selection. One must focus on the interactions that occur within an organism and between the organism and its environment, inclusive of other organisms; natural selection is about adaptively significant changes in organism-environment *relationships*. But this begs the question: what factors are responsible for bringing about changes in organism-environment relationships? The answer, of course, is many things. It could be a functionally significant mutation, a chromosomal transposition, a change in the physical environment, a change in one species that affects another species, or it could be a change in behavior that results in a new organism-environment relationship. In fact, a whole sequence of changes may ripple through a complex pattern of relationships. For instance, a climate change might alter the ecology, which might induce a behavioral shift to a new habitat, which might encourage an alteration in nutritional habits, which might precipitate changes in the interactions among different species, resulting ultimately in the differential survival and reproduction of alternative morphological characters and the genes that support them. (An excellent in vivo illustration of this causal dynamic can be found in the longitudinal research program in the Galápagos Islands among "Darwin's finches." See Grant and Grant [1979, 1989, 1993, 2002] and Weiner [1994].)

To underscore this conceptualization of natural selection, which is rather more subtle than the shorthand characterizations that are often found in the literature, I will provide one more example. English land snails (*Cepaea nemoralis*) are subject to predation from thrushes, which have developed the clever habit of capturing the snails and then breaking open their shells with stones. Accordingly, a behavioral innovation (including tool use) in one species became a cause of natural selection in another species. However, two additional factors, one genetic and the other ecological, have also influenced the course of natural selection in *C. nemoralis*. It happens that these snails are polymorphic for shell banding patterns, which provide varying degrees of camouflage. The result is that the more cryptic genotypes have been less intensively preyed upon than those that are more visible. However, at the ecological level the pattern of predation by thrushes (and the frequencies of the different snail genotypes) varies greatly because the thrush populations, being themselves subject to predators, display a marked preference for well-sheltered localities. So, paradoxically, the snails are generally much less subject to predation in more open areas (see B. Clarke 1975).

The cardinal point in these examples is this: it is the functional (bioeconomic) effects or consequences of various organism-environment pattern-changes, insofar as they may impact on differential survival, that constitute the causes of natural selection. Another way of putting it is that causation in evolution also runs backwards from our conventional view of things; in evolution, functional effects are also causes. It is an iterative process. To use Ernst Mayr's (1965) well-known distinction, it is the "proximate" functional effects that result from any change in the organism-environment relationship that are the causes of the "ultimate" (transgenerational) selective changes in the genotype, and the gene pool of a species. (It should be noted in passing that this dynamic is analogous to E. L. Thorndike's famous law of effect in psychology, which forms the backbone of the behaviorist learning paradigm.)

This is where synergistic phenomena fit into the picture. Cooperative interactions in nature that produce positive functional consequences, however they may arise, can become units of selection that differentially favor the survival and reproduction of the parts (and their genes). In other words, it is the proximate advantages (the payoffs) associated with various synergistic interactions (in relation to the particular organism's needs) that constitute the underlying cause of the evolution of cooperative relationships and complex organization in nature. To put it baldly, functional synergy is the ultimate cause of cooperation (and complexity) in living systems, not the other way around. (As an aside, it is similar to the way in which market forces are said to work in human economies; if a widget sells, more widgets are likely to be produced for sale. If not, the widget will soon go extinct.)

This bioeconomic theory of cooperation and complexification in evolution is particularly relevant to symbiosis and sociobiology. The hypothesis is that the synergies that may result from cooperative behaviors are the very cause of the systematic evolution of those behaviors over time, via their impacts on differential survival and reproduction. Moreover, many of these evolutionary changes originate—and are initially adopted—at the behavioral level (as illustrated above with respect to the thrushes that prey on English land snails). To repeat, as Ernst Mayr (1960) long ago observed, behavioral innovations are often the "pacemakers" of evolutionary change. In C. H. Waddington's words, "It is the animal's behavior which to a considerable extent determines the nature of the environment to which it will submit itself and the character of the selective forces with which it will consent to wrestle" (1975, p. 170).

As noted in chapter 2, the idea that behavioral innovations might be a significant cause of evolutionary change can be traced back to Lamarck. Darwin also alluded to the idea in *The Origin of Species*. At the turn of the twentieth century, a movement among evolutionists of that day known as organic selection theory was developed in an effort to highlight the creative role of behavior in evolutionary change. It was subsequently buried by Weismannism and, much later, was resurrected, downgraded, and renamed the Baldwin effect by George Gaylord Simpson. Waddington himself developed a variation on this theme in the 1950s, which he dubbed "genetic assimilation." And Ernst Mayr has repeatedly argued the case for behavior as a cause of evolution in his various writings (reviewed in Corning 1983, 2003; also see Plotkin 1988; P. P. G. Bateson 1988; P. P. G. Bateson et al. 1993; J. H. Campbell 1994; Avital and Jablonka 1994, 2000; Deacon 1997; Weber and Depew 2003; cf. Skinner 1981).

Evidence for the Synergism Hypothesis

The evidence for the presence of synergy at every level of living systems is compelling. Here I will simply mention a few of the highlights.

Synergy was most likely involved in the very origins of life. Indeed, it is an implicit premise in every one of the various formal hypotheses that have been proposed for the earliest steps in the evolutionary process, from Eigen and Schuster's (1977, 1979) hypercycles to Szathmáry and Demeter's (1987) stochastic corrector model and Wächtershäuser's (1988, 1990) surface metabolism model. All share the common assumption that cooperative interactions among various component parts played a central role in catalyzing living systems.

DNA, the basic molecule of life, also utilizes synergy. Among other things, the double-stranded, antiparallel backbone, or scaffolding, of each giant DNA molecule hangs together only because there are covalent bonds that "glue" together the atoms of its constituent phosphate and deoxyribose molecules. By the same token, the vital role of DNA in biosynthesis is made possible by a highly coordinated division of labor between three different forms of RNA—the messenger RNA that makes copies of the relevant DNA sequence, the transfer RNA that assembles the appropriate amino acids, and the ribosomal RNA that lines up the amino acids in the proper order for assembling a protein.

Similarly, at the level of the genome, it goes without saying that genes do not act alone, even when major single-gene effects are involved. An example (noted above) is the so-called homeobox gene complex, which is responsible for defining the basic body plan for a wide range of organisms, from insects to humans. As was also noted above, the human genome sequencing project has established, among other things, that there are 1,195 distinct genes associated with the human heart, 2,164 with white blood cells and 3,195 with the human brain (Little 1995).

The origin of chromosomes, likewise, may have involved a cooperative or symbiotic process (see Maynard Smith and Szathmáry 1993). Sexual reproduction, one of the major outstanding puzzles in evolutionary theory, is also a cooperative phenomenon, as the term is used here. Although there is still great uncertainty about the precise nature of the benefits, it is assumed that sexual reproduction is, by and large, a mutually beneficial joint venture.

As we move up the ladder of complexity, we find further variations on the theme of functional cooperation. Once upon a time bacteria were considered to be mostly loners, but this is no longer the case. It is now recognized that large-scale, sophisticated cooperative efforts—complete with a division of labor—are commonplace among bacteria and can be traced back at least to the origin of the so-called stromatolites (rocky mineral deposits) that were constructed by bacterial colonies some 3.5 billion years ago (J. A. Shapiro 1988; J. A. Shapiro and Dworkin 1997; Margulis 1993). Shapiro suggests that bacterial colonies can be likened to multicellular organisms.

Complex eukaryotic cells (some of them ten thousand times the size of a bacterium) can also be characterized as cooperative ventures—obligate federations that may have originated as symbiotic unions (parasitic, predatory or perhaps mutualistic) between ancient prokaryote hosts and what have now become cytoplasmic organelles, particularly the mitochondria, the chloroplasts and, possibly, eukaryotic undulipodia (cilia) and certain internal structures that may have evolved from structurally-similar spirochete ancestors (Margulis 1993).

Support for the Synergism Hypothesis

Over the past decade or so there has been a growing appreciation for the role of synergy in the natural world. Some explicit applications of the synergy concept include Kondrashov's (1982, 1988) hypothesis regarding the basis of sexual reproduction, which relies on synergistic linkages between deleterious mutations. Similarly, Maynard Smith and Szathmáry's (1993) theory of the origin of chromosomes postulates a synergistic relationship among primordial genes, as noted above. Szathmáry (1993) also utilizes the concept in a model derived from metabolic control theory which suggests that, under some conditions, two mutations affecting a metabolic pathway could act synergistically. Gary Rosenberg (1991) postulates a necessary role for "synergistic selection" in the evolution of warning coloration (aposmatism) in marine gastropods. Synergy occurs when a potential predator has multiple "distasteful" encounters with the same morph, which enhances the joint selective value for each bearer. (See the further discussion of this issue in the contributions by Guilford and Cuthill 1991; Tuomi and Augner 1993.) Hurst (1990) suggests that parasite diversity in a given cell or organism may be more burdensome than a similar quantity of uniform types, because various synergistic interactions among different parasites may enhance their mutual effects. Hurst proposes that diploidy, multicellularity and anisogamy may all be anti-parasite mechanisms; they might serve to reduce parasite diversity.

Leo Buss (1987), as noted above, utilizes the concept of synergy (or what he calls "synergisms") in a broader theoretical context, as an explanatory principle in connection with the evolution of metazoa and "higher" units of selection. Though he never explicitly defines the term, his usage is idiosyncratic; he equates synergy with positive, or mutually beneficial relationships between lower and higher levels of organization, or wholes and parts, as contrasted with conflicts between levels. "The organization of any unit will come to reflect those synergisms between selection at the higher and the lower levels which permit the new unit to exploit new environments and those mechanisms which act to limit subsequent conflicts between the two units" (1987, p. viii). (For more recent treatments of this issue, see Michod 1997, 1999; also Frank 1998, 2003.)

Another positive development is the growing number of field research programs in the behavioral sciences, especially in behavioral ecology, that are explicitly looking for, and finding, synergy. For instance, Gordon (1987) observed "synergistic interactions" among three major activities in colonies of red harvester ants (*Pogonomyrmex barbatus*) in response to various perturbations. Santillán-Doherty and his colleagues (1991), in a study of stump-tailed macaques (*Macaca arctoides*), found nonlinear synergistic effects among three variables—kinship, sex, and rank—in shaping the behavioral interactions among the animals in their study population. And Packer and Ruttan (1988) also explicitly recognized the role of synergy in cooperative hunting. They observed that, when individual hunting success is already high, there is little to be gained by cooperating. Cooperation depends upon synergy—an increase in the average individual feeding efficiency through joint efforts. "An increase in hunting success with group size therefore indicates synergism from cooperation, whereas a decrease indicates some form of interference [negative synergy]" (1988, p. 183). Some other examples of synergy include the following:

- Nest construction in the social wasp (*Polybia occidentalis*) is a complex activity requiring the coordination of various tasks. To study the bioeconomics, Robert Jeanne (1986), conducted a comparative study of small versus large colonies, as well as the nest construction technique used by social wasps versus the less efficient method of solitary wasps. Jeanne found that small colonies required almost twice as many worker-minutes to complete the same amount of construction (due mainly to materials handling inefficiencies that larger colonies could minimize). In addition, he was able to determine that social wasps could collect and process a given amount of nest material with 2.6 times fewer foraging trips than were required by solitary wasps (with the added advantage that the social foragers were able to reduce their exposure to predators in the field). In other words, the synergies here were measurable.

- Marzluff and Heinrich (1991) tested the hypothesis that immature common ravens (*Corvus corax*) form social groups (in contrast with breeding adults that are territorial) in order to gain access to defended carcasses. They found that groups ranging from nine to twenty-nine immature birds were significantly more likely to overcome adult carcass defenders and were able to feed at higher rates than were smaller groups or solitary individuals. The group benefits resulted from a combination of reductions in the neophobia of the foragers and the reduced aggression of adult defenders as group sizes increased.

- In a comparative study of reproduction during a single breeding season among southern sea lions (*Otaria byronia*), Campagna and others (1992) observed that only 1 of 143 pups born to gregarious, group-living females died before the end of the season, as compared to a 60 percent mortality among the 57 pups born to solitary mating pairs. Pups in colonies were protected from harassment and infanticide by subordinate males and were far less likely to become separated from their mothers and die of starvation.

A number of other theorists have implicitly recognized synergistic effects in their studies and analyses without using the term explicitly. Thus, Page and Robinson (1991) refer to "non-additive inter-individual effects" in relation to possible genetic influences on the division of labor in honeybees. Bell (1985) focuses on the non-additive functional efficiencies that arise with specialization and a division of labor in Volvocales and, in his analysis, invokes the reasoning of Adam Smith. Hoogland (1981) stresses that there is a strong relationship between group-size in prairie dog colonies and both functional improvements in the detection of predators and decreased individual scanning activity—efficiencies that are synergistic.

There are also many quantitative, cost-benefit studies (in addition to those mentioned above) that tacitly support the Synergism Hypothesis. To cite a few examples: In birds, Ligon and Ligon (1982) analyzed the communal nesting and extensive helping behaviors among green woodhoopoes (*Phoeniculus purpureus*), both among closely related and unrelated birds. They found that this behavior pattern markedly increased the woodhoopoes' likelihood of survival and reproductive success in an East African environment characterized by a severe shortage of suitable nest sites. A similar pattern was identified by Clarke (1989) in the bell miner (*Manorina melanophrys*). (But note also the more strongly kin-oriented pattern observed in other woodhoopoe populations by Du Plessis 1993.) Parker and others (1994) used DNA fingerprinting to document that food sharing in feeding aggregations of common ravens (*Corvus corax*) in the forests of western Maine was not primarily kin-directed. Møller (1987) analyzed various trade-offs (costs and benefits) of colonial nesting in swallows (*Hirundo rustica*) and concluded that the costs and benefits of coloniality varied markedly with such factors as group size, the frequency of predation, exposure to parasites, and so forth.

In the same vein, Mumme and his coworkers (1988) were able to conduct a comparative cost-benefit analysis of a fifteen-year data set comparing joint-nesting and solitary acorn woodpeckers (*Melanerpes formicivorous*). The data indicated that communally nesting females experienced a fitness trade off: lower average annual reproduction in exchange for higher year-to-year survival rates. In a later study of the Florida scrub jay (*Aphelocoma c. coerulescens*), Mumme (1992) showed that the presence of nonbreeding helpers in experimental groups correlated with lower predation and higher nestling survival rates than was the case with control groups that were denied helpers. And Haig and others (1994), utilizing a DNA analysis with 224 red-cockaded woodpeckers (*Picoides borealis*), found that helping behaviors involved a variety of related and unrelated birds and that there was no direct benefit to the helpers from "extra matings." (J. L. Brown 1987, in a book-length synthesis on communal breeding and helping behaviors in birds, provided additional evidence, although he also observed that, as a rule, unrelated helpers do not seem to work as hard as close kin.)

Recent discoveries that many insect colonies consist of multiple queens or multiple patrilines have presented a challenge to the long-standing inclusive fitness explanation for social insects (a thesis that can be traced all the way back to Darwin). For instance, Queller and others (1988) observed that swarm-founding neotropical wasp colonies (*Parachartergus colobopterus*) may have multiple queens, sometimes numbering in the hundreds, and yet the level of relatedness and inbreeding is low. Similarly, Strassman and coworkers (1994) compared allozyme polymorphisms in incipient social wasps of the subfamily Stenogastrinae and estimated that the average relatedness among colony members in one of the best-studied species (*Liostenogaster flavolineata*) was 0.22, the lowest so far reported for any primitively eusocial insect. And M. P. Scott (1994) found that, in the burying beetle (*Nicrophorus tomentosus*), competition with flies (as well as conspecific groups) promotes communal breeding among unrelated males and females. As Breed (1988) points out, genetic models predict that reduced relatedness among colony members should have a divisive, if not fatal, effect on colony functioning. Nevertheless, eusocial species do exist and are obviously successful—if less than perfectly integrated.

Furthermore, Sherman and colleagues (1988) hypothesized that genetic diversity within social hymenoptera may have a previously unrecognized group-level advantage as a buffer against parasites and pathogens. This hypothesis was subsequently supported in a study of the bumblebee (*Bombus terrestris*) (Shykoff and Schmid-Hempel 1991). In addition, a series of reports by Robinson and Page (1988), Page and others (1989), and Page and Robinson (1991) have supported the hypothesis that the genetic differences observed within honey bee colonies (*Apis mellifera*) can be correlated with performance differences among workers with respect to the division of labor and to ecological variations. In other words, the genetic composition of the colony may reflect downward causation in relation to colony-level functional (bioeconomic) needs; natural selection in this domain may operate on the parameters of the colony as a dynamic system—see below.

Supporting evidence for this hypothesis was also found in both honey bees (*Apis mellifera*) and dwarf honey bees (*Apis florea*) by Oldroyd and his coworkers (1992a, 1992b, 1994). (See also Woyciechowski 1990.) Likewise, Rissing and coworkers (1989) discovered in a field study of the colonial leaf-cutting ant (*Acromyrmex versicolor*) that co-foundresses were unrelated and yet the colonies exhibited specialization without apparent conflict. These researchers concluded that intense between-colony competition and brood raiding provided a group-level selection pressure in favor of such behaviors. (See also the analysis by Mesterton-Gibbons and Dugatkin 1992.)

In social carnivores, Packer and Pusey (1982) observed that breeding coalitions of African male lions included nonrelatives much more commonly than kin selection theory would predict. And Scheel and Packer (1991) found a similar pattern in the hunting and cub-guarding behaviors of female lions. In primates, J. Moore (1984) reviewed and reanalyzed the earlier studies of Goodall, Teleki, McGrew, and others on meat sharing in chimpanzees, a pattern whose potential costs and benefits turned out to be surprisingly complex and were not unambiguously associated with inclusive fitness. Stanford (1992) studied allomothering in capped langurs (*Presbytis pileata*) and found that it could best be interpreted as a low-cost behavior that benefits both related and non-related recipients. And, in the evening bat (*Nycticeius humeralis*), G. S. Wilkinson (1992) documented an extensive pattern of communal nursing of pups that was not preferentially directed to kin.

As noted above, synergy is also implicit in Egbert Leigh's several discussions of how groups are able to contain or override individual advantages for the good of the group (Leigh 1971, 1977, 1983, 1991). To repeat, Leigh argues that, if the potential payoffs (synergies) for each of the participants are high enough, this may provide a sufficient incentive for them to impose government in the common interest. Some examples cited by Leigh include: selection for honest meiosis and the elimination of segregation distorters from diploid genomes; the purging by honeybee workers of eggs laid by other workers rather than the queen (whose offspring represent the products of multiple matings and are genetically more closely related to the workers); the self-regulating division of labor and activity cycles in honeybee hives; the generally harmonious cooperative relationships between eukarytic cells and their endosymbiotic organelles; the suppression by leaf-cutter ants (*Atta*) of reproductive activity in their symbiotic fungi, except when colonizing a new nest; anisogamy in eukaryotes and the transmission of organelles and other cytoplasmic factors exclusively via the maternal line.

Symbiogenesis and the Synergism Hypothesis

A particularly important development in support of the Synergism Hypothesis during the course of the past decade has been the rediscovery of symbiogenesis as a major cause of evolutionary change and complexification. The origin of this hypothesis traces back to an obscure school of nineteenth and early twentieth-century Russian botanists, most notably A. S. Famintsyn (1907a, 1907b, 1918) and K. C. Merezhkovsky (1909, 1920). In fact, it was Merezhkovsky (1920) who coined the term *symbiogenesis*, which he defined as "the origin of organisms through combination and unification of two or many beings, entering into symbiosis" (p. 65). These and others of the Russian school correctly inferred that the chloroplasts in eukaryotic cells are endosymbionts and they also recognized the symbiotic character of lichens. However, their hypothesis was presented as an alternative to Darwin's theory and was generally ignored or rejected in the West. Later, in the 1920s and 1930s, another Russian theorist, B. M. Kozo-Polyansky (1924, 1932) recognized the compatibility between Darwinism and the symbiogenesis hypothesis. As Kozo-Polyansky observed: "The theory of symbiogenesis is a theory of selection relying on the phenomenon of symbiosis" (1932, p. 25). However, Kozo-Polyansky's works were also not known or appreciated in the West; they were published only in Russian and had the misfortune to appear at the height of the Stalinist era. (We are indebted to Liya Khakhina of the Russian Academy of Sciences for her efforts to bring this work to our attention, and for her translations of key passages. See Khakhina 1979; 1992; also Margulis and McMenamin 1993.)

Meanwhile, the American biologist Ivan Wallin (1927) independently advanced a similar hypothesis. He made the "rather startling proposal" (as he candidly acknowledged) that bacteria might be "the fundamental causative factor" in the origin of species (p. 8). Claiming that mitochondria could be grown independently of their host cells (a dubious proposition), his theory was widely rejected by his peers and was soon forgotten. (Even Wallin himself dropped the subject.) However, the endosymbiotic theory of eukaryotes, and the more general theory of symbiogenesis in evolution, was revived once again by Lynn Margulis, beginning in the 1970s. (See especially Margulis 1970, 1981, 1993; also Margulis and Sagan 1986, 1995.) At first widely discounted, the endosymbiosis hypothesis gradually gained recognition over the years as supporting evidence accumulated, and it is now widely recognized as an important source of evolutionary complexification.

As noted in chapter 2, the significance of symbiogenesis in relation to the Synergism Hypothesis is that these creative processes have constituted an important subset of the total universe of synergistic phenomena that have played a causal role in the evolution of complexity. (We will discuss symbiogenesis further in chapter 4). However, the concepts of synergy and symbiosis are not equivalent. The term *symbiosis* is also of Greek origin; it means "living together." It was introduced into biology as a technical term by the pioneering German mycologist Anton de Bary (1879), who employed it to denote the living together of "dissimilar" or "differently named" organisms in lasting and intimate relationships. De Bary's focus was on relationships, and the paradigmatic example, both in de Bary's time and ever since, is lichens (although de Bary also included in his definition what would now be called parasitic relationships). Today, there seem to be a number of conflicting definitions of symbiosis in the literature. Among other things, this reflects important differences of opinion about how the subject matter of the field should be defined, and about which phenomena should be included. Adding to the confusion is the fact that symbiologists are not always consistent in practice even with their own definitions.

Nevertheless, there seems to be general agreement that symbiosis refers to *relationships* of various kinds between biological entities and the functional processes that arise from those relationships. Synergy, on the other hand, refers to the interdependent functional effects (the bioeconomic payoffs) of symbiosis, among other cooperative phenomena. In short, all symbioses produce synergistic effects, but many forms of synergy are not the result of symbiosis. Accordingly, synergy is a room without walls in terms of which kinds of cooperative relationships are applicable; combined effects of all kinds and at every level of living systems are relevant; indeed, the term can even accommodate such unconventional but important biological phenomena as animal-tool "symbioses," not to mention the relationships between humans and their technologies. Synergy can also comfortably handle both mutualistic and parasitic effects, as well as various asymmetrical distributions of costs and benefits and even cooperative effects that defy the conventional categories, as noted above. By focusing on cooperative effects of all kinds, synergy is thus a more pan-disciplinary and inclusive term.

But, in any case, the concept of synergy focuses our attention on the functional effects produced by cooperative interactions of all kinds, including symbioses. This is of great theoretical importance because in evolution it is the functional effects produced by the "interactors" (to use David Hull's term once again) that are the "target" of natural selection, not the relationships per se.

Synergistic Selection Revisited

Maynard Smith's use of the synergy concept is also supportive of the Synergism Hypothesis and deserves a special note. Well known for his introduction of game theory models into evolutionary theory (among other contributions), Maynard Smith (1982a) coined the term *synergistic selection* (as mentioned above). It is more or less synonymous with D. S. Wilson's (1975, 1980) concept of "trait group selection" and a similar formulation by

Matessi and Jayakar (1976), which sought to account for the evolution of altruism without the need for inclusive fitness theory. The general approach involved temporary (functional) interactions among non-relatives in non-reproductive groups. The key feature of the synergistic selection model, according to Maynard Smith, was a fitness gain to interacting altruists that was greater than the gain to an altruist and a nonaltruist.

Maynard Smith discussed the concept of synergistic selection further in a 1983 paper. Again, he paralleled Wilson's trait group selection model, identifying non-additive interaction effects (labeled "r" in his equations) as the critical factor. And again, he assumed that the interaction involved altruism. Subsequently, Queller (1985) elaborated on Maynard Smith's ideas in an analysis of inclusive fitness theory, where he proposed that synergistic effects might provide an alternative to altruism as an explanation for the evolution of social behaviors. Queller suggested the use of a coefficient of synergism ("s") to reflect any joint effects produced by cooperators.

In Maynard Smith's 1989 textbook on evolutionary genetics, there is a significant shift of focus. Here he follows Queller's lead and moves the concept out of the classical population genetic framework and into game theory, with its emphasis on finding an ESS (evolutionarily stable strategy). No longer is synergistic selection associated with altruism; the stress is on cooperation as a class of behaviors with a variety of potential payoff distributions. Now synergy (relabeled "s") is defined as the non-additive payoff increment to cooperating partners. Maynard Smith concludes that, if the synergistic increment is greater than the cost, the behavior will be an ESS (i.e., if $1n + 1n = 3n$ or more). Although inclusive fitness is not required for such interactions to occur, he suggests that relatedness could be a significant facilitator, especially in initiating cooperation.

As noted in chapter 2, Maynard Smith and Szathmáry also make liberal use of the synergy concept in their two volumes on the evolution of complexity, *The Major Transitions in Evolution* (1995) and *The Origins of Life* (1999). Moreover, their detailed study of the process of biological complexification in evolution is consistent in its overall vision with the more explicit conceptualization in *The Synergism Hypothesis*. Finally, thanks to David Sloan Wilson (1975, 1980; also D. S. Wilson and Sober 1989, 1994; Sober and Wilson 1998) and a growing number of coworkers, group selection—for thirty years a pariah in evolutionary theory—has been resuscitated on a new foundation. At first, Wilson too assumed that one of the cooperators was an altruist. However, his argument has been strengthened by the more recent realization that group selection can also include mutualistic, win-win forms of cooperation that provide differential reproductive advantages to all concerned. (See the more expansive discussion of the group selection controversy in chapter 2.)

Testing the Synergism Hypothesis

As noted above, the Synergism Hypothesis and the concept of synergistic selection (or functional group selection), like the concept of natural selection, represents an umbrella term for a broad category of causal influences. It is not a discrete mechanism or concrete causal agent. The causes of synergistic selection, like those of natural selection, are always situation-specific. Therefore, it is not possible to devise a single, decisive test of the Synergism Hypothesis. To our knowledge, nobody has ever succeeded in doing so for natural selection, either. Rather, the Synergism Hypothesis directs our attention to the combined effects produced by things that work together, or cooperate. Accordingly, it can be tested in much the same way that the role of natural selection is routinely tested, with hypotheses and analytical tools that are appropriate to the particular context.

One important method for verifying the role of synergy in a given case was first suggested by Aristotle in the *Metaphysics* (1961, 1041b11–31), to my astonishment. To paraphrase Aristotle's wording, many parts may be needed to make a whole, yet the loss of even a single part may be sufficient to destroy it. (I refer to this methodology as "synergy minus one.") Thus, we need only to remove one of the major parts from any living system (or any human technology, for that matter) and observe the consequences. As a thought experiment, imagine what would happen if some of the constituent amino acids were removed from the homeobox complex in the homeotic genes during morphogenesis, or if the transfer RNA were removed from the machinery of reproduction, or if the mitochondria were removed from a eukaryotic cell, or the gut bacteria from a termite, or the sub-majors from an army ant colony, or the beak from one of Darwin's finches, or the vowels from the words in table 1 above; or the cyanobacteria from the Igorot's rice-terrace system, or the wheel from an automobile.

Of course, there are also a great many cases where "synergy minus one" merely attenuates the overall effects, with consequences that might only be measurable in statistical terms. Thus, the removal of one member from a school of dwarf herrings might only marginally affect the probability that any of the remaining members will be eaten by a barracuda. And the loss of one member from a coalition of male lions, or chimpanzees, might or might not tip the scales in subsequent confrontations. On the other hand, if you remove one oarsman from a varsity eight, the chances are that the remaining seven will lose the race. (In chapter 7, where I discuss the phenomenon of "devolution" in human societies, it will be argued that the fate of many civilizations in the past may have been sealed by some variant of the synergy-minus-one scenario.)

A second method for testing hypotheses about synergistic effects involves comparative studies of various kinds. Raven's (1992) comparison of the functional differences between lichen symbionts and other asymbiotic forms provides one illustration. Other examples mentioned above include Bell's (1985) comparative study of size-effects and functional differentiation among the Volvocales; Jeanne's (1986) study of colony-size effects in social wasps; Marzluff and Heinrich's (1991) study of group-size effects in ravens; Campagna and coworkers' (1992) comparison of pup survival rates in sea lions; Stander's (1992) comparative study of lion hunting behaviors; Ligon and Ligon's (1982) analyses of helping behaviors in woodhoopoes; the observations of Parker and others (1994) of food sharing in ravens; the studies by Mumme and his coworkers (1988) of joint nesting behaviors; the group-level advantages of genetic diversity in bumblebees postulated by Shykoff and Schmidt-Hempel (1991); and Stanford's (1992) study of allomothering in capped langurs. Again, if synergistic effects are real and measurable, then it should be possible to demonstrate the differences that they make in a given context. In fact, the literature in such "hard sciences" as biochemistry, physiology and pharmacology provides a wealth of examples.

Game theory offers a third method for testing various hypotheses about synergy, as suggested above. Game theory models are especially useful in analyzing facultative relationships where the synergistic effects can be quantified and the costs and benefits can be allocated in various ways among the "players."

One objection to this theory might be the charge that there is nothing new here; it might be argued that innumerable theorists through the centuries have recognized that wholes are more than the sum of their parts. Yes, but . . .! To reiterate, what *is* new here is the idea that the functional effects produced by living systems (their synergies) are the very cause of their existence, their reason for being. As noted above, in evolutionary processes, causation often works backwards from our conventional view of things; in evolution, functional effects are also causes. It is the functional effects of various kinds (in a given environment) that determine the differential survival of the genes, and structures, and behaviors that are responsible. Hence, it is the synergies that are the cause of cooperation in nature, not the other way around. Equally important, this theory is by no means unchallenged, or undisputed. A major alternative theme advanced by many theorists over the years—from Lamarck in the eighteenth century to Herbert Spencer in the nineteenth century, biologist Stuart Kauffman in the 1990s, and the well-known science writer Robert Wright in his recent book *Non Zero: The Logic of Human Destiny* (2000)—is that there is an inherent, deterministic trend in evolution toward greater complexity, which allows us to predict future developments. In other words, the evolution of complexity is seen as an autonomous self-organizing process. I disagree (see chapter 4).

Conclusion: A Science of Relationships

What are some of the implications? First, the term *synergy* could serve as a *lingua franca* for the cooperative/emergent/interactional effects that are observed and studied by various disciplines. By removing a language barrier, the term could facilitate cross-disciplinary communication and understanding.

Second, by directing our attention to context-specific historical relationships and interactions, rather than mechanisms or reductionist "laws," Holistic Darwinism encourages a multileveled, multidisciplinary research and theory that is free from the intellectual shackles of nineteenth century Newtonian physics. Furthermore, in contrast with the bloodless mathematical caricatures that are blind to the functional properties of the phenomenal world, this paradigm draws our attention to the functional aspect of cooperative effects. As noted above, concepts with broad applicability to many different kinds of phenomena may play an important theoretical role in the sciences. The synergy concept provides a framework for integrating the research in various disciplines that may be relevant for understanding the broader causal role of cooperative phenomena in nature and evolution. To borrow the "preowned" parable of the blind men and the elephant, if we are ultimately to make sense of the whole, we will need to pool our discoveries about *both* the parts and the whole.

All scientific concepts are inescapably Procrustean and selective—highlighting certain aspects of the phenomenal world to the exclusion of others. None can be all things to all scientists. The ultimate test is fruitfulness. By that standard, the concept of synergy would seem to hold promise. Among other things, it offers a theoretical framework that, like the concept of natural selection, can provide a focus for explaining a major aspect of the evolutionary process, namely, the evolution of organized complexity. Indeed, an invigorated science of synergy would shine a spotlight on a fundamental property of the phenomenal world. Equally important, because it is pan-disciplinary (and egalitarian) in its methodological implications, the concept of synergy could provide a useful bridge between various specialized disciplines. If synergy can provide functional advantages elsewhere in the phenomenal world, why not also within the scientific enterprise itself.

— ∽ —

From nature's chain whatever link you strike
Tenth, or ten thousandth, breaks the chain alike.

—Alexander Pope

SUMMARY: A thesis advanced in this book is that synergy of various kinds has played a significant creative role in evolution; it has been a prodigious source of evolutionary novelty. It has been proposed that the functional (selective) advantages associated with various forms of synergistic phenomena have been an important factor in the progressive evolution of complex systems over time. Underlying the many specific steps in the complexification process, a common functional principle has been operative. Recent mathematical modeling work in complexity science, utilizing a new generation of nonlinear dynamical systems models, has resulted in a radically different vision. It has been asserted that spontaneous, autocatalytic processes, which are held to be inherent properties of nature and living matter itself, may be responsible for much of the order found in living systems and that natural selection is merely a supporting actor. A new "physics of biology" is envisioned in which emerging natural laws of organization will be recognized as being responsible both for driving the evolutionary process and for truncating the role of natural selection. This chapter introduces the issue and discusses a possible relationship between the Synergism Hypothesis and self-organizing phenomena.

4

Synergy versus Self-Organization in the Evolution of Complex Systems

Synergy versus Self-Organization

It has always seemed to me ironic that we are surrounded and sustained by synergistic phenomena, yet we do not, most of us, seem to appreciate its importance; we take its routine miracles for granted. Indeed, synergy is literally everywhere around us and within us; it is unavoidable. Here are just a few examples:

- Water has a unique set of combinatorial properties that are radically different from those of its two constituent gases. But if you simply mix the two gases together without a catalyst such as platinum, you will not get the synergy.
- Our written language, with well over three hundred thousand words, is based on various combinations of the same twenty-six letters. Recall (from chapter 3) how the letters "o," "p," and "t" can be used to make "top," "pot," "opt," and "p.t.o." (paid time off). But if you remove the vowel, there will be no pattern recognition in the reader's mind.
- About two thousand separate enzymes are required to catalyze a metabolic web, like glycolysis in living organisms. But if you were to remove just one of the more critical enzymes, say hexokinase, the process would not go forward.
- As noted earlier, bricks can be used to make a great variety of useful structures—houses, walls, factories, jails, roads, watchtowers, fortifications, and even kilns for making more bricks. The brick is truly a synergistic technology. But without mortar and human effort (and a plan), you will have only a pile of bricks.

- A modern automobile is composed of roughly fifteen thousand precisely designed parts, which are derived from some sixty different materials. But if a wheel is removed, this incredible machine will be immobilized.
- Economist Adam Smith's classic description in *The Wealth of Nations* (1964) of an eighteenth century pin factory is often cited as a paradigm of the division of labor. Smith observed that ten laborers, by dividing up the various tasks associated with making pins, were able collectively to produce about forty-eight thousand pins per day. However, Smith opined that if each laborer were to work alone, doing all of the tasks independently, it was unlikely that on any given day the factory would be able to produce even a single pin per man.

These mundane forms of magic are not magic at all, of course, but a fundamental characteristic of the material world; things in various combinations, sometimes with others of like kind and sometimes with very different kinds of things, are prodigious generators of novelty. To repeat, the Synergism Hypothesis posits that synergistic effects of various kinds have been a major source of creativity in evolution. The Synergism Hypothesis asserts that it was the functional (selective) advantages associated with various forms of synergy that facilitated the evolution of complex, functionally organized biological and social systems. In other words, underlying each of the many particular steps in the complexification process, a common functional principle has been at work. It is quintessentially a bioeconomic theory of complexity.

The recently developed theories of self-organization would seem to be "orthogonal" (opposed) to this functionalist, selectionist theory. Mathematical modeling work in biophysics, utilizing a new generation of nonlinear mathematics, has produced a radically different hypothesis about the sources of biological order. As articulated by Stuart Kauffman (1993, 1995, 2000), much of the order found in nature may be spontaneous (autocatalytic)—a product of the generic properties of living matter itself. Kauffman envisions a new physics of biology in which the emerging natural laws of organization will be recognized as being responsible both for driving the process and for truncating the role of natural selection. Natural selection in Kauffman's paradigm is viewed as a supporting actor (see below).

The Evolution of Complexity: A Theoretical Challenge

Complexity seems of late to have become a buzzword. There have even been popular books chronicling the research and theory that have burgeoned in this category (Lewin 1992; Waldrop 1992). Nevertheless, the underlying theoretical challenge is not new. Attempts to explain the origins and evolution of living systems can be traced back at least to the Old Testament. Even a concept as fashionable as autocatalysis can be found in the writings of Aristotle, the first great biologist, who developed what has become an enduring theme in western natural science. Aristotle postulated an intrinsic directionality, or unfolding process in nature (entelechy). Aristotle also inspired the concept of an ascending ladder of perfection, or hierarchy, that later came to be associated with the Latin term *scala naturae* (Granger 1985; Lovejoy 1936).

At the beginning of the nineteenth century, the French naturalist Jean Baptiste de Lamarck postulated a natural tendency toward continuous developmental progress in nature, energized by what he called the "power of life" (Lamarck 1963). Likened by Lamarck to a watch spring, it involved the idea that living matter has an inherent developmental energy.

Orthogenetic theories of evolution reached an apogee of sorts during the nineteenth century with the multivolume, multidisciplinary magnum effort of Herbert Spencer, who was considered by many contemporaries to be the preeminent thinker of his era. Spencer formulated an ambitious "universal law of evolution" (as he called it) that spanned physics, biology, psychology, sociology, and ethics. In effect, Spencer deduced society from energy by positing a cosmic progression from energy to matter, to life, to mind, to society, and, finally, to complex civilization. "From the earliest traceable cosmical changes down to the latest results of civilization," he wrote in "The Development Hypothesis" (Spencer 1892), "we shall find that the transformation of the homogeneous into the heterogeneous is that in which progress essentially consists." Among other things, Spencer maintained that homogeneous systems are less stable than those that are more differentiated and complex. (It is worth noting that, while Spencer viewed this progression as spontaneous in origin, he also believed that it was sustained by the fact that more complex forms are functionally advantageous.) As noted earlier, Spencer also coined the term *superorganism* (see chapter 8).

There have been many less-imposing vitalistic and orthogenetic theories since Spencer's day, ranging from Henri Bergson's *élan vital* to Hans Driesch's *entelechy*, Pierre Teilhard de Chardin's *omega point*, Pierre Grassé's *idiomorphon*, and Jean Piaget's *savoir faire*. However, during the past century, Darwin's theory of natural selection cast a long shadow over various autocatalytic theories. Darwin seemed to be rebutting Lamarck and Spencer directly when he wrote in *The Origin of Species*, "I believe in no fixed law of development" (1968, p. 318). And again, later on, Darwin stated, "I believe . . . in no law of necessary development" (1968, p. 348). One of the formulators of the so-called modern synthesis (which predated Neo-Darwinism), Theodosius Dobzhansky, put the matter succinctly: "Natural selection has no plan, no foresight, no intention" (1975, p. 377).

A striking illustration is the eye, that revered object of nineteenth century natural theology. We now know that the eye did not unfold deterministically or arise full-blown. It developed independently on perhaps forty different occasions in evolutionary history, utilizing at least three different functional principles—the pinhole, the lens, and multiple tubes. Nor do all the eyes of a similar type work in the same manner. (More recent work in the new science of genomics suggests some possible genetic continuities—i.e., the reuse of conserved genes—as well.)

While the evidence for natural selection as a directive agency in evolution is overwhelming, many theorists over the years have felt that the Neo-Darwinian synthesis is inadequate, by itself, to account for the undeniable progressive trend from the primordial chemical soup to simple, one-celled prokaryotes and eukaryotes, and, ultimately, to large, complex, socially organized mammals. The evolution of complexity has seemed to require something more than random point mutations in an amorphous "gene pool." The long-term trend toward greater complexity (in tandem with the many examples of stasis) seems to suggest the presence of some additional mechanism or mechanisms. Some years ago, biologist C. H. Waddington articulated these doubts with characteristic bluntness: "The whole real guts of evolution—which is how do you come to have horses and tigers and things—is outside the mathematical theory" (quoted in D. E. Rosen 1978, p. 371).

More broadly, the question is: Why does complexity exist? Why have various parts aggregated over time into larger, more-complex wholes? And why have many wholes differentiated into various specialized parts? For that matter, what *is* complexity? And, in the context of modern biology, what *are* wholes and parts? The accumulating data on mutualism, parasitism, colonialism, social organization, coevolution and the dynamics of ecosystems have revealed many nuanced interdependencies and have blurred the supposedly sharp demarcation lines between various biological units.

Physicist Larry Smarr (1985) has pointed out that complexity is, in reality, a multidimensional, multidisciplinary concept; there is no one right way to define and measure it. A mathematician might define it in terms of the number of degrees of freedom in computational operations. A physicist might be concerned with the number and frequency of interactions in a system of interacting gas molecules. The systems theorists of the 1960s were fond of using the rubric (suggested independently by mathematicians Andrei Kolmogorov and Gregory Chaitin) of "algorithmic complexity"—the size of the smallest mathematical description of system behavior. Social scientist Herbert Simon (1962) advocated the use of a hierarchical measure—the number of successive levels of hierarchical structuring in a system, or what biologist G. Ledyard Stebbins (1969) characterized as "relational order." For obvious reasons, biologists have traditionally preferred such biologically relevant measures as the number of parts (e.g., cells), types of parts (e.g., cell types), or the number of interdependencies among various parts. In recent years there have also been a number of efforts to define complexity in relation to thermodynamics, entropy, and information (see especially Wicken 1987; Haken 1988b; Brooks and Wiley 1988; Weber et al. 1988; Salthe 1993). (See also chapter 12.)[1]

John Tyler Bonner, in his well-known book on the evolution of complexity (1988), suggests that biological (and, by extension, social) complexity should also be defined in terms of the functional nature of living systems. What is most salient about biological systems is not just the number of parts, or even the number of interconnections among the parts, Bonner argues, but the division of labor (and the combining of capabilities) that results; this is the distinctive hallmark of biological complexity. In other words, biological complexity should be associated with the functional synergies that it produces.

In recent years there has also been increasing acceptance of the views of biologists Ludwig von Bertalanffy (1950, 1967), W. Ross Ashby (1960, 1956), C. H. Waddington (1962, 1968), Paul Weiss (1971), and others that biological complexity is also characterized by cybernetic properties; it is not just ordered but also organized (see also Wiener 1948; Powers 1973; J. G. Miller 1995). That is to say, biological (and social) systems are distinctive in being (a) goal-oriented (or teleonomic), (b) hierarchically organized, and (c) self-regulating (they display processes of feedback control) as well as being uniquely self-developing and self-determining. The physical chemist Engelbert Broda (1975) stressed the functional imperatives: "The more the division of labor was developed [in evolution], the more important became intercellular and interorganismal communication and control. Hence, for an understanding of more complicated systems, thermodynamics and kinetics must increasingly be supplemented by cybernetics, by applied systems analysis" (p. 141). In hindsight, Broda could have added molecular and intracellular communication and control to the list of biological processes with cybernetic properties.

One other distinctive feature of complex, living systems is that they cannot be fully understood, nor can their evolution and operational characteristics be fully explained, by an exclusive focus either on the system as a whole or on the component parts. As noted earlier, both chemist Michael Polanyi and physicist Philip Anderson (among others) stressed that, in the process of constructing a complex, living system, the causal dynamics are in fact multileveled. On the one hand, the properties of the whole are constrained and shaped by the properties of the parts, which in turn are constrained and shaped by the lower-level properties of their constituent raw materials and by the laws of physics and chemistry. To a devout reductionist, this is a truism that modern science daily reconfirms.

On the other hand, the extreme reductionist argument that an understanding of the parts fully explains the whole leads to what C. F. A. Pantin called the "analytic fallacy." To repeat, Polanyi, Anderson and others point out that a whole also represents a distinct, irreducible level of causation that harnesses, constrains, and shapes lower level parts and which may, in fact, determine their fates. In effect, wholes may become both vessels and selective fields for the parts—and may even come to exercise hierarchical, cybernetic control over the parts.

Moreover, wholes can do things that the parts cannot do alone. An automobile cannot be fully understood, or its operation explained, by separate descriptions of how each part works in isolation. Not only is the design of each part affected by its role and relationship to the whole, but its performance and functional consequences may only be comprehensible in terms of its interaction with other parts and the whole (see Corning 1983, 2003; Haken 1973, 1977, 1983; also, compare the concept of "interactional complexity" in Wimsatt 1974).

Thus an automotive engineer must always look both upward and downward (and horizontally) in the hierarchy of causation when trying to comprehend the operation of any part. And the same applies to the students of living systems. Evolution has produced several emergent levels of wholes and parts. Furthermore, the power and impact of these emergent wholes has greatly expanded over the course of time; complexity has been at once a product of evolution and a cause of evolution (an important point to which we will return below).

Until a few years ago, the Neo-Darwinian explanation, while subject to vigorous debate over the details, was essentially uncontested; it was assumed that the trend toward biological complexification was somehow functionally driven. But the nascent science of complexity has challenged the selectionists' hegemony; non-Darwinian vitalistic/orthogenetic theories—now respectably clothed in a new wardrobe of nonlinear dynamical system models—are again in vogue. We will consider this issue briefly here and in more detail in chapter 12.

Self-Organization and Complexity

"Self-organization" is almost as much of a buzzword these days as is "complexity." However, it is hardly a newly discovered phenomenon. Aristotle enshrined it in his classic metaphor about the growth of acorns into oak trees. The pioneering nineteenth century embryologists, such as Karl Ernst von Baer, also appreciated, and observed, self-organization in the process of morphogenesis. But more important, self-organization is also compatible with Darwin's theory. Modern Neo-Darwinians, following the lead of Francisco Ayala (1970), Theodosius Dobzhansky (1974), and Ernst Mayr (1974a,b), have generally associated self-organization with Colin Pittendrigh's term *teleonomy* (evolved purposiveness) and the concept of an internal "program" (Roe and Simpson 1958). In this formulation, self-organization has been equated with the mechanisms of cybernetic self-regulation and feedback. Self-organization is viewed as being a product of, and subordinate to, natural selection.

Darwin also categorically rejected the idea of an inherent energizing or directive force in evolution, as mentioned earlier. However, it is important to note that the theory of evolution via natural selection does not stand or fall on this issue, so long as any autocatalytic processes are (a) of a material nature, (b) empirically verifiable and (most important) (c) subject to testing for their functional (fitness) consequences in relation to survival and reproduction.

In their new book, *Self-Organization in Biological Systems* (2001), biologist Scott Camazine and his coauthors usefully define *self-organization* as a term that encompasses a broad array of pattern-forming physical processes—from sand dunes to fish schools. Self-organizing systems are distinctive in that they are *not* organized by some outside force or agency; the causal dynamics are internal to the parts (or participants) and their interactions (see Yates et al. 1987; Whitesides and Grzybowski 2002). This seems straightforward enough, and it presents no inherent conflict with Darwinian theory (differential selection among functional variants) or the bedrock principle that coded instructions in the genome play a major directive role in the ontogeny and life history of living systems. Nor does this conflict with the idea that the genome might use simple rules or procedures, and co-opt basic physical principles, to generate structural complexity.

Nevertheless, many enthusiasts have assumed that self-organization and natural selection are opposed to one another and that evidence for self-organization somehow diminishes the role of natural selection. A case in point is biologist Brian Goodwin (1994), who cites the marine alga *Acetabularia* as an example of a living organism that seems to follow a simple mathematical principle as it develops a stalk and various branches from a single giant cell. However, as John Maynard Smith (1998) has pointed out, genetic instructions and feedback are also involved in the growth of *Acetabularia*. Moreover, it is likely that the evolution of *Acetabularia* was influenced by differential selection among functional variants. So *Acetabularia* does not undermine Darwinian theory.

In a similar vein, entomologist Robert E. Page and philosopher Sandra D. Mitchell (1998) proposed that a division of labor can emerge from living systems without the influence of natural selection. This claim overlooks the fact that the emergent effects produced by a division/combination of labor are subject to differential selection in relation to the synergies they produce. They are not exempted from natural selection. Camazine and his colleagues summed it up nicely: "There is no contradiction or competition between self-organization and natural selection. Instead, it is a *cooperative 'marriage'* [their emphasis] in which self-organization allows tremendous economy in the amount of information that natural selection needs to encode in the genome. In this way, the study of self-organization in biological systems promotes orthodox evolutionary explanation, not heresy" (Camazine et al. 2001, p. 89).

In this light, let us consider the important and highly visible contributions of biophysicist Stuart Kauffman, beginning with *The Origins of Order: Self-Organization and Selection in Evolution* (1993). (See also Jantsch 1980; Brooks and Wiley 1988; Brooks et al. 1989; Csányi 1989; Salthe 1993.) Not only is this magnum opus an imposing, even daunting, guided tour of the rapidly developing science of complexity but it goes well beyond the claims of other workers in this theoretical vineyard. In fact, Kauffman's thesis is nothing less than subversive to natural selection theory. Natural selection is not the primary source of biological organization, he asserts, but a supporting actor that fine-tunes a self-organizing natural world. As Kauffman puts it: "Much of the order found [in nature] is spontaneously present . . . Such order has beauty and elegance, casting an image of permanence and law over biology. Evolution is not just 'chance caught on the wing.' It is not just a tinkering of the ad hoc, of bricolage, of contraption. It is emergent order honored and honed by natural selection" (1993, p. 644).

Kauffman states that the origins of order are to be found in the generic properties of living matter itself, which he characterizes as an "invisible hand." Natural selection is portrayed as being subordinate to these self-organizing principles. "Vast order abounds [in nature] for selection's further use" (1993, p. 235). Indeed, Kauffman suggests that biological order may exist sometimes "despite" natural selection. He speaks of how the new physics of biology requires us to view natural selection as being highly constrained by the natural laws of organization. Natural selection is "privileged" to improve upon the imminent order that exists in biological systems.

Some of Kauffman's results are not particularly controversial because they build on a long tradition in biology that has had as its focus the elucidation of various laws, constraints, and emergent properties associated with complex systems. The assertion that natural selection is constrained by the laws of physics and thermodynamics challenges no orthodoxy. What Kauffman adds to this tradition are some law-like constraints associated specifically with the dynamics of complex systems.

On the other hand, Kauffman's overarching conclusion *is* controversial, even gratuitously so. The proposition that autonomous, autocatalytic processes are the primary sources of order in nature, and that natural selection merely fine-tunes the results, represents a radical reformulation of evolutionary theory. Yet this conclusion is not the ineluctable result of the work Kauffman so carefully explicates. It is, as he acknowledges, a "bold leap" beyond the models (and a limited body of empirical support). Nor, as we noted above, is his hypothesis new. The vision of evolution as being self-propelled, or as a self-determined unfolding process, is a venerable theme in natural philosophy, tracing its roots at least to Empedocles.

For the record, it should also be noted that, in his most recent book, *Investigations* (2000), Kauffman moves toward the middle-ground. He now acknowledges an interplay between self-organization and selection. More important, he focuses on the role of "autonomous agents"—that is, organizations of matter, energy, and information that are active participants in what he characterizes as an emergent, "creative" process (sound familiar?). Yet Kauffman is still searching for overarching laws. He speculates that there may be a "fourth law of thermodynamics"—an inherent trend in nature toward greater diversity and "the persistent evolution of novelty in the biosphere." He also stresses what he calls "collectively autocatalytic" phenomena.

However, Kauffman's "fourth law" can be fully accounted for in mainstream Darwinian terms as an outcome of the inherent variability in living organisms (and their environments), which is subject at all times to differential survival and reproduction based on context-specific functional criteria. Darwin characterized it as a law of variation. As for the postulate of self-organization, the evidence is overwhelming that biological organization is predominantly controlled by purposeful "control information" (see chapter 14). To repeat, self-organizing processes may facilitate and introduce economies into the process of constructing living organisms, but the results are always subject to the final editorial "pruning" of natural selection. Self-organization survives only if it works in relation to the ongoing problem of survival and reproduction.

Kauffman's image of autonomous agents is also troubling. As a general rule, living organisms are hardly autonomous. Their basic purpose has been "preprogrammed" by evolution. They are shaped and constrained by the functional control information contained in the genome. They are subject at all times to the inescapable challenges associated with survival and reproduction. And they are enmeshed in a more or less complex system of interdependencies and feedbacks, both with other organisms and with their environments. Thus, autonomy in living organisms is a matter of degree, and it is in any case an emergent (functional) product of evolution via natural selection, not a free lunch. It is subject at all times to differential selection.

Natural Selection versus Self-Organization

Many years ago, Theodosius Dobzhansky voiced what still stands as the most important scientific objection to such orthogenetic and nonselectionist visions. The basic problem, he noted, is that these theories implicitly downgrade the contingent nature of life and the basic problem of survival and reproduction. In fact, they explain away the very thing that requires an explanation: "No theory of evolution which leaves the phenomenon of adaptedness an unexplained mystery can be acceptable" (1962, p. 16). There's the rub. Order is not a synonym for adaptation, and adaptation in nature depends on functional design.

Nor can the need for adaptation be so lightly dismissed as many orthogenetic theorists do. For one thing, energy is not a free good; it must be captured and converted to various organic uses. (This point is developed further in chapter 13.) Also, the thermodynamic processes and structures that characterize living systems are never autonomous (independent of environmental contingencies). It matters a great deal whether these systems are located on land or in water, in the Arctic or in the tropics, in an oxygenated or an anoxic environment and whether they are subject to competition, or predators, or parasites. Life in the real world is always contingent and context dependent. Indeed, orthogenetic theories are often obtuse or cavalier about the prevalence of extinctions in evolutionary history. Finally, "information" and "structure" are not functional equivalents; what get tested in the environment are the properties of the structures themselves, not their informational content. (Again, see the discussion of control information in chapter 14.)

Self-organization is an undisputed fact, and the case for autocatalysis, especially in the early stages of evolution, is **compelling**. Indeed, Kauffman's work strengthens the case. His formulation suggests that life may have "crystallized" initially in a collective phase transition leading to connected sequences of biochemical transformations—an interesting alternative to the hypercycle concept. (Elsewhere, Kauffman likens the process to a set of pegs scattered on a floor that are gradually tied together to form a net.) If this scenario is correct, life may have an "innate holism" (synergy); life began as an integrated, emergent property of complex systems of polymer catalysts. Also, it may have been more easily achieved than we have heretofore imagined. However, Kauffman's vision does not extend very far up the phylogenetic tree, to the point where morphology, functional design, and a division of labor begin to matter.

Equally important, much of Kauffman's case, despite its admirable rigor, rests on the models themselves—on hypotheses that still require testing. At this point, some are little more than mathematical "just-so" stories. The jury is still out on a key question: To what extent are the models isomorphic with the dynamics of the real world? What is the relationship between these computer-driven equations and concrete, feedback-driven cybernetic systems with specific functional properties and requirements? Do the *quantitative* numerical relationships in the models map to the *qualitative* functional interactions (and logic) within and between living systems?

This issue is of crucial importance. In its original (Darwinian) formulation as a functional theory of evolution, natural selection referred to those functional effects (adaptations) of all kinds, and at various levels of biological organization, that, in each successive generation, influence, if not determine, differential survival and reproduction. It is the functional effects produced by a gene, or a genic "interaction system" (Sewall Wright's felicitous term), or a genome, or a phenotype, or an interdependent set of genomes (symbionts, socially organized groups, coevolving species) in relation to the contingencies of survival and reproduction that constitute the directive (causal) aspect of natural selection. Neither randomness (strictly speaking), nor incrementalism, nor even competition (strictly defined) are indispensable. Natural selection can also be a party to synergistic autocatalytic processes. It can be a party to discontinuous (catastrophic) symbiotic functional fusions. And it can be a party, as well, to novelties that create new niches and mitigate competition. Natural selection becomes a co-conspirator whenever an innovation has functional consequences for survival and reproduction. Likewise, natural selection reconfirms existing adaptations in each new generation via normalizing or stabilizing selection.

In this broader, functional conceptualization, which was clearly articulated by the developers of the so-called "modern synthesis"(contrary to some slanderous caricatures), natural selection is superordinate to all proximate forms of functionally significant causation, including those that may be autocatalytic and self-ordering (see J. S. Huxley 1942; Dobzhansky 1937, 1962, 1970; Simpson 1967; S. Wright 1968–1978; Stebbins 1969; E. Mayr 1963, 1982, 2001; Stebbins and Ayala 1985). In this formulation, natural selection often serves as an editor or a censor.

Accordingly, Kauffman's models, and comparable efforts, beg the question: Do the dynamical attractors in a Boolean network model really represent autonomous self-ordering processes? Or do they perhaps model stable combinations of polymers, genes, cells, or organisms that, in the real world, would be likely to be favored by natural selection? The answer may be that they do both. The ordering observed in evolution may have been a trial-and-success process in which the stable attractors identified in dynamical system models also happen to simulate functionally viable synergistic combinations—the material entities that must survive in the real world. In this vision, natural selection is not simply the interior decorator who was brought in to hang the curtains after the house was built. From the very outset, natural selection was posing the question to both the architect and the builder: Does it work?

Several conclusions can be drawn from this important body of theory and research and from the convergent efforts of many other workers in this area. First, it would appear likely that order existed prior to selection and arose through autonomous, self-organizing processes; natural selection did not create order *ab initio*. It may only have bestowed a blessing on it. Nor was natural selection the exclusive designer of biological organization later on. (Indeed, I argue that the products of evolution have themselves become increasingly important co-designers over time, though often inadvertently, in a process that has been interactive.)

A second conclusion, one that will also be discussed further below, is that wholes may have been more fundamental biological entities than parts in terms of the process of biological complexification. In the earliest stages of evolution, the parts, in fact, had no meaning, no directional consequences, and no selective value until they were combined into functional units. Classical competitive selection began to play a shaping role when the first self-organized reproducing wholes appeared, and it intensified in proportion to the increasing organizational and functional capabilities of various combinations of parts. It was the emerging functional capacities for replication, metabolism, damage repair, mobility, predation, defense—all of which are products of more complex organization—that accelerated the evolutionary arms race.

Finally, we can discern, at an early stage of the evolutionary process, a principle that will be elaborated upon below, namely: "competition via cooperation." Cooperation is not a peripheral survival strategy in a world governed by competition. The synergy resulting from cooperative interactions of various kinds provides the functional *raison d'être* for biological organization and for the observed evolutionary trend toward greater complexity. In many, but obviously not all, cases, synergy has provided more complex forms with a competitive edge. Let us turn then to a consideration of the role of synergy (and symbiosis) in the evolution of complexity. Though mentioned earlier, symbiosis deserves further comment in this regard.

Symbiogenesis in Evolution

Symbiosis in general, and mutualism in particular (to say nothing of the broader concept of synergy), represented at best a minor theme in biology until recently, and it certainly played no role in mainstream evolutionary theory. Some theorists considered symbiosis to be a myth; others viewed it merely as a small class of anomalies or oddities that in no way challenged the dominant Neo-Darwinian synthesis (basically, competition and mass selection among point mutations in individual genes); still others recognized that mutualistic, cooperative relationships might provide organisms with a competitive advantage under some circumstances (even Darwin appreciated that), but it was assumed that such phenomena were relatively rare.

During the past two decades, however, a dramatic change has occurred. There has been an upsurge in research and theorizing about symbiosis, mutualism, cooperation, and even synergy. Among the major developments were the following:

- A growing acceptance of the endosymbiotic theory of eukaryote evolution, indisputably one of the major benchmarks in biological complexification;
- A flowering of research and publications on symbiosis across more than a dozen subdivisions of biology along with a growing number of courses and

textbooks on symbiosis, a new international journal called *Symbiosis*, and
several conferences devoted to the subject;

- A parallel upsurge of interest in mutualism and coevolution among ecologists, with special reference to the application of cooperative versions of the
  well-known Lotka-Volterra equations, as well as various cost-benefit models;
- A recognition of the importance of cooperation among hard-core Neo-
  Darwinians in general and sociobiologists in particular, in part due to the
  iterated prisoner's dilemma model developed by Robert Axelrod and
  William Hamilton (Axelrod and Hamilton 1981; Axelrod 1984; Axelrod
  and Dion 1988; Lima 1989) and the plethora of other models and research
  that this landmark effort spawned (see Gintis 2000a; also chapter 6);
- Work in Europe on symbiosis and the emergence of a new academic sub-
  specialty, endocytobiology, which is devoted to cellular symbioses
  (Schwemmler and Schenck 1980; Schwemmler 1989);
- As noted earlier, the belated discovery in the west of an entire school of
  symbiogenesis theorists, dating back to a group of late nineteenth and early
  twentieth century Russian botanists (but including also a few advocates in
  the west), who advanced the hypothesis that symbiosis has played a major
  causal role in evolution (according to some, in opposition to natural selection) (Khakhina 1979, 1992; Margulis and McMenamin 1993);
- And in a culmination of this process, as noted earlier, a landmark international conference in 1989 on symbiosis as a source of evolutionary innovation, at which some twenty participants documented the ubiquity of
  symbiosis and developed the case for symbiogenesis as a significant factor in
  evolution (Margulis and Fester 1991).

Though much work still remains to be done by symbiologists on the
costs, benefits, and evolutionary implications of symbiosis, a number of tentative conclusions are possible.

First, there can be no doubt that the synergies associated with symbiosis
have played a significant role in the evolution of complexity. Symbiosis is
implicated in the emergence of the first living organisms, as well as in early
mergers among primitive prokaryotes possessing complementary functional
specializations, the development of complex nucleated cells, the emergence
of land plants and ruminant animals, the colonization of various aquatic and
terrestrial environments, and the coevolution over time of both aquatic
and terrestrial ecosystems.

A second conclusion is that symbiosis is clearly a robust phenomenon.
Not only is it found within and between all five kingdoms of living organisms, but functionally similar kinds of symbiosis have evolved repeatedly

and independently within (and across) various taxa. For instance, the lichen taxon, which has no common ancestor, includes sixteen of the forty-six recognized orders of ascomycetous fungi (as well as some basidiomycetous and conidial fungi) and over thirty types of algae and cyanobacteria (Hawksworth 1988). By the same token, cyanobacteria, which are photo-synthesizers, form symbiotic relationships not only with fungi but also with some algae, bryophytes, aquatic ferns (*Azolla*), cycads, and angiosperms (*Gunnera*) (Ahmadjian and Paracer 1966). Similarly, many species of cleaner fish and cleaner birds provide services opportunistically for an array of fish, alligators, crocodiles, iguanas, elephants, rhinos, and many other species of herbivores. The independent evolution of multiple symbiotic relationships has also been documented in corals, legumes, the gut symbionts of ruminant animals, and hydrothermal vent species, among others.

A third conclusion is that symbiosis has provided many opportunities for organisms to occupy ecological niches that would not otherwise have been viable. The lichens that are often the first living forms to colonize a barren environment provide an example. And so do the richly populated coral communities that are frequently found in what would otherwise be unproductive tropical waters. Likewise, gut symbiosis is indispensable for the niche occupied by ruminant animals. (The hydrothermal vent species, which occupy a unique ocean floor niche, were also noted earlier.)

It should also be evident that symbiosis may serve a number of highly survival-relevant purposes, including defense and protection, nutrition, mobility, and reproduction. The character of these relationships can also vary widely across various parameters including duration, persistence, speci-ficity, universality, level of dependency, type and level of integration, mode of transmission, and the distribution of costs and benefits. For instance, many flowering plants are completely dependent on their animal symbionts for reproductive assistance, but the relationships are transitory and promis-cuous. By contrast, some cattle egrets, oxpeckers, and other bird species may form more or less permanent attachments to one or a few ruminant animals but may not depend for their nutritional needs on the animals' parasites; they may be diversified ground foragers as well.

Another point is that symbiosis typically precipitates the coevolution of various facilitative adaptations—morphological structures, chemical sub-stances, communications modalities, and even cybernetic regulatory mech-anisms. There are, for example, the specialized organs that accommodate the light-emitting bacteria in luminescent pony fish (Ruby and Morin 1979); the honeydew secretions that aphids produce for their ant hosts (E. O. Wilson 1975); the hollow thorns and glycogen-rich nectar that attract *Pseudomyrmex* ants to *Acacia* plants where, in return, the ants provide protection from harmful parasites and disperse the plants' seeds (Ahmadjian and Paracer 1966); the nectar and honey guides produced by flowering plants; and the bicarbonate ions in the saliva of ruminants that help to maintain an acceptable pH for gut symbionts (Smith and Douglas 1987). Perhaps most significant is the evidence that host symbionts may regulate the growth and reproduction of their partners, presumably in the interest of preserving the integrity of the whole (Smith and Douglas 1987). As noted earlier, the reproductive rates of symbionts must, of necessity, be synchronized (Margulis and McMenamin 1993).

It is also evident that symbiosis frequently involves more than two species and may constitute some of the key building blocks of an ecosystem. One of the most extraordinary examples is the single-celled eukaryotic protist, *Mixotricha paradoxa* (Margulis and McMenamin 1993; E. Mayr 1974a). In fact, each cell represents an association of at least five different types of organisms. In addition to the host cell, there are three surface symbionts, including large spirochetes, small spirochetes, and bacteria. The function of the large spirochetes, if any, is not clear; they may even be parasites. However, the hairlike small spirochetes, which typically number about 250,000 per cell, provide an unusually effective propulsion system for the host through their highly coordinated undulations, the control mechanism for which is still obscure. Each of these spirochetes, in turn, is closely associated with another surface symbiont, a rod-shaped anchoring bacterium. Finally, each *Mixotricha* host cell contains an endosymbiont, an internal bacterium that may serve as the functional equivalent of mitochondria, removing lactate or pyruvate and producing ATP.

What makes this partnership all the more extraordinary is the fact that *Mixotricha* is itself an endosymbiont. It is found in the intestine of an Australian termite, *Mastotermes darwinensis*, where it performs the essential service of breaking down the cellulose ingested by its host. Indeed, these and other symbionts may constitute more than half the total weight of the termite.

Perhaps the most impressive form of multiple symbioses, though, can be found in coral communities. A single coral reef may encompass millions of organisms from dozens of different plant and animal species, many of which are symbiotic with one another as well as with the coral outcropping itself. The coral provides oxygenated water and shelter. The plants and animals consume the oxygen, plankton, and organic debris and deposit calcium to build the coral. In addition, there are many kinds of symbioses between the creatures that are associated with the corals—among others, clams and algae, crabs and sea anemone, fish and sea anemone, shrimp and sea anemone, and sea urchins and fish. The functions associated with these relationships include nutrition, protection from predators, mobility, mutual defense, and parasite removal (Perry 1983).

Among the implications of symbiosis for the science of complexity, three should be mentioned at this point.

1. Fusion, the functional integration of various elements, parts, or organisms via symbiosis, is one of the major mechanisms of complexification. However, the pattern can be extremely varied and the functional consequences are neither straightforward nor fixed.

2. Evolutionary complexification via fusion or integration has a distinctive causal dynamic in which behavioral changes (broadly defined) precipitate new options for selection. These changes are sustained in the short term by proximate mechanisms—direct rewards or reinforcements (real-time payoffs and feedback). However, the ultimate cause of the process—natural selection—is the fitness consequences of the short-term functional effects. To reiterate, in evolutionary change, effects may also be causes. (More on this point later on.)

3. This dynamic suggests a different approach to the modeling of evolutionary change. In effect, a coupling of two (or more) separately evolved genomes creates a new selective unit that may or may not compete directly with other organisms or with nonsymbiotic siblings. Yet, so far as I know, there has been little if any effort to incorporate symbiotic relationships into formal models of evolution, or of complexity. For instance, the word symbiosis does not even appear in the index to Kauffman's 1993 volume. And his NK models of correlated fitness landscapes are designed to model alternative epistatic interactions among the genes in a single genome.

A final point, to reiterate what was said earlier, is that symbiosis, or *functional integration*, represents only one of two very different modes of evolutionary complexification. Under the broader umbrella of synergy, there are also the multifaceted processes of *functional differentiation* and *elaboration*, which occur at many different levels and involve a very different sort of causal dynamic. We will consider this mode of complexification further below.

Synergy and Evolution

As noted earlier, synergy exists in many different forms. Indeed, it defies efforts to develop an exhaustive typology. Some of the more common and functionally important categories (although these are not all mutually exclusive) include synergies of scale, threshold effects, phase transitions, emergent phenomena, functional complementarities, augmentation or facilitation (e.g., catalysts), joint environmental conditioning, cost-and-risk-sharing, information sharing, collective decision making, a division of labor (or, better said, a combination of labor), animal-tool symbioses, and convergent fortuitous combinations. (These categories are described in more detail in Corning 1995, 2003.) They encompass both integrative and differentiating dynamics.

It should also be noted that the synergies associated with functional differentiation may take either of two forms. One type involves the disaggregation of a single complex task into a set of specialized subtasks. The human immune system—one of the marvels of nature—provides a microlevel illustration. As described by immunologists Ivan Roitt (1988) and Gustav Nossal (1993), the system that defends us against the enormous number and variety of potentially pathogenic microbes in our environment consists of at least nine different types of mechanisms, some of which are localized at vulnerable places in the body and act more or less independently while others range throughout the body and are highly interactive with other elements of the system. Perhaps most impressive is our acquired immune system, which consists of a widely dispersed network of primary and secondary organs (the thymus and bone marrow, and the lymph nodes, spleen, and tonsils) that orchestrate a highly coordinated defense of the body using an array of functionally specialized cells and molecules that are distributed via our lymphatic and circulatory systems. These specialized agents include, among others, antigen-presenting cells that identify antigens, major histocompatibility complex (MHC) molecules that display antigen pieces (peptides), T lymphocytes that "read" the antigen peptides and signal other components of the system (such as macrophages), and B lymphocytes that produce a prodigious variety of antibody proteins. (And this is only an abbreviated outline of a much more complex story.)

The second type of functional differentiation arises from the fact that the survival problem is usually multifaceted, and that the various subtasks associated with survival and reproduction may be allocated to specialists of various kinds (e.g., army ants). In either case, functional differentiation in turn creates a need for cybernetic regulation; the parts must be coordinated to achieve the objectives of the whole. Accordingly, the synergies associated with functional differentiation (or symbiotic integration) create selective contexts that in turn favor the evolution of cybernetic regulatory mechanisms (see below).

To repeat a key point, synergy can produce a variety of measurable, quantifiable benefits. Thus, information sharing by weaver birds can measurably reduce individual energy expenditures for foraging, and the huddling behavior of emperor penguins measurably reduces individual energy expenditures for thermoregulation. Other examples include Bonner's (1988) observation that aggregates of myxobacteria, which move about and feed en masse, secrete digestive enzymes that enable them to collectively consume much larger prey. Similarly, Schaller (1972) found that the capture efficiency (captures per chase multiplied by one hundred) and the number of multiple kills achieved by his Serengeti lion prides increased with group size—although a later study by Caraco and Wolf (1975) found that these results were dependent on the size of the prey.

In the highly social African wild dog (*Lycaon pictus*) population, overall kill probabilities in hunting forays were found to be vastly superior (between 85 and 90 percent) to those achieved by less social top carnivores (Estes and Goddard 1967). Kummer (1968) found that collective defense in hamadryas baboons (*Papio hamadryas*) is highly successful and reduces the net risk to each individual troop member. Ligon and Ligon (1978, 1982) analyzed the remarkable communal nesting behavior of the green woodhoopoe (*Phoeniculus purpureus*) and discovered that the extensive pattern of helping behaviors, even among unrelated individuals, markedly increased their likelihood of survival and reproductive success in their harsh Kenyan environment. Partridge (1982) and his colleagues have shown that fish schooling, which may include active forms of cooperation, is highly adaptive for the individual members. For instance, evasive maneuvers utilized by dwarf herring against predatory barracudas dramatically reduces the joint risk of being eaten. And H. O. von Wagner (1954) observed that the Mexican desert spiders (*Leiobunum cactorum*) cluster together in the thousands during the dry season, enabling them to avoid dehydration.

In all of these cases, and in countless human analogues, there were synergies—cooperative economies—that could not otherwise be achieved. However, as noted earlier, synergy is always context-specific and contingent. Consider again the examples cited above. Weaver birds have nothing to gain from information sharing when food is plentiful and widely distributed; huddling behavior by emperor penguins is not functional—and is not done—during the warm summer months; myxobacteria would find it dysfunctional to feed in large aggregations if their food sources were all small and widely dispersed; African lions would do better to hunt small, slow-moving prey alone; if wild dogs were ruminants, sociality would most likely not provide any nutritional benefits; collective defense by hamadryas baboons is relevant only because there are dangers to defend against; in more salubrious environments, green woodhoopoes would probably not find it advantageous to feed unrelated nestlings; dwarf herring might not find it advantageous to school if there were not barracudas about; and desert spiders have nothing to gain by congregating during the wet season.

Economic activity in human societies exhibits many of the same properties and synergies. One important distinction has to do with the role of technology (which is often likened to a form of symbiosis) in driving the evolution of human cultures. Thus, for example, a native Amazonian using a steel ax can fell about five times as many trees in a given amount of time as his ancestors could with stone axes. Likewise, a farmer with a horse can plow about two acres per day, while a farmer with a modern tractor can plow about twenty acres per day. One New Guinea horticulturalist can produce enough food to feed himself and about four or five other people; an American farmer can produce enough to feed forty-five to fifty people. And when the Mobil Oil Corporation purchased a Thinking Machines CM-5 computer a few years ago to replace its existing supercomputer, the time (and cost) required to process a major batch of seismic data dropped from about twenty-nine weeks and $2.8 million to ten days and $100,000.

A second point has to do with the nature of evolutionary causation and the causal role of synergistic phenomena. Earlier it was asserted that synergistic effects have played a significant role in the well-documented emergence of more complex systems, both in nature and in human societies. (We use the rubrics of size, functional differentiation, interdependence, and hierarchical ordering as our measuring rods.) That is, the functional effects produced by synergistic phenomena of various kinds have been an important "mechanism" of complexification.

Any factor that precipitates a change in functional relationships—that is, in the viability and reproductive potential of an organism or the pattern of organism-environment interactions—represents a potential cause of evolutionary change. It could be a functionally significant gene mutation, it could be a chromosomal rearrangement, a change in the physical environment, or (most significant for our purpose here) a change in behavior. In fact, a sequence of changes may ripple through an entire pattern of relationships: Thus, a climate change might alter the ecology, which might induce a behavioral shift to a new environment, which might lead to changes in nutritional habits, which might precipitate changes in the interactions among different species resulting, ultimately, in morphological changes and speciation.

What, then, are the sources of creativity in evolution? There are many different kinds, but the role of behavioral changes as a pacemaker of evolutionary change—mentioned earlier—should be emphasized. To quote an authority on the subject, Ernst Mayr (1960, pp. 373, 377–78),

> A shift to a new niche or adaptive zone requires, almost without exception, a change in behavior . . . It is very often the new habit which sets up the selection pressure that shifts the mean of the curve of structural [or functional] variation . . . With habitat selection playing a major role in the shift into new adaptive zones and with habitat selection being a behavioral phenomenon, the importance of behavior in initiating new evolutionary events is self-evident . . . Changes of evolutionary significance are rarely, except on the cellular level, the direct result of mutation pressure.

However, this model also begs the question: What causes behavioral changes? While this is a vastly complicated subject, one important underlying principle can be identified. In fact, behavioral changes often involve proximate causal "mechanisms"—the immediate rewards and reinforcements that psychologist E. L. Thorndike (1965) enshrined in his famous "law of effect," which forms the theoretical backbone of behaviorist psychology. At the behavioral level, in other words, there is a *proximate* selective agency (in Ernst Mayr's terminology) at work that is analogous to natural selection. Moreover, this "mechanism" is very frequently the initiating cause of the *ultimate* changes associated with natural selection (see Corning 1983; Plotkin 1988; P. P. G. Bateson 1988; Weber and Depew 2003; cf. Skinner 1981).

This is where the phenomenon of functional synergy (and the subcategory of symbiosis) fits into the evolutionary picture: It is the immediate, bottom-line payoffs of synergistic innovations in specific environmental contexts that are the cause of the biological/behavioral/cultural changes that, in turn, lead to synergistic, longer-term evolutionary changes in the direction of greater complexity, both biological and cultural/technological.

An Illustration from Adam Smith

Consider this illustration, involving another well-known example from *The Wealth of Nations*. Adam Smith drew a comparison between the transport of goods overland from London to Edinburgh in "broad-wheeled" wagons and the transport of goods by sailing ships between London and Leith, the seaport that serves Edinburgh. In six weeks, two men and eight horses could haul about four tons of goods to Edinburgh and back. In the same amount of time, a merchant ship with a crew of six or eight men could carry two hundred tons to Leith, an amount that, in overland transport, would require fifty wagons, one hundred men and four hundred horses.

The advantages of shipborne commerce in this situation are obvious. Indeed, shipment over water has almost always been an advantageous form of long-distance transport, as many different societies have demonstrated historically. But the causal explanation for Smith's paradigmatic example is not so obvious. In part it involved a division of labor and the merging of an array of different human skills; in part it involved the fairly sophisticated technology of late eighteenth century sailing vessels; it also required the capital needed to finance the construction of the ships; it required a government that permitted and encouraged private enterprise and shipborne commerce (including the protection afforded by the British navy); it also required a market economy and the medium of money; in addition, it required an unobtrusive environmental factor, namely, an ecological opportunity for waterborne commerce between two human settlements located (not coincidentally) near navigable waterways with suitable tidal currents and prevailing winds.

In other words, the causal matrix involved a synergistic configuration of factors that worked together to produce a favorable result. And the result—which played an important role in the rise of the British Empire—represented a significant step in the ongoing process of technological, economic, and societal evolution. However, it should be reiterated that, if any major ingredient were to be removed from the recipe, the result would not have occurred. Take away, say, the important component technology of iron smelting. Or, in like manner, take away the power supply from Jurassic Park (the movie). Synergistic causation is always configural, relational, and interdependent; the outcomes are always codetermined.

The relationship between synergistic effects and the evolution of complexity should now be more apparent. The process of complexification in evolution has been closely linked to the production of novel, more potent forms of synergy. That is, the *differentiation* and/or *integration* of various parts, coupled with the emergence of cybernetic regulation and the development of hierarchical controls, has been driven by the "mechanism" of functional synergy; synergistic effects of various kinds have been a primary cause of the observed trend toward more complex, multifunctional, multileveled, hierarchically organized systems. Furthermore, the same causal agency is applicable both to biological complexification and to the evolution of complex human societies—though (quite obviously) both the sources of innovation and the selective processes involved differ in some important respects.

Returning to another point raised earlier, we can now also see why it may be said that, at least in the process of evolutionary complexification, wholes have been more important units of selection than parts (see S. Wright 1980). It is wholes of various sorts that produce the synergies that then become the objects of positive selection (i.e., differential survival and reproduction). Thus, synergistic relationships of various kinds, and at various levels of organization, have been important units of evolution. To repeat, the Synergism Hypothesis is a theory about the *relationships* among biological phenomena; it is a theory about the causal role of relationships. Synergistic combinations, whether they arise through an integration of various parts (symbioses) or through the differentiation and specialization or elaboration of an existing whole (or, for that matter, through various agglomerative processes with synergistic outcomes), may provide a competitive advantage. Measurable proximate functional benefits may translate into measurable ultimate selective benefits—thus the slogan "competition via cooperation."

Synergistic Partnerships

Biologist David Sloan Wilson and philosopher Elliott Sober (1989), in the course of an argument for a multileveled view of evolution, provided an elegant example. They noted that many species of beetles in the family Scolytidae have adopted the strategy of tunneling under the bark or into the heartwood of various trees to create "galleries" for laying their eggs and protecting their larvae. Normally, these invaders would be thwarted by the trees' defensive measures (which include filling the cavities with resin). However, the beetles are able to overcome the trees' defenses with the assistance of a symbiont—a pathogenic fungus that kills the wood in the vicinity of the gallery. Meanwhile, a separate community of nonpathogenic fungi and yeast bacteria are also deposited by the beetles, and these symbionts produce a thick lining for the gallery that, among other things, serves as a food supply for the beetle larvae. Not only are these symbionts functionally interdependent but, significantly, the beetles have also evolved a specialized structure, called a mycangium, that enables them to carry their symbionts with them when they mature and leave their natal galleries.

In this example, there is a de facto partnership—a functionally interdependent unit whose survival and reproductive success is a product of the joint contributions of each of the partners.

One other especially potent example may also be helpful here. The African honey guide is a bird with a peculiar taste for bees' wax, a substance that is even more difficult to digest than cellulose. Moreover, in order to obtain bees' wax, the honey guide must first locate a hive, and then attract the attention of the African badger (ratel) (*Mellivora capensis*) and enlist it as a co-conspirator. The reason is that the ratel has the ability to attack and dismember the hive, after which it will reward itself by eating the honey while leaving the wax. However, this unusual example of cooperative predation between two different species in fact depends upon a third, unobtrusive co-conspirator. It happens that honey guides cannot digest bees' wax. They are aided by parasitic gut bacteria that produce an enzyme that can break down wax molecules. So this improbable, but synergistic, feeding relationship is really triangular. And, needless to say, the system would not work if any of the partners, for whatever reason, withdrew (Bonner 1988).

What makes this example of synergy particularly apropos for our purpose is the fact that the African honey guides also form symbiotic partnerships with humans, the nomadic Boran people of northern Kenya. (It is, of course, only one of many examples of animal-human partnerships.) Biologists Hussein Isack and Hans-Ulrich Reyer (1989) conducted a systematic study of this behavior pattern some years ago and quantified the synergies. They found that Boran honey-hunting groups were approximately three times as efficient at finding bees' nests when they were guided by the birds. They required an average of 3.2 hours to locate the nest compared with 8.9 hours when they were unassisted. The benefit to the honey guides was even greater. An estimated 96 percent of the bees' nests that were discovered during the study would not have been accessible to the birds without their human partners, who used tools to pry them open. (The bird-human partnership is also aided by two-way communications—vocalizations that serve as signals.)

Synergy and Self-Organization Revisited

What is the relationship, then, between synergy and self-organization? In fact, these two paradigms may not be contradictory but complementary. The process of evolutionary complexification may well have had autocatalytic aspects and certain inherently self-organizing properties that were independent of Darwinian selection processes, at least initially. But the wholes that resulted ultimately had to be functionally efficient as well. They had to pass the test of fitness. And, in fact, the most significant thing about organization, however it arises, is the synergy it produces. Thus, synergy is found at the heart of self-organizing phenomena; in effect, synergy may be the functional bridge that connects self-organization and natural selection in complex systems.

However, it should be stressed that synergy per se is not the same as *functional synergy* in terms of the problems of survival and reproduction, and self-organization is not equivalent to functional organization. Since there is no theoretical restriction on how synergy may arise in evolution, the only issue is whether or not self-organizing phenomena are exempted from, or conform to, the imperatives of functional viability; are these self-organized synergies compatible with the functional requirements for survival and reproduction, or do they exist despite natural selection, as we have defined it here? I believe that, for the most part, it will prove to be the case that autocatalytic and self-organizing phenomena are also subject to the editorial screening of natural selection. Thus, to reiterate the point made above, functional synergy may be the bridge that connects self-organization and natural selection. Prigogine's dissipative structures, Eigen and Schuster's hypercycles and Kauffman's dynamical attractors—insofar as they exist in the phenomenal world—can also be expected to produce synergies that are subject to differential selection in relation to their functional (adaptive) fitness.

In sum, a fully comprehensive theory of evolution must encompass both self-organization and selection, not to mention drift, historical contingencies, the laws of physics, and what the Nobel prize-winning geneticist Francis Crick called "frozen accidents."

It should also be stressed that a common source of confusion in the contemporary literature on self-organization has to do with a widespread failure to differentiate between two radically different kinds. One form of self-organization is nonpurposive in nature and should properly be called "self-ordering," while the other form is ends-directed; it has a systemic "purpose." The latter implies functional design—adaptations (and structures) that are either directly or indirectly products of natural selection (cf. Banerjee et al. 1990). Indeed, organic forms of self-organization rely on instructions and feedback, whereas self-ordering processes do not.

"Self-determination," likewise, is often conflated with self-ordering and self-organization. However, as I interpret the term, self-determination involves a phenomenon that transcends *both* the mechanism of spontaneous self-ordering and of self-organization via natural selection (and instructions). Self-determination implies a degree of autonomy: the ability to (a) establish goals, (b) make choices (decisions), and (c) exercise control over the conditions that are required to actualize those choices.

Theodosius Dobzhansky, one of the twentieth century's leading evolutionists, noted that:

> Purposefulness, or teleology, does not exist in nonliving nature. It is universal in the living world. It would make no sense to talk of the purpose or adaptation of stars, mountains, or the laws of physics. Adaptedness of living beings is too obvious to be overlooked. . . . Living beings have an *internal*, or natural teleology. Organisms, from the smallest bacterium to man, arise from similar organisms by ordered growth and development. Their internal teleology has accumulated in the evolutionary history of their lineage. (Dobzhansky et al. 1977, pp. 95–96)

All self-organization (as defined here) has internal teleology, but self-determination implies some degree of freedom, the potential for creativity and innovation and the ability to exercise a measure of self-control over the process of adaptation. Self-determining systems can actualize their purposiveness in ways that can contribute significantly to the dynamics of evolutionary change, as we noted earlier (chapters 2 and 3) in relation to the Baldwin effect and Mayr's pacemaker concept. We will have more to say on this below.

Cybernetics and Evolution

A fundamental characteristic of self-determining systems is that they are also cybernetic systems. A distinguishing feature of cybernetic systems is that they are controlled by the *relationship* between endogenous "goals" and the external environment. Consider this problem: When a rat is taught to obtain a food reward by pressing a lever in response to a light signal, the animal learns the instrumental lever-pressing behavior and learns to vary its behavior patterns in accordance with where it is in the cage when the light signal occurs so that, whatever the animal's starting position, the outcome is always the same. Now, how is the rat able to vary its behavior in precise, purposeful ways so as to produce a constant result? Some behaviorists postulated environmental cues that modify the properties of the main stimulus acting on the animal and so modify the animal's behavior. But this is implausible. It requires the modifying cues to work with quantitative precision on the animal's nervous system; these cues are hypothetical and have never been elucidated; and, most important, this model cannot deal with novel situations in which the animal has had no opportunity to learn modifying cues. A far more parsimonious explanation is that the animal's behavior is purposive: The rat varies its behavior in response to immediate environmental feedback in order to achieve an endogenous goal (food), which in this case also involves a learned subgoal (pressing the lever).

The systems theorist William T. Powers (1973), in a landmark study, showed that the behavior of such a system can be described mathematically in terms of its tendency to oppose an environmental disturbance of an internally controlled quantity (figure 3). That is to say, the system will operate in such a way that some function of its output quantities will be nearly equal and opposite to some function of a disturbance in some or all of those environmental variables that affect the controlled quantity, with the result that the controlled quantity will remain nearly at its zero point. The classic example is a household thermostat.

Needless to say, the model described above is greatly simplified and portrays only the most rudimentary example—a homeostatic system. More complex cybernetic systems are obviously not limited to maintaining any sort of simple and eternally fixed steady state. In a complex system, overarching goals may be maintained (or attained) by means of an array of hierarchically organized subgoals that may be pursued contemporaneously, cyclically, or seriatim. Furthermore, homeostasis shares the cybernetic stage

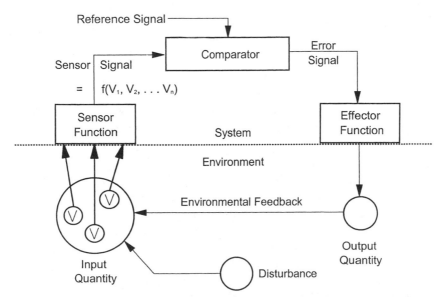

Figure 3. A model cybernetic control system. Redrawn from W. T. Powers, "Feedback: Beyond Behaviorism" *Science* 179 [1973]: 351–56.

with homeorhesis (developmental control processes) and even teleogenesis (goal-creating processes).

It is also important to note that cybernetic processes are not limited to only one level of organization. Over the past decade or so, we have come to appreciate the fact that they exist at many levels of living systems. They can be observed in, among other things, morphogenesis (J. A. Shapiro 1991, 1992; Thaler 1994), cellular activity (Hess and Mikhailov 1994), and neuronal network operation, as well as in the orchestration of animal behavior. Also, it should be noted that the cybernetic model encompasses processes that conform to Haken's paradigm of "distributed control." Some examples include bacterial colonies (J. A. Shapiro 1988), cnidarians (Mackie 1990), honeybees (Seeley 1989), army ants (Franks 1989), and, of course, humans. Another way to put it is that several levels of feedback processes exist in nature, and complex organisms such as mammals—and especially socially organized species—are distinctive in their reliance on the higher level controls (see Corning 1983; Kline 1995). (For a history of feedback-control mechanisms in human technology, which date back to antiquity, see O. Mayr (1970).)

It should also be noted that cybernetic control processes may produce results that resemble Boolean dynamical attractors, but they are achieved in a very different way. By the same token, the cybernetic model, properly applied, calls into question the hypothesis (e.g., J. E. Lovelock 1990) that the biosphere is controlled by automatic nonteleological feedback relationships. Without some internal reference signal (teleonomy), there can be no feedback control, although there can certainly be self-ordered processes of reciprocal causation at work, or perhaps Darwinian processes of coevolution and stabilizing selection. Indeed, the existence of systemic purposiveness (teleonomy) is what distinguishes organisms (and "superorganisms") from ecosystems (see chapter 8; also Wilson and Sober 1989). The mere fact of functional interdependence is insufficient to justify the use of an organismic/cybernetic analogy.

With the emergence and increasing scope of cybernetic self-control, a subtle but important dividing line was crossed in evolution; self-organization was augmented by self-determination. Accordingly, a fundamental challenge for autocatalytic, self-ordering theories of evolution is this: Can hierarchical, cybernetic controls evolve spontaneously (i.e., without reference to their functional properties and performance)? Stuart Kauffman (1993, p. 202) suggests as much. He makes the surprising claim that Boolean networks are "functionally equivalent" to cybernetic regulatory systems. However, this does not seem possible because (a) the causal factors underlying the two types of processes are obviously very different, and (b) the critical property of teleonomy—constant outcomes that are achieved or maintained by variable, feedback-driven behavior—do not appear to be present in these models. Once again, a quote from Dobzhansky may be relevant here:

> The origin of organic adaptedness, or internal teleology, is a fundamental, if not the most fundamental problem of biology. There are essentially two alternative approaches to this problem. One is explicitly or implicitly vitalistic. Organic adaptedness, internal teleology, is considered an intrinsic, immanent, constitutive property of all life. However, like all vitalism, this is a pseudo-explanation; it simply takes for granted what is to be explained. The alternative approach is to regard internal teleology as a product of evolution by natural selection. Internal teleology is not a static property of life. Its advances and recessions can be observed, sometimes induced experimentally, and analyzed scientifically like other biological phenomena. (Dobzhansky et al. 1977, p. 96)

Dobzhansky did not live to witness the recent discoveries in molecular biology that are revolutionizing our conception of evolution. It is becoming evident that even DNA acts in purposeful, feedback-dependent ways not only to control morphogenesis, but, more important, to shape the dynamics of natural selection itself (Cairns et al. 1988; J. A. Shapiro 1991, 1992; Thaler 1994). To quote Thaler: "The environment not only selects among preexisting variants, it also interacts with the organism in sophisticated ways to generate the variation on which selection acts. . . . The components exist for feedback between the generators of genetic diversity and the environment that selects among variants" (p. 225).

In any event, the evolutionary emergence of self-determination over the course of time has had two implications: One is that self-determining processes have gained increasing ascendancy over the blind processes of autocatalysis, mutations, and natural selection. And the second is that, as noted earlier, the partially self-determining organisms that are the products of evolution have come to play an increasingly important causal role in evolution; they have become co-designers of the evolutionary process.

The Pacemakers of Evolution

Recall our earlier discussion of the pacemaker role of behavior in evolution. It is now widely recognized that teleonomy, or purposiveness, is an important property of the behavior of living systems with roots that can be traced far back in evolutionary history (see Rosenbleuth et al. 1943; Ayala 1970; E. Mayr 1974b; Dobzhansky et al. 1977; Corning 1983). Even primitive *E. coli* bacteria, planaria (flatworms), and various insects (drosophila flies, ants, bees, etc.) can adapt and learn novel responses to novel situations and even, in some cases, engage in creative problem-solving.

An unambiguous illustration involves the honey bee's aversion to alfalfa, whose flowers possess spring-loaded anthers that deliver a sharp blow to any bee that attempts to enter. Experienced bees normally avoid alfalfa altogether, but modern, large-scale agricultural practices sometimes leave the honeybee with the choice of alfalfa or starvation. In such situations, the bees have learned to avoid being clubbed by foraging only among flowers where the anthers have already been tripped or by eating a hole in the back of the flower to reach the nectar (Pankiw 1967; Reinhardt 1952). Many other examples are cited in Corning (2003).

In recent years it has become clear that the learning capabilities of animals go well beyond the simplistic behaviorist paradigm. They include specific learning predispositions, selective attention, stimulus filtering and selection, purposive trial-and-error learning, observational learning, and even capabilities for cost-benefit estimates, risk-assessments and discriminative choice-making.

Thus it may be appropriate to deploy the notion of *teleonomic selection* (or neo-Lamarckian selection) to characterize the proximate "mechanism" of value-driven, self-controlled behavioral changes. As the evolved products of evolution have gained greater power to exercise teleonomic control over their relationships to the environment (and to each other), natural selection has become a dog that is increasingly wagged by its tail. Teleonomic selection has become an important instigator of evolutionary change and complexification.

One example of this causal pattern is the evolution of giraffes, which is frequently cited in elementary biology textbooks as an illustration of the distinction between Lamarckian and Darwinian evolution. Evolutionists like to point out that the long necks of modern giraffes are not the product of stretching behaviors that were somehow incorporated into the genes of their short-necked ancestors (as Lamarck posited). Instead, natural selection favored longer-necked giraffes once they had adopted the "habit" of eating tree leaves. And that's the point. A change in the organism-environment relationship among ancestral giraffes, occasioned by a novel behavior—a teleonomic selection—precipitated a new selection pressure for morphological change.

A contrasting example involves one of Darwin's Galapagos Islands finches, the so-called woodpecker finch (*Cactospiza pallidus*). In order to excavate the bark of trees in search of insect larvae, this remarkable bird has been able to circumvent the need to evolve the kind of long, probing tongue that is characteristic of the classic mainland woodpeckers by inventing a digging tool—a cactus spine or small twig that it holds lengthwise in its beak and carries from tree to tree. In other words, a creative behavioral adaptation has enabled *C. pallidus* to circumvent what would otherwise have resulted in a selection pressure for morphological change.

Teleonomic selection is also implicated in the process of evolutionary complexification. Many of the synergistic/symbiotic phenomena that were described above most likely were the result initially of behavioral innovations—ranging from the earliest bacterial colonies to eukaryotes, lichen symbioses, coral communities, land plants, ruminant animals, and the division of labor in socially organized insects and mammals. Synergy provided the proximate rewards, or payoffs, and natural selection affected the appropriate longer-term biological changes.

A similar linkage between synergy and self-determination (teleonomic selection) can be observed in the evolution of human societies. For a detailed review, see Corning (1983, 2003), Hallpike (1986), A. W. Johnson and Earle (1987), J. P. Scott (1989), Durham (1991), and Howells (1993); also see the work of economist Brian Arthur (1988, 1990) and others on the role of positive feedback in economic evolution and the important work on self-determination in human psychology by Deci and Ryan (1985).

One example, from the California Gold Rush era, illustrates not only the role of technological innovation in human evolution but also the economic and organizational (social cybernetic) concomitants. Over a five-year period, from 1848 through 1853, the ontogeny of gold-mining technology in effect recapitulated our entire technological phylogeny up to that time. Within the first year, the classic model of individual prospectors wading in mountain streams with tin pans was largely supplanted by three-man teams using shovels and "rocker boxes," an innovation that also increased the quantity of material that could be processed in a day from ten or fifteen buckets to more than one hundred buckets, or at least twice as much per man. Shortly thereafter, the wooden sluice made its appearance. Though it required six- to eight-man teams (with an associated ownership and management structure), a sluice could handle 400 to 500 buckets of material per day, or about twice as much per man as a rocker box.

When hydraulic mining was introduced in 1853, teams of twenty-five or more men were required to process and haul the materials and manage the water pumps, hoses, etc., that were used to blast away the faces of entire hillsides. A relatively large amount of capital was also needed and an organization was required to manage the technology and the large workforce. However, the amount of material processed daily also jumped to one hundred tons or more. Again, the functional consequences of synergistic phenomena can be measured and quantified.

Many other examples could be cited, but perhaps this one will suffice to illustrate some of the major features of the process of complexification in human evolution. The process has included purposive innovation, cybernetic social control, the production of synergistic (cooperative) effects, teleonomic selection, and, in its train, microevolutionary biological changes via natural selection. Much of our early history as a species remains shrouded and subject to varying interpretations as to the particulars, but the overall pattern described above seems valid. Indeed, supporting evidence can be found among contemporary human populations living in extreme environments—deserts, the arctic, high altitudes—where distinct cultural and morphological adaptations have followed the teleonomic selections associated with migration into these environments.

One other aspect of self-determination in evolution should be mentioned here. It has to do with the role of information. As Robert Rosen (1985) has pointed out, information is one of the most widely used, exhaustively analyzed, and theoretically muddled concepts in all of science. In physics (and electronics), the classic Shannon-Weaver formulation—the quantity of binary bits associated with a given communications transaction—has provided a convenient and durable measuring rod (Shannon 1948; Shannon and Weaver 1949). But this also skirts the issue of how to define information. Furthermore, in living systems, the quantity of information is very often less relevant than the *quality*; all information is not created equal. Nor is there any single unit of information that can give quantitative precision to the concept where living systems are concerned. Information takes different forms at different levels of biological organization. Although there have been numerous efforts in recent years to develop a theoretically useful definition for the life sciences (see especially Rosen 1985; Weber et al. 1988; Banerjee et al. 1990; Salthe 1993), none has won a consensus.

The definition I favor is perhaps the most unorthodox of all. I have proposed (1983, 1992, 2001a, 2001b) that information does not, in fact, exist; in reality it is an umbrella concept like natural selection that we use to characterize certain properties, or functional aspects, of a wide variety of phenomena associated with the construction (ontogeny, phylogeny) and operation of thermodynamic/cybernetic systems. (Physicists, of course, have a very different view of the term.) The cybernetics pioneer, Norbert Wiener (1948), equated information with the degree of organization (or negative entropy) in a cybernetic system. However, I prefer to define cybernetic "control information" in terms of its functional role, to wit: *the capacity to exercise cybernetic control over the acquisition, disposition, and utilization of matter/energy by living systems.* If energy is defined as the capacity to do work, control information is defined as the capacity to control the capacity to do work. We will explore this concept in depth in chapter 14.

In any event, information in the cybernetic sense is vitally linked to the future of our capacity for self-determination and self-control, both individually and collectively. Much will depend upon the evolving technologies of communications. But, equally important, much will depend upon the capacity of the sciences (and the humanities) to contribute to the development of more powerful and effective information. The future of self-determination, then, is also tied to the acquisition (and application) of useful knowledge.

It may well be the case that the primordial origins of complexity in evolution were rooted in self-ordering processes; living systems may indeed have been energized and catalyzed by spontaneous biophysical and biochemical activity. However, the trajectory of evolution has moved the causal dynamics inexorably away from autocatalytic phenomena toward purposive and functional phenomena. Accordingly, the future lies with self-determination—that is, information-based, purposive innovations. And it can safely be predicted that new forms of synergy will play a central role in shaping our future evolution as a species, for good or ill.

— ⟡ —

The emergent is unlike its components in so far as these are incommensurable, and it cannot be reduced to their sum or their difference.

—G. H. Lewes

SUMMARY: Despite its current popularity, emergence is a concept with a venerable history and an elusive, ambiguous standing in contemporary evolutionary theory. This chapter briefly recounts the history of the term and details some of its current usages. Not only are there radically varying interpretations about what emergence means, but so-called "reductionist" and "holistic" theorists have very different views about the issue of causation. However, these two seemingly polar positions are not irreconcilable. Reductionism, or detailed analysis of the parts and their interactions, is essential for answering the "how" question in evolution—how does a complex living system work? But holism is equally necessary for answering the "why" question—why did a particular arrangement of parts evolve? In order to answer the "why" question, a broader, multileveled paradigm is required. The reductionist approach to explaining emergent complexity has entailed a search for underlying "laws of emergence." In contrast, the Synergism Hypothesis focuses on the economics—the functional effects produced by emergent wholes and their selective consequences. It will also be argued that emergent phenomena represent, in effect, a subset of a much larger universe of combined effects in the natural world; there are many different kinds of synergy, but not all synergies represent emergent phenomena.

5

The Re-Emergence of Emergence: A Venerable Concept in Search of a Theory

Introduction

If "complexity" is currently the buzzword of choice for our newly minted millennium—as many theorists proclaim—"emergence" seems to be the explication of the hour for how complexity has evolved. Complexity, it is said, is an emergent phenomenon. Emergence is what self-organizing processes produce. Emergence is the reason why there are hurricanes, and ecosystems, and complex organisms like humankind, not to mention traffic congestion and rock concerts. Indeed, the term is positively awe-inspiring. As physicist Doyne Farmer observed: "It's not magic . . . but it *feels* like magic" (quoted in Waldrop 1992).

Among other things, emergence has been used by physicists to explain Bénard (convection) cells, by psychologists to explain consciousness, by economists and investment advisors to explain stock market behavior, and by organization theorists to explain informal networks in large companies. Indeed, a number of recent books view the evolutionary process itself as an emergent phenomenon. But what is emergence? What does it explain, really? And why is it so readily embraced, in spite of its opacity, by reductionists and holists alike? There are very few terms in evolutionary theory these days—not even *natural selection*—that can command such an ecumenical following.

Though emergence may seem to be the "new, new thing"—from the title of the recent bestseller by Michael Lewis about high technology in Silicon Valley—in fact it is a venerable term in evolutionary theory that traces back to the latter nineteenth and early twentieth centuries. It was originally coined during an earlier upsurge of interest in the evolution of wholes, or, more precisely, what was viewed unabashedly in those days as a "progressive" trend in evolution toward new levels of organization culminating in mental phenomena and the human mind. This long-ago episode, part of the early history of evolutionary theory, is not well known today, or at least not fully appreciated. Nonetheless, it provides a theoretical context and offers some important insights into what can legitimately be called the re-emergence of emergence.

The Origin of Emergence

According to the philosopher David Blitz in his authoritative history of emergence entitled, appropriately enough, *Emergent Evolution: Qualitative Novelty and the Levels of Reality* (1992), the term *emergent* was coined by the pioneer psychologist G. H. Lewes in his multivolume *Problems of Life and Mind* (1874–79). Like many post-Darwinian scientists of that period, Lewes viewed the evolution of the human mind as a formidable conundrum. Some evolutionists, like Alfred Russel Wallace (the codiscoverer of natural selection), opted for a dualistic explanation. The mind is the product of a supernatural agency, he claimed. But Lewes, following the lead of the philosopher John Stuart Mill, argued that, to the contrary, certain phenomena in nature produce what he called "qualitative novelty"—material changes that cannot be expressed in simple quantitative terms; they are emergents rather than resultants. To quote Lewes:

> Every resultant is either a sum or a difference of the cooperant forces; their sum, when their directions are the same—their difference, when their directions are contrary. Further, every resultant is clearly traceable in its components, because these are homogeneous and commensurable. . . . It is otherwise with emergents, when, instead of adding measurable motion to measurable motion, or things of one kind to other individuals of their kind, there is a cooperation of things of unlike kinds. . . . The emergent is unlike its components in so far as these are incommensurable, and it cannot be reduced to their sum or their difference. (p. 413)

Years earlier, John Stuart Mill had used the example of water to illustrate essentially the same idea: "The chemical combination of two substances produces, as is well known, a third substance with properties different from those of either of the two substances separately, or of both of them taken together" (quoted in Blitz 1992, p. 77). However, Mill himself had an illustrious predecessor. In fact, both Mill and Lewes were resurrecting an argument that Aristotle had made more than two thousand years earlier in a philosophical treatise, later renamed the *Metaphysics*, about the significance of wholes in the natural world. (I alluded to it in earlier chapters and it was quoted in the frontispiece.) Aristotle wrote: "The whole is something over and above its parts, and not just the sum of them all . . ." (1961, book H, 1045:8–10). (We will return to Aristotle's famous catchphrase later on.) So the ontological distinction between parts and wholes was not exactly a new idea in the nineteenth century. The difference was that the late Victorian theorists framed the parts-wholes relationship within the context of the theory of evolution and the challenge of accounting for biological complexity.

The basic quandary for holistic theorists of that era was that evolutionary theory as formulated by Darwin did not allow for radically new phenomena in nature, like the human mind (presumably). As every first-year biology student these days knows, Darwin was a convinced gradualist who frequently quoted the popular canon of his day, *natura non facit saltum*— nature does not make leaps. (The phrase appears no less than five times in *The Origin of Species*.) Indeed, Darwin rejected the very idea of sharp discontinuities in nature. In *The Origin,* Darwin emphasized what he called the "Law of Continuity," and he repeatedly stressed the incremental nature of evolutionary change, which he termed "descent with modification." Darwin believed that this principle applied as well to the evolution of the "mind." In the *Descent of Man*, he asserted that the difference between the human mind and that of "lower" animals was "one of degree and not of kind" (1874, vol. I p. 70).

Many theorists of that era viewed Darwin's explanation as unsatisfactory, or at least incomplete, and emergent evolution theory was advanced as a way to reconcile Darwin's gradualism with the appearance of "qualitative novelties" and, equally important, with Herbert Spencer's notion (elaborating on Lamarck) of an inherent, energy-driven trend in evolution toward new levels of organization. Emergent evolution had several prominent adherents, but the leading theorist of this school was the comparative psychologist and prolific writer, Conwy Lloyd Morgan, who ultimately published three volumes on the subject, *Emergent Evolution* (1923), *Life, Spirit and Mind* (1926), and *The Emergence of Novelty* (1933). (Other theorists in this vein included Samuel Alexander, Roy Wood Sellars, C. D. Broad, Arthur Lovejoy, William Morton Wheeler, and Jan Smuts, a one-time Prime Minister of South Africa.)

The main tenets of Lloyd Morgan's paradigm will sound familiar to modern-day wholists: quantitative, incremental changes can lead to qualitative changes that are different from, and irreducible to, their parts. By their very nature, moreover, such wholes are unpredictable. Though higher-level, emergent phenomena may arise from lower-level parts and their actions, there may also be "return action," or what Lloyd Morgan also called "supervenience" ("downward causation" in today's parlance). But most important, Lloyd Morgan argued that the evolutionary process has an underlying progressive tendency, because emergent phenomena lead in due course to new levels of reality.

It was a grand vision, but what did it explain? As Blitz observes, it was not a causal theory. "Emergent evolution related the domains studied by the sciences of physics, chemistry, biology, and psychology—a philosophical task not undertaken by any one of them—but did not propose mechanisms of change specific to any one of them—a scientific task which philosophy could not undertake"(1992, p. 100). Indeed, Lloyd Morgan ultimately embraced a metaphysical teleology that portrayed the evolutionary process as an unfolding of inherent tendencies, which he associated with a creative divinity (shades of Spencer, Henri Bergson, Pierre Tielhard de Chardin, and other orthogenetic and vitalistic theorists, not to mention some of today's complexity theorists).

Jan Smuts's visionary book *Holism and Evolution* (1926) also deserves mention, not least because he was apparently the one who coined the term *holism* (from the Greek word for wholes). But, like Lloyd Morgan, Smuts advanced a concept of "holistic selection" that was pointedly non-Darwinian in nature. Smuts posited an inherent developmental tendency in evolution—an underlying whole-making "force." As Smuts put it:

> The creation of wholes, and ever more highly organized wholes . . . is an inherent character of the universe. There is not a mere vague indefinite creative energy or tendency at work in the world. This energy or tendency has specific characters, the most fundamental of which is whole-making. . . . Wholeness is the most characteristic expression of the nature of the universe in its forward movement in time. It marks the line of evolutionary progress. And Holism is the inner driving force behind that progress. (1926, p. 99)

In sum, emergent evolution in the hands of Lloyd Morgan, Smuts, and others of that era was orthogenetic; it was not really a scientific theory, though the boundary line was not so sharply delineated back then. But far more damaging to the cause of emergent evolution was the rise of the science of genetics in the 1920s and 1930s and the triumph of an analytical, experimental approach to biology. In its most strident form, reductionism swept aside the basic claim of emergent evolutionists that wholes had irreducible properties that could not be fully understood or predicted by examining the parts alone. Critics like Stephen C. Pepper, Charles Baylis, William McDougall, Rudolph Carnap, and Bertrand Russell claimed that emergent qualities were merely epiphenomena and of no scientific significance. Russell, for instance, argued that analysis "enables us to arrive at a structure such that the properties of the complex can be inferred from those of the parts" (Russell 1927, pp. 285-286). The reductionists conceded that it was not currently possible, in many cases, for science to make such inferences and predictions, but they believed that this shortcoming was a reflection of the state of the art in science and not of some superordinate property in nature itself. In time, it was said, reductionism would be able to give a full accounting for emergent phenomena.

The Submergence of Emergence

Under this theoretical onslaught, the doctrine of emergent evolution went into a prolonged eclipse, although it never succumbed completely to the promissory notes proffered by the reductionists. During the decades that followed, the Aristotelian argument that wholes have distinctive, irreducible properties re-emerged in several other venues (though often with different terminology). In the 1930s, for example, embryologist Joseph Needham advanced the idea of "integrative levels" in nature and argued for "the existence of [different] levels of organization in the universe, successive forms of order in a scale of complexity and organization" (1937, p. 234). A decade later, the biologist Julian Huxley, a principal architect of the "modern synthesis" in evolutionary biology, sought to define evolution as "a continuous process from star-dust to human society." Among other things, Huxley asserted that "now and again there is a sudden rapid passage to a totally new and more comprehensive type of order or organization, with quite new emergent properties, and involving quite new methods of further evolution" (J. S. Huxley and T. H. Huxley 1947, p. 120). Biologist Alex B. Novikoff also defended the idea of emergent levels of reality in a much-cited 1945 article in *Science* entitled "The Concept of Integrative Levels in Biology."

The growth of the new science of ecology in the 1930s also stimulated an interest in whole systems and macro-level relationships. Among the pioneer ecologists—such as Charles Elton, A. G. Tansley, Raymond Lindeman, G. Evelyn Hutchinson, and others—there was much talk about how the natural world is an integrated economy, a biological community and even, for some theorists, a "quasi-organism" (Tansley 1935). Ironically enough, the seminal concept of an ecosystem—which has since become a centerpiece of modern ecology—was originally conceived by Tansley in the context of his belated conversion to reductionism. "Wholes," he wrote, "are in *analysis* nothing but the synthesized actions of the components in associations" (1935, p. 289). (For an in-depth history of ecology, see Donald Worster's *Nature's Economy: A History of Ecological Ideas*, 1977.)

A much broader reaffirmation of the importance of wholes in nature occurred in the 1950s with the rise of general systems theory. Inspired especially by the writings of biologist Ludwig von Bertalanffy (1949, 1968), the systems movement was to that era what complexity theory is today, and the Society for General Systems Research, founded in 1956, provided an interdisciplinary haven for the beleaguered band of holistic theorists of that era. (The organization was later renamed The International Society for the Systems Sciences). Indeed, the Society's yearbook—*General Systems*—was a beacon (and a treasure-trove) for the systems movement for more than a generation. It included the contributions of such luminaries as Kenneth Boulding, Ralph Gerard, Anatol Rapoport, H. Ross Ashby, Heinz von Foerster, Russell Ackoff, Stafford Beer, Donald T. Campbell, Herbert Simon, George Klir, Robert Rosen, Lawrence Slobodkin, Paul Weiss, James Grier Miller, and many others. (Herbert Simon's 1962 article on "The Architecture of Complexity" was seminal, along with Paul Weiss's 1969 article on "The Living System: Determinism Stratified"; see below.)

A Re-Emergence

It is difficult to attach a date to the re-emergence of emergence as a legitimate, mainstream concept, but it roughly coincided with the growth of scientific interest in the phenomenon of complexity and the development of new, non-linear mathematical tools—particularly chaos theory and dynamical systems theory—which allow scientists to model the interactions within complex, dynamic systems in new and insightful ways. Among other things, complexity theory gave mathematical legitimacy to the idea that processes involving the interactions among many parts may be at once deterministic yet for various reasons unpredictable. (One oft-noted constraint, for instance, is the way in which initial conditions—the historical context—may greatly influence later outcomes in unforeseeable ways.)

One of the benchmarks associated with the re-emergence of emergence was the work of Nobel psychobiologist Roger Sperry (1964, 1969, 1991)—noted earlier—on mental phenomena and the role of what he was the first to call "downward causation" in complex systems like the human brain. Sperry also employed Lloyd Morgan's term, *supervenience*.

Meanwhile, in physics Herman Haken and his colleagues broke new ground with "synergetics"—the science of dynamic, "cooperative" phenomena in the physical realm (though he later ventured into neurological and cognitive phenomena as well). Over the past thirty-odd years, synergetics has produced a large body of wholistic theory (Haken 1973, 1974, 1977, 1983, 1988a, 1990). Likewise, the Nobel physicist Ilya Prigogine's work in non-equilibrium thermodynamics, especially his concept of "dissipative structures," represented yet another wholistic approach to the rise of complexity in nature (Prigogine 1978, 1980; Prigogine and Nicolis 1971; Prigogine et al. 1972a, 1972b; Nicolis and Prigogine 1977, 1989).

In the United States, much of the recent work on the subject of emergence has been fueled by the resources and leadership of the Santa Fe Institute. Beginning in the mid-1980s, the Instititute's annual Proceedings have contained many articles related to this subject, and a number of the scholars associated with the Institute have published books on complexity and emergence. (See especially Kauffman 1993, 1995, 2000; Casti 1995, 1997; Holland 1995, 1998; also, Lewin 1992 and Waldrop 1992.) Kauffman, as noted earlier, theorizes that life is an emergent phenomenon in the sense that it represents a "spontaneous crystallization" of pre-biotic molecules that can catalyze networks of reactions. Life is a collective property of a system of interacting molecules, says Kauffman: "the whole is greater than the sum of its parts"(1995, pp. 23-24). Likewise, Holland published an entire book devoted to the subject, entitled *Emergence: From Chaos to Order* (1998).

What Does Emergence Mean?

Despite the recent proliferation of writings on the subject, it is still not clear what the term denotes or, more important, how emergence emerges. One problem is that the term is frequently used as a synonym for "appearance," or "growth," as distinct from a parts-whole relationship. Thus, one of the dictionaries I consulted defined the term strictly in perceptual terms and gave as an example "the sun emerged from behind a cloud." Even the Oxford English Dictionary, which offered four alternative definitions, gives precedence to the version that would include a submarine which submerges and then re-emerges.

It is not surprising, then, that the overwhelming majority (close to 100 percent) of the new journal articles on "emergence" and "emergent" that are identified each week by my computer search service involve such subjects as the emergence of democracy in Russia, the emergence of soccer as a school sport in the U.S., the emergence of the Internet, the emergence of mad cow disease, and the like. I have deliberately played on this conflation of meanings in this chapter to illustrate the point, but even avowed complexity theorists commonly use the term (perhaps unwittingly) in both ways. Thus, the subtitle of Mitchell Waldrop's book *Complexity* (1992) is *The Emerging Science at the Edge of Order and Chaos*.

Unfortunately, some theorists seem to take the position that emergence does not exist if it is not perceived; it must be apparent to an observer. But what is a "whole"—how do you know it when you see it, or don't see it? And is the mere perception of a whole—a gestalt experience—sufficient, or even necessary? John Casti (1997), like Lewes and Morgan, associates emergence with dynamic systems whose behavior arises from the interaction among its parts and cannot be predicted from knowledge about the parts in isolation. "The whole is bigger than the sum of its parts," echoes editor Michael Lissack (1999) in the inaugural issue of the new journal *Emergence*. John Holland (1998), by contrast, describes emergence in reductionist terms as "much coming from little" and imposes the criterion that it must be the product of self-organization, not centralized control. Indeed, Holland tacitly contradicts Casti's criterion that the behavior of the whole is irreducible and unpredictable. Holland's approach represents reductionism of a different kind—more like Herbert Spencer's search for a universal "law" of evolution than Bertrand Russell's focus on identifying all the parts.

Perhaps the most elaborate recent definition of emergence was provided by Jeffrey Goldstein (1999) in the inaugural issue of *Emergence*. To Goldstein, emergence refers to "the arising of novel and coherent structures, patterns and properties during the process of self-organization in complex systems." The common characteristics are: (1) radical novelty (features not previously observed in the system); (2) coherence or correlation (meaning integrated wholes that maintain themselves over some period of time); (3) A global or macro level (i.e., there is some property of "wholeness"); (4) it is the product of a dynamical process (it evolves); and (5) it is "ostensive"—it can be perceived. For good measure, Goldstein throws in supervenience—downward causation.

Goldstein's definition is hardly the last word on this subject, however. One indication of the ambiguous status that the term currently holds in complexity science is the discordant dialogue that occurred in an on-line (Internet) discussion of the topic hosted by the New England Complex Systems Institute (NECSI) during December 2000 and January 2001. Here are just a few abbreviated (and paraphrased) excerpts:

- Emergence has more to do with concepts and perceptions.
- Emergence arises when an observer recognizes a "pattern."
- Perception is irrelevant—emergence can occur when nobody is there to observe it.
- The mind is an emergent result of neural activity.
- In language, meaning emerges from combinations of letters and words.
- A society is an emergent, but it is in turn composed of emergent collections of cells.
- When water boils and turns to steam, this is emergence—something new in the macro world emerges from the micro world.
- Temperature and pressure are emergents—macrolevel averages of some quantity present in microlevel phenomena.
- Emergence involves a process. Thus, economists can say that a recession emerges.
- It's like a dynamical attractor, or the product of a "deep structure"—a preexisting potentiality.
- Another participant responded to this with "I don't know what a deep structure is, but it feels good to say it."
- Still another objected that dynamical attractors are mathematical constructs—they say nothing about the underlying forces.
- Emergence requires some form of "interaction"—it's not simply a matter of scale.
- Others disagreed—if the properties of the whole can be calculated from the parts and their interactions, it is not emergence.
- Emergents represent rule-governed creativity based on finite sets of elements and rules of combination.
- Emergence does not have logical properties; it cannot be deduced (predicted).
- Another participant replied, "maybe not, but once observed, future predictions are possible if it is deterministic."
- Another discussant asserted that a "very simple example" is water, and its properties should in principle be calculable by detailed quantum-level analysis.
- A discussant familiar with quantum theory disagreed—given the vast number of "choices" (states) that are accessible at the quantum level, one would, in effect, have to read downward from H2O to make the right choice.
- Yet another discussant pointed out that quantum states are always greatly affected by the boundary conditions—the environment.
- Finally, one discussant disputed the entire concept of emergence—it's all in the eye of the beholder—if we cannot even know that there is a real world, that hydrogen and oxygen actually exist, how can we "know" what they do in combination?

In short, contradictory opinions abound. There is no universally acknowledged definition of emergence, nor even a consensus about such hoary (even legendary) examples as water. And if emergence cannot be defined in concrete terms—so that you will know it when you see it—how can it be measured, or explained? As Jeffrey Goldstein noted in his *Emergence* article, "emergence functions not so much as an explanation but rather as a descriptive term pointing to the patterns, structures or properties that are exhibited on the macro-scale" (1999, p. 58). Editor Michael Lissack, in his own inaugural *Emergence* article, acknowledged that "it is less an organized, rigorous theory than a collection of ideas that have in common the notion that within dynamic patterns there may be underlying simplicity that can, in part, be discovered through large quantities of computer power . . . and through analytical, logical and conceptual developments . . ." (1999, p. 112). (Well, not always. On this point, see below.)

Redefining Emergence

How can we sort all of this out? The place to start, I believe, is with the more inclusive concept of synergy. Since this concept has been treated in depth in earlier chapters, here I will be brief. To repeat, synergy, broadly defined, refers to *the combined (cooperative) effects that are produced by two or more particles, elements, parts or organisms—effects that are not otherwise attainable.* In this definition, synergy is not "more" than the sum of the parts, just different (as Aristotle himself long ago argued). Furthermore, there are many different kinds of synergy.

Accordingly, some of the confusion surrounding the term *emergence* might be reduced (if not dissolved) by limiting its scope. Rather than using it loosely as a synonym for synergy, or gestalt effects, or perceptions, etc., I would propose that emergent phenomena be defined as a "subset" of the vast (and still expanding) universe of cooperative interactions that produce synergistic effects of various kinds, both in nature and in human societies. In this definition, emergence would be confined to those synergistic wholes that are composed of things of "unlike kind" (following Lewes's original definition). It would also be limited to "qualitative novelties" (after both Lewes and Lloyd Morgan)—i.e., unique synergistic effects that are generated by functional complementarities, or a combination of labor. In this more limited definition, all emergent phenomena produce synergistic effects, but many synergies do not entail emergence.

In other words, emergent effects would be associated specifically with contexts in which constituent parts with different properties are modified, re-shaped or transformed by their participation in the whole. Thus, water and table salt are unambiguous examples of emergent phenomena. And so is the human body. Its 10 trillion or so cells are specialized into some 250 different cell types that perform a vast array of important functions in relation to the operation of the whole. Indeed, in biological systems (and in technological wholes like automobiles), the properties of the parts are very often shaped by their functions for the whole. On the other hand, in accordance with the Lewes/Morgan definition, a sand pile or a river would not be viewed as emergent phenomena. If you've seen one water molecule you've seen them all.

Must the synergies be observed in order to qualify as emergent effects, as some theorists claim? Most emphatically not. The synergies associated with emergence are real and measurable, even if nobody is there to observe them. And what about the claim that emergent effects can only be the result of "self-organization"? Is this a requirement? Again, emphatically not. As noted in the last chapter, self-organization is another academic buzzword these days that is often used rather uncritically. However, to repeat, there is a fundamental distinction between self-organizing processes (or, more precisely, what should be called "self-ordering" processes) and wholes that are products of *functional organization* (as in organ systems). Living systems and human organizations are largely shaped by "instructions" (functional information) and by cybernetic control processes. They are not, for the most part, self-ordered; they are predominately organized by processes that are "purposeful" (teleonomic) in nature and that rely on "control information." (See also chapters 4 and 12 to 14.)

Consider this example. A modern automobile consists of some 15–20,000 parts (depending on the car and how you count). If all of these parts were to be thrown together in one great "heap" (a favorite word of Aristotle), they could be described as "ordered" in the sense that they are not randomly distributed across the face of the earth (or the universe, for that matter). Nevertheless, they do not constitute a car. They become an organized, emergent phenomenon—a useable "whole"—only when the parts are assembled in a very precise (purposeful) way. As a disorganized heap, they are indeed nothing more than the sum of the parts. But when they are properly organized, they produce a type of synergy (emergent effects) that the parts alone cannot.

In this light, let us return briefly to the NECSI Internet discussion. As defined here, emergence has nothing to do with concepts, patterns, or appearances (despite the conflated usage of the term in everyday language). The mind is indeed an emergent phenomenon, but steam is not. Some emergent phenomena may be rule-governed, but this is not a prerequisite; much of it is also instruction-governed. A water molecule is also an emergent phenomenon, but the debate over whether or not the whole can be predicted from the properties of the parts in fact misses the point. Wholes produce unique combined effects, but many of these effects may be co-determined by the context and the interactions between the whole and its environment(s). In fact, many of the properties of the whole may arise from such system-environment interactions. This is preeminently the case with living systems.

The Laws of Emergence?

This conclusion, and the fundamental distinction that was drawn above between emergent phenomena that are self-ordered and the many products of purposeful organization (functional design) also has important theoretical implications, I contend. Indeed, this distinction goes directly to the heart of the reductionist-holist debate about the properties of wholes (and how to explain them) that traces back to the nineteenth century, and it poses a direct challenge to the contemporary search for "laws" of emergence and complexity in evolution.

Holland, in his recent book on emergence, acknowledges that this newly fashionable term remains "enigmatic"—it can be defined in various ways. Nevertheless, he believes that some general laws of emergence will ultimately be found. Holland asks: "How do living systems emerge from the laws of physics and chemistry. . . . Can we explain consciousness as an emergent property of certain kinds of physical systems?"(1998, p. 2). Elsewhere he speaks of his quest for what amounts to the antithesis of the entropy law (the Second Law of Thermodynamics)—namely, an inherent tendency of matter to organize itself. Holland illustrates with a metaphor. Chess, he says, is a game in which "a small number of rules or laws can generate surprising complexity." He believes that biological complexity arises from a similar body of simple rules.

In fact, there have been many variations on this basic theme in recent years, with numerous theorists invoking inherent self-organizing tendencies in nature. Francis Heylighen and his colleagues (1999) claim that evolution leads to the "spontaneous emergence" of systems with higher orders of complexity. Mark Buchanan (2000) discerns a "law of universality" in evolution—from our cosmic origins to economic societies—as a consequence of self-organized criticality (after Per Bak and his colleagues). As noted in chapter 4, Stuart Kauffman in his latest book (2000) speaks of a new "fourth law of thermodynamics"—an inherent organizing tendency in the cosmos that counteracts the entropic influence of the Second Law. Steve Grand (2001) views the emergence of networks as a self-propelled, auto-catalytic process. Albert-László Barabási (2002) invokes "far reaching natural laws" that, he believes, govern the emergence of networks. And Niels Gregersen and his contributors (2002) see an "innate spontaneity" in the emergence of complexity. All of these grand visions can be called reductionist in the sense that they posit some underlying, inherent force, agency, tendency or "law" that is said to determine the course of the evolutionary process, or some important aspect; emergence is thus treated as an epiphenomenon.

Edward O. Wilson also speaks in reductionist terms about emergent phenomena. In his discipline-defining volume, *Sociobiology: The New Synthesis* (1975), Wilson proclaimed that: "The higher properties of life are emergent" (p. 7). He also referred to a "new holism" that would avoid what he called the "mysticism" of past holists, such as Lloyd Morgan and William Morton Wheeler. Wilson did not elaborate on this theme in his volume, but in his more recent book, *Consilience: The Unity of Knowledge* (1998), he endorses what he characterizes as the "strong form" of scientific unification. His "transcendental world view," as he puts it, is that "nature is organized by simple universal laws to which all other laws and principles can be reduced" (p. 55). "The central idea of the consilience world view is that all tangible phenomena, from the birth of stars to the workings of social institutions, are based on material processes that are ultimately reducible, however long and tortuous the sequences, to the laws of physics" (p. 226). Wilson claims that an emergent phenomenon such as the human mind can, in theory at least, be reduced to its constituent parts and their interactions. Of course, he concedes, "this would require massive computational capacity," but he derides the claim that the mind and other such "wholes" cannot be understood by reductionist analyses alone. He calls this notion a "mystical concept" (quoted in Miele 1998, p. 79).

In a similar vein, Francis Crick, in a 1994 book, explains: "The scientific meaning of emergent, or at least the one I use, assumes that, while the whole may not be the simple sum of its separate parts, its behavior can, at least in principle, be *understood* from the nature and behavior of its parts *plus* the knowledge of how all these parts interact [italics in original]" (p. 11). He illustrates with an example from elementary chemistry. The benzene molecule is made of six carbon atoms arranged in a ring with a hydrogen atom attached to each. It has many distinctive chemical properties, but these can be explained, he claims, in terms of quantum mechanics. "It is curious that nobody derives some mystical satisfaction by saying 'the benzene molecule is more than the sum of its parts.' . . ."

Nobody can gainsay the fact that a great deal has been learned about how nature and living systems work through the use of reductionist methods in science, and surely there is much more to come. There may indeed be many law-like patterns at different levels and in different domains of the natural world. But the water example given above (chapter 3) illustrates why there are ultimate limits to reductionism, and why holistic, systems approaches (and even systems-environment approaches) are also essential for understanding "organized" biological wholes. We can see why this is the case by revisiting some of the views expressed above.

First, consider Holland's chess analogy. Rules, or laws, have no causal efficacy; they do not in fact "generate" anything. They serve merely to describe regularities and consistent relationships in nature. These patterns may be very illuminating and important, but the underlying causal agencies must be separately specified (though often they are not). But that aside, the game of chess illustrates precisely why any laws or rules of emergence and evolution are insufficient. Even in a chess game, you cannot use the rules to predict "history"—that is, the course of any given game. Indeed, you cannot even reliably predict the next move in a chess game. Why? Because the system involves more than the rules of the game. It also includes the players and their unfolding, moment-by-moment decisions among a very large number of available options at each choice point. The game of chess is inescapably historical, even though it is also constrained and shaped by a set of rules, not to mention the laws of physics. Moreover, and this is a key point, the game of chess is also shaped by teleonomic, cybernetic, feedback-driven influences. It is not simply a self-ordered process; it involves an organized, purposeful activity.

Similar limitations and biases can be seen in some of the other recent writings on emergence. Thus, for example, Barabási (2002) speaks of a "law' of network development, but the process he describes in effect amounts to a Darwinian theory of networks. He tells us that the "fittest" nodes—based on the context and their functional properties—will expand and become the biggest, and most central, at the expense of other nodes. Likewise, Steven Johnson, in his book *Emergence* (2001), cites ant behavior as a model for spontaneous self-organization in nature. But this is inaccurate. In fact, the behavior of the ants is highly purposeful, even though the "machinery" of cybernetic control may be distributed; ant-behavior is instruction-driven, not law-driven. Finally, in his newest book, Kauffman repeatedly hints at "laws" of evolution but concedes these are yet to be found. As noted in chapter 4, he now recognizes two other important causal agencies in evolution—"autonomous agents" (a.k.a. living organisms) and natural selection! "Self-organization mingles with natural selection in barely understood ways . . ." (2000, p. 2).

Harold Morowitz (2002), in his newest book, comes closer than most theorists of this genre to a view that is compatible with the Darwinian paradigm. Recognizing that variability is inherent in the living world at every level, Morowitz posits that there are "pruning rules" that shape the forms that arise out of the many possibilities in evolution. The problem here is that this formulation implies that there are preexisting selection criteria, rather than recognizing that selection is historically defined and context-specific. Biological evolution is an open-ended trial-and-success process. Indeed, Morowitz finds himself in sympathy with Teilhard de Chardin (and many others) in believing that there is "something deeper" in the "orderly unfolding" of the universe.

As for Edward Wilson's reductionist claim that we lack only sufficient computational capacity to elucidate the workings of the human mind, the problem with this formulation is that the human mind is not a disembodied physical entity, or a mass-produced machine with interchangeable parts. Each mind is also a product of its particular "history"—its distinct phylogeny, its unique ontogeny and its ongoing, moment-by-moment interactions with its environment(s). Molecular biology and neurobiology—however important to our understanding of mental phenomena—can only illuminate some of the many levels in the life of the mind. As for all the rest of the causal hierarchy, unfortunately we are not omniscient and never will be.

Equally important, there is a major theoretical segue involved in the modernized version of reductionism espoused by Wilson, Crick and others. In its nineteenth and early twentieth century incarnation, reductionism meant an understanding of the "parts"—period. Modern-day reductionists, by contrast, speak of the parts *and* their interactions. But the interactions among the parts (and between the parts and their environments) *are* "the system." The whole is not something that floats on top of it all. So this cannot properly be called reductionism; it is systems science in disguise. Indeed, the interactions among the parts may be far more important to the understanding of how a system works than the nature of the parts alone. For example, we now have a relatively complete map of the human genome. Yet we still have only a sketchy idea of how the genome produces a complete organism. The great challenge for molecular biology in this century will be to do systems science at the molecular level. (For a path-breaking volume on systems biology, see Kitano 2001.)

Evolution as a Multilevel Process

To repeat a point made earlier, though reductionism will no doubt continue to play a vital role in helping us to understand "how" organized systems (emergent phenomena) work in nature, a number of theorists, including this author, have argued that a multileveled selectionist approach is necessary for answering the "why" question—why have emergent, complex (living) systems evolved over time? (See Corning 1983, 1995, 1996a, 2003; Maynard Smith and Szathmáry 1995, 1999; D.S. Wilson 1997a,b; Sober and Wilson 1998; see also chapters 1 to 3.)

To reiterate, Holistic Darwinism, and the multileveled approach to complexity, is based on the cardinal fact that the material world is organized hierarchically. What the reductionist claims overlook is the fact that new principles, and emergent new capabilities, arise at each new level of organization in nature. (Again, the water example in chapter 3 provides an illustration.) A one-level model of the universe based, say, on quantum mechanics and the actions of quarks and leptons, or energy flows, or whatever, is therefore totally insufficient. This point was argued with great clarity and erudition many years ago in a landmark essay, cited above, by the biologist Paul Weiss entitled "The Living System: Determinism Stratified." "Organisms are not just heaps of molecules," Weiss pointed out (1969, p. 42). They organize and shape the interactions of lower-level "subsystems" (downward causation), just as the genes, organelles, tissues and organs shape the behavior of the system as a whole (upward causation). Furthermore, one cannot make sense of the parts, or their interactions, without reference to the combined effects (the synergies) they produce.

The arguments advanced in the two landmark articles in *Science*, mentioned earlier, also bear repeating. In "Life's Irreducible Structure" (1968), chemist Michael Polanyi pointed out that each level in the hierarchy of nature involves "boundary conditions" that impose more or less stringent constraints on lower-level phenomena, and that each level operates under its own, irreducible principles and laws. And the Nobel physicist Philip Anderson, in "More is Different"(1972), noted that one cannot start from reductionist laws of physics and reconstruct the universe. "The constructionist hypothesis breaks down when confronted with the twin difficulties of scale and complexity. . . . At each level of complexity entirely new properties appear."

Accordingly, emergent phenomena in the natural world involve multilevel systems that interact with both lower- and higher-level systems—or "inner" and "outer" environments, in biologist Julian Huxley's characterization. Furthermore, these emergent systems in turn exert causal influences both upward and downward—not to mention horizontally. (If determinism is stratified, it is also very often "networked.") The search for "laws" of emergence, or some quantum theory of living systems, is destined to fall short of its goal because there is no conceivable way that a set of simple laws, or one-level determinants, could encompass this multilayered "holarchy" and its inescapably historical aspect. (I will have more to say on this crucial point in chapter 12.)

The Two Faces of Janus

The writer Arthur Koestler, in his landmark 1969 volume *Beyond Reductionism: New Perspectives in the Life Sciences* (coedited with J. R. Smythies), deployed a metaphor that was meant to convey the idea that both reductionism and holism are essential to a full understanding of living systems. Janus—the Roman god of entries, exits, and doorways—has traditionally been portrayed as a head with two faces that are looking in opposite directions—both in and out, past and future, forward and back . . . and, for Koestler, upward and downward. Emergence (at least as defined here) is neither a mystical concept nor is it a threat to reductionist science. However, a holistic approach to emergence also has a major contribution to make. In accordance with the Synergism Hypothesis, it is the synergistic effects produced by wholes that are the very cause of the evolution of complexity in nature. In other words, the functional effects produced by wholes have much to do with explaining the parts. (To repeat, it is synergy that explains cooperation in nature, not the other way around.) In this light, perhaps the time has come to embrace the full import of Koestler's famous metaphor; in fact, both faces of Janus are indispensable to a full understanding of the dynamics of the evolutionary process. For an elaboration on this theme, see Koestler's *Janus: A Summing Up* (1978).

— ∽ —

Politics is one of the unavoidable facts of human existence. Everyone is involved in some fashion at some time in some kind of political system.

—Robert A. Dahl

SUMMARY: In earlier chapters it was proposed that the functional (selective) advantages associated with various forms of synergistic phenomena have played an important role in the evolution of complex systems over time. Underlying the many steps in the complexification process, a common functional principle has been operative. A major codeterminant of this process, however, has been the parallel evolution of cybernetic (communications and control) processes (and systems). Cybernetic control processes are an indispensable concomitant in living systems of all kinds. This chapter will briefly discuss how the Synergism Hypothesis relates specifically to the evolution of cybernetic systems (i.e., political systems) in human societies.

6

Synergy, Cybernetics, and the Evolution of Politics

Introduction

The tumultuous political events of the past decade or so have, among other things, compelled political scientists to rethink some of their long-established concepts and analytical constructs. One example is *political development,* a term that has traditionally been associated with the optimistic post–World War II scenario in which "developing nations" were said to be following "industrial societies" into a final stage of "postindustrial society" that would, presumably, be permanently embalmed in stable democracy and some variant of the traditional "balance of power"—or terror. That smug scenario has been deflated by a sequence of events which suggests that the modern nation-state may itself be a transient phenomenon, a stepping stone on the way to something larger, or smaller, or both—or perhaps neither.

One indication of a sea change in the discipline is a growing interest in the concept of "political evolution." This seemingly innocuous linguistic shift is not merely a fad or a borrowed metaphor, but the reflection of a fundamental paradigm shift. It represents a reconceptualization of macrolevel political change. In this nascent new paradigm—which has yet to be fully articulated, much less agreed upon—political development can be viewed as analogous to political "engineering"—that is, the construction of a viable political process or structure more or less from preexisting plans. ("Nation-building" is the much used—and abused—current buzzword for this process.) Political evolution, on the other hand, is located at the creative cutting-edge, where old problems are solved with new techniques or new forms of organization, and where new problems are brought under political control. Following Charles Darwin's broad definition of evolution as "descent with modification," political evolution may also include systemic reconfigurations, reorderings and even "devolution" (see chapter 7). Accordingly, it can be said that political development is to political evolution as ontogeny is to phylogeny (see Corning and Hines 1988).

Some years ago, political scientist Kenneth Waltz (1975) drew a useful distinction between a political system and what he called a "political market," the latter being a collection of independent actors in a "framework" of forces, with varying relationships and varying degrees of interaction. For reasons that will become evident below, I prefer the term *political ecosystem*. But in any event, the distinction is an important one, especially in international politics, because it highlights the fact that political evolution also involves irreversible historical changes in the character of the global system. However, it should also be emphasized that political evolution is not simply another name for political history. To the contrary, it connotes a patterned process whose causal dynamics are amenable to theoretical generalizations, to causal theories.

One implication of an evolutionary perspective is that short-range issues (say, the future of Eastern Europe or the prospects for the European Union) can usefully be viewed within a much broader theoretical framework. An evolutionary paradigm can—and should—encompass, among other things, the evolution of other social species, the five-million-year process of human evolution, and the evolution of complex human societies and polities over the past ten thousand years or so—long before the modern nation-state was even conceived. In addition, as we shall see, an evolutionary paradigm must account for the propensities of human nature and the opportunities, constraints, and imperatives in the natural environment along with the traditional social, economic, and political variables. Such a paradigm provides a far richer perspective for theorizing about political change in the immediate past, present, and future. (I will elaborate on this contentious point below.)

Theories of Political Evolution

Over the past two decades, a number of political scientists have become conversant with this broader evolutionary paradigm. Equally important, a variety of hypotheses have been advanced to explain the process of political evolution, either as a whole or in part. For example, Gary R. Johnson (1995) proposes what could be called a sociobiological hypothesis to account for the origin of human polities. Politics, in his view, is derived from reproductive competition, and the advancement of various cooperative efforts is seen as secondary, in terms of the functional basis of government, to the containment of individual conflicts. Johnson also adopts the sociobiological assumption that there are only three bases for social organization, all of them rooted in individual reproductive interests: altruism (or nepotism toward closely related individuals), reciprocity, and exploitation. Nepotistic "kin selection," he argues, was the "primary force" responsible for establishing societies and polities.

A variation on this theme is what could be called the "ethological model." A particularly flagrant example was advanced by anthropologists Lionel Tiger and Robin Fox in their provocative popularization, *The Imperial Animal* (1971). What Tiger and Fox did, and with a certain relish, was to equate politics in human societies with dominance competition in the natural world. Thus politics is "a world of winners and losers." The political system, they claimed, is synonymous with a "dominance hierarchy." At first glance, it may seem that Tiger and Fox were promoting the Machiavellian vision (seconded by such modern-day theorists as Hans Morgenthau) that politics is essentially a struggle for power. As the character O'Brien famously put it in novelist George Orwell's masterpiece, *1984*, "power is not a means; it is an end. . . . The object of power is power" (1984, p. 266) Yet Tiger and Fox also recognized that dominance competition in nature has a purpose. It is related to competition for scarce resources—nest sites, food, and especially obtaining mates. Tiger and Fox concluded that "the political system is the breeding system" (p. 25). For a more recent treatment of this hypothesis, see Somit and Peterson (1997). An application to the evolution of warfare can also be found in Thayer (2004).

A number of other theorists have adopted a game theory approach to political evolution. The pioneer in this area was biologist John Maynard Smith (1982b), who was the first to apply classical game theory models to the problem of explaining social evolution. Maynard Smith's focus was on the strategies pursued by individuals within a defined population, and his objective was to identify adaptive strategies for the members of the population as a whole that could not be invaded or replaced by exploitative strategies. Such robust strategies were then characterized as being "evolutionarily stable."

An important variation on this approach, with direct implications for political evolution, was developed by political scientist Robert Axelrod and biologist William Hamilton (1981) (also see Axelrod 1984). As noted earlier, this involved a revised version of the famous two-person prisoner's dilemma game, incorporating a number of more realistic assumptions about the nature of the game and the players. Axelrod and Hamilton then proceeded to conduct a tournament among a number of their colleagues. The winning strategy, submitted by systems scientist Anatol Rapoport, was called tit-for-tat (cooperate initially, then respond to whatever the other player does in subsequent rounds), and it proved to be remarkably effective as a generator of cooperative behaviors among individual players. Among other things, it was found that (1) cooperation can get started even in a world that may also favor "defectors"; (2) it can also thrive in an environment where many other strategies are also being tried; and (3) it can resist invasion by less cooperative strategies. "Thus," Axelrod concluded, "the gear wheels of social evolution have a ratchet" (1984, p. 20). (See also the broader concept of a cultural "ratchet effect" in Boesch and Tomasello [1998].)

Among the outpouring of more recent game theory approaches to politics, the path-breaking work of Elinor Ostrom and her colleagues on the structural basis for collective action is of particular importance (see Ostrom 1990, 1998, 2000). Much theoretical analysis and empirical research have been devoted to illuminating the dynamics of political development and evolution from the ground up—that is, in terms of the influences that shape individual participation in government in different contexts. (I will have more to say on this below.)

In contrast with these microlevel approaches, other political theorists have focused on political evolution at the most inclusive macrolevel. George Modelski is concerned with the evolution of the global political system over the past thousand years. He is well known for a theory of long cycles in world history, which he conceptualizes as a learning process (Modelski 1987). In his earlier work, he envisioned waves of innovation coupled with a recurrence of wars and periods of political hegemony in apparently repetitive patterns. More recently, Modelski (1994b) has adopted a more explicitly evolutionary paradigm. He now characterizes global history as a process involving structural change and directionality along a steady path. He also speaks of the mechanisms of "variation, and innovation, cooperation and reinforcement" (p. 2) Yet, at the same time, Modelski envisions the global process as an "unfolding" according to an "inner logic," and he quotes the systems-theorist-cum-evolutionist Ervin Laszlo: "Evolution is not an accident but occurs whenever certain parametric requirements have been fulfilled" (Modelski 1994b, p. 13).

In his most recent iteration, developed in collaboration with Tessaleno Devezas, Modelski and his coauthor speak of modeling a Darwin-like evolutionary process (Devezas and Modelski 2003). But this is not so. In fact, these theorists develop a complex mathematical model and conduct an analysis of historical data to derive a deterministic dynamic that predicts the eventual emergence of a global political system; it is now 80 percent complete, they say. Indeed, their analysis is replete with the buzzwords of complexity theory—self-organized criticality, multilevel cascades, major transitions, a dynamic regime between order and chaos, etc. (I will critique this model below.)

Another variation on this macrolevel approach to political evolution was proposed some years ago by Gebhard Geiger (1988), though his theoretical focus was confined to the Weberian transformation of small face-to-face societies into large-scale, hierarchical, bureaucratic states (macrostructures). Geiger was concerned with explaining the evolution of political power (i.e., specialized instruments of centralized control that are endowed with the ability to use force). He argued that this transition requires a theory that goes beyond Neo-Darwinian inclusive fitness models because these explanations are not sufficient to account for various factors in real-world human politics. Specifically, he claimed that hierarchical organizations in human societies are not an adaptation and are not designed to engender mutual benefits for their members. Accordingly, Geiger proposed that the explanation for such political macrostructures lies in an extension of the theory of self-organizing dynamical systems. That is, the properties of "natural self-organization" were postulated by Geiger to engender structural stability in a dynamical system, including large-scale polities.

There are also various coercive theories of political evolution. The so-called "warfare hypothesis" is a perennial favorite, dating back at least to Thucydides's great treatise, *History of the Peloponnesian War* (1962). Another classic on this subject is von Clausewitz's *On War* (1968). Darwin, Herbert Spencer, and an assortment of nineteenth and early twentieth century social Darwinist writers also singled out the role of warfare in human evolution. More recently, anthropologist Robert Carneiro (1970) advanced a theory of war based on "environmental circumscription"—a refinement of pioneer sociologist William Graham Sumner's "man-land ratio"—to explain the formation of early states. Carneiro's theory is concerned with the relationship between populations and resource constraints, particularly arable land.

There is also the "balance of power" hypothesis of, among others, Arthur Keith (1949), Robert Bigelow (1969), and sociobiologist Richard Alexander (1979), who calls it an "imbalance of power" theory. The core idea is that, over time, human polities have grown progressively larger, *primarily* in order to strengthen themselves against other human groups. Alexander envisions a three-stage process: (1) the formation of multimale bands mainly for protection against large predators, (2) a combination of defense against predation and group hunting, and (3) a combination of antipredation, group hunting, and competition and conflict with other human groups. Moreover, as populations grew larger, warfare with other groups came to predominate over other forms of cooperation. Warfare, Alexander claims, is both the necessary and sufficient cause of large-scale human societies. An elaboration of this thesis can be found in Thayer (2004). For broad reviews of the literature on warfare, see Corning (2001a), van der Dennen (1991, 1995), and Corning and van der Dennen (2005). Also, see below and chapter 8.

In this chapter, I will update a radically different theory of political evolution, one that dovetails with the larger effort, described in earlier chapters, to account for the evolution of complex biological and social systems. To be specific, this theory is a special case of the Synergism Hypothesis. The key elements of this theory are the concept of synergy and the utilization of a cybernetic model of biological, social, and political systems and processes. The Synergism Hypothesis and cybernetics were discussed in earlier chapters, so we will focus here on the evolution of politics. For more in-depth treatments, see Corning (1971a, 1971b, 1974, 1977, 1983, 1987, 1996b, 2003, 2004b).

Defining Politics

How does the Synergism Hypothesis relate to the evolution of political systems? We begin with the perennial problem of defining *politics*. Charles Evans Hughes, a distinguished Chief Justice of the United States, was indiscreet enough in his pre–Supreme Court days to remark that "the Constitution is what the judges say it is." In like manner, or so it seems, politics is whatever political scientists and political anthropologists say it is. And, not surprisingly, there seem to be almost as many definitions of politics as there are theorists. The problem is that any given definition may rule in, or out, certain kinds of phenomena, or perhaps stress only one aspect of a multifaceted class of phenomena.

Nevertheless, a choice must be made. Political scientist Robert Dahl has written that a definition is, in effect, "a proposed treaty governing the use of terms" (1970, p. 8). The treaty I advocate defines politics as isomorphic with social cybernetics: *A political system is the cybernetic aspect, or "subsystem," of any socially organized group or population. Politics in these terms refers to social processes that involve efforts to create, or to acquire control over, a cybernetic subsystem, as well as the process of exercising control.*

This definition is not original. The term *cybernetics* can be traced to the Greek word *kybernetes*, meaning steersman or helmsman, and it is also the root of such English words as "governor" and "government." In the nineteenth century, the French scientist André Ampère took to using the term *cybernetics* as an equivalent for *politics*. More recently, the term has been employed by political scientists Karl Deutsch (1963), David Easton (1965), and John Steinbruner (1974), among others. The cybernetic model is also widely employed by life scientists, engineers, and systems scientists. For an up-to-date and authoritative sourcebook on cybernetics, see the *International Encyclopedia of Systems and Cybernetics*, edited by Charles François (2004).

As noted earlier, the single most important property of a cybernetic system is that it is controlled by the relationship between endogenous goals and the external environment. Recall the work of William T. Powers (1973), who has shown that the behavior of a cybernetic control system can be described mathematically in terms of its tendency to oppose an environmental disturbance of an internally controlled quantity (see chapter 4, especially figure 3). The system operates in such a way that some function of its output quantities is nearly equal and opposite to some function of a disturbance of any of the environmental variables that affect the controlled quantity, with the result that the controlled quantity remains nearly at its zero point. In this model, endogenous "purposes" and "feedback" play a key role in controlling the behavior of the system.[1]

What is the justification for dehumanizing politics and converting the multifarious real-world processes into an abstract model? One advantage is that it reduces the many disparate cases to an underlying set of generic properties that transcend any particular institutional arrangements, not to mention the motivations and perceptions of the actors who are involved. The cybernetic definition is also functionally oriented. It is focused on the processes of goal setting, decision making, communications, and control (including the all-important concept of feedback), which are in fact indispensable requisites for all purposeful social organizations. Indeed, cybernetic regulatory processes exist in families, football teams, business firms, and at all levels of government.

In this paradigm, the struggle for power—or dominance competition in the argot of ethology—is relevant and may even affect the Darwinian fitness of the participants, but this aspect is subsidiary to the role of politics qua cybernetics in the operation of any social system. Equally important, power struggles are a subsidiary aspect of the explanation for why such systems evolve in the first place. Social goals (goals that require two or more actors) and anticipated or realized outcomes are the primary drivers.

Another advantage of this definition is that it enables us to view human politics as one variant among an array of functionally analogous (and sometimes even homologous) cybernetic regulatory processes which are found in all other socially organized species—from bacterial colonies to army ants to wolf packs—and in all known human societies, including, by inference, our group-living, protohominid ancestors of more than five million years ago. Though there are great differences among these species, and among societies, in how political/cybernetic processes are organized and maintained, both the similarities and the differences are illuminating. This is illustrated in figure 4, a pointed variation on Powers's model in chapter 4 that is intended to model a modern government.

Thus, a cybernetic definition of politics is grounded in a biological—and functional—perspective and is related, ultimately, to the biological problem of survival and reproduction in, and for, organized societies. Politics in these terms can be viewed as an evolved biological phenomenon that has played a significant functional role in the evolutionary process; political evolution has been inextricably linked to the synergies that have inspired the progressive evolution of complex social systems.

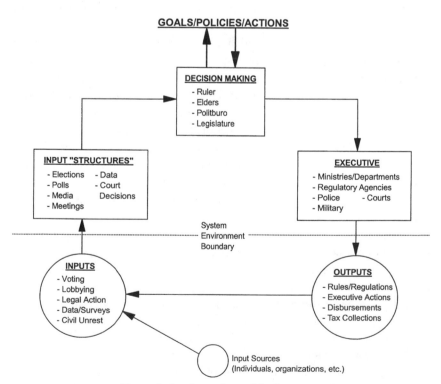

Figure 4. A cybernetic model of government.

Within this biopolitics-cum-cybernetics paradigm, political evolution refers to the invention, and perhaps diffusion, of novel cybernetic mechanisms, processes, and systems. Many of these are directly or indirectly related to the survival, and thus to the Darwinian fitness, of the members of a social group, but in modern human societies many others obviously are not. Some of these cybernetic innovations may lead to greater complexity in an existing system or to the emergence of a new level of cybernetic control (the Port Authority of New York and the European Union are two examples during the past century), but this need not always be so. In fact, some innovations involve an analogue of adaptive simplification in biological evolution. The substitution of a regulated market for a bureaucratic command system might be an example.

Explaining Political Evolution

How, then, do we account for the evolution of political systems, both historically and in the sometimes puzzling contemporary cases? For example, how do we account for the collapse of the Soviet empire, which, as political scientist Kenneth Jowitt points out, "was not supposed to happen"? Or, for that matter, how can we account for the recent "Balkanization" of the Balkans?

In *The Synergism Hypothesis* (Corning 1983), a chapter was devoted to what was called an "interactional paradigm," which was really a synthesis of various interdisciplinary paradigms that have been put forward over the past two decades. Here I can only provide a sketch of that causal framework. In brief, the pattern of causation in something as complex and variegated as the evolution of human societies requires a framework that is multidisciplinary, multileveled, configural (or relational), functional, and cybernetic. It involves geophysical factors; biological and ecological factors; an array of biologically based human needs (and derivative psychological and cultural influences) that must be attended to, as well as organized economic activities and technologies (broadly defined); and, of course, political processes, all of which interact with one another in a path-dependent, cumulative historical flux (see figure 5). It is preeminently a coevolutionary process.

This framework compels us to focus explicitly on the many codetermining factors that, in each case, interact synergistically, rather than trying to single out some monolithic causal variable that is ultimately destined to fall short. Also, it requires the recognition that the process of political evolution is always situation-specific, even when there may be invariances and recurrent patterns of covariance within the total configuration of factors.

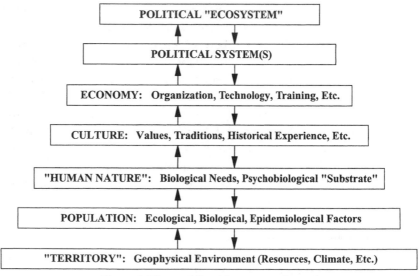

Figure 5. The multilevel interactional paradigm.

(The development of evolutionary economics over roughly the past decade has introduced a similar perspective into economic theory. See chapters 9 and 10.)

Some of these variables are obvious to political scientists. They involve the staples of conventional political analyses. But other variables are not always appreciated, or may be treated as constants. One case in point is fresh water resources, which have played a key role (necessary but not sufficient) in codetermining both the locations and the rise and decline of various civilizations—not to mention the conflicts between them. Thus, recent research has indicated that a major climate change precipitated the sudden collapse of the Akkadian empire in ancient Mesopotamia around 2200 BC (H. Weiss et al. 1993). Climate changes have also been implicated in the fall of the Mayan civilization and of Teotihuacán.[2] In fact, a major challenge for any theory of political evolution is that it must be able to account not only for progressive innovations and complexifications but also for "regressive" changes, for the episodic rise and decline of political systems. Two examples, one of each kind, will perhaps suffice to illustrate the synergistic nature of such changes. (I will have more to say on this issue below and in chapter 7.)

The rise of the Zulu nation in the nineteenth century provides an instructive example of progressive political evolution (see Gluckman 1940, 1969; D. R. Morris 1965). Until the early 1800s, the people of mainly

Bantu origin who inhabited what became known as Zululand (part of the modern South African province of Kwa-Zulu Natal) consisted of a disorderly patchwork of cattle-herding and minimally horticultural clans that frequently warred with one another. The most common casus belli were disputes over cattle, rights to grazing lands, and water rights. The ensuing combats were usually brief, for the most part involving prearranged pitched battles at a respectable distance between small groups of warriors armed with assegai (a lightweight, six-foot throwing spear) and oval cowhide shields. Injuries and fatalities were usually few.

As the human and cattle populations increased over time, resulting in "environmental circumscription" (Carneiro's term), there was a corresponding increase in the frequency and intensity of warfare among the clans until a radical discontinuity occurred in 1816, when a twenty-nine-year-old warrior named Shaka took over the leadership of the Zulu clan. Shaka immediately set about transforming the pattern of Natalese warfare by introducing a new military technology involving disciplined phalanxes of shield-bearing troops armed with short hooking and jabbing spears designed for combat at close quarters.

Shaka's innovation was as great a revolution in that environment as was the introduction of the stirrup and gunpowder into European warfare. After ruthlessly training his ragtag army of some 350 men, Shaka set out on a pattern of conquests and forced alliances that quickly became a juggernaut. Within three years, Shaka had forged a nation of 250,000 people, including a formidable and fanatically disciplined army of about 20,000 men (who were motivated in part by Shaka's decree that they were not allowed to marry until they were blooded in battle). Shaka's domain had also increased from about 100 square miles to 11,500 square miles. There was not a tribe in all of black Africa that could oppose the new Zulu kingdom, and, in short order, Shaka began to expand his nation beyond the borders of his people's traditional lands.

The further evolution and ultimate downfall of the Zulu nation at the hands of the Europeans in the latter part of the century is another chapter. What is significant here is the profound structural and functional changes— changes involving the superposition of an integrated political system—which occurred among the Zulu by virtue of decisive political entrepreneurship stimulated by population pressures and coupled with synergistic changes in military techniques and organization. Again, the causal process was configural and interactional, with cybernetic control processes being an integral part of the synergies that resulted. Moreover, these synergies were positively reinforcing (in the classical behaviorist sense) as well as providing positive feedback in the strict cybernetic sense.

The classic example of political decline is the Roman Empire, which recent scholarship suggests involved, among other things, a nexus of populational, economic, and political causes. (For a more detailed analysis, see Corning [2003]; also, see the discussion of devolution in chapter 7.) The explanation begins, ironically, with a population explosion. In 400 BC, there were only about 150,000 adult citizens on the entire Italian peninsula; as late as 70 BC, there were about 500,000 citizens and about the same number of slaves and freemen in metropolitan Rome, according to the Roman census. But by 28 BC, the number of Roman citizens had reached about 4 million, three-quarters of whom, it is thought, were by then living in the provinces.

Meanwhile, a profound shift was occurring in the Roman economy. The rapid population increases created a growing dependence on overseas food imports—especially grain from Sicily and Egypt—to a considerable extent due to a conversion of domestic agriculture to large-scale, export-oriented, slave-based *latifundia*. At the same time, Rome's once-thriving export markets for manufactured goods declined as the provinces learned to make Roman products more cheaply at home. Unfavorable trade balances eventually led to inflation and a debasement of the currency.

To cope with this imbalance, Rome began to place greater tax burdens on its empire, ostensibly to support the military legions and civil servants that were supposed to provide protection and maintain law and order but who came to be perceived by the locals as being there to support the tax collectors. The rest of the story is complicated, but this important configuration of changes (exacerbated by a stagnation of investment and enterprise, serious structural weaknesses in the political system, and some other factors not mentioned here) fundamentally altered the cost-benefit calculus for many Roman subjects and thus undermined the synergies that had been responsible for Rome's ascendancy.

There was nothing deterministic or orthogenetic about these two evolutionary episodes. Nor can any monolithic causal variable encompass them. The causal matrix in each case involved a dynamic mix of interacting factors located at several levels of causation—from geophysical to ecological, biological, technological, and political. Numerous factors worked together, synergistically, in a relationship of mutual and reciprocal causation to bring about the destruction of the Roman Empire. As the eminent classical scholar Charles Alexander Robinson observed, "The problem of the decline of the Roman Empire will probably be debated as long as history is studied, for it was a complex phenomenon in which many factors interacted, not one of which can be singled out as the prime cause" (1951, p. 611).

It would appear that a similar configuration of factors worked together to undermine the Soviet empire. Ironically, a reduction in Russia's historic sense of vulnerability to external attack was one of the factors that served to weaken the perceived need for the empire. When this was coupled with a disastrous war (Afghanistan), an upwelling of internal demands for dissolution, and the need to reform a sclerotic internal economy, among other factors, the calculus of perceived costs and benefits of the empire was altered for those who had the power to defend it.

Testing the Theory

As noted above, some political evolutionists have proposed linear or cyclical theories of political evolution, with postulates of law-like or orthogenetic properties. The synergism-cum-cybernetics theory, by contrast, posits an open-ended, historically contingent functional explanation that, far from being vacuous and untestable, permits a number of falsifiable propositions. Several of these were advanced in *The Synergism Hypothesis* (Corning 1983) and were discussed in detail there (see also Corning 1987). Here I will mention three in particular:

1. This theory would be falsified if any purposeful, socially organized group could be found where cybernetic (political) processes were nonexistent.
2. This theory would be falsified if any case could be found where cybernetic (political) processes existed in the absence of any perceived functional benefits, or positive synergies, for one or more of the participants.
3. This theory predicts that there will always be a strong correlation between the size and complexity of any social organization and the size and complexity of its cybernetic subsystem; the theory could be falsified if this expected correlation proved to be weak or nonexistent.

To be sure, there may be short-term lags in the predicted relationships, but they should nevertheless be valid over the longer term. In other words, allowing for exceptions that don't disprove the rule, political systems do not exist in the absence of functional synergies, and functional synergies in social organizations do not exist without correlative cybernetic control processes. That is as close to a testable law of political evolution as I believe we will get.

One way of evaluating this theory may be in terms of the light it can shed on the various hypotheses about political evolution that were briefly described above. Although a detailed analysis is not possible here, a few observations are in order.

First, consider Gary Johnson's hypothesis about what might be called primeval politics. Kin selection/inclusive fitness (that is, reproductive self-interest) may very well have provided both a constraint and a window of opportunity for the earliest steps in hominid social organization. However, kin selection is not a *sufficient* explanation. Reproductive self-interest is universal in nature and is always a constraint on social behavior. Something more is needed to explain why some species have exploited various modes of cooperative behavior while others have not. Indeed, why have so many species engaged in symbiotic relationships with other species in total disregard for their biological relatedness? There is also a biological imperative that, for many species, counteracts the constraining influence of inclusive fitness—namely, the deleterious effects of inbreeding depression and the pressure to outbreed (Thornhill 1993). This genetically based contradiction to the selfish-gene hypothesis impelled our hominid ancestors to regularly exchange their genes with those of non-kin.

More important, inclusive fitness theory is manifestly unable to account for the most revolutionary aspect of human social evolution—the fact that our ancestors broke through the inclusive fitness barrier and launched a trend toward ever larger, more functionally differentiated and complexly organized societies composed predominantly of non-kin. How have human polities succeeded in vastly expanding the nature and scope of their cooperative activities? As Geiger correctly pointed out, something more is needed to account for more recent human evolution, and for politics. But does Geiger's alternative model meet the need?

Geiger's basic premise is that bureaucratic regimes cannot be explained in terms of kin selection. However, the self-organizing model that he proposes is not the only alternative available. A functionalist, selectionist model based on the Synergism Hypothesis is another alternative. As pointed out in chapter 4, a self-organizing process may or may not be compatible with a functionalist explanation. But, as Geiger formulates his hypothesis, it amounts to a null hypothesis for the synergy-cum-cybernetic model; if it can be shown that large-scale bureaucratic states evolved independently of any functional goals, or consequences, the Synergism Hypothesis would be falsified. I do not think this is likely.

As pointed out in chapter 2, there is also a structural deficiency in the sociobiological model of social behavior; it appears to exclude the entire category of mutualism—win-win forms of cooperation and teamwork that are not dependent on genetic relatedness. One variation on this theme concerns what I call "corporate goods." I can provide only an abbreviated discussion of this concept here.

The "Corporate Goods" Model

In the corporate goods model (which can include any number of players), the participants may contribute in many different ways to a joint product (say the capture of a large game animal or the manufacture and sale of an automobile). However, unlike collective goods, or public goods that are indivisible and must be equally shared (even, possibly, with nonparticipants and cheaters), corporate goods can be divided in accordance with various principles, rules), or contracts. The division of the spoils is thus not preordained, as is the case with the payoffs in most game theory models; in other words, the payoff matrix can be manipulated at will. Indeed, the question of how the goods are divided up may be crucially important in determining if the game will be played at all. If this sounds familiar, even commonplace, it is because corporate-goods games are, in fact, ubiquitous in human societies. It is the predominant form of economic organization in a complex society. Yet, surprisingly enough, this everyday phenomenon has not been treated systematically either in sociobiology (evolutionary psychology) or in game theory, to my knowledge.

Some other problems with the conventional game theory paradigm—and with Neo-Darwinism—should also be noted. For instance, there are many cases in the natural world where the alternative to a win-win cooperative effort is not zero (the lowest possible value in a game theory payoff matrix) but death. If you were a small animal faced with the prospect of confronting a large predator, cooperative defense might be the only logical choice. Cheating would be self-defeating. Another problem is that game theory models have not, as a rule, dealt with multiple interests, where cooperation in one area—say mutual grooming—may also affect cooperation in other areas like hunting, meat sharing, coalition-building, or mutual defense. Nor does game theory capture the sometimes complex interplay between the costs and benefits associated with various choices or strategies.

A further problem, inherent in the game theory paradigm and in Neo-Darwinism generally, is that it is particularly insensitive to synergies of scale—the many cases where collective action produces combined effects that would not otherwise be possible. Biologist Lee Dugatkin (1999) cites an example (based on some research by Susan Foster) involving the collective behavior of the wrasse, a tropical reef fish that preys on the abundant supply of eggs produced by the much larger sergeant-major damselfish. Because female damselfish aggressively defend their nests, no single wrasse, nor even a small group, can overwhelm the damselfish's defenses. However, very large groups can do so and are rewarded with a gourmet meal of damselfish caviar.

Since success in raiding a damselfish's nest can only be achieved by a large group of wrasse acting in concert, it is an unambiguous example of a synergy of scale. Dugatkin calls this "byproduct mutualism" (an incidental byproduct of individual actions), but this is a misnomer. If an animal will only engage in a dangerous activity (like mobbing a predator) in concert with other animals and will reliably choose flight when it is alone, such collective behaviors are not simply statistical artifacts.

One other mode of cooperation in the natural world should also be mentioned, namely reciprocity. One well-studied subcategory of this behavior is called "indirect reciprocity." It involves a class of cooperative actions that do not seem to bear any relationship at all to reproductive fitness. For instance, helping behaviors among unrelated individuals—say meerkat babysitters or the helpers at the nest in various bird species—appear to be an evolutionary puzzle. What do the helpers gain from this? Some years ago, biologist Richard Alexander (1987) developed the concept of indirect reciprocity as a possible explanation. Alexander's argument was that, in a stable, ongoing network of cooperators, a donor might ultimately receive a fair return indirectly for some helping behavior if it later became the recipient of some other member's generosity. In effect, it formalized the old expression "what goes around comes around."

Much more thought and analysis has been devoted to this phenomenon in recent years, and the consensus seems to be that indirect reciprocity may well be a factor in sustaining socially organized species, independently of kinship (see especially Boyd and Richerson 1989; Mumme et al. 1989; Mesterton-Gibbons and Dugatkin 1992; Nowak and Sigmund 1998a, 1998b; Gintis 2000a; Clutton-Brock et al. 2001; Clutton-Brock 2002; Barclay 2004). Significantly, this phenomenon seems to occur under the conditions that, most likely, also characterized the evolution of the human species.

Also important is the burgeoning literature on "strong reciprocity" as a cooperation-enhancing mechanism, including the work of Gintis (2000a, 2000b); Fehr and Gächter (2000a, 2000b, 2002); Henrich and Boyd (2001); Sethi and Somanathan (2001); Fehr et al. (2002); M. E. Price et al. (2002); Bowles and Gintis (2004); de Quervain et al. (2004); and others. The term *strong reciprocity* refers to what amounts to an altruistic (not directly self-serving) propensity to punish cheating or violations of social norms. Closely related to this is the expanding body of work on fairness as a facilitator of cooperation in humankind (see especially Gergen 1969; Greenberg and Shapiro 1971; Kahneman et al. 1986a, 1986b; Rabin 1993; Fehr and Schmidt 1999; Gintis 2000a, 2000b; Nowak et al. 2000; Henrich et al. 2001; Fehr and Fischbacher 2002, 2003; Sigmund et al. 2002). Another closely related development is the growing body of work on the evolution and role of group-serving norms in securing cooperation in human societies (see especially Axelrod 1986; Sethi and Somanathan 1996, 2001; Ostrom 2000; Boyd and Richerson 2002; Young 2003; Gowdy 2004a; Richerson and Boyd 2004).

The Ethological Model Revisited

In this light, we can perhaps better appreciate the more coherent and sophisticated ethological model of politics that has been articulated in a succession of works by the primatologist Frans de Waal, beginning with his *Chimpanzee Politics: Power and Sex among Apes* (1982). See also de Waal (1989, 1996) and Harcourt and de Waal (1992). Drawing on his own extensive research among captive chimpanzees, as well as the many long-term field studies of these animals by other primatologists, de Waal offered a deeper, richer perspective on the issue. The struggle for power and influence is ubiquitous among these animals, he acknowledged. From the animals' motivational perspective, this may well be an end in itself. And, yes, the dominant animals may gain advantages in terms of such things as nesting sites and breeding privileges. But there is much more to dominance behaviors than this. The competition for status very often involves coalitions and alliances; it is often a group process rather than an individualistic, Hobbesian war. Moreover, there is much evidence that social constraints on dominance behaviors are common, both in these and other social animals; coalitions sometimes form to thwart the actions of a dominant animal. And in bonobos (pygmy chimpanzees), a loose female hierarchy seems to form the organizational backbone of the group; females often band together to constrain an aggressive male (de Waal 1997). (Also highly relevant is the evidence for what Boehm [1996, 1997, 1999], calls an "egalitarian syndrome" in small-scale human societies like hunter-gatherers.)

More important, stable dominance hierarchies in chimpanzees and other social animals also have functional importance for the group—maintaining peace, arbitrating disputes, limiting destructive competition, mobilizing collective action, and even defending the group against outside threats. The intense interdependence of social animals like chimpanzees and bonobos also leads to a degree of reciprocity and generosity such as food sharing. Indeed, as de Waal (1996, pp. 9, 102) points out, we also need to ask, ". . . what's in it for the subordinate?" His answer was, "The advantages of group life can be manifold . . . increased chances to find food, defense against predators, and strength in numbers against competitors. . . . Each member contributes to and benefits from the group, although not necessarily equally or at the same time . . . Each society is more than the sum of its parts." In other words, cooperative social groups may produce mutually beneficial synergies.

More recent studies of chimpanzees, bonobos, orangutans, and other socially organized species also suggests that interpersonal social relationships and interactions can be very complex and that cultural influences can also play an important part (see especially de Waal 1989, 1996, 1999, 2001). In fact, there may even be a degree of democratic participation in various group decision-making processes (Conradt and Roper 2003). Nor does one size fit all. The dynamics may differ from one group to the next, or even within the same group over time (see especially Kummer 1968, 1971; E. O. Wilson 1975; Lopez 1978; Strum 1987; Dunbar 1988; Wrangham et al. 1994; Boesch and Tomasello 1998; Whiten et al. 1999; Hammerstein 2003; van Schaik et al. 2003).

Accordingly, in the modern version of the ethological model, dominance behaviors may take on the functional attributes of leadership, and a dominance hierarchy may provide a framework for organizing various cooperative activities, including a division (combination) of labor (see Corning 1983, 2004b; cf. Masters 1989; Grady and McGuire 1999; Rubin 2002). Such organized political systems are characterized by overarching collective goals, decision making, interpersonal communications, social control processes, and feedback. To repeat, political systems are cybernetic systems. De Waal (1982, p. 213), invoking Aristotle, concluded with the following: "We should consider it an honour to be classed [along with chimpanzees] as political animals." For the record, this is also consistent with Aristotle's usage, as political scientist Larry Arnhart points out. Aristotle applied the term to any socially organized species that cooperates in jointly pursuing various aspects of the survival enterprise, from honeybees to wild dogs and killer whales. For obvious reasons, Aristotle placed humans at the pinnacle of this category.

Synergy Goes to War

As for the warfare hypothesis, organized collective violence is, for better or worse, a significant source of synergy in human societies. There are, for instance, the threshold synergies associated with the number of combatants, the technological synergies that are embodied in the weapons of war, and the synergies associated with a military division of labor—say the five-thousand-person crew of a modern aircraft carrier. Warfare provides a preeminent example of the slogan "competition via cooperation." There is also much evidence—ethnographic, archeological, historical, and even ethological—that warfare has played a major role in human history. But is warfare the necessary and sufficient cause of large-scale human polities, as Alexander claimed?

In the first place, given the grave risks and potential costs to the participants (in most cases), one needs to ask, why do wars occur? What might be the offsetting benefits? In fact, there is a very extensive body of scholarly literature on the causes of war. Among others, see Russett (1972); Choucri and North (1975); Waltz (1979); Zinnes (1980); Bueno de Mesquita (1981); Gilpin (1981); Corning (1983, 2001a, 2001b); van der Dennen (1991, 1995); Keegan (1993); Eibl-Eibesfeldt and Salter (1998); Corning and van der Dennen (2005).[3] Some of the overarching conclusions found in this literature are as follows: Wars can involve all of the levels in the causal framework illustrated in figure 4: resource constraints (or opportunities), population growth, psychological predispositions, cultural factors, technologies and economic activities, the dynamics of political processes (and political entrepreneurship), and relationships with other human populations. Conflict between the small, kin-based groups of early hominids were most likely related to competition for needed resources: water holes, favorable sleeping sites, herds of game animals, etc. Later on, as the economies of human societies evolved, the sources of conflict expanded to include control over such critical resources as firewood and obsidian, and, still later, arable land and strategic rivers; it is one of the ironies of human evolution that every new technology has created a new resource dependency. Oil is an obvious contemporary case in point.

Wars have not always been *necessary* to resolve such conflicts. Many conflicts have been resolved by peaceful means, and there have been many examples of peaceful coexistence. However, warfare has often been the preferred solution to conflict situations. Nor have wars been a *sufficient* explanation. Without the many other factors that have contributed to the evolution of complex societies, warfare would not have carried human populations much beyond the level of the Yanomamö or the Dugum Dani. Indeed, if conflict between groups were sufficient, there should also be nation-states composed of chimpanzees, which (we noted earlier) are quite "warlike" (Goodall 1986). To repeat, the very absurdity of this idea highlights the fact that many other factors worked together to further the process of complexification in human polities. These factors included (to name a few) bipedalism, binocular vision, opposable thumbs, tool making, an omnivorous diet, group hunting, the development of symbolic language (and, later on, writing), increased intelligence, the domestication of fire and of various animal species (dogs, camels, elephants, horses, etc.), herding, agriculture (and the agricultural surpluses which enabled human populations to expand), and division of labor. Moreover, many of these factors entailed the elaboration of functionally based modes of social cooperation as well as instrumentalities of political control and coercion. On this point, see especially Gurr (1988). Again, if any one of these codetermining factors, say bipedalism or language or agriculture, were to be removed from the recipe, the outcome would not have been the same.

In sum, warfare has been a frequently used adaptive strategy for human societies, an instrumentality that has mostly been related to the broader problem of survival and reproduction for human groups. The implication of a "synergy model" of warfare is that it is predominantly, though not exclusively, a synergy-driven functional activity—even when the outcomes do not fulfill the combatants' expectations. (This is not meant to condone wars but to explain them.) Furthermore, the causal matrix is far more complex than any prime-mover theory can encompass. The process of societal/political evolution is rooted in a configuration of biologically based human needs, of which physical security and access to needed resources are essential but not the entire story. Warfare has been an important facet of human evolution, but it has been shaped in turn by other aspects of the collective survival enterprise. For instance, it is obvious that technology has been one of the major influences in the evolution of warfare, but many military technologies were initially invented for nonmilitary purposes and were only later adapted to warfare: gunpowder, steel making, telephones, automobiles, and airplanes come readily to mind.

Finally, a word is in order about such deterministic theories as that of Geiger (1988) and Modelski and Devezas (2003). The underlying problem with these paradigms, in brief, is that past historical trends cannot safely be used to predict the future; to put it simply, a trend is not a law. If any of the codetermining variables that contributed to the historical trajectory of the past should be altered—what I have termed "synergy minus one"— then the entire process may be disrupted. To cite a specific case in point, there is growing evidence that the past ten thousand years have been a period of exceptional climatic stability globally and that instability—rapid, drastic climate changes—has been the rule over the past several million years (see especially Alley 2000; Corning 2004a; Stipp 2004). As will be emphasized in chapter 7, political devolution has also been a frequent occurrence and is essential for understanding our past history and future prospects. In sum, devolution falsifies deterministic theories.

Conclusion

Does the Synergism Hypothesis have any heuristic value? The answer, I believe, is yes. For one thing, it implies the use of more expansive, multi-leveled, and multivariate (and multidisciplinary) analyses to comprehend systematic political change, with a focus on the functional relationships among the variables and not merely on their additive statistical properties. A second implication is that a more sustained effort should be devoted to elucidating the bioeconomics (and economics) of synergy—the concrete, measurable consequences, or payoffs, of cooperative phenomena that may serve to sustain or undermine the correlative cybernetic (political) processes (see Corning 1996b, 2003).

One especially important test of this theory relates to the role of politics in the broad saga of human evolution, a process that may have taken six million years and is still ongoing. The accumulating evidence suggests that this process included three distinct transitions. The first, and in many ways the most important, of these involved a shift by our remote ancestors from an arboreal existence to a terrestrial mode of adaptation. This momentous change, I have argued elsewhere (Corning 2003), was accomplished with a behavioral package that included sociocultural and political synergies and a crucially important tool/weapon symbiosis. The second transition, which entailed a dramatic "hominization" (a suite of major anatomical developments), was the result of a synergistic new pattern of social behaviors including potent new tools, systematic hunting, the exploitation of fire, the adoption of home bases, the invention of a more elaborate division/combination of labor, and more elaborate political organization. Finally, the worldwide diaspora that resulted in the replacement of archaic *Homo sapiens* and Neanderthals with modern humans, beginning about fifty thousand years ago, was also a synergistic cultural and political phenomenon as larger groups with more advanced technology and organization overwhelmed other hominid populations, not to mention many other megafauna, in a global spasm of extinctions. In each of these major transitions, moreover, functional synergy and political/cybernetic processes are likely to have played an important part.

This theoretical paradigm also invites us to utilize more fully the insights we have gained about the evolution of complexity in the natural world. As noted earlier, two major modes of functionally based complexification have been evident in the broader process of biological evolution: (1) symbiotic partnerships, or mergers of various kinds, that have precipitated new forms of synergy and new functional capabilities; and (2) autogenous differentiation, specialization, and elaboration resulting in new forms of functional synergy and new cybernetic processes. Indeed, if social organization based on an inclusive fitness model may provide an appropriate framework for explaining the earliest phase of human evolution, it may well be the case that a symbiogenesis model (cooperative partnerships among unrelated individuals) best fits the revolutionary changes in human societies since the Paleolithic period. One must, of course, avoid using facile analogies as a substitute for rigorous analysis. But in this case the biological analogy directs our attention to phenomena in human societies that may be variations on a basic evolutionary theme rather than being borrowed metaphors.

A further test of the Synergism Hypothesis is whether or not it has any predictive value. What does it predict about the future? Deterministic theories, which are rooted in a venerable pre-Darwinian tradition, permit firm predictions because they assume that the process is governed by some directive influence or force (extrinsic or intrinsic) that ultimately determines (controls) the trajectory or outcome of the process. Currently, there are a number of theorists who fit this category (as noted above). Darwinian theories, by contrast, are grounded in the assumption that the process is historically contingent and chaotic in the sense that the overall pattern is held to be fully determined but not predictable. Indeed, the evolution of self-determination in the human species has introduced a unique source of creativity into the evolutionary process, which orthogenetic theories necessarily discount (see chapter 4).

But if the Synergism Hypothesis does not allow one to make unequivocal predictions about the future course of political evolution, it is possible to make a number of conditional if-then predictions based on an understanding of the causal factors involved and of the relationships among them. For example, one can predict that, if global economic interdependency continues to increase (as the ongoing developments in technology, industry, and international trade portend), cybernetic mechanisms and mechanisms of regulation and governance will evolve apace. The many existing areas of international regulation—postal, aviation, oceanic, telecommunications, and others—will be augmented by new economic, financial, and monetary regimes and, perhaps, by more formalized and legitimated enforcement powers. However, there are many contingent "ifs" that represent preconditions. For instance, various unpredictable, cataclysmic events could intervene—plagues, global climate changes, nuclear catastrophes, or political upheavals of various kinds. The configurations of synergies that provide the motivation for political innovation are always historically rooted. (The future of world politics will be explored in somewhat greater depth in chapter 8.)

By the same token, one can also predict that the apparently contra-dictory countertrend toward political disaggregation/fragmentation will continue insofar as the functions/synergies provided by the traditional nation-states—including, preeminently, national security—decline or are superseded. The apparent paradox dissolves within the context of the propositions cited above. (Again, we will explore the subject of devolution further in chapter 7.)

Finally, there is the challenge of testing further the cybernetic theory itself. A better causal theory of political evolution may in due course be advanced. But, in the meantime, I would hope that this one will be given serious consideration.

— ℰℐ —

These climatic events were abrupt, involved new conditions that were unfamiliar to the inhabitants of the time, and persisted for decades or centuries. They were therefore highly disruptive, leading to societal collapse . . .

—Harvey Weiss and Raymond S. Bradley

SUMMARY: Devolution is a political buzzword these days. But what does devolution mean? How can we measure it? And, most important, how do we explain it? As discussed in chapter 6, one corollary of the Synergism Hypothesis is the proposition that all teleonomic systems require cybernetic control processes which, in human societies, are typically referred to as political systems, management systems, or governments. Accordingly, it is postulated that the fate of any cybernetic control process in a living system is ultimately contingent upon the underlying functional effects that the system produces; functional synergies are the very cause of the differential selection and survival of complex systems and their cybernetic subsystems. It is argued here that the phenomena often referred to as cases of devolution provide an important opportunity to test this theory. A causal explanation of sociopolitical systems should be able to account not only for various progressive trends but also for the many cases in which regression or collapse occurs. Some studies related to political devolution will be discussed, and the arguments for competing hypotheses will be considered. A major example of political devolution will also be invoked in support of this theory.

7

Devolution as an Opportunity to Test the Synergism Hypothesis and the Cybernetic Theory of Political Systems

The Vicissitudes of Devolution

Devolution is a political buzzword these days as empires, nations, bureaucracies, and even business firms collapse, divide, downsize, outsource, and in various ways become less than they once were.

In the political sphere, the term *devolution* is commonly used in two different ways. On the one hand, it is associated with the current trend in western countries toward reducing or relinquishing the central government's role (power and resources) in various social programs and services—welfare, education, health care, railroads, public utilities, and the like. States and provinces (and even members of the private sector) are being granted greater responsibility for these functions.

On the other hand, devolution is also widely used in connection with a broader political trend that involves the breaking up of entire polities such as nation-states and empires. Devolution in this sense often involves the redrawing of political boundaries. Whole populations may be divided into new political units. Thus, the British Empire has devolved into the British Commonwealth; the Soviet Union is long gone (though the situation bears watching); the old Yugoslavia experienced a bloody dismemberment (more on this below); the United Kingdom is in the process of devolving as we speak; and there was recently a near miss in Canada when the issue was put to a vote in Quebec. (Whether or not political devolution will become a longer-term trend remains to be seen.)

Yet, paradoxically, in the biological and social sciences the very concept has lately become taboo; for many biologists and anthropologists in particular, devolution is redolent of orthogenesis—the view that evolution has an inherent directionality toward some form of improvement or perfection. As noted earlier, many nineteenth- and early twentieth-century evolutionists claimed that there has indeed been a broad, progressive trend in evolution that, needless to say, culminated in humankind. In this paradigm, devolution amounts to a setback or a deviation from the main course. There are hints of this orthogenetic vision in Aristotle, but it was more clearly enunciated by Jean Baptiste de Lamarck, Herbert Spencer, and a veritable host of their intellectual progeny during the twentieth century. For instance, anthropologist Robert Carneiro, following Spencer, defined cultural evolution as a directional change toward greater complexity, while devolution, to him, connotes a temporary step backward, a regression (Carneiro 1972, 1973). He predicted that, provided we can avoid a nuclear catastrophe, "a world state cannot be far off" (Carneiro 1978, p. 219). It is a matter of decades or centuries, not millennia, he declared. (We saw a more recent example of this genre in the work of Modelski and Devezas in chapter 6.)

The critics of orthogenesis contend that this conception of the evolutionary process is fundamentally flawed and wishful thinking. *Progress* is unavoidably a value-laden term that imposes external criteria on a process that is not, in fact, guided or pointed in some specific direction. Darwin's theory of evolution is deeply opposed to deterministic, all-encompassing causal theories like Herbert Spencer's universal law of evolution and the many similar formulations, from Teilhard de Chardin's omega point to Ilya Prigogine's thermodynamic law of evolution. Darwinian evolution has no hidden agenda. It is governed by adaptation to the immediate context, or local circumstances, and any observed trends are artifacts of past evolutionary history.

Progress in Evolution

These criticisms are well-taken. However, some anti-progressives have thrown out the baby with the bath water; they deny, or at least downgrade, the reality and significance of cumulative, functionally based (naturally selected) trends in evolution. It is perfectly legitimate and proper to recognize that there have in fact been specific directional trends of various kinds over the course of evolutionary history that are not the products of orthogenesis, vitalism, thermodynamics, or, for that matter, random accidents (a "drunkard's walk" in the vivid metaphor of Stephen Jay Gould 1996).

This is not to say that such trends are irreversible; they are at all times contingent. But they can properly be labeled progressive in relation to some specific functional criteria, and, in many cases, these criteria involve functional (economic) improvements such as greater efficiency, lower costs, higher yields, and greater reliability. Indeed, a great many traits in complex organisms (from the four nucleotide bases that constitute the genetic alphabet to the homeobox gene complex, nucleated eukaryotic cells, and endoskeletons) represent evolutionary inventions that have been conserved over countless generations. Accordingly, *devolution, adaptive simplification, regressive evolution,* and similar terms may imply nothing more sinister than the reversal of a clearly defined functional trend of some sort.

To illustrate this point, compound, image-focusing eyes with some 2.5 million photoreceptors and complex neural processing systems, which are capable of rendering full-color, stereoscopic, "motion-picture" images of the surrounding environment, are functionally superior to a single photoreceptor cell or even a small cluster of light-sensing cells behind a small pinhole. There has been evolutionary progress in the form of cumulative functional improvements over time in the eyes of certain lineages with respect to clearly defined functional criteria. However, this has not been the product of a unilinear trend.

Conversely, devolution in the sense of the loss of some functional trait has been a common occurrence in the course of evolutionary history. There are many examples: the loss of eyesight in cave-dwellers; the stubby wings of flightless birds; the atrophied forelimbs of kangaroos; hair loss in naked mole rats (and humans); the loss, in humans, of the ability to synthesize ascorbic acid; the loss of mitochondria in some eukaryotic protists; and the surrender of some 254 genes in the chloroplasts of land plants, resulting in a loss of the ability to synthesize some forty-six proteins that their free-living "cousins" can produce (Margulis and Sagan 1995).

Accordingly, one of the major contingent, reversible trends in evolutionary history has been an overall increase in biological complexity. For the purpose of explaining biological and social complexity, we will utilize the functional criteria that are widely employed both in biology and the social sciences (and control engineering). These criteria do not by any means exhaust the possible ways of measuring complexity in living systems, but they are significant because they are associated with important functional attributes and capabilities in nature as well as in social systems. These functional criteria consist of the following: the number of parts in the system; the number of different specialized roles or functions performed by those parts (or "functional differentiation," to use a Spencerian term); and the number of cybernetic feedback loops involved (a direct indicator of cybernetic communications and control relationships as well as the functional interdependencies among the parts). (See also chapter 4.)

Applying these criteria to living organisms, it could be argued that humans are not the most complex forms to walk (or swim) on Earth. Dinosaurs were and blue whales are obviously a great deal larger than humans. A 150-pound human has an estimated 10^{13} cells of about 250 different types. A blue whale weighs about 425,000 pounds (roughly 2,830 times as much as a human) and has an estimated 2.8×10^{15} cells. The number of different cell types in blue whales has not been determined, to my knowledge, but it is unlikely that there would be a great many more or many fewer than the number of cell types in humans. On the other hand, if one counts the functional specializations that occur within each cell type, the number of discrete functional tasks performed by human cells is vastly greater. The human brain alone has an estimated one hundred billion neurons that perform an immense number of different information processing, communications, and control tasks. So, if these finer-grained functional criteria are used, humans are unquestionably at the pinnacle of morphological complexity.

An obvious analogy in human societies would be the number of different types of workers in a large corporation—say, General Motors. If you only differentiate between blue collar and white collar workers, or hourly and salaried employees, you will find only a small number of different worker types (two). But if you differentiate in terms of the specific task each employee performs, the total number is vastly larger—in the thousands. Although the use of more fine-grained functional criteria to define biological (and biosocial) complexity obviously presents a major research challenge, it also introduces a more sophisticated way of measuring the capabilities of the whole.

But more to the point, these functional criteria provide a useful common metric for defining complexity in living systems, both in the natural world and in human societies. And this, in turn, has facilitated the development of a causal theory to account for the evolution of complex systems in nature and humankind. To reiterate, the Synergism Hypothesis represents an economic theory (broadly defined) of organized complexity in evolution. The hypothesis, in a nutshell, is that it is the selective advantages arising from various forms of functional synergy that account for the directional trend toward greater complexity in evolution. Over the course of evolutionary history, a common functional principle has been operative; synergy of various kinds has been the common denominator, so to speak, in the process of evolutionary complexification—from eukaryotic cells to bacterial colonies and human societies.

An important corollary of the Synergism Hypothesis, as discussed in chapter 6, is that cybernetic processes (goal setting, decision making, communications, control activity, and feedback) are a necessary concomitant of organized biological complexity. To repeat, cybernetic processes are found at all levels in living systems, from genomes to animal societies, and the fate of these control processes is intimately tied to the underlying functional synergies that the systems produce. In human societies, these systems are typically referred to as political systems, management systems, and government—though every family, every football team, and every factory also has one. Accordingly, a political system can be defined as being the cybernetic aspect, or subsystem, of any socially organized, goal-oriented group or population. Politics in these terms refers to social processes that involve efforts to create, or to acquire control over, a cybernetic subsystem, as well as the process of exercising control.

Testing the Synergism Hypothesis

Some ways of testing this theory of complexity—and the corollary theory of political complexity—were discussed in chapter 3 and chapter 6. Here we will consider further the methodology, which traces back to Aristotle, that I call "synergy minus one." The term was inspired by the recordings that were popular a few years ago called "Music Minus One," which allowed a singer or instrumentalist to fill in the missing part.

This methodology involves experiments or thought experiments in which a major part is removed from the whole and the consequences are then documented. Thus, to reiterate, it is not hard to imagine what would happen if a major gene were to be removed from the homeobox gene complex, or if the mitochondria were removed from a eukaryotic cell, or the gut bacteria from a termite, or the submajors (porters) from an army ant colony, or a wheel from an automobile, or the water supply from a human settlement. For that matter, we know very well what happens to a modern industrial society when the electrical power grid shuts down.

Accordingly, the Synergism Hypothesis is also highly relevant to the problem of explaining macrolevel political devolution because it predicts that the specific causes are likely to vary from one case to the next but that the disruption of any one major element of the full "package" of basic survival requisites for a human population may prove fatal to the system as a whole. For the Easter Islanders, the decisive factors were (apparently) the exhaustion of their wood supply and soil depletion. For the Ik, it was a drought. For the Moriori, it was a genocidal invasion. For the aboriginal Australians, the South African San people, and many Native American civilizations, it was imported disease epidemics. And for a large number of

Mesopotamian civilizations, according to the theory proposed by Harvey Weiss and his colleagues, a severe, sustained region-wide drought that occurred around four thousand years ago most likely devastated and de-populated almost simultaneously many otherwise thriving Middle Eastern societies— along with their political systems (H. Weiss et al.1993; H. Weiss 1996; H. Weiss and Bradley 2001). (The fate of the Akkadians, in particular, was also mentioned in chapter 6.) As Harvey Weiss and Raymond Bradley put it in a major *Science* article on this issue, these societal collapses were "an adaptive response to otherwise insurmountable stresses" (2001, p. 609). (One of the anonymous reviewers for this volume also reminded me of anthropologist Colin Turnbull's classic 1972 study of the Ik, which documented what can happen to an adapted culture that is suddenly forced to change under extreme stress conditions.)

In short, if synergy refers to the combined effects produced by wholes, the removal of even a single major part could have a negative effect on the performance of the whole and may even be catastrophic. Thus, if political-cum-cybernetic control systems arise in order to facilitate the operation of complex, synergistic systems at all levels of social organization, then the fate of the political system is necessarily tied to the functional viability of the system and its parts.

Political Devolution Defined

The term *political devolution* can be defined in a number of different ways. It could refer to reduced complexity, or it could mean the complete collapse, dissolution, or physical extinction of a population. Likewise, it could refer to a voluntary disaggregation or to an externally imposed or coerced change.

Here the focus will be limited to the cybernetics—systems of decision making, communications, and collective control among various groups and populations. If the progressive evolution of greater political complexity is associated with the communications and control processes that are necessary concomitants of being able to mobilize people and resources for one or more collective purposes—from group hunting to cooperative foraging, large-scale farms, manufacturing enterprises, and military defense and offense against other groups—then the converse involves a decline or collapse of a cybernetic (political) system and its capabilities. In these terms, political devolution can either be voluntary or coerced. It can involve only a limited functional decline, or it can be accompanied by the physical disappearance of a population. In any case, the hypothesis is that both the development and the dismemberment of any political (cybernetic) system is ultimately determined by the "economics"—its relationship to the production of various functional synergies.

Many forms of political devolution involve the termination of a system that was only temporary and narrowly focused (even ephemeral) to begin with. The research literature on primates and social carnivores provides many examples: temporary coalitions of lions, hyenas, or chimpanzees that coordinate individual efforts for the purpose of joint predation, or for collective defense or offense against another group, or to compete with other males for mating privileges, or even to contain and resist a dominant animal. In these cases, devolution occurs when the job is done.

The ethnographic research literature on human societies is also replete with apt examples of ephemeral political systems. One of the most famous examples involves the Great Basin Shoshone of the American Southwest. Until very recently, the Native Americans who inhabited this dry, harsh environment survived mainly by foraging in small family groups for various plant foods—nuts, seeds, tubers, roots, berries, and the like. Occasionally, however, a number of these families would gather into larger groups, numbering seventy-five or more, when there were opportunities for a large-scale rabbit hunt (or sometimes an antelope hunt) under the leadership of a "rabbit boss." These joint ventures involved highly coordinated efforts with huge nets, rather like tennis nets only hundreds of feet long, that were used to encircle and capture large numbers of prey. Yet, when the hunt was completed and the prey were consumed, the family groups would disperse once again (Steward 1938; A. W. Johnson and Earle 1987).

In a similar vein, the Native Americans of the North American Great Plains were legendary for their massive summer encampments. Dozens of small foraging bands, each with fifty members or less, would congregate each year into tribes numbering in the thousands under a tribal council and a chief, who organized and directed various tribal activities, especially the annual buffalo hunt (Carneiro 1967).

There are also a great many examples of ephemeral political systems in contemporary human societies. When the basketball game is over, the team members leave and go home for the night; when the stage show is over, the actors disperse; and when the collective response to a local disaster has achieved its immediate objectives, the ad hoc political system that arose to coordinate the efforts of various agencies (fire, police, repair services, shelter and food distribution services, etc.) will be disbanded. Such systems have been studied in depth by political scientist Louise Comfort (1994a, 1994b, 1998).

Similarly, in the business world, there are innumerable joint ventures and partnerships between separate firms, ranging from the many that are short-term and single purpose (say, a consortium of banks that underwrite a corporate merger) to those that are multifaceted and enduring (such as a parts supplier for a major automobile manufacturer). Some joint ventures are highly successful, while others are abject failures and are quickly abandoned. In either case, devolution is a common occurrence in the private sector as well, whenever the underlying functional need that inspired it no longer exists. The frequent downsizing of corporate conglomerates during the past two decades provides one obvious example. By the same token, ethnographers have documented innumerable military alliances between various bands, tribes, chiefdomships, and states (in the anthropologists' terminology) that have lasted only as long as there was a common enemy to be resisted—or attacked.

However, the most spectacular cases of political devolution, historically, have involved the overarching systems that are associated with the "collective survival enterprise"—that is, the procurement or protection of the basic survival requisites for a human society (see below and chapters 10 and 11).

Studies of Devolution

There is, needless to say, a long tradition of scholarship on the political devolution of human societies, from Edward Gibbon's *Decline and Fall of the Roman Empire* to the writings of Oswald Spengler, Arnold Toynbee, Herbert Simon, various systems theorists, catastrophe theorists, chaos theorists, and, of course, many modern-day environmentalists (the Club of Rome and the Limits to Growth theorists come to mind). There is even a specialized area of engineering, called failure analysis, that encompasses social system failures as well.

Especially important, however, are the data and case studies on political devolution that are found in the research literature in anthropology, archeology, and ancient history. The examples are, of course, plentiful; a great many societies have downsized, disaggregated, or disappeared over the millennia. Some were defeated on the battlefield and were put to the torch. Others disappeared mysteriously. Still others seem to have been burdened by a complicated nexus of destructive factors—a negative synergy. (Rome was mentioned in chapter 5.) By the same token, in some cases the society's central locations were completely depopulated while in other cases the population continued to grow in succeeding centuries, albeit under new management. The list of relevant case studies includes, among many others, the Mayans, the Incas, the Aztecs, the Olmec, the inhabitants of Teotihuacan, the Anasazi, the Hohokam, the Sumerians, the Babylonians, the Akkadians, the Hittites, the Minoans, the people of Mohenjo-Daro, the Easter Island-ers, the Moriori, the Tasmanians, the Maasai, the members of the Hawiian and Zulu kingdoms, the inhabitants of Han China, the Carthaginians, and, of course, the Ancient Romans.

Five of the many studies that are related to this subject are particularly relevant here. One is Robert Edgerton's 1992 book *Sick Societies*. Edgerton's overall focus is the problem of adaptation in human societies. He debunks the Panglossian notion held by some anthropologists that human soci-eties/cultures are generally well adapted and that every cultural practice, no matter how bizarre it may seem, is adapted for the society in which it is found. In other words, Edgerton rejects the commonplace argument that, because of our cultural blinders—or racist prejudices—we just don't under-stand other societies.

On the contrary, Edgerton argues, there are a great many practices that are objectively harmful to individuals and, in some cases, to entire societies. Some of these practices even imperil biological survival. To cite one example, the Bena Bena of the New Guinea highlands suffer from a shortage of protein, yet they have a taboo against eating the chickens (or chicken eggs) that are plen-tiful in their environment. Other clear-cut historical examples of serious maladaptation, according to Edgerton, include the Nuer, the Tasmanians, the Siriono, the Mayans, the Montegrano (Itialian farmers), and perhaps such communal organizations as the Shakers and the Oneida Community.

Nevertheless, many sick societies seem to thrive and continue to grow in numbers. Why? The short answer is that the maladaptive practices reported by Edgerton (and others) may or may not seriously threaten the viability of the society. Edgerton ultimately adopts three "self-evident" criteria for cul-tural maladaptation at the societal level: (1) the outright failure of a popu-lation to survive; (2) a context in which a sufficient number of the pop-ulation are deeply enough dissatisfied with the status quo to threaten the viability of the society and its institutions; and (3) a cultural practice that severely impairs the physical or mental health of a population so that its members cannot adequately meet their own needs or maintain their social and cultural system (1992, p. 45).

Edgerton's study is supported by two major anthropological studies of societal collapses. One is the edited volume by Norman Yoffee and George Cowgill, *The Collapse of Ancient States and Civilizations* (1988), which includes eleven detailed case studies and analyses from Rome to Mesoamerica and Han China. The editors also draw a clear distinction between political decline/collapse and the collapse of a civilization, although they are a bit vague about exactly what these terms delineate.

In any case, Yoffee and Cowgill's most important overall conclusion is that every collapse included in their study was different in character. No consistent pattern could be found, and there are no evident "prime movers" that propelled the political decline; each case was unique. Although a variety of contributing factors could be identified—poor leadership, trade disruptions, climate changes, government corruption, inflation, etc.—many of the examples utilized in their volume seemed to involve what Rice Odell is quoted by the editors as calling a "synergistic result" of a combination of factors, rather than a single decisive coup. Yoffee also stresses political scientist Herbert Kaufman's description (in his chapter for the volume) of a "downward spiral" of mutually harmful endogenous and/or exogenous factors.

Rome is one of the prime examples that the editors and contributors cite. To augment the old saying, Rome was neither built nor destroyed in a day. The sack of Rome by Alaric in 410 AD, and its ignominious aftermath, culminated several centuries of progressive decline involving a complex nexus of ecological, economic, social, and political factors. No doubt this is one reason why the fall of Rome is a source of endless fascination—and endless scholarship. Rome provides a relatively well-documented example of a multifactored, "dysergistic" process, but it is not unique. For a more in-depth analysis of Rome's rise and decline incorporating recent scholarship and new insights, see chapter 6 and Corning (2003).

Tainter's Theory

In sharp contrast to Yoffee and Cowgill's approach, Joseph Tainter's *The Collapse of Complex Societies* (1988), a formidable single-authored synthesis, represents an attempt to develop a broad explanatory principle for political devolution. Tainter attempted to support his thesis with material drawn from twenty case studies from various historical eras in both old and new world settings.

Complex human civilizations, Tainter points out, are "fragile, impermanent things," (p. 1) and a study of the many known examples of societal collapse can, he says, illuminate the underlying principles that govern both their rise and their decline. Tainter's objective, then, is to offer a general explanation for why such reversals of fortune have occurred over the course of human history.

Tainter begins by noting that there have been at least eleven specific themes (not mutually exclusive or free of overlaps) that various theorists have invoked to account for sociopolitical collapses: (1) depletion or denial of a major resource, (2) the establishment of a new resource base, (3) the occurrence of an insurmountable catastrophe, (4) an insufficient response to some challenge, (5) the actions of other societies, (6) intruders, (7) class conflicts or elite mismanagement, (8) social dysfunction, (9) mystical factors, (10) a chance concatenation of events, and (11) economic factors.

However, Tainter disagrees with these theorists. He finds all of their explanations insufficient, except perhaps as contributing factors. Tainter's key proposition is that the collapse of a complex sociopolitical system will predictably occur when there are "declining marginal returns" (p. 191)— when the economic costs of additional investments in complexity outweigh the additional benefits. In effect, Tainter's theory represents an alternative to the Synergism Hypothesis; it is based on an internal economic calculus relating to the costs and benefits of complexity in the political system itself.

Unfortunately, there are some technical problems with the theory. First, Tainter does not define the term *complexity* in such a way that one can measure it, and in his accompanying discussion he blurs the distinctions between complexity, inefficiency, bloat, and the sheer number of workers. Indeed, a full-fledged collapse in his terms differs from a decline only in its relative suddenness and rapidity, not in its concrete, measurable consequences. Nor does Tainter give us any measuring rod for devolution, or even a surrogate indicator. Likewise, we are not given any way of measuring either the inputs to, or the outputs from, greater complexity—that is, the marginal value. It also begs the question: marginal value to whom? Bureaucrats? A political elite? An underclass of slave laborers?

But most important, the theory does not accord well with more recent findings related to this issue. Even if, for the sake of argument, complexity in Tainter's terms could be defined and measured, a marginal value relationship would, at best, constitute but one variable—neither necessary nor sufficient to explain the many historical instances of political devolution. As we noted earlier, there is now strong evidence that, in many cases, precipitous sociopolitical collapses were directly attributable to such exogenous variables as conquests, epidemics, key resource depletions, and drastic environmental changes independently of any discernable political dynamic. Conversely, there are many other cases in which political devolution has occurred when the mission was accomplished; there was no longer a need and no further potential for realizing positive synergies.

However, it should also be noted that, in a recent, jointly authored article (Allen, Tainter, and Hoekstra 1999), Tainter's paradigm shifted away from diminishing marginal returns to the system and to a broader economic calculus associated with the marginal returns to the total population and the economy from "extracting resources" or other societal benefits.

This iteration represents a major change; it is now much more compatible with the Synergism Hypothesis, where the burden of maintaining a political system is weighed against the underlying functional objectives of the system. Thus, according to the synergism-cum-cybernetics paradigm, even a bloated, inefficient army will continue to be publicly supported if it effectively deters potential invaders. However, the converse is far less likely to be the case.

Jared Diamond's "Package" Approach

Much can also be learned from Jared Diamond's important study, *Guns, Germs, and Steel* (1997). Diamond's work is focused on explaining the rise of large, complex civilizations over the past thirteen thousand years or so, but his explanatory framework is also relevant to the converse problem of explaining political devolution and collapse.

Very briefly, Diamond takes up the forbidding challenge of explaining not only how and why the evolutionary trend toward societal complexity occurred in humankind but also why it happened where and when it did and why it did not happen elsewhere or elsewhen. A key aspect of Diamond's approach, one that directly contradicts some of the deepest metatheoretical assumptions of the social sciences, is that one cannot explain these fundamentally historical phenomena in terms of some context-free, deterministic (law-like) mechanism. The evolutionary process, including the evolution of humankind, is inescapably historical in nature; context-dependent factors have played a crucial role in the process. What is required, Diamond says, is "a science of history" (p. 421). (For a more extended treatment of this formulation, see Corning 2003.)

Accordingly, each major breakthrough in the evolution of complex societies, as well as each replication in some other geographic venue, was the result of a site-specific, synergistic nexus—a convergence of many "ultimate" and "proximate" factors (terms Diamond uses in a different sense than evolutionary biologists do). Diamond does not use the term *synergy*. He refers to a "package" of contributing factors. But the meaning is the same; each instantiation involved a combination of necessary and sufficient elements (see figure 6).

The development of food production and the surpluses that resulted were the key to it all, Diamond argues, but this in turn depended upon many other factors. One important precursor was the prior emergence of anatomically modern humans with language skills and sophisticated cultural resources by about 50,000 BP. Another factor was the decline and

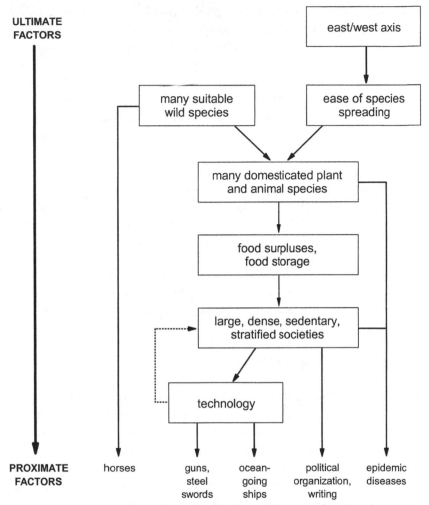

Figure 6. Jared Diamond's evolutionary "package." Redrawn from Jared Diamond, *Guns, Germs, and Steel: The Fates of Human Societies* (New York: W. W. Norton, 1997).

mass extinction of many of the large megafauna, upon which evolving humans had depended, coupled with a rise in human population levels. This demand-supply imbalance created increasing pressure to find suitable supplements to the standard hunter-gatherer diet. The fortuitous co-location of key founder crops, especially emmer wheat (which could be domesticated with a single gene mutation), together with legumes and animal husbandry (which allowed for a balanced diet) meant that the Fertile Crescent was the most likely location for a breakthrough that could sustain a large, sedentary population. Equally important, though, were such cultural inventions as food storage, draft animals, record keeping, and complex political organizations. Needless to say, this brief summary can hardly do justice to a much more elaborate synthesis.

Finally, a word is also in order regarding Jared Diamond's most recent book, *Collapse: How Societies Choose to Fail or Succeed* (2005), a work that is even more relevant to the issue here. Diamond's thesis is that societal collapses, historically, have seldom been attributable to any one factor and that the choices a society makes may determine the outcome. Willful environmental damage by human actors has often played a major role, Diamond argues, along with climate changes, hostile neighbors, and the loss of trading partners. In the examples Diamond selects to analyze in depth, his multifactored approach seems quite plausible. However, there are other cases where climate changes were severe, prolonged, and very likely decisive—for instance, with the Akkadians and the Old Kingdom in Egypt. Likewise, there have been many cases where conquest by an aggressive and vastly superior neighbor was the decisive factor and could not realistically have been avoided.

Rising to the Challenge of Decline

Nevertheless, I wholeheartedly support a synergistic "package" approach to understanding what is required to sustain a human population. What Diamond omits is a more complete inventory of what is both necessary and sufficient to sustain a human society and its members over time. This is where the survival indicators (basic needs) framework can be of use. The thesis, in a nutshell, is that the survival enterprise can be "operationalized" as a concrete analytical paradigm in terms of an array of at least fourteen basic needs that, to a first approximation, provide the specifications for the survival and reproduction (adaptation) of any given individual, and it can be scaled up for an entire human population. All of these fourteen basic needs are prerequisites for the continued viability of a human population, and if any of these needs is not met, the associated political system will be threatened and may collapse. (The full delineation of this biologically-grounded paradigm and a discussion of its use as an analytical tool can be found in chapter 11.[1])

Furthermore, the challenge of meeting these basic needs entails a multileveled hierarchy of causal factors, as illustrated in figure 5 (see chapter 6). This hierarchy is defined (somewhat arbitrarily) in terms of the span of cybernetic control—from "the piling up of little purposes" in ecosystems (to borrow a term from Lynn Margulis and Dorion Sagan [1995]) to the potentially enormous destructive power of large-scale, modern political systems. The main point of this graphic, however, is to underscore the fact that many different factors interact in complex ways to affect the fate of a human population and its political system. (Note especially that the causal arrows in figure 5 point in both directions.)

Although space does not permit a detailed discussion of this paradigm, perhaps we can illustrate it with a reference to a recent case study. I am referring to the Balkanization (literally) of the former Yugoslavia in the 1990s. It could be argued that there were no obvious survival threats to the population of that country; their basic needs were provided for. Yet, on closer inspection, both the political union forged by Josip Broz Tito earlier in the twentieth century and its recent disaggregation were driven by deep underlying conflicts and, ultimately, survival-related concerns.

A key to understanding the progressive evolution-devolution of Yugoslavia lies in the fact that it represented an uneasy, even forced, alliance—under a charismatic and dictatorial leader supported by a small communist elite—among historically antagonistic smaller political units with cross-cutting ethnic, religious, and linguistic animosities (Slovenia, Croatia, Bosnia-Herzegovena, Serbia, Montenegro, Macedonia, Kosovo, and Vojvodina). Tito had united them against perceived external survival threats (common enemies) during World War II. However, the country never became fully integrated or interdependent economically, and its political institutions were jury-rigged. Indeed, it remained a relatively backward country, with extremes of wealth and poverty that became worse during the communist era. For instance, per capita income in the north was six times that of the south, on average, at the time of the breakup.

Hence, when at long last Tito was gone and the major external threats to the population dissolved (the collapse of the Soviet Union was the final knell), so did Yugoslavia's functional foundation—its *raison d'être*. In the process, historical hatreds and tensions among the nation's constituent ethnic/nationalist/religious groups reemerged and became a serious internal threat. The dynamic of devolution at the national level was exacerbated by a process of political mobilization and conflict among its parts; eventually these antagonisms erupted into bloodshed, wanton brutality, and ethnic cleansing. As the casualties mounted, it became increasingly the case that physical survival was at stake for the various parties in this conflict (however senseless it may have seemed), and the political process came to be driven by this life-and-death imperative.

How could this tragedy have been avoided? Setting aside the egregious failures of leadership and other contributing factors, true economic integration (interdependence), a more equitable division of the economic pie, and/or a new external menace might have succeeded in holding this jury-rigged nation together. But this is debatable and hotly debated (see McFarlane 1988; Akhavan and Howse 1995; C. Bennett 1995; Rogel 1998).

In Tainter's original theory, the process of political collapse was viewed as being governed by an internal cost-benefit calculus related to the burden of complexity itself. This can hardly account for what has happened to Yugoslavia. In contrast, the Synergism Hypothesis posits that the fate of a political system is determined by the underlying functional processes (the synergies) to which it is related. Again, it is the functional synergies that are ultimately responsible both for the progressive evolution of more complex political systems and, in their absence, for the reverse dynamic of political devolution. In the absence of a functional basis for unity and a growing sense of a zero-sum relationship between nationalist groups (fanned by political leaders and the local media), Yugoslavia was likely to devolve. And, in this case, the dissolution process was hastened by ethnic conflicts that were inflamed and exacerbated by the political regime itself.

The Devolution of the United States

A more benign, peaceable example of political devolution—theoretically significant because it exemplifies the many systems that are created to meet a defined, short-term goal—can be found in, of all places, the United States. Although the image of "big government" and election campaign rhetoric portraying the federal government as a bloated bureaucracy has been a recurring theme in American politics over the past two decades, the reality is quite different if one contrasts the size and scope of the federal government, and the level and intensity of cybernetic control over the population, in 1944 (at the height of World War II) and in 1994 (fifty years later).[2]

World War II is now only a dim memory, and the generation that fought the war is mostly gone. However, the conversion of the United States from a depression-plagued peacetime economy with a pitifully small military (350,000 personnel in 1939) to a huge war machine (the "arsenal of democracy") with 11.4 million uniformed military personnel and 3.3 million civilian employees (compared to less than one million employees in 1939) is well documented. And these numbers do not include the many millions of Americans who became involved in war production work (17 million new jobs were created during the war, a 34 percent increase in the labor force) or the 10 million organized civilian volunteers of various kinds. In short, the war produced a radical economic, political, and military transformation. The national mobilization at every level of society and the degree of regimentation and control exerted over the population and the economy were totally unprecedented in the U.S., before or since. To be sure, this massive undertaking succeeded only because the population was united against two formidable enemies and (by and large) willingly accepted the sacrifices and constraints that were imposed. Nevertheless, the changes were radical and convulsive.

Over a six-year period, the American military establishment inducted, trained, clothed, housed, and fed a total of 15 million soldiers, sailors, and airmen, including several million who were shipped overseas to fight on far-flung battlefronts. In addition, the American Lend-Lease Program provided (and delivered to its various allies, despite losses to enemy submarines) food and war matériel amounting to a total of $50 billion (or $435 billion in 1995 dollars). This part of the war effort alone dwarfs the lunar space program or the Desert Storm operation against Iraq. Indeed, the avalanche of wartime production generated, among other things, a cornucopia of statistics: 30,000 airplanes; 87,000 warships of all types; 88,000 tanks; 400,000 artillery pieces; 634,000 jeeps; 2.7 million machine guns; 7.3 million five-hundred-pound bombs; 25 billion rounds of .30-caliber ammunition; 57.5 million wool undershirts; 519 million pairs of socks; 116 million pounds of peanut butter; 15.6 million shaving brushes; and 106.5 million tent pins, among many other items.

Needless to say, it is not feasible to directly measure the cybernetic aspects of this vast enterprise—the total volume of decision-making, communications, and control activities by the federal government during 1944 (or in any other year for that matter). Although archival materials and historical accounts do exist—in abundance—the task of tabulating them is so overwhelmingly large that it is obviously not practicable. Instead, we must rely on some surrogate statistics that, it is argued, are highly correlated with the relevant cybernetic processes. For instance, total federal government employment, including military personnel, went from approximately 1 percent of the total population in 1939 to 10 percent in 1944. The federal budget, likewise, went from $9 billion in 1939 (10 percent of the GNP) to $98.4 billion (46.8 percent of GNP). Meanwhile, the percentage of the economy that was directly engaged in war production went from less than 5 percent to over 40 percent.

The impact of the war on the American economy and population in cybernetic terms are also well documented. There were tight controls on prices, wages, rents, profits, raw materials, manufacturing activities, construction, transportation services, merchant shipping, and more. Some twenty major consumer items were strictly rationed, including gasoline, heating oil, meat, butter, sugar, tires, shoes, and coffee. Many other items became scarce or simply disappeared from store shelves, including liquor, soap, cigarettes, stockings, burlap, and cotton, because available supplies were diverted for military use or the raw materials were used for military goods. Cars and other major appliances were also unavailable during the war; the manufacturers of nonessential consumer goods were mostly recruited for war production work. The news media were also heavily censored, as were all overseas letters, and the scientific and educational establishments were both enlisted for war work of various kinds. The budget for the Office of Scientific Research, for example, went from $74 million in 1940 to $1.6 billion in 1945.

To oversee and manage this mobilization process, an alphabet soup of government agencies was created on a crash basis. The Office of Price Administration, with 5,500 local boards and 60,000 employees, was the most intrusive. However, the War Production Board, the Office of Civilian Defense, the Office of War Information (censorship and propaganda), the Office of Defense Transportation, the Public Health Service, and several other agencies collectively redirected the entire economy and society. For instance, there was a huge increase in the need for overland transportation during the war. But fuel rationing drastically reduced the usage of trucks and cars. So people turned to using trains, and this put the nation's railroad system under tremendous pressure. By 1945, passenger mileage alone had jumped to three times the prewar level. The agency responsible for coping with this need was the newly created Office of Defense Transportation, which, in effect, commandeered the nation's complex network of privately owned rail companies for the duration.

But perhaps the most significant indicators of the increased level of federal government control over the economy were the changes that occurred in the tax system. For the first time in U.S. history, the government mandated that income tax payments were to be withheld from paychecks and forwarded by employers directly to the Treasury. Taxes were also drastically increased (partly to finance the war but also as one means, among others, of drawing excess consumer demand out of the economy); the top (marginal) tax rate jumped to a confiscatory 94 percent. Federal government tax receipts in 1940 were $2.7 billion. In 1944 they had increased to $35.4 billion, more than thirteen times the prewar level.

Devolution by Design

Even before the war was over, the U.S. government began planning for reconversion to a peacetime economy. A special concern was how to meet the pent-up demand for consumer goods, from automobiles to washing machines, without causing runaway inflation. (Despite the level of high taxes, liquid assets waiting to be spent had increased from $50 billion in 1941 to $140 billion in 1944.) So industries that were expected to experience a rapid surge in demand after the war were given a priority in shifting out of war production work. In this and many other areas, the government deliberately planned for a demobilization and downsizing (and a devolution of the federal government's role) that was not only successful but, despite the Cold War that followed, never reverted to anything approximating the broad scope and pervasive power that was exercised during World War II.

Fifty years after the war ended, the statistics tell the story. Federal employment in 1994, including the military, amounted to 1.53 percent of the total U.S. population, versus 10.7 percent during the war. In fact, the total number of civilian and military personnel in 1994 represented less than one-third of the number in 1944. Despite the perceptions of most Americans, federal employment was only one-half of a percentage point higher in 1994 than it was in 1939. Likewise, total federal government outlays as a percentage of GDP amounted to 21.1 percent, less than half the 1944 percentage (46.8 percent) and roughly equivalent to the percentage in 1939 after subtracting transfer payments for Social Security, welfare, and the like (plus interest on the national debt). See table 3 for a summary.

Moreover, the declines in federal employment, expenditures, and taxes were correlated with a drastic reduction in the degree of governmental control over the economy after the war. Again, the statistics that are available must serve as surrogates.

Conclusion

As these data show, the political devolution that occurred in the U.S. after World War II fulfilled the theoretical expectation that political devolution can be the result either of success or failure. From a functional, synergistic perspective, this duality is not at all paradoxical. It was a direct consequence

Table 3. A comparison of the federal government's scope in 1939, 1944, and 1994.

Indicator	1939	1944	1994
U.S. population (millions)	131	138	260
Federal government employees (millions)	0.97	3.34	2.97
Military personnel (millions)	0.35	11.40	1.61
Government and military as a percentage of population (%)	1.00	10.70	1.53
GNP/GDP (billions)	$90.5	$210.1	$6931.4
Government outlays (billions)	$9.0	$98.4	$1416.7[a]
Outlays as a percentage of GNP/GDP (%)	10.0	46.8	21.1[a]
Outlays in constant (1995) dollars (billions)[b]	$98.9	$855.7	$1503.8[a]

Sources: *Statistical Abstract of the United States*, 1953, 1997; *Historical Statistics of the United States*
[a] 1994 outlays include transfer payments of $319.6 billion for Social Security, $144.7 billion for Medicare, $37.6 billion for veterans' benefits, and $214 billion for welfare, plus $203 billion in interest on the national debt, or 63 percent of the total.
[b] Utilizes the Consumer Price Index as a deflator.

of the disappearance of the underlying functional need, which was clearly survival related. No other theory that we are aware of can reconcile this seeming paradox.[3]

Accordingly, none of the long list of vanished polities, past or present, conforms to any rule—except one. The rule is this: if a collective survival enterprise and its political system are unable to secure one or more of the basic needs for its members (or is no longer needed for that purpose), the regime will in due course be threatened with collapse or be replaced. This is not exactly a revelation, but the framework of basic needs and the hierarchy of causation outlined above makes the argument more explicit (and testable) and enables us to see why all of the impressive scholarship on this issue has failed to identify a universal doomsday scenario. There is none.

In summary, history matters. But so do the imperatives of survival and reproduction. Our basic biological needs profoundly shape our cultures, whether we are consciously aware of this fact or not. (This point will be explored in depth in chapters 10 and 11.) And the synergy-minus-one test identifies and makes explicit the rationale that societies and their rulers/leaders generally utilize, often implicitly, to prioritize their problems and allocate resources—whether it be a tsunami, a disease epidemic, a military threat, a drought, the depletion of a key resource, or an internal threat to the regime and those who depend upon it for their survival and well-being. To update the old saying about the squeaky wheel getting the grease, if a tire goes flat during a trip your priorities will drastically change. You will need to make an unplanned stop to repair or replace it. Likewise, if the history of the human species has been marked by many unplanned political failures as well as successes, the record suggests that the future will hold more of the same. This is not a counsel of despair but a call to acknowledge and prepare for the formidable challenges that future generations will inevitably face.

— ☙ —

All kinds of creatures are alike in so far as each exhibits co-operation among its components for the benefit of the whole; and this trait . . . is common also to societies.

—Herbert Spencer

SUMMARY: The so-called organismic analogy, which has graced social and political theory (off and on) ever since Plato, has reemerged in evolutionary biology in recent years as a way of characterizing key properties of social organization in the natural world—although Herbert Spencer's term *super-organism* is the preferred moniker these days. (Biologists often give credit to one of their own, William Morton Wheeler, but Wheeler's writings appeared several decades later.) As Spencer himself argued, the organismic analogy is justified by the existence of common functional properties at higher levels of biological organization, including especially "functional differentiation" and "integration" with respect to overarching, collective goals or objectives; there is a functional analogy between organisms and superorganisms. More important, superorganisms may also constitute a distinct unit of selection (and adaptive change) in the evolutionary process. Without exception, however, superorganisms are also dependent upon cybernetic (communications and control) processes—or governance—as discussed in chapters 6 and 7. Accordingly, human superorganisms (and their political systems) are not *sui generis* but are variations on a major evolutionary theme. Indeed, it is likely that social organization played a key part in human evolution and in the rise of civilization. A modern human society represents an elaboration upon an ancient hominid survival strategy. It is, quintessentially, a "collective survival enterprise" (as I call it). This perspective casts a different light on the ongoing process of cultural evolution and the much-debated prospects for global governance, or political devolution—or both. The current, dualistic trend toward both more and less inclusive superorganisms may continue as the economic and political topography and the functional needs for governance continue to evolve. A global superorganism may well be emerging even as traditional nation-states are devolving. However, some deep problems currently exist.

8

Synergy and the Evolution of Superorganisms: Past, Present, and Future

Introduction

Superorganisms are again in vogue. Of course, they were never really threatened with extinction, like some endangered species. The reality of biological organization above the level of individual organisms, including but certainly not limited to human societies, remains undiminished. Indeed, the list is still expanding (see Currie 2001).

What has changed over the years is scientific fashion—and theoretical dogma. Although the term *superorganism* has a venerable pedigree, it became a pariah among biologists during the middle years of the twentieth century and was widely criticized as an inappropriate, even mystical metaphor. Of course, this was an era of strident reductionism in evolutionary biology. Any relationship that was more encompassing than the cells and organs of an individual organism (and for some, only individual genes) was treated by many theorists as an epiphenomenon of little or no evolutionary significance. Recall biologist George Williams, in his legendary critique of group selection, *Adaptation and Natural Selection* (1966), who characterized superorganisms as a figment of a "romantic imagination" (p. 220) and "not an appreciable factor in evolution" (p. 8).

Many other theorists of that era followed Williams's lead (e.g., Dawkins 1976, 1982; West-Eberhard 1975). West-Eberhard called the term "undesirable," since it implied selection at a higher level than that of an individual. However, this "head-in-the-bag denial" of functional organization and selection at the social level (to borrow an epithet from Richard Dawkins) was ultimately unsustainable. (See chapter 2; also Corning 1983, 2003; D. S. Wilson and Sober 1989; Hölldobler and Wilson 1990; Ghiselin

1997.) Williams himself has since backed away from this theoretical salient (Williams 1992).

But if the term *superorganism* is now being deployed once again without qualms, it is still very uncertain what it means, much less what theoretical status, if any, it commands. In fact, it has been invoked in a variety of incompatible ways. Here I will try to address these issues. I believe the term can be defined with some precision, and that it can be accounted for within the framework of the Synergism Hypothesis. Not only do superorganisms represent an important aspect of the evolutionary process generally but they have had an important role to play in the ongoing evolution of human societies, or so I will argue. First, however, some historical perspective is in order.

The Organismic Analogy

The so-called organismic analogy is hardly of recent coinage. In fact it can be traced back to the very roots of Western social and political theory—at least to Plato in the *Republic* (and perhaps to his mentor, Socrates) more than two thousand years ago. According to Plato, the "political body" or *politea,* is distinct from a herd or a mob in that it is organized to provide for the collective needs and wants (and ultimately "the good life") of the citizenry through a division of labor and an exchange of goods and services. In a much-quoted passage, Plato declared: "We must infer that all things are produced more plentifully and easily and of better quality when one man does one thing which is natural to him and does it in the right way, and leaves other things" (1985, 370c). But as Plato well understood, the trade-off for these benefits was mutual dependence, much like the parts of a body. Plato specifically rejected the assertion of the Sophists, like Antiphon, that a society was merely a facultative social contract between autonomous, independent individuals. As Plato stressed, the body politic involves functional interdependence. The loss of a part could cause harm or even be fatal to the whole.

This functional analogy, one of the most profound (and relentlessly debated) insights in political thought, was echoed by (among others) Aristotle in the *Politics*, Marsilio of Padua (he called the state a "living being"), St. Thomas Aquinas, various Enlightenment theorists, the French physiocrats, Auguste Comte, Edmund Burke, Emile Durkheim, and a bevy of so-called structural-functionalists among twentieth century sociologists and anthropologists. James Lovelock (1993), principal author of the controversial Gaia hypothesis in ecology, identifies James Hutton, the father of geology, as the first modern scientist to use the term *superorganism*, in 1788. Hutton wrote, "I consider the Earth to be a superorganism and its proper study should be by physiology." Nevertheless, it was the nineteenth century polymath Herbert Spencer, in the first volume of his massive, three-volume *The Principles of Sociology* (1897), who popularized the term and applied it to social organization. Like Plato and Aristotle before him, Spencer was quite clear about where the organismic analogy began and ended:

> Let it once more be distinctly asserted that there exist no analogies between the body politic and a living body, save those necessitated by that mutual dependence of parts which they display in common. Though, in the foregoing chapters, sundry comparisons of social structures and functions to structures and functions in the human body, have been made, they have been made only because structures and functions in the human body furnish familiar illustrations of structures and functions in general. . . . These, then, are the analogies alleged; community in the fundamental principles of organization is the only community asserted. (vol. I, p. 592)

In the twentieth century, the superorganism metaphor became the "dominant theme" in the biological literature on social behavior from 1911 until about 1950, according to Edward O. Wilson (1971, p. 317). Some biologists credit the distinguished entomologist of that era, William Morton Wheeler (1928), as the originator of the term. Wheeler applied the concept specifically to insect societies.[1] On the other hand, a contemporary colleague, biologist Warder C. Allee, adopted the original Spencerian definition. Allee, who wrote extensively on the role of cooperation in nature (see especially Allee 1931, 1938), focused on the functional commonalities in all forms of social organization.[2] However, the superorganism metaphor eventually fell out of favor and was even discredited in many quarters during the 1960s, when group selection theory and the "quasi-mystical vision" (in E. O. Wilson's pejorative phrase) of a society as a whole that evolves independently of its members was for a time attacked and rejected.

Superorganisms *Redivivus*

Yet it was Wilson himself who braved the wrath of the reductionists by reintroducing the *superorganism* term, somewhat tentatively, into his discipline-defining text, *Sociobiology: The New Synthesis* (1975). In a chapter devoted to colonial species that are intermediate between loose societies and complex, fully integrated organisms, such as the siphonophores (e.g., the *Physalia*, or

Portuguese man-of-war), as well as various myxobacteria, slime molds, flagellates, sponges, flatworms, coelenterates (like jellyfish), and others, Wilson cautiously broke the taboo: "The very term *colony* implies that the members are physically united, or differentiated into reproductive and sterile castes, or both. When the two conditions coexist in an advanced stage, the 'society' can be viewed equally well as a superorganism, or even an organism." (1975, p. 383). In other words, Wilson was utilizing Spencer's bedrock definition; a superorganism exists when there is a differentiation of functions (functional specialization) within a social aggregate, with the additional proviso that one of these specializations must relate to reproduction (a root characteristic of organisms).

Another, more visible effort to revive the superorganism concept occurred in 1989, when group selection advocates David Sloan Wilson and Elliott Sober published an article on "Reviving the Superorganism" in the *Journal of Theoretical Biology*.[3] (A defense of the concept in my 1983 book, *The Synergism Hypothesis*, was not so widely noticed.) Wilson and Sober returned to Spencer's definition, asserting that "the hallmark of an organism is functional organization. . . . We define a superorganism as a collection of single creatures that together possess the functional organization implicit in the formal definition of organism" (p. 339). For Wilson and Sober, the key property was not reproductive specialization, as Edward O. Wilson asserted, but the susceptibility of the whole to group selection. Superorganisms exist when functional interdependence (a "shared fate," in David Sloan Wilson's term) affects natural selection. "Not all groups and communities are superorganisms, but only those that meet the specified (and often stringent) conditions," Wilson and Sober wrote (p. 343).

One year later, in a landmark volume on *The Ants* coauthored by Bert Hölldobler and Edward O. Wilson (1990), and in a subsequent popularization (1994), Hölldobler and Wilson resurrected Wheeler's usage and applied the superorganism concept specifically to ant colonies:

> The amazing feats of the weaver ants and other highly evolved species comes not from the complex actions of separate colony members but from the concerted actions of many nestmates working together. . . . One ant alone is a disappointment; it is really no ant at all. . . . The colony is the equivalent of the organism, the unit that must be examined in order to understand the biology of the colonial species. (1994, p. 107)

After describing the complex organization of leafcutter ant colonies, Hölldobler and Wilson conclude that these colonies "do precisely the right thing for their own survival. Guided by instinct, the superorganism responds adaptively to the environment" (1994, p. 122).

Emboldened by Hölldobler and Wilson's unapologetic usage, biologists Robin Moritz and Edward Southwick two years later published a book-length monograph on *Bees as Superorganisms: An Evolutionary Reality* (1992). (See also the study of paper wasps, *Ropalidia marginata*, by Gadagkar [2001].) Not only was their treatment systematic and their documentation compelling, but Moritz and Southwick took pains to define the term "superorganism" (as they were using it) with some care. Following Wilson's earlier lead, they argued that superorganisms must consist of "organisms arranged in at least two non-uniform types and differentiated into sterile and reproductive organisms with different functions" (p. 4). The term should not be used as a synonym for sociality per se, they admonished. Other criteria for superorganism status, Moritz and Southwick claimed, included being (usually) sedentary, the ability to maintain colony homeostasis, being able to defend (or disguise) the colony, and having large numbers. These were not prerequisites, however. As Moritz and Southwick put it, in the end only one feature really counts: "It makes absolutely no sense invoking such a definition if natural selection does not act upon the superorganism itself. . . . In many of the insect societies, however, selection on the colony level seems to override selection at the individual level"(p. 6).

In other words, Moritz and Southwick embraced the definition of David Sloan Wilson and Elliott Sober; a superorganism exists when the whole is a unit of selection. (As an aside, the distinction between individual and group or colony selection is not so clear-cut as it might appear. Individuals are, after all, the vessels that carry the genes and reproduce themselves. However, individual survival may also depend upon the combined efforts of a group or colony. In some cases, selection may favor individual traits that are advantageous to the group but detrimental to the individual—i.e., altruistic. In other cases, though, a trait may be selectively advantageous both to the individual and the group, as I noted earlier.)

A sure sign that the resuscitated superorganism concept has regained legitimacy is that it is being used these days in more promiscuous ways. Science writer Roger Lewin (1996), among others, associates it with ecosystems, like peat bogs. Peat bogs the world over tend to be convergent in their patterns of development and succession, and such regularities suggest a deep structure of interdependencies. Lovelock (1993) goes even further. He claims that Gaia, the global ecosystem that includes all forms of life, along with the earth's geology and climate, is a vast interdependent superorganism. Lovelock maintains that the biosphere cannot be explained otherwise than as the product of a self-regulating, homeostatic, global feedback process. (In his most recent iteration, Lovelock [2003] adopts a more emergent/selectionist stance, however.) Only somewhat less encompassing is the proposition advanced by Léo Mathieu and Sorin Sonea (1995) that bacteria form a giant, interacting global superorganism. Setting aside the contentious debate over the merits of these assertions, is it legitimate to classify ecosystems, or the biosphere, as superorganisms? "It depends what you mean by superorganism, and who you ask," Lewin (1996, p. 31) points out.

Indeed, some theorists go so far as to associate the superorganism concept with virtually any complex system in the living world. Biologist Brian Goodwin, a critic of Neo-Darwinism, defines any rule-governed, emergent system as a superorganism. Philospher Sandra Mitchell equates it with "strange attractors"—the order that can emerge spontaneously from the internal dynamics of a complex system (cited in Lewin 1996). But perhaps the most far-removed, if not far-fetched, usage of the term is in the 1993 book by Gregory Stock, *Metaman: The Merging of Humans and Machines into a Global Superorganism*. According to Stock, the coming global superorganism includes not only humans but our crops, machines, buildings, communications transmissions, and other human artifacts. *Metaman* is another example of the current genre of global futurist books (Howard Bloom's breathtaking *Global Brain: The Evolution of Mass Mind from the Big Bang to the 21st Century* also falls into this category.) However, Stock's superorganism is pure metaphor; it has no scientific grounding, needless to say, and one caustic critic pointed out that it amounted to a neologism for civilization.[4]

Resurrecting Spencer's Usage

I occupy a position somewhere between the strict constructionists, who would confine the term to eusocial species with reproductive specialists, and the expansive, Whole Earth Catalog approach. I prefer to define a superorganism, in the spirit of Spencer (and of Plato before him), as *a behavioral system (social system) in which there are interdependent, coordinated actions with respect to one or more collective, goal-related activities*. The key point is that superorganisms are emergent phenomena (see chapter 5). They are able to do collectively what individual organisms cannot; they produce synergistic effects.

Among other things, this definition decouples the concept from reproductive specialization, which is after all only one manifestation of the broad principle of functional differentiation in nature. Equally important, it relaxes (but certainly does not reject) the requirement that a superorganism can only exist if it is subject to group selection. In human cultures, after all, there are many examples of social organizations that may or may not be relevant for selection, depending upon the context. Indeed, a strict selectionist definition renders the term problematic in relation to the ongoing process of cultural evolution (which, after all, was the focus of Spencer's usage). Although it is very likely that group selection played a role in human evolution (as noted earlier), many functionally important aspects of modern human societies are irrelevant for selection. Consider this paradox: a society mobilized to fight a war against a genocidal neighbor might justifiably be called a superorganism, even by a strict selectionist standard, but if a truce is signed the superorganism would technically vanish, even if the peace were being maintained by standing armies (a balance of power). A purely functional, Spencerian definition of superorganisms avoids such sand traps.

Another advantage in using a more permissive, functional definition of the term is that it encompasses a range of social interdependencies that would otherwise be excluded, both in the natural world and in human societies. Under E. O. Wilson's sociobiological definition, many forms of symbiosis would not count, even when they have important selective consequences—such as ruminant animals and their gut symbionts, or ants and their aphid partners, or the vitally important association between many land plants and various mycorrhizal fungi (Lewis 1991). Some forms of lichens reproduce via a specialized thallus that produces symbiotic diaspores, but others do not. So does this mean that only some lichens are superorganisms? What about eukaryotic cells, whose mitochondria reproduce independently? Should they be classified only as symbionts, not organisms, as Lynn Margulis tauntingly suggests? The exclusion of many symbiotic partnerships from superorganism status is especially unfortunate in light of the fact that, as discussed earlier, symbiogenesis has been one of the major evolutionary pathways to greater organismic complexity.

From a Spencerian perspective, all goal-oriented functional organization at the social level, both within and between species, represent superorganisms. With the exception of some human societies, the overwhelming majority of superorganisms have survival-related functions and thus influence (and are influenced by) natural selection. However, this begs the question: Are ecosystems (or Gaia) superorganisms? And what about emergent complexity in general? Here I will side with the strict constructionists who insist that ecosystems are not, in fact, integrated systems at all. An ecosystem is a reification of a more or less elaborate network of both facultative and obligatory relationships among separate species. It involves an interplay of separate purposes—often competitive, frequently cooperative, sometimes commensal, and interdependent to varying degrees. The distinction here is analogous to the one that economists draw between firms and markets.

All superorganisms—from army ants to the U.S. Army—also have cybernetic systems—that is, goals, communications, and control (and feedback) processes that guide the behavior of the system. This is a bedrock functional property of superorganisms that is not always invoked by biologists, but I believe it is a valid criterion and a useful indicator (see chapter 6). To be sure, a cybernetic system does not necessarily have a centralized controller, or governor, as in Norbert Wiener's (1948) classic model (as pointed out in chapter 4). There is increasing evidence, most notably in insect societies, that cybernetic control can also be distributed among the colony members, and that individuals can adjust their behavior in such a way that colony-level regulation is a collective, emergent effect (see especially Franks 1989; Seeley 1989; Moritz and Fuchs 1998; Detrain et al. 1999; and Gordon 1999). As Detrain and his associates (1999, p. xiii) have noted: "Classically, problem solving is assumed to rely on the knowledge of a central unit which must take decisions and collect all pertinent information. However, an alternative method is extensively used in nature: problems can be collectively solved through the behavior of individuals, which interact with each other and with the environment." Lovelock (1990) claims that Gaia qualifies as a superorganism because it has a global, cybernetic feedback system. But Lovelock uses the term *feedback* in a loose, even metaphorical way. There may well be interactive, self-regulating cycles at work in nature, but these cannot entail feedback *sensu stricto* unless there is an evolved internal goal-directedness that is governed by "control information." Control information is teleonomic in character and is therefore a product of natural selection; it is not a statistical artifact or an embedded property of nature (see chapter 14). In other words, not all homeostasis has cybernetic properties.

Accordingly, some important features of superorganisms, *sensu* Spencer, include the following:

- The term refers only to a limited (and variable) aspect of any given social aggregation, namely, organized, interdependent functional activities. As Spencer (and Plato before him) suggested, the concept of a superorganism is also closely linked to what economists have traditionally called a division of labor—though in many cases it could be more accurately termed a combination of labor.[5]
- A particular group's superorganismic properties may change over the course of time. For example, some of the coalitions and alliances that occur in various primate species are stable and long-lasting, whereas many others are

ephemeral, short-lived, and situation-specific (de Waal 1982, 1996). Yet, regardless of its duration, each of these groupings has an implicit collective goal and exhibits the coordination of individual behavior toward the realization of that goal, whether it be group defense, group hunting, dominance competition, rivalry over mating privileges, or the like.

- The properties of superorganisms can vary widely. They may have one or multiple levels of organization; there may be a single objective or several overlapping purposes; also, there may be varying degrees of interdependency among the participants, and various asymmetries in the distribution of costs and benefits.

- Superorganisms may or may not be composed of genetically related individuals, either in nature or in human societies. Nor are they required to exhibit reproductive specializations. To repeat, some superorganisms are based on genetic altruism, but many more involve what I refer to as "egoistic cooperation." In effect, the window for social cooperation and superorganisms is much larger than is implied by the classical models in sociobiology.

- Without exception, superorganisms also exhibit cybernetic control processes—or government in a broad, generic sense. Functional communications play a central role, as they do in organisms. However, control may be exercised by means of carrots, or sticks, or self-organized "volunteerism" (so to speak). In any event, the surest indicator of the emergence or demise of a superorganism is the presence, or absence, of cybernetic control processes (as discussed in chapters 6 and 7).

- Finally, to reiterate, superorganisms are predominantly concerned with various aspects of the ongoing survival enterprise—meeting one or more basic survival and reproductive needs through collective efforts. In other words, the underlying purpose of organisms and (most) superorganisms is homologous. (We will relate this crucial point to human societies later on.)

Examples of Superorganisms

Superorganisms as defined here can be found in all parts of the living world. Here are just a few examples:

- Bacteria invented superorganisms. The so-called stromatolites—rocky domes of various sizes that litter seashores from the Bahamas Islands to Shark Bay, Australia—are often referred to as "living rocks." However, stromatolites are actually composed of layers of bacteria (or algae) and inorganic materials (sand, gravel, mud, and other particles) deposited by wave action over many years and cemented together by bacterial activity. Some of these structures are still living, but many others may be ancient fossils. What makes them important examples of superorganisms is that the resident communities use collective action to create a structure that can resist the destructive effects of wind, water, waves, and ultraviolet (UV) radiation. The top layer of the stromatolite typically consists of dead cells that serve the community by shielding the next layer—the energy-producing photosynthesizers—from the damaging effects of UV radiation. Below that are layers of producers and consumers, along with various channels to allow for the circulation of nutrients, enzymes, and wastes. Under that are layers of limestone and the hardened corpses of many previous generations. And when the sediment at the crown becomes too thick, the next layer below will migrate upward, adding a new level to the structure. The stromatolite community also has a sophisticated chemistry, utilizing both anaerobic (oxygen-free) and aerobic processes (see J. A. Shapiro 1988; J. A. Shapiro and Dworkin 1997; Margulis 1993).

• Similar characteristics are found today in the thick bacterial carpets and scums known as "microbial mats" or "biofilms" that can be found almost everywhere in the world, though they lack the hard structure of the stromatolites. Many of these "slime cities," as science writer Andy Coghlan (1996) has dubbed them, include a mix of different strains that collaborate in various ways and coordinate their efforts via precise chemical signals. To repeat, microbiologist James Shapiro (1988) has likened bacterial colonies to multicellular organisms.

• Superorganisms are also found in Volvocales. The Volvocales are a primitive order of colonial algae that exhibit a great variety of somatic and reproductive patterns, and sizes. In a detailed study by Bell (1985), it was noted that the largest of the *Volvox* species, which are visible to the naked eye, also display a division of labor (or, more precisely, a combination of labor) between a multicellular soma and segregated, encapsulated germ cells. Comparative analyses have suggested some of the functional benefits. First, it appears to facilitate growth and result in a much larger overall size. It also results in more efficient reproductive machinery (namely, a larger number of smaller germ cells). Bell hypothesizes that in this case larger overall size also results in a greater survival rate. It happens that these aquatic, planktonic algae are subject to predation from filter feeders, but there is an upper limit to the prey size that their predators can consume. Integrated, multicellular colonies are virtually immune from predation by filter feeders. (I refer to this as a "synergy of scale.")

• By any definition, the leafcutter ants of the genus *Atta* are among the most spectacular examples of a superorganism. E. O. Wilson, who has studied them extensively, reports that the leafcutters are true agriculturalists. They grow fungi (mushrooms) in extensive underground chambers. The fungi are provisioned with great masses of fresh vegetation that is harvested by hordes of workers that bring it back to the nest in a complex division of labor that involves several categories (and sizes) of workers and a highly orchestrated effort. Each new colony is founded by a new queen that digs a nest and personally nurtures the first few workers. Eventually, however, the queen retires to a lifetime of producing many millions of daughters, most of whom become workers and soldiers. As Hölldobler and Wilson (1994, p. 121) observe, leafcutter ants have validated "the idea of the ant colony as a tightly regulated unit, a whole that indeed transcends the parts."

- Socially organized mammalian species often create temporary superorganisms. The study of hunting behavior in lions by Stander (1992) speaks to the assertion by some theorists that cooperative hunting behaviors may only be fortuitous effects. Stander's data for 486 group hunts by lions (*Panthera leo*) in Namibia displayed a clear pattern of coordinated roles among "wings" and "centres," or "drivers" and "catchers," with morphologically differentiated individuals commonly occupying the same role. Stander concluded: "cooperative hunts were more successful than non-coordinated group hunts . . . and hunting success was further improved when lionesses hunted in their preferred stalk categories" (p. 452). (Some previous studies of group hunting had characterized as cheating behaviors what Stander interprets instead as, at least in some cases, an unrecognized role in a division of labor.) Similar coordinated hunting behaviors have also been observed in other species as well—including wolves (Mech 1981, 1988), dolphins and whales (Würsig 1989), and ravens, among others.

Below are two more detailed examples of superorganisms.

The Naked Mole Rats

One of the most thoroughly documented examples of a superorganism in nature, by any definition, is the naked mole rat (*Heterocephalus glaber*). The naked mole rat is an African rodent species that lives in large underground colonies (usually numbering 75–80 but sometimes over 200). They subsist by eating plant roots and succulent tubers. Affectionately dubbed "sabretoothed sausages" because they are hairless and have two outsized front teeth that they use for digging, the naked mole rats represent a particularly significant example of a division/combination of labor in mammals. In fact, these odd-looking animals utilize specialized worker "castes" and a pattern of breeding restrictions that is highly suggestive of the social insects.

Typically (but not always), the breeding is done by a single queen, with other reproductively suppressed females waiting in the wings. The smallest of the nonbreeders, both males and females, engage cooperatively in tunnel digging, tunnel cleaning and nest making, as well as transporting the colony's pups, foraging for food, and hauling the booty back to strategic locations within the colony's extensive tunnel system. (One investigator, Robert A. Brett, found a mole rat "city" in Kenya that totaled about two miles of underground tunnels and occupied an area equivalent to twenty football fields.) Biologist Paul Sherman and a group of researchers who have studied these animals extensively (Sherman et al. 1991, 1992), wrote the following description of the mole rats' tunnel-building activities:

> The animals line up head-to-tail behind an individual who is gnawing [with its outsized, powerful front teeth] on the earth at the end of a developing tunnel. Once a pile of soil has accumulated behind the digger, the next mole rat in line begins transporting it through the tunnel system, often by sweeping it backward with its hind feet. Colony mates stand on tiptoe and allow the earthmover to pass underneath them; then, in turn, they each take their place at the head of the line. When the earthmover finally arrives at a surface opening, it sweeps its load to a large colony mate that has stationed itself there. This "volcanoer" [so named because its actions appear to an observer outside to produce miniature volcanic eruptions] ejects the dirt in a fine spray with powerful kicks of its hind feet, while the smaller worker rejoins the living conveyor belt. (Sherman et al. 1992, p. 78)

The vital and dangerous role of defense in a mole rat colony is allocated to the largest colony members, who respond to intruders like predatory snakes by trying to kill or bury them or by sealing off the tunnel system to protect the colony. This mole rat militia will also mobilize for defense against intruders from other colonies.

Why do mole rats utilize this highly cooperative survival strategy? Eusociality is relatively rare in nature, and the traditional view has been that a haplodiploid reproductive pattern provides a genetic facilitator. But this is obviously not the case with mole rats, which are diploid. (Indeed, it seems that haplodiploidy is neither necessary nor sufficient; all species of Hymenoptera are haplodiploid, but most are not eusocial; on the other hand, all termites are eusocial and diploid.) Sherman and colleagues (1992, p. 78) provide a bioeconomic (synergistic) explanation for the mole rat strategy: "We hypothesize that naked mole rats live in groups because of several ecological factors. The harsh environment, patchy food distribution and the difficulty of burrowing when the soil is dry and hard, as well as intense predation, make dispersal and independent breeding almost impossible. By cooperating to build, maintain and defend a food-rich subterranean fortress, each mole rat enhances its own survival" (see also Sherman et al. 1991). (Although it is not stressed in the mole rat research literature, another critically important facilitator is a cooperative relationship—and synergy—between the mole rats and endosymbiotic bacteria that are able to break down the cellulose in succulent tubers.)

If the bioeconomics—the functional synergies—provide an important part of the explanation for the naked mole rat survival strategy, the political (cybernetic) aspects are equally important, and are also well documented. As is the case with many other socially organized species, naked mole rats exhibit a combination of self-organized cooperation (preprogrammed individual volunteerism) and orchestrated social controls that are policed by various coercive means. The control role of the breeding queen is of central importance. The queen is usually the largest animal in the colony (size usually determines the dominance hierarchy), and she aggressively patrols, prods, shoves, and vocally harangues the other animals to perform their appointed tasks. Indeed, it has been observed that her level of aggressiveness varies with the relative urgency of the tasks at hand. In addition, the queen acts to suppress breeding and reproduction on the part of non-queen females, who are always ready to take over that role. (Occasionally other females are allowed to share the breeding function with the queen; why this is so is not known.) The queen also intervenes frequently in the low-level competition that goes on among colony members over such things as nesting sites and the exploitation of food sources. And when the reigning queen dies, there is a sometimes-bloody contest among the remaining females to determine her successor.

All of this control activity is facilitated by an elaborate communication system that includes seventeen distinct categories of vocalizations—alarms, recruitment calls, defensive alerts, aggressive threats, breeding signals, and so forth. In fact, the mole rats' communication system rivals that of some primate species in its level of sophistication. Thus, a naked mole rat colony may be characterized as a superorganism with a superordinate system of cybernetic control—government. In accordance with the theory of politics discussed in chapter 6, in mole rat colonies functional synergy and cybernetic processes go hand in hand.

Multiple Synergies in Meerkats

Not only do superorganisms benefit from many different kinds of synergy but very often a group enjoys multiple synergies—a synergy of synergies.

One illustration can be found in meerkats. Long before Walt Disney discovered the meerkats (the Afrikaner name for mongoose) and gave them a featured role in the movie *The Lion King*, these highly gregarious small mammals were using the synergy principle in various ways to cope with the challenge of survival in the Kalahari Desert and other marginal areas in southern Africa. Renowned for their ability to stand tall on their hind legs and scan the horizon while using their long tails to form a tripod for balance, meerkats live in elaborate underground burrows with multiple-family groups of up to thirty or more animals.

Among other synergies, meerkats benefit from huddling closely together for warmth during the cold desert nights (environmental conditioning); they also hunt collectively and will jointly defend their burrows with noisy displays and threatening charges (synergies of scale); they take turns standing sentry duty to watch out for predators, such as hyenas, jackals, and eagles (cost sharing); and they use various signals, including sharp warning cries when danger appears, to communicate with their companions (information sharing). There is even a rough division of labor. The adult males are primarily responsible for defending the burrow and its dozen or more entrances, while the females and immature males share in nurturing the infants. In addition, meerkats economize on building and maintaining their burrows by sharing their quarters with noncompetitive solitary yellow mongooses and social ground squirrels (a cost-sharing symbiosis; see Macdonald 1986; Doolan and Macdonald 1996a, 1996b, 1997).

Are Human Societies Superorganisms?

Can a human society also be called a superorganism? By Edward O. Wilson's definition, the answer is "probably not"; in his terms, it is only loosely analogous. However, there is one possible loophole. All sexually reproducing species, after all, have a degree of specialization between males and females. This is augmented in human societies by the various functional roles that the males play in provisioning, defense, and so on. Beyond that, there are both historical and contemporary cases in human societies where reproductive activities are either highly restricted or skewed, ranging from the princely harems and royal eunuchs of ancient times to celibate nuns and priests even today. By David Sloan Wilson's definition, on the other hand, the answer regarding human societies is a more equivocal "yes *and* no." Organized human societies have frequently, though not always, constituted units of differential survival and reproduction in competition with other societies. The accumulating evidence suggests that group selection may have played a significant role in human evolution, but group selection has not been the case always and everywhere in human history.

However, if one uses the more liberal Spencerian definition of a superorganism, the answer to the question above is unambiguously "yes." All human societies are, with rare pathological exceptions, synergy-producing superorganisms. Indeed, complex modern societies are multileveled superorganisms—beginning with nuclear families and including volunteer groups, schools, churches, community organizations, and the plethora of economic, military, political, and legal institutions. Moreover, the underlying purpose of human societies, without exception, is homologous with that of other superorganisms in nature. Going back to Plato's taproot definition, a human society represents, at bottom, a combination of labor with respect to the array of basic needs that are directly related to biological survival and reproduction (again, see chapter 11). A human society constitutes, quintessentially, a "collective survival enterprise." It represents a variation on a common evolutionary theme. To be sure, many of the subordinate superorganisms in complex modern societies are not directly survival-related—from basketball teams to symphony orchestras, movie production companies, and cruise ship crews, among others. Nevertheless, these culturally evolved superorganisms utilize the same underlying organizational principles, as Spencer observed, and they depend upon more or less complex cybernetic communications and control processes.

The Synergism Hypothesis and Cultural Evolution

Superorganisms, as defined here, are a subset of a broad spectrum of complex systems in the natural world, from eukaryotic cells to blue whales to the Boeing aircraft corporation. Accordingly, superorganisms can be accounted for within the framework of the Synergism Hypothesis.

To reiterate, the Synergism Hypothesis seeks to account for the evolution of complex systems in the living world, including human societies (see chapter 3). It represents an economic theory (broadly defined) of organized complexity in evolution. The hypothesis, to repeat, is that it is the selective advantages arising from various forms of functional synergy that account for the directional trend toward greater complexity in evolution. Over the course of evolutionary history, a common functional principle has been operative; synergy of various kinds has been the common denominator, so to speak, in the process of evolutionary complexification—from eukaryotic cells to bacterial colonies to human societies.

Chapter 2 briefly described the role of synergy in human evolution, and chapter 6 discussed the correlative evolution of political (cybernetic) processes. (A more elaborate discussion can be found in Corning 2004b.) Here we will consider specifically the evolution of complex human civilizations. The explosive rise of technologically sophisticated human societies since the Paleolithic has inspired many prime mover theories. (For extensive reviews, see Corning 1983, 2003.) Herbert Spencer deserves credit for developing the first full-blown modern theory, as mentioned earlier. To repeat, in his multivolume *Synthetic Philosophy*, an outpouring of works that spanned nearly forty years and influenced many other theorists of his era, Spencer formulated a "Universal Law of Evolution" that encompassed physics, biology, psychology, sociology, and ethics. In effect, Spencer deduced society from energy by positing a sort of cosmic progression from energy (characterized as an external and universal force) to matter, life, mind, society, and, finally, complex civilizations. Spencer defined evolution as a process characterized by "a change from an indefinite, incoherent homogeneity to a definite, coherent heterogeneity through continuous differentiations [and integrations]" (1892, p. 1).

Increasing complexity provides functional advantages, Spencer argued, but the "proximate cause of progress" in human societies was the pressure of population growth—the Malthusian dynamic: "It produced the original diffusion of the race. It compelled men to abandon predatory habits and take to agriculture. It led to the clearing of the earth's surface. It forced men into the social state; made social organization inevitable and has developed the social sentiments. It has stimulated men to progressive improvements in production, and to increased skill and intelligence. It is daily pressing us into closer contact and more mutually-dependent relationships" (1852, p. 501). (On one key point, Spencer was quite wrong. As noted earlier, humankind evolved in social groups, for reasons that had to do with positive economic benefits.)

Unfortunately, Spencer became a victim of his association with the social Darwinists (who were inspired by some of his early, more radical writings), and he became a pariah among twentieth century social scientists. For a time, he was a virtual non-person. Nevertheless, Spencer inspired a passel of subsequent prime mover theories. For instance, anthropologist Leslie White (1949, 1959), picking up on the Spencerian notion that progress is closely associated with the ability to harness and control energy, developed what he called the "Basic Law of Evolution." In White's words, "culture advances as the amount of energy harnessed per capita per year increases, or as the efficiency or economy of the means of controlling energy is increased, or both." (L. A. White 1959, p. 56). Calling himself a "cultural determinist," White claimed that culture evolves independently of our will: "We cannot control its course, but we can learn to predict it" (L. A. White 1949, p. 39; also pp. 330, 335).

Other Prime Mover Theories

Another modern-day prime mover theory invokes population growth, although Spencer's prior claim to this idea is generally not acknowledged. In the 1960s, anthropologist Esther Boserup (1965) proposed that population growth might have played a key role in the development of agriculture. Don Dumond (1965) focused on the relationship between population growth and cultural evolution in general. But it was Mark Nathan Cohen (1977), in a closely reasoned book-length treatment, who adopted the most Spencerian posture. Calling population growth the "cause of human progress," he asserted that population pressure is an "inherent" and "continuous" causal agency in cultural evolution. "Rather than progressing, we have developed our technology as a means of approximating as closely as possible the old status quo in the face of ever-increasing numbers" (p. 285).

Unfortunately, this explanation is too simple. All species have the potential for exponential growth and all species ultimately have limits. Humans are not unique in this regard. Not only do human societies practice various means of birth control to limit population growth but various external factors, from wars to diseases, droughts, and famines (as Malthus so kindly pointed out), may impose severe population constraints. More important, human populations do not grow in a vacuum; they grow only in favored locations and at propitious times, when the wherewithal exists in the natural environment for their sustenance and growth. And this in turn has depended on favorable environments and specific adaptations (human technologies).[6]

Social conflict—internal or external—is also frequently touted as the "engine" of cultural evolution, and there is certainly good reason to believe that violent confrontations between human groups have ancient roots (as discussed in chapter 6). But many theorists have claimed that warfare alone can also account for the evolution of civilization, from hunter-gatherers to advanced nation-states. Darwin, Spencer, and a host of social Darwinists stressed social conflict to varying degrees, but some theorists have gone much further. They attribute cultural evolution to our supposed aggressive and acquisitive instincts (shades of Raymond Dart). Sir Arthur Keith, with his *A New Theory of Human Evolution* (1949), was probably the first and least-known theorist of this genre, while the writings of Konrad Lorenz

(1966), Robert Ardrey (1966, 1976), and Robert Bigelow (1969), among others, caused something of a furor in the latter 1960s and 1970s. (Some, like Bigelow, stressed the complementary role of cooperation as well.)

As noted earlier, the well-known biologist Richard Alexander (1979) took perhaps the strongest position on this issue. In his so-called balance of power scenario, the process of cultural evolution is seen as being driven by competition between human groups, which in turn is an expression of inclusive fitness maximizing behavior. In other words, warfare is a form of reproductive competition by other means. Whereas various economic hypotheses are neither necessary nor sufficient explanations for large-scale societies, Alexander argued, warfare *is* both necessary and sufficient.

As pointed out in chapter 6, it is obvious that organized warfare has been a major source of synergy in human societies. The evidence is overwhelming that warfare has played a significant role in shaping the course of recorded human history. For instance, a major study of this issue some years ago examined twenty-one cases of state development, ranging in time from 3000 B.C. to the nineteenth century A.D. It was found that coercive force was a factor in every case and that outright conquest was involved in about half of them (Corning 2001).

However, I argue that warfare is neither the necessary nor sufficient cause of complex societies. If warfare involves grave and possibly fatal risks to the combatants, we need to probe more deeply into why wars occur. In fact, there is a vast research literature on this subject, spanning several academic disciplines, which supports at least one unambiguous conclusion. Warfare is itself a complex phenomenon with many potential causes and many different consequences. Wars cannot be treated as simply the expression of an instinctive urge or an uncontrollable external pressure. There are too many anomalies, and too many problems with any monolithic theory. Why is it that some quite warlike societies did not evolve into nation-states? Why did some societies achieve statehood and then subsequently collapse or even disappear? And why did the first pristine states appear during a very small slice of time in the broader epic of evolution, within a few thousand years of one another at most? Finally, there are the cases in which population pressures were relieved by increased trade or an intensification of subsistence technologies (again, see Corning 2001).

Technology as a Prime Mover

Technology has been another popular candidate for the role of prime mover in cultural evolution. Nobody would dispute the fact that technology has played a major role in the process, with synergies that are very often quantifiable. !Kung San hunter-gatherers living in the African Kalahari desert in the 1960s extracted 9.6 calories of energy from the environment for every calorie expended, according to the classic study by anthropologist Richard Lee (1968). By contrast, an American of the 1960s returned 210 calories for every calorie invested. Since Americans worked twice as many hours as their Kalahari counterparts, they secured forty-six times as many calories per person.

Many other synergies of this kind are documented in the research literature of human ecology (see especially Salisbury 1973). We noted earlier that a native Amazonian using a steel axe can fell about five times as many trees in a given amount of time as could his ancestors using stone axes, and chain saws add literally hundreds of multiples to the lumberjack's bottom line. Similarly, a shotgun is at least two to three times more efficient than a bow and arrow at bagging game on the hoof. An early farmer with a horse and wooden moldboard plow could turn over about one acre a day. His modern-day counterparts, with specially bred work horses and steel plows, can do at least two acres, and a farmer with a tractor and modern farm machinery can plow twenty acres per day, and sometimes much more.

Recall also Adam Smith's textbook case study of technological progress in *The Wealth of Nations* (1964). Smith did a comparison between the transport of goods overland from London to Edinburgh in broad-wheeled wagons and by sailing ships from London to Leith, the seaport that serves Edinburgh. In six weeks, two men and eight horses could haul about four tons of goods to Edinburgh and back, Smith found. In the same amount of time, a ship with a crew of six or eight men could carry two hundred tons to Leith, a load that, if transported overland, would require fifty wagons, one hundred men, and four hundred horses. (Other examples can be found in Corning 2003.)

One problem with elevating technology to the status of a prime mover is that it is not a force, or a "mechanism." It is not even confined to tools or machines. It is really an umbrella concept—a broad label that we use to identify the immense number of techniques we have devised for earning a living and reproducing ourselves. At bottom, the term refers to human activities involving the use of various inventions, behaviors, tools, objects, or even other organisms that have been appropriated, developed, or fabricated to serve human purposes. Some technologies are mainly a matter of deploying knowledge and skills. Thus, many agricultural practices—the use of dung as a fertilizer, crop rotation, interplanting, controlled watering regimes, among many more—are very important technologies. Likewise, many of our common plant and animal food products are the result of countless generations of selective breeding (genetic engineering) for various desired properties—size, texture, color, nutritional content, disease resistance, and the like. Similarly, domesticated animals are, in essence, some of humankind's oldest and most important technologies.

Many other human technologies involve the more or less skillful manipulation of objects in the environment. We already mentioned the role of fire, one of our earliest and still most vital technologies. The techniques required to gather, process, and cook various plant foods also played an important role in our evolution. The use of pits, dead falls, cul-de-sacs, and other stratagems for capturing game were very likely among the early hominid food-getting technologies. The diversion of water for irrigation purposes was a critically important step in the development of large-scale agriculture. So were dams, walls, fences, weirs, and many other early cultural innovations. In other words, technology is not really some external agency; it is a synergistic relationship involving human knowledge, human skills, and various external objects.

A second key point about technology is that it almost always requires organized cooperative activities by humans—what Karl Marx called "relations of production." The Boeing Company, for instance, in 2001 had 42 major facilities, 200,000 employees, and some 10,000 suppliers—many of them major corporations in their own right—that are scattered throughout North America and, indeed, the world. A Boeing 747 is the product of a vast cooperative effort. A third point is that every technology is embedded in a specific environment. It is enmeshed, so to speak, in the historical context; it is not a separate, autonomous agency but is always part of a larger economic and cultural system. More important, both the natural environment and the historical/cultural venue exert an important causal influence; they are codeterminants.

Technological innovations have the following properties in common: (1) they arise from human needs and human purposes in a specific historical context; (2) they utilize but also modify past cultural and technological attainments; (3) they are interdependent parts of a larger synergistic system; (4) they involve highly purposeful, goal-oriented development processes, as well as many progressive improvements over time; and (5) they are subject to a neo-Lamarckian selection process, in that the outcomes are ultimately epiphenomena—the combined result of many individual user choices among the available options. There is at least a tacit benefit-cost calculation associated with each individual decision, though many other cultural influences may also contribute.

Synergy and Superorganisms

This last point is critically important. It provides the linkage back to the Synergism Hypothesis and our theory of human evolution. In the final analysis, it is the synergies that determine the emergence and diffusion of a new technology; it is the payoffs that induce the positive selection of each innovation, in accordance with the backwards logic discussed earlier. The wellspring of cultural innovation is organized intelligence, but it is the functional effects—the synergies—that shape the selection process. So the Synergism Hypothesis is applicable to the ongoing process of cultural evolution—to the evolution of superorganisms—as well. There is nothing predestined about this process, any more than there is a deterministic directionality in the natural world. Moreover, each succeeding generation in effect reevaluates the technologies, social institutions, and practices that it inherits. A given technology is sustained over time by a cultural analogue of what is known in population genetics as "stabilizing selection," just as various functional improvements over time are products of "directional selection" within and across each new generation of users. By the same token, the many cases in which an older technology is supplanted could be likened to adverse selection in nature. However, as Richerson and Boyd (1999) point out, synergy is not enough. Complex societies have also developed "workarounds" to compensate for the lack of the social instincts that are needed to reinforce large-scale cooperation. Such instincts are essential in small face-to-face societies and in other social species, such as leafcutter ants. In comparison with ant colonies, Richerson and Boyd argue, a complex human society is at best a "crude superorganism."[7]

The main problem with prime mover theories is that they don't work. They may highlight important influences but they are manifestly inadequate—perhaps necessary but certainly not sufficient—to explain cultural evolution. This is especially apparent when you begin to ask historical questions. Why did a particular breakthrough happen when and where it did, and not at some other time or place? Nor can prime mover theories account for the manifest influence of other important movers. More important, societies do not change in some automatic way or follow a unilinear path. Often the path may lead downhill; prime mover theories have no explanation for what was referred to in chapter 7 as "devolution."

Accordingly, the Synergism Hypothesis posits that the evolution of complex human superorganisms has entailed an incremental, contingent trial-and-error-cum-trial-and-success process that, at every step, entailed a synergistic nexus of factors—from geophysical and ecological factors to an array of basic resources (starting with fresh water), plus technology, economic and political organization, interactions of various kinds with other human populations (from trade to war), and much more. Recall again the discussion surrounding figure V in chapter 6, where many codetermining factors were portrayed as exerting both upward and downward causation in a path-dependent historical process.

The Future of Superorganisms

So what are the future prospects for large-scale human superorganisms? Is global government in the cards, or the tea leaves, as many theorists over the years have predicted (and many others have scorned and debunked)? Or is devolution the wave of the future? On the one hand, there have been some important progressive (albeit controversial) developments in recent years: the World Trade Organization (WTO), the North Atlantic Free Trade Association (NAFTA), and further progress in the ongoing evolution of the European Union (EU). At the same time, devolution has been occurring at various levels as well.

Paradoxical as it may seem, both trends may be real; a dualistic process involving both fragmentation and superordinate integration may be at work. Political scientist James Rosenau (1999) has characterized this concurrent process as "fragmegration." Rosenau's neologism may or may not have "legs," but, by whatever name it is called, this apparent duality is not at all puzzling or inexplicable in light of the Synergism Hypothesis. These shifts reflect changes in the underlying functional bases for existing political superorganisms at various levels, and especially for the system of sovereign nation-states that, we should remember, evolved only within the last few centuries. There is nothing sacred or immutable about them. They arose in relation to specific functional needs—economic and military—and in a specific set of environments (natural, cultural, economic, and political). As both the underlying needs and the larger global context continue to evolve over time, so will the instrumentalities of governance—of cybernetic control.

Consider the striking contrast between the collapse of the Soviet Union, on the one hand, and the continued progress toward a "United States of Europe" (in Winston Churchill's appellation). Each of these two historic political shifts involves very complex, multidimensional processes, to be sure, but the underlying causes in both cases were changing sets of functional needs and shifting economic and political landscapes. Moreover, these changes were clearly recognized and accepted both by the relevant political elites and by a broad cross-section of the affected populations.[8]

Indeed, a political system will persist only so long as people believe that it exists and act accordingly (and are willing to use force to defend it). As political scientists Yale Ferguson and Richard Mansbach (1999) point out, modern nation-states are much overrated in terms of their ability to exercise control over vast areas of economic and social life, not to mention the well-being of their citizens. Some nation-states that exist on paper (or on world maps) are currently in a state of civil war, or are near collapse, or have ceased to exist except as a legalistic formality. Others are confronted by a variety of both internal and external challenges to their authority; their sovereignty is being corroded. For better or worse, the new World Trade Organization has become a focal point and instrumentality for this corrosive process. (Of course, the United States represents a possible exception to this trend. In the wake of the 9/11 terrorist attack, the United States has asserted itself in the global arena with new vigor, but its current political posture may not last. Grave economic problems lie just ahead.)

Political scientist Joseph Nye (2002), in a perceptive new analysis, utilizes the metaphor of a three-dimensional chess game to characterize the emerging multilevel pattern of power relationships in the post-9/11 global arena. On the top chessboard, where military power is deployed, the game is largely unipolar. Military spending by the United States is higher than the next eight countries combined, and its military capabilities are vastly superior to those of any potential challenger. Nonetheless, there are severe constraints on American hegemony. Terrorists employing suicidal tactics are only one example. The boundaries of domestic political support in the United States for foreign military engagements also remain uncertain at this point, but it certainly does not amount to a carte blanche for reckless adventures.

The second level of the chessboard, according to Nye, is economic power, and here the United States must operate in a multipolar game in which its ability to act unilaterally is severely constrained and gradually diminishing. OPEC has long served as an example of this trend. The trade war that erupted in the early years of this century over the U.S. decision to impose steel import tariffs, and the European appeal to the World Trade Organization for sanctions, underscores this new reality. The game on this chessboard promises more of the same as time goes on. Finally, the bottom level of the chessboard involves a widely dispersed pattern of transnational actors—multinational business firms, communications media, banks, nongovernmental organizations, terrorists, hackers, and so forth—that, in many cases, operate without significant governmental constraints on their global activities. This chessboard also can be expected to continue expanding, and governance will require transnational measures. What makes this three-dimensional chess game a formidable challenge to the traditional nation-states is that the chessboards are also vertically connected and influence one another. Political power is, for better or worse, a multileveled game.

What are the prospects, then, for the emergence of a global superorganism? We certainly do not lack for predictions about the future of world politics (reviewed in Modelski and Thompson 1999). Some theorists, the so-called realists, see the future as an extension of the traditional "balance of power"—a shifting pattern of competition and conflict between nation-states and empires (e.g., Kenneth Waltz 1993, and Samuel Huntington 1996). There has been much (overblown) talk recently about the new American empire. But other analysts see ecological and demographic turmoil ahead, with unfathomable political disorder as a likely consequence (e.g., Paul Kennedy 1993). Immanuel Wallerstein (1996) posits in ineluctable political dynamic that is driven by six intertwined change vectors that will lead ultimately to a new global political configuration, while Francis Fukuyama (1992) foresees the triumph and coming hegemony of liberal democracy. Finally, George Modelski and William R. Thompson (1999) detect an overall pattern of long cycles in international power relationships, while Devezas and Modelski (2003) envision a deterministic march to a world state.

But all this begs the question: how do we get from here to there? The Organization for Economic Cooperation and Development (OECD) (1995), in a detailed study that compared the dynamics associated with multilateral trade patterns with the process of regional political integration, drew a number of important conclusions that also seem relevant to the prospects for global integration (see also Hewson and Sinclair 1999; Modelski 1999). Some of the highlights of the OECD study were as follows:

- Transnational expansion has been occurring, and will continue to occur, in multiple dimensions: trade, investment, ownership, communications, production, markets, and more. Moreover, this trend has occurred to a considerable extent without political integration and will continue to do so.
- Even without integration, there have been a plethora of bilateral and multilateral trade agreements between nations that have served to establish rules, regulate trade patterns, adjudicate conflicts, compensate various parties, and so forth. Over one hundred such agreements were negotiated between 1965 and 1995, and the new World Trade Organization has greatly accelerated this process.

- Though limited trade agreements give the appearance of preserving the autonomy of the participants, in fact they often shade into encroachments on sovereignty and amount to incremental steps toward transnational governance. (This trend is clearly evident in the actions of the World Trade Organization.)

- Political integration as an end in itself is a non-starter. However, integration can be a facilitator for concrete economic and political objectives. As a practical matter, integration may be as much a psychological as an instrumental step, though the psychological benefits should not be underrated. These may include a greater sense of interdependence and cooperation, more willingness to abide by the rules, and more sensitivity to matters of equity (it enlarges the "playing field"), among other things.

- However, there may also be many practical advantages to integration—synergies of scale in production and markets; reduction of tariffs and other artificial trade barriers; uniformities in rules and regulations; elimination of structural impediments (such as internal monopolies), and more.

Finally, the OECD study concluded that market forces must play a crucial role in catalyzing and building support for political integration. Political motives alone are not enough, though political objectives can greatly augment the economic rationale. In other words, a synergistic confluence of various economic objectives and political constituencies is desirable.

A global superorganism, if it emerges, will be the product of an incremental, mutualistic, cooperative effort. As Harrison (1995) stresses, the participants in this process will ultimately find that it is advantageous to, in effect, pool their sovereignty and place it under the control of an entity that is shared in common yet also has a significant degree of independence. It will work because it is in the mutual self-interest of the participants.

In short, a global superorganism will succeed only as a cooperative process in which the pooling of sovereignty on a global level is viewed as mutualistic and is not seen as the imposition of a regime that serves the interests of only one power, or one segment of the world community, namely the rich countries and multinational firms. The record of the WTO to date has been a disappointment, and the nascent global trade regime has become a focal point of controversy (Amsden 2002; Faux 2002; Green 2002). Many of the developing countries feel, with justification, that the new order has been destructive to their interests rather than beneficial. This is a recipe for ultimate failure . . . and worse, a formula for economic and political chaos. A new global social contract is urgently needed that is focused not on free market capitalism as an end in itself, or on expanding markets for multinational firms, but on fostering global economic development. This is the most urgent priority, and the weight of history going back even to the Roman Empire and Periclean Athens attests to its transcendent importance.

Beyond that, there is much more to be done to build the infrastructure of a global superorganism, including a unified financial architecture (the dream of former U.S. Treasury Secretary Robert Rubin), an effective environmental regime, a global public health system (increasingly urgent in the age of bioterrorism and AIDS), a legal regime, and, longer term, an effective global security system that does not depend upon the precarious balance of power between existing states or the hot-and-cold exertions (some welcome, others not) by the world's only remaining superpower mainly in the pursuit of its own self-interest.

There is reason to believe that the potential for achieving positive synergies of various kinds at the global level is increasing, as are the potential costs for not moving toward this goal. In the aftermath of the attack on the World Trade Center, along with the wars in Afghanistan and Iraq and the continuing conflict between Israel and the Palestinians, deep political rifts and animosities have come to the surface. We cannot ultimately defeat terrorists armed with suitcase-sized weapons of mass destruction unless we also address the roots of their anger. Nor can we defeat the heroin poppy growers and the drug traders if their survival depends upon catering to the robust market that supports this rampant habit. Equally important, the global environmental and demographic threats are real, and they have life-and-death implications. This inescapable interdependency—this "entangled bank" in Darwin's term—is the great challenge of the twenty-first century. Workable, win-win economic and political instrumentalities at the global level remain underdeveloped, and the will to construct them has been piecemeal, unfocused, and even resisted. What the world sorely needs is another Jean Monnet (father of the European Union) with a practicable vision for a global superorganism. But even if such a leader steps forward, a global superorganism can only arise if we have the collective will, and wisdom, to implement it—and that is a very big "if" indeed.

PART II

Bioeconomics and Evolution

— ☙ —

The life of man in society, just as the life of other species, is a struggle for existence, and therefore it is a process of selective adaptation.

—Thorstein Veblen

SUMMARY: Many contemporary social scientists still adhere to the self-serving claim that human societies are *sui generis*—autonomous entities that transcend the constraints and vicissitudes of the natural world. It is said that the theory of evolution, and biology, may provide useful analogies and exert a limited influence on humankind, but human behavior, by and large, has been shaped by cultural forces. "Economic man" is therefore a bundle of idiosyncratic "tastes" and "preferences." This view is directly challenged here, beginning with a review essay that is focused on two major books associated with the new interdiscipline of evolutionary economics, one by economist Geoffrey Hodgson and the other by business consultant Michael Rothschild. In very different ways, each is a leading advocate for a relatively new paradigm in economics that has borrowed Darwinian evolutionary concepts and adopts an evolutionary view of economic change. I will argue that evolutionary economics, as currently framed, still falls short of a truly integrated bioeconomics. A radically different bioeconomic paradigm will be developed in the course of this and the next two chapters.

9

Evolutionary Economics: Metaphor or Unifying Paradigm?

Introduction: "The Economy of Nature"

To our preliterate ancestors, untutored in academic economics but well attuned to the vicissitudes of living in the late Pleistocene, the basic problems that they confronted—along with all other living things—were survival and reproduction. Earning a living in the "economy of nature" (to borrow the classic term—see chapter 10) was a relentless, inescapable, and somewhat unpredictable imperative.

Accordingly, the dirt-under-the-fingernails *weltanschauung* of these long-gone hominids included certain universal bioeconomic principles—which could have been documented for posterity had there been a perspicacious, time-warped economist available to take notes. These bioeconomic principles included the following, among others:

- The survival problem is always context-specific. The parameters of the problem for each organism differ, depending on its particular characteristics, the nature of its often-changing environment, and, most important, the precise organism-environment *relationship*.
- Energy, and access to relevant cybernetic control information about how to capture and utilize it, are two universal requisites for survival and reproduction, along with an array of other, more diverse survival needs.
- The time and energy that any organism has at its disposal to meet its needs is always limited and must be utilized relatively efficiently—or else.
- All species deploy specific strategies and tactics for earning their living. Some work alone while others form symbiotic partnerships (mutualistic or parasitic). Some survive by capturing sunlight and a few other basic

215

nutrients while others survive by preying on the photosynthesizers. Some reproduce asexually while others form sexually reproducing partnerships in close-knit social communities. Some use various body parts as tools while others fabricate tools as needed from various raw materials. Some hoard their information while others are inveterate gossips. And so it goes.

- Ecological competition is a common feature of the bioeconomic realm, but so also are interdependency, cooperation, symbiosis, and the division of labor. Moreover, competition is not the fundamental organizing principle in the economy of nature, as many theorists have asserted. The touchstone is the problem of earning a living and reproducing—adaptation—and both competition and cooperation are subsidiary phenomena. They are contingent "survival strategies." In fact, many species are practiced in the art of avoiding direct competition.

- Finally, all species exploit the synergy principle in one way or another. The phenomenon of synergy provides a way of conjuring economic leverage from an almost endless variety of cooperative effects.

"Academic Scribblers"

If none of these principles sound like conventional economics, it is because economists have erected a science that does not bear much relationship to the biological fundamentals. Thanks to the diligent efforts of two centuries of "academic scribblers" (to borrow economist John Maynard Keynes's self-deprecating caricature), including the scribblings of a passel of Nobel Prize winners, we have come to view economies, by and large, the way professional economists tell us to.

An economy, so they say, is about competition, markets, prices, preference functions, marginal utilities, rational choices, demand and supply relationships, equilibrium conditions, and so forth. Moreover, we are specifically admonished not to utilize any a priori concept of biological "needs," either as bedrock measuring rods or to make predictions about economic behaviors. We can only deal (scientifically) with the infinitely variable tastes and revealed preferences of "economic man" (and woman). In other words, economic behavior depends on the things economists study. And the obvious shortcomings of economics as a predictive science, we are told, can be attributed to the discipline's growing pains. In time, better analytical tools, better models, and better sources of data will lead inexorably to greater predictive power. Or, alternatively, various imperfections in the real world, if corrected, will make it conform more closely to the economists' models.

In the halcyon days of the social sciences, after World War II, such a conceit seemed plausible, at least in technologically advanced industrial societies. An ever-smaller proportion of the population was required for food production while a growing number of consumers enjoyed a high standard of living and a generous supply of discretionary income. Keynesian economics held out the hope that major economic depressions were a thing of the past. Like the rest of us in those days, economists were prone to echo Plato's observation in the *Republic* that, if the original purpose of human societies was to provide for "mere life," we had progressed to the point where it was now possible to focus on "the good life." Indeed, progress was an article of faith in western societies, socialist and capitalist alike. As a 1950s General Electric commercial put it, "progress is our most important product." Not only could progress be taken for granted, it didn't really need to be explained; it was self-evident. And the residual problems of advanced economies required only some fine-tuning. Accordingly, Darwinism and the biological paradigm were widely viewed as being irrelevant because, it was said, survival was no longer a problem; economic societies had transcended "nature, red in tooth and claw."

If the worldview of the postwar era was a part of the problem, the metatheoretical framework of economics (and the other social sciences)—their model of how to do science—was equally at fault. As both Geoffrey Hodgson (1993) and Michael Rothschild (1990) document persuasively in their two important but very different volumes on evolutionary economics, neoclassical economics, whose roots can be traced back to Adam Smith and the Scottish School, was based on a deeply flawed methodology—in fact, a metaphor borrowed from Newtonian physics (see also Gowdy 1994, 2000, 2004a, 2004b). The metaphor was that of a self-equilibrating mechanism that operates according to discoverable economic laws. Moreover, most economists of the day favored (and probably still do) what the legendary economic theorist Joseph Schumpeter characterized as "methodological individualism," the assumption that the workings of an economy as a whole can be reduced to a simple sum of the actions of each individual participant. Adam Smith's immortal metaphor said it all and, in the bargain, provided subsequent generations of economic realists with a justification for personal greed: "In spite of their natural selfishness and rapacity [men] are led by an invisible hand . . . to advance the interest of society" (1984 p. 184).

But perhaps the most perverse aspect of neoclassical economics, not to mention the other social sciences, was the methodological premise associated with cultural determinism and value relativism. It was widely held that, in searching for the relevant causal forces in human societies, the social sciences did not need to look beyond the realm of social phenomena, or perhaps the "laws" of Skinnerian behaviorist psychology. Indeed, the very notion of autonomous human purposes, goals, or creativity as causal factors in economic life was all but banished as unscientific. Frequently quoted was the dogma of the French sociologist Emile Durkheim in *The Rules of Sociological Method* (1938, pp. 104, 110), a bible for several generations of social scientists: "Every time that a social phenomenon is directly explained by a psychological phenomenon, we may be sure that the explanation is false. . . . *The determining cause of a social fact should be sought among the social facts preceding it and not among the states of individual consciousness* [italics in original]."

This comfortable worldview began to crumble in the 1970s and 1980s. First, there was the intrusion of some uncomfortable biological facts: the population bomb, in Paul Ehrlich's explosive metaphor, the oil shocks and increasing concern about resource depletion, the growing menace of environmental pollution, and, more recently, the ominous threat of global warming. It has become abundantly evident—as a matter of survival if not social justice—that economies are not always self-equilibrating, although the die-hard advocates of *laissez-faire* doubtless remain unpersuaded.

Then there was the challenge of the biological sciences, especially the rise of sociobiology, which asserted that biological (and biopsychological) facts also play a role in human behavior (how much has been the subject of rancorous debate). But the most important development by far has been ontological. There has been a growing recognition that a scientific paradigm based on Newtonian physics is not even applicable to modern physics, much less the social sciences. Many social scientists have come to recognize that a more appropriate metaphor for economics, if one is necessary, should be biological and evolutionary, and Geoffrey Hodgson's book, *Economics and Evolution* (1993), may well be the very best treatment of this subject to date. This chapter will briefly consider Hodgson's argument—and his survey of the theoretical terrain—and follow it with a brief discussion of Rothschild's book. But first, some historical perspective is in order.

The Origins of Evolutionary Economics

Technically, the roots of evolutionary economics could probably be traced back to the Reverend Thomas Malthus in his progenitive *Principles of Political Economy* (1820). However, the modern (twentieth century) inspiration for an evolutionary approach to economics, it seems, was the seminal work of the American economist Armen Alchian (1950). Alchian was the first postwar economist to adopt an explicitly selectionist approach to economic change. He postulated that profitability, as opposed to profit maximization, was the fundamental criterion for the survival of a business firm. "Those who realize *positive profits* are the survivors; those who suffer losses disappear [emphasis in original]" (p. 213). Alchian's survivalist approach was similar to the Nobel economist Herbert Simon's (1957) famous "satisficing" principle, which was developed some years later.

Alchian's work, though influential, failed to spark a broader movement within the discipline. The very few selectionist writings that appeared in the scholarly literature of the social sciences over the next two decades were predominantly developed by anthropologists and sociobiologists. All this changed with the publication of Richard Nelson and Sidney Winter's path-breaking book, *An Evolutionary Theory of Economic Change* (1982). The effect was catalytic. However, their approach was very conservative. They did not reach outside of the orthodox framework of economic explanation, and their analytical focus was the population of business firms in a complex society. In effect, their paradigm amounted to a borrowed set of analogies—a methodological frame-shift—rather than a theoretical reconceptualization. To quote (selectively) from their introduction:

> Our use of the term *evolutionary theory* to describe our alternative to orthodoxy . . . is above all a signal that we have borrowed basic ideas from biology, thus exercising an option to which economists are entitled. . . . We have already mentioned one borrowed idea that is central to our scheme—the idea of economic "natural selection." . . . Supporting our analytical emphasis on this sort of evolution by natural selection is a view of "organizational genetics"—the process by which traits of organizations, including the traits underlying the ability to produce output and make profits, are transmitted through time. . . . [However] we disavow any intention to pursue biological analogies for their own sake, or even for the sake of progress toward an abstract, higher-level theory. . . . We are pleased to exploit any idea from biology that seems helpful in the understanding of economic problems. . . . We also make no effort to base our theory on a view of human nature as a product of biological evolution, although we consider recent work in that direction to be a promising departure. . . . (1982, pp. 9–11)

For better or worse, Nelson and Winter's metaphorical approach became the standard model for the emerging new field of evolutionary economics. It did not fundamentally challenge the core neoclassical assumption of "economic man," nor did it invite the importation of biologically based causal influences (genotypes and phenotypes) into economic analyses, although this barrier is now beginning to crumble.

Hodgson's Critique

All this and more is reviewed in Hodgson's landmark 1993 volume, and in several of his subsequent works that will be mentioned later on. For starters, *Economics and Evolution* is a work of deep and meticulous scholarship that, among other things, methodically reconstructs the history of evolutionary thinking in economics, from Malthus to Marx, Engels, Spencer, Marshall, Menger, Veblen, Schumpeter, and Hayek. Many of these theorists are cast in a new light. However, Hodgson's purpose is not merely to summarize their work but to critique their ideas in light of our modern understanding of biological evolution.

I will mention a few highlights. Hodgson makes a strong case against treating Marx and Engels as evolutionary theorists. Marx and Engels had a wholly inadequate understanding of Darwin's theory and, worse yet, used highly questionable sources to rebut Darwinism. Their core conception of a dialectical clash between economic classes that are being swept along in a deterministic trajectory was in actuality a theory constructed within the mechanistic (law-driven) Newtonian tradition, even though it purported to explain an historical process. The causal dynamics were thus deeply anti-thetical to Darwin's theory of evolution as a cumulative process of contingent, incremental change via selection for functional, adaptive properties.

Hodgson argues that the Reverend Thomas Malthus, in addition to playing an inspirational role in Darwin's own evolutionary thinking, should properly be considered a founding father of evolutionary economics. In various writings, including his major textbook on the *Principles of Political Economy* (1989), Malthus portrayed economic life as a dynamic process that is driven by the biological fundamentals—particularly population growth and the "means of subsistence." In contrast with the many economic theorists who have envisioned an equilibrium condition as the natural or ultimate state of humankind, Malthus's "dismal science" (as a contemporary writer, Thomas Carlyle, called it) was based on the premise of unending conflict, struggle, and change.

Herbert Spencer, at once one of the towering figures of nineteenth century social science and a virtual nonentity for most of the twentieth century (even though his ideas have been freely expropriated by others), is given due credit and fair criticism, for the most part. Hodgson rightly finds fault with Spencer's orthogenetic view of evolution as a law-like, progressive developmental process leading to greater complexity and harmony and the withering away of the state—a vision Spencer shared, ironically, with Marx (albeit with some important theoretical differences). On the other hand, Hodgson perpetuates some negative judgments about Spencer's work that are not entirely justified.

For instance, he holds Spencer to account for not appreciating the entropy law (the second law of thermodynamics), which, he claims, contradicts Spencer's views about the functional advantages of structural heterogeneity. In fact, the second law does not contradict this conclusion; heterogeneity (complexity) may well be associated with functional improvements (though not always—see chapters 1 through 3). There is also empirical evidence of the benefits of progressive complexification in cultural evolution. Indeed, anthropologist Robert Carneiro (1967), in an elegant analysis, added credibility specifically to Spencer's views on the relationship between population size and cultural complexity by demonstrating that there is a mathematical relationship (approximately two-third power) between village size and village structural complexity in an ethnographic sample of forty-six societies.

Of particular concern is Hodgson's caricature of Spencer's views on cultural evolution. He attributes to Spencer the crude notion that culture can only evolve insofar as humans are changed biologically, an idea that demeans a theorist of Spencer's analytical power and encyclopedic breadth. To the contrary, Spencer should be given credit, along with Darwin, for appreciating the subtly *interactional*, coevolutionary nature of human biological and cultural evolution (for more on Spencer, see Corning 1982).

Marshall, Schumpeter, Hayek, and Veblen

Hodgson's analysis of Alfred Marshall, one of the giants of the discipline, is particularly incisive. Although much has been made of Marshall's statement in his classic text, *The Principles of Economics* (1890), that economics is "a branch of biology broadly interpreted," in fact Marshall never moved beyond a static, mechanistic paradigm; the core of his work was equilibrium-oriented and a promised companion volume on the dynamics of economic change never materialized. There are some suggestive ideas and intriguing leads in *The Principles*—allusions to variation and selection processes, the use of an organismic analogy to characterize economic development, an appreciation of the role of organism-environment interactions. Nevertheless, as Hodgson shows, Marshall's biology was Spencerian, not Darwinian, and it remained, as Hodgson puts it, "a promise unfulfilled." Marshall's insights were later ignored, and his example was not pursued by others. Subsequent generations of economists (for instance, Marshall's influential follower Pigou) turned instead for their inspiration to the hard science of Sir Isaac Newton or the energetics of Walras, Jevons, Pareto, and others.

Joseph Schumpeter and Friedrich Hayek, two of the twentieth century's leading economic theorists, are also found wanting in Hodgson's analyses. In Hodgson's view, Schumpeter is unjustly given credit nowadays as a progenitor of evolutionary economics. Schumpeter's evolutionism was in reality built on Léon Walras's dualistic vision of a general equilibrium punctuated by revolutionary creativity—a vision that was somewhat reminiscent of Marxism (and of punctuated equilibrium theory in evolutionary biology). While Schumpeter did use the term *evolution,* he did not use it in connection with the Walrasian dynamic. A close reading of his works suggests that he meant it to be nothing more than a synonym for general change. Indeed, Schumpeter even objected to the drawing of any specific analogies between biological and economic evolution.

In contrast, as Friedrich Hayek's theoretical views evolved over the years, he became increasingly enamored of the evolutionary paradigm. Early on, his work was a throwback to the neoclassical free-market tenets of Adam Smith and the Scottish School. The problem, as Hodgson (1993) observes, is that Hayek's Newtonian model of economic life relies on "methodological individualism." In an insightful critique, Hodgson shows that there is no justification for treating the individual actor as a "black box" that responds mechanistically to whatever values, tastes, or preferences are poured into it. Such insularity about the causal dynamics of human behavior managed to preserve the autonomy of the discipline at the cost of disconnecting *Homo economicus* from the real world.

Hayek was no more at fault in this respect than a host of other economic theorists, but his problems were compounded by an ill-conceived attempt to preserve an individualistic, self-equilibrating free-market model while, at the same time, grafting onto it a vaguely selectionist superstructure that is focused on the evolution of rules, institutions, and even whole societies. In this paradigm, individual actions are viewed as having functional significance for more inclusive wholes, which Hayek asserted are not, after all, reducible to their parts. Hayek never seemed to appreciate, much less reconcile, the profound theoretical contradiction that his evolving theory produced. Worse yet, Hayek's loose-jointed evolutionism was not at all Darwinian (his cavalier treatment of Darwin was a travesty), nor was it in touch with the voluminous (and growing) scholarly literature on the dynamics of cultural evolution. Finally, Hayek's psychology seemed to be gratuitous, as if plucked from thin air to support an a priori ideological position. Hodgson minces no words: "Unfortunately, these are not unique cases of a casual attitude to sources and scholarship in Hayek's work" (1993, p. 160).

If many of the pioneer economists fare poorly under Hodgson's scrutiny, Thorstein Veblen, an early twentieth century American theorist, provides an example of a pleasant surprise—indeed, a revelation. Though Veblen's evolutionism was little appreciated or emulated, either in his own day or until quite recently, in point of fact he got it right, even though his theoretical framework was never fully fleshed out (nor could it be, given the relatively primitive biology, psychology, and economic science of his day).

"Why is economics not an evolutionary science?" Veblen (1919, p. 56) asked in a famous article originally published in 1898. Economics, he argued, should be focused on explaining evolution and change, rather than fixing its gaze on the illusion of a static equilibrium. Veblen was also unique in fully grasping and utilizing the fundamental elements of Darwin's theory (variation, heredity/reproduction, and natural selection), and he diligently sought to develop analogous principles for socioeconomic evolution, with an emphasis on institutions as key units of selection, along with individual "habits of thought." In his classic work on *The Theory of the Leisure Class* (1899), Veblen penned a summary of his vision (quoted in part in the epigraph above) that sounds very Darwinian: "The life of man in society, just as the life of other species, is a struggle for existence, and therefore it is a process of selective adaptation. The evolution of social structure has been a process of natural selection of institutions" (p. 188).

Hodgson also points out that Veblen, unlike many other economic theorists, before or since, featured the distinctively purposeful (teleonomic) aspect of socioeconomic evolution, particularly the role of innovation and technological development. Although Hodgson claims that, in this respect, Veblen departs (somewhat) from a strictly Darwinian model by adopting a Lamarckian view of how cultural traits are transmitted, in fact Hodgson's interpretation reflects a widespread misunderstanding of evolutionary theory, an important point to which we will return. (See also the discussion of Lamarck in chapters 2 and 3.) Noting that many of Veblen's insights are now in the process of being rediscovered, often without an appreciation for their heritage, Hodgson (1993, p. 188) concludes: "Veblen should thus be placed among the founding figures of modern evolutionary economics." (Hodgson elaborates on this conclusion in more recent writings.) Veblen emphasized a selectionist dynamics and cumulative causation over time and, equally important, the role of emergent phenomena—especially social institutions—and the downward causation they may exercise on individual behaviors. The implications of an economic theory that recognizes behavioral influences other than an individualistic, rational calculus—i.e., multilevel causation—are potentially far-reaching. For more on this issue, see Hodgson (1998a, 1998b, 1999, 2001, 2004) and Gowdy (1994, 2004a, 2004b).

Hodgson's Paradigm

The reward for taking Hodgson's guided tour through the pantheon of economic theory is that it provides a firm foundation for his ultimate objective—to clarify the implications of an evolutionary paradigm and to lay out some guiding principles. Though there is a great deal of merit in what Hodgson has to offer, I will argue that it was ultimately too timid and severely constrained by the mindset of his discipline (but see below).

Hodgson sets the stage with a frontal attack on the foundations of neoclassical theory. "Crisis," he observes, is "an over-used word" (1993, p. 3). Nevertheless, the problems at the very core of the discipline are so serious that the entire edifice needs to be rebuilt. The neoclassical paradigm is totally at odds with the underlying reality of a contingent, irreversible historical process. Neither the mechanistic assumptions of equilibrium theory nor those of classical rationality are sustainable. Furthermore, the reductionist assumption that self-seeking actors, if left unfettered, will produce social and economic order, and even optimal results, is obviously untenable. Chaos theory and a variety of other nonlinear approaches to modeling dynamic processes lend support to the contention that economic life displays historicity—a sensitivity to initial conditions, path dependency, directional trends that are associated with positive feedback loops, and, not least, an enormous accumulation of baggage from past cultural and economic development. The reality of human purposes and human choices also contradicts the orthodox model; the "black box" turns out to be a Pandora's box. "Real world economic phenomena have much more in common with biological organisms and processes than with the mechanistic world of billiard balls and planets," Hodgson (1993, p. 24) notes.

Hodgson makes short work of the accusation that an evolutionary paradigm implies social Darwinism, or an endorsement of dog-eat-dog competition. Nor does it imply biological determinism, what some critics have called "vulgar sociobiology." In a brilliantly argued section on "Problems for Dr. Pangloss," Hodgson also rebuts the charge that an adaptationist/selectionist paradigm implies a process that inevitably leads to improvement, not to mention some form of optimization or perfection.

The problem with Hodgson's approach, in a nutshell, was in his follow-through. At this point in his own personal evolution, Hodgson was insisting on the autonomy of culture and economic life. Hodgson adopts and vigorously defends the so-called dual-inheritance model of cultural evolution—the predominant view among biologically oriented social scientists that biological and cultural evolution are separate processes that must be understood and explained in their own terms. Hodgson concedes that there may be interactions between the two processes. Anthropologist William Durham's (1991) coevolution model stresses this aspect, as does the work of Boyd and Richerson (1985), and Richerson and Boyd (2004); also see part I of this volume and Corning (1983, 2003). But the idea that culture and economic life are somehow an expression of (or are shaped by) our biologically based needs, drives, and capacities is avoided by Hodgson, and many other social scientists, who have felt compelled to defend the social sciences against the imperialism of sociobiology.

Accordingly, the logical, even necessary implication is that the theory of biological evolution should be treated only as a metaphor. There may be loosely analogous processes of variation, inheritance/reproduction, and selection, but these should be viewed merely as heuristic tools. Indeed, Hodgson assigns an entire chapter to the task of trying to show how metaphors have a perfectly respectable scientific pedigree (take natural selection, for instance) and how, in effect, the adoption of an evolutionary approach in economics would simply replace one metaphor with another that is more appropriate to the subject matter. The challenge, then, is to identify the relevant analogues. Hodgson's preference, following Veblen, is to treat "habits" and "institutions" as the units of selection. (An interesting example, involving vehicular traffic conventions, can be found in Hodgson and Knudsen 2004a.)

Although this approach is politically safe and certainly useful—economies do, after all, have many emergent, systemic properties—in my view it is ultimately insufficient. It amounts to a revival and repackaging of institutional economics. This may help to reinvigorate a beleaguered scholarly tradition, but Hodgson did not move the argument much beyond where Nelson and Winter had planted their flag in 1982. If the ultimate objective is to get us closer to an understanding of the underlying causal dynamics of economic processes and economic change, it is essential to include the sources of human motivation and choice (including our choice of habits for that matter). Despite Hodgson's commendable plea for reestablishing the role of human purposes and goals in economic science, in the end he opted to keep the lid on the black box.

In several more recent writings, Hodgson has devoted considerable effort to making the case for a revival of Veblenian institutional economics (see especially Hodgson 1999, 2001a, 2003a, 2003b, 2004a, 2004b; Hodgson and Knudsen 2004b). In *How Economics Forgot History* (2001b), Hodgson defended the notion that economic theory must be historically rooted and sensitive to the specific context. This approach was commonplace in the nineteenth century but disappeared into a Newtonian fog after World War II. In a companion to this volume, *The Evolution of Institutional Economics* (2004b), Hodgson develops an in-depth case for resurrecting the tradition in economic theory that focuses on institutions and their interactions, both with one another and with individual actors. Also, Hodgson has moved toward a more multileveled, interactional perspective. In theory, causal influences at all levels, from genes to political institutions, may influence the course of economic evolution, Hodgson agrees. Indeed, Hodgson even criticizes Veblen's normative focus on an "instinct for workmanship" and acknowledges: "If there is a normative standard by which to judge the adequacy of social institutions, then it is in terms of their capacity to sustain or enhance the means of human life" (2004, p. 201). Unfortunately, this sounds rather tentative, and he does not pursue its implications. But then, neither did Veblen, despite his firm embrace of Darwinism.

Hodgson also addresses two currently contentious theoretical issues in his more recent writings. One involves the distinction between so-called Lamarckian selection (the transmission of acquired characters between generations) and Darwinian selection (differential survival based on functional differences). In contrast with Nelson and Winter, among others, Hodgson posits that a Darwinian mode of selection is the primary dynamic of change at the institutional level in human societies (see also Hodgson 2001c; Knudsen 2001; Vromen 2001; Wilkins 2001).

The second issue concerns what goes under the heading "Universal Darwinism." First proposed by biologist Richard Dawkins (1983) (though it was presaged by earlier theorists), the notion is that certain fundamental causal processes can be said to characterize all forms of evolution, namely, variation, replication, and differential selection (see also Dawkins 1989; Plotkin 1994).

At the risk of giving offense, I believe that the somewhat scholastic, even tedious debates over these two issues in recent years could easily have been avoided. In the case of the Lamarckian/Darwinian selection issue, the unequivocal answer is that both are relevant—even in biological evolution. To understand why this is so requires an appreciation of the fact that behavioral and cultural innovations in the natural world routinely get transmitted between generations via social learning and that these behavioral traits may even serve as pacemakers that indirectly influence biological evolution. Of course, the direct inheritance of acquired characters does not, in fact, occur, but this was only part of what Lamarck proposed. (See the discussion of a Darwinized neo-Lamarckian selection in chapters 2 and 3; also see the more expansive discussion of this issue in Corning 2003.) However, "habits" and other cultural artifacts do not get transmitted mindlessly between generations, as a rule, without reference to their functional consequences for the bearer and the recipient; habits do change after all, so they are subject to both stabilizing (or normalizing) selection and to directional selection over time. In other words, there is no categorical distinction between biological evolution and cultural or economic evolution. In nature and human societies alike, there has been an interplay (a coevolution) between these processes, though the means of transmission between generations differs.

Likewise, on the subject of Universal Darwinism, this issue might have been defused by making a simple terminological shift. Universal Darwinism sounds imperialistic—as if biologists are threatening to take over the social sciences (again). In light of the history of social Darwinism and the ideological connotations that are, unavoidably, associated with Darwin's name among many social scientists, it is gratuitous and unnecessary to insist on calling it Universal Darwinism. Not only does the term *universal selectionism* have a more ecumenical, pan-disciplinary ring to it but it can be defined in a way that is consonant with Universal Darwinism.

Rothschild's Paean to Capitalism

Michael Rothschild's best-selling popular book *Bionomics* (1990), which is a classic of its genre, also deserves consideration here. (In passing it should be noted that Rothschild's title is a pretentious neologism that overlooks the established term *bioeconomics* while inappropriately excising the *oikos*, or household part of the original Greek term; we are left with something that means, roughly, "life management.") A distinctive feature of Rothschild's volume is that he provides an array of vivid illustrations that are meant to draw out analogies between the economy of nature and human economies. In so doing, however, Rothschild unwittingly helps to make the case for the more radical thesis that human economies are not simply metaphorical ecosystems—with capitalism as the engine of economic evolution. (The hardcover edition of his book had the unfortunate subtitle "The Inevitability of Capitalism.") At the risk of being politically incorrect, I contend that it will ultimately prove more fruitful to view human economies as variations and elaborations on a common theme that is rooted in the process of biological evolution itself. (I will expand on this point below and in the next two chapters.)

Rothschild takes pains—rather too vehemently—to denounce social Darwinists and sociobiologists, whom he lumps together in his eagerness to distance himself from the taint of biological determinism:

> In their view, human culture is not parallel to, but an extension of, human genetic information. For them, the tree of cultural evolution grows from genetic roots. In bionomics, genes and knowledge are not connected. . . . Our genes do not program us to become capitalists. Capitalism is simply the process by which technology evolves. By way of analogy, bionomics argues that, on a day in–day out basis, biologic and economic life are organized and operate in much the same way. . . . Though the analogy between genetic and technologic evolution is powerful, it is not perfect. (p. xiii)

To my knowledge, nobody has claimed otherwise. And has anyone, in the past century, seriously asserted that capitalism is programmed into our genes? (Even if such things have been asserted in the popular press, this hypothesis would make an easy target if it were presented to anyone who has been trained in behavioral genetics.) A more serious criticism is that there is also a bit of ideological sleight of hand at work here. On the one hand, Rothschild asserts that an economy as a whole bears only an analogous relationship to the biological realm. On the other hand, capitalism is justified as a natural phenomenon.

Rothschild then proceeds to illustrate his thesis—an unabashed paean for the free market system—with a wonderful selection of examples, for this is a very well researched and superbly written volume. As he shifts focus repeatedly from human economies to the economy of nature, Rothschild compares humans with other species in a variety of ways. For instance, he shows how comparable are the uses of information in biological and economic evolution. Yet he conspicuously passes over the closer analogies and even homologies between humans and other species, especially the primates, in terms of how they are able to acquire and utilize information through learning and social interactions.

Likewise, he discusses the dynamics of biological evolution as it has been reconstructed for trilobites and juxtaposes it with the latest thinking about the evolution of *Homo sapiens*, which he describes without embarrassment as an interactive process in which behavioral/cultural/technological developments have gone hand in hand with anatomical changes over time. He also details the many parallels between the division of labor in eukaryotic cells and human factories, yet he avoids mention of the even closer parallel between the division of labor in human societies and, say, army ant colonies or naked mole rat societies, with regard to such group-level functions as defense, food procurement, and reproduction. (Perhaps because it does not suit his thesis, Rothschild illustrates but does not specifically point to the fact that complex human societies resemble *both* ecosystems and organisms to varying degrees.)

Rothschild's treatment of the cost-benefit calculus embedded in economic processes, particularly in relation to energetics, is especially noteworthy. His illustrations range from hydrothermal vent species to Bernd Heinrich's landmark studies of bumblebee economics to the fascinating corporate history of Cub Foods. He also highlights the vital role of real cost reductions in economic evolution. For instance, he estimates with carefully documented calculations from historical data that, over three centuries of machine power evolution, real costs (in constant dollars) have declined from about $6,000 per horsepower for the original Newcomen engine to $3 per horsepower for a modern automobile engine. Similarly, data for the American economy between 1910 and 1986 reveal that retail egg prices (in constant dollars) declined an astounding 80 percent. Yet Rothschild glosses over the fundamental commonalities in animal and human energetics, as stressed by the pioneering economist Nicholas Georgescu-Roegen in his thermodynamically oriented work (see chapter 10). Indeed, Georgescu-Roegen is not even referenced in Rothschild's volume.

Finally, Rothschild is well enough schooled in the literature of evolutionary theory, ethology, and behavioral ecology to recognize the importance of mutualism, symbiosis, and social organization in the natural world, and to identify analogues in human societies. He writes: "Avoiding head-on competition—in the wild and in the marketplace—leads to diversity, which, in turn, promotes interdependence. Mutually beneficial relationships, common among species in nature, are echoed in business, where the vast majority of affiliations are based on mutual profitability." And yet, later he insists on our uniqueness: "Human beings *are* different from all other creatures. We are conscious beings. As social animals we are socially conscious. . . . We choose to form communities for mutual aid, support, and sharing. As a species we have always done so. Indeed, our capacity to cooperate may well be our most powerful adaptive trait." In sum, Rothschild makes the case, compellingly, for a biological paradigm in economics. He fails, conspicuously, to convince us that the analogy is only skin deep and of little theoretical consequence.

Explaining Economic Evolution

The fundamental question for an evolutionary economics is this: What are the causes of both the *continuities* and the *changes* that we can observe in economic life? (Darwin termed it "descent with modification.") What are motivations that drive the choices that are made by individuals and institutions? Whereas Hodgson was more concerned, at least in his 1993 volume, with the sources of continuity (heredity) and with identifying economic analogues of genes, Rothschild addressed the dynamics of change; his focus was the accumulation and modification of economic processes, especially technology and its handmaiden, information. Economic evolution was portrayed by Rothschild as being, fundamentally, a process of cumulative learning, which can be captured quantitatively in the so-called learning curve (or "experience curve" in the argot of contemporary business consultants). Economic enterprises are characterized by Rothschild as organized intelligence, and organizational learning over the course of time (a process that is often accelerated by the synergies that collective problem solving can produce) is identified as the primary cause of economic evolution.

In other words, Rothschild is talking about the *evolution* of institutions, among other things. Yet once again Rothschild contaminates his argument by insisting on the uniqueness of humans. In what amounts to a non sequitur, he observes that the learning curve, and our accumulation of organizational knowledge, is what sets us apart from Edward Tolman's maze-learning laboratory rats. Other species can only improve their economic performance by changing their genes, Rothschild claims, whereas humans alone can change their technologies, a statement that is flatly wrong. Rothschild cites several key references pertaining to the evolution of behavior and culture in other species, but evidently he didn't read them carefully enough to discover the broad consensus view among the students of animal behavior that an organism's phenotype is generally the product of an *interaction* between its genotype and its specific environment, and that many adaptive behaviors arise through learning and even cultural transmission (see Corning 1983, 2003; also chapters 1 through 8).

Despite his insights about the workings of a complex economy (the fruit of several years of battlefield experience in the business world as a lawyer and consultant), Rothschild's Panglossian bottom line constitutes, in my opinion, a serious flaw. He concludes that the experience curve of technology and economic organization, when given its head in a free market environment, will ultimately negate the Malthusian dynamic and "obliterate" the central myth of the dismal science (not to mention John Stuart Mill's law of diminishing returns). This, to put it bluntly, is a Pollyanna look-alike. It entails an optimistic gamble that we would be foolhardy to take; we won't have to wait until the next ice age to see Rothschild's thesis disproved.

In effect, both of these metaphorically minded economic evolutionists put the human species into its own theoretical Disney World. In Rothschild's case, the claim that humans alone can invent new adaptive strategies is totally at odds with the extensive evidence that other organisms are able do so as well, although humans obviously excel in this respect. In Hodgson's case, his insistence on the autonomy of the economic realm as opposed to the more qualified—and theoretically challenging—concept of interdependence amounts to the same thing. (Hodgson did use the phrase "partial autonomy" at least once, but he obviously did not stress it or pursue its theoretical implications, either in 1993 or subsequently.) In rejecting the fundamental, bottom-line homology between nature's economy and that of humans (I'm not denying the obvious differences), these biologically oriented economic theorists perpetuate the conceit that has for so long obfuscated the basic paradigm shared by all living species—namely, the problems of survival and reproduction in an uncertain, historically constrained, contingent world.

Finally, it should be noted that, in addition to the two outstanding volumes that were reviewed here, there have been a number of other important contributions to evolutionary economics within the past decade or so (see Witt 1992, 2003; Baum and Singh 1994; Gowdy 1994, 2004a, 2004b; Vromen 1995; Nelson 1996; Metcalfe 1998; Aldrich 1999; Mokyr 1999; Dosi et al. 1998, 2000; Dosi 2000; Potts 2000; and the Hodgson works cited earlier). However, the outlook for this emerging interdiscipline is still guarded. In a recent survey of the state-of-the-art in evolutionary economics (Laurent and Nightingale 2001, p. 10), the editors concluded: "The question becomes one of whether evolutionary theories explain economic phenomena better than alternative theories . . . whether evolutionary economics is worth exploring. The answer to that must await . . . development of mature evolutionary theory."

Another Alternative: Bioeconomics

What is the alternative to these metaphorical visions of economic life? Ironically, Hodgson himself pointed the way. To quote Hodgson's own loose definition of economics: "Economics should be the study of the social relations and processes governing the production, distribution and exchange of the *requisites of life* [italics added]" (1993, p. 8). I could not have said it better. Economics is concerned, first and foremost, with meeting the ongoing biological needs of human societies and their members. To repeat, a human society is, at bottom, a "collective survival enterprise"; we are deeply dependent upon one another for our survival needs, as Adam Smith himself noted. Furthermore, the survival problem remains a pressing daily concern. Despite our propensity for self-denial, survival is directly or indirectly associated with most of the economic activity of humankind worldwide. Consider, for instance, the fact that humans spend about one-third of their lives sleeping, and that a substantial share of our economy is devoted to providing for this basic biological need in various ways, from bedrooms to hotel rooms, sleeping bags, and sleeping pills. Or, consider how much economic activity is devoted to keeping the human body within a comfortable temperature range—thermoregulation. (More on this in chapter 11.)

As suggested earlier, in order to distinguish between this very different economic paradigm and more conventional approaches, I prefer to call it "bioeconomics." Needless to say, this term is hardly new. However, I use it in a distinctive way. (There is no universally accepted definition of bioeconomics; it's still up for grabs.) I will define the term further in the next chapter and will provide some historical background. Here let me stress that, as I use it, the term *bioeconomics* refers to an empirically grounded definition of what is the basic problem for the human species. Bioeconomics is not about scarcity, or efficiency, or profitability, or habits, or what have you. It's about biological survival and reproduction. If there is a progenitor for this approach, the most obvious candidate is the "dismal" economist himself, Thomas Malthus. Arguably, it could be said that I am proposing to return economic theory to its historical roots—to the very paradigm that inspired Darwin himself.

Much of the work these days in evolutionary economics is concerned with describing and explaining how economies grow, change, and perhaps decline. These are important issues. However, bioeconomics as I define it is centrally concerned with how economies meet our basic needs—or fail to do so. Moreover, this paradigm is multileveled and interactional in its approach to the causal dynamics. Again, this approach was originally detailed in a chapter of my earlier book, *The Synergism Hypothesis: A Theory of Progressive Evolution* (1983). It was part of a broader analytical framework that I characterized there as an "interactional paradigm." (See also chapter 6, especially figure 5, and chapters 10 and 11 in this volume.) In general, the focus of bioeconomics, as I define it, is on the relationship between our biologically based motivational substrate and the vast range of learned and culturally molded behaviors (and institutions) through which these motivations are expressed. This area was hardly terra incognita in 1983, but there has been much additional progress since then. (A recent computer search for the period from 1985–1995 identified over two hundred book-length studies devoted to exploring this bio-psycho-social interface, and many of these works are syntheses of the burgeoning research literature.) In other words, we already know quite a bit about what goes on inside the "black box." There is no scientific justification for keeping the lid on it. Indeed, the new field of behavioral economics and the work in experimental economics, not to mention a well-established scholarly tradition in welfare economics, reflect efforts by economists themselves to address the challenge of understanding better how human nature and complex economies interact.

A few additional points should be added about the implications of a bioeconomics paradigm. There is more to come in the next two chapters:

- First, this approach is focused on *both* the many and varied causes and the survival consequences of behavior—economic and otherwise—including the ways in which proximate biopsychological factors influence the ongoing cultural and economic evolution of human societies. This contrasts with the dual inheritance and coevolutionary models that focus on the replicators—the mechanisms of information storage, transmission, and change over time. The objectives of these other approaches are not contradictory but complementary. (Needless to say, this approach also contrasts with the simplistic determinism of vulgar sociobiology.)

- To reiterate, a key aspect of this paradigm is the concept of a multileveled hierarchy of causation—from the physical environment to the most inclusive political entities, including several levels of emergent biological and social wholes that are at once partially independent and interdependent; complex processes of both upward and downward causation are continuously at work. Indeed, the causal dynamics are usually configural and interactional in nature; they have synergistic properties. Accordingly, in addition to the creative activities singled out by Rothschild (and many other economists), the process of evolutionary change also includes such important variables as population growth, epidemics, environmental challenges and opportunities, and, not least, resource availabilities. (To cite one well-known example, England's adoption of coal as an energy source at the dawn of the Industrial Revolution was hastened by a drastic depletion of its supply of firewood.)

- In this framework, a distinction is drawn between various units of selection as passive repositories of information storage and replication and the units of selection in causal/functional terms at the proximate, phenotypic level (what I refer to as neo-Lamarckian selection). In this paradigm, selection is a process that goes on at multiple levels—among goods and services, tools and technologies, resources and raw materials, physical locations and ecosystems, individuals and households, volunteer organizations and business firms, markets and economic sectors, governmental entities, and even whole societies. The search by some social scientists for a universal replicator analogous to the gene (e.g., memes) is misguided. The problem with "Universal Darwinism" as a monolithic theory of cultural/economic evolution is that it involves both multiple (interacting) levels of selection and a plethora of different agencies, or objects—from ideas and techniques to diverse physical and social structures. Furthermore, the precise relationships and interactions among the various kinds and levels of selection are immensely complex. To cite one example, in Ireland there are stone walls that have been used continuously to enclose grazing lands for cattle for over 6,000 years. The walls have been passed down between generations of herdsmen, but they can hardly be called memes; they bear no functional resemblance whatsoever to replicators. (For more on this contentious issue, see the extended discussion in Corning 2003, and chapter 6.)

- Finally, this framework also comfortably accommodates purposiveness in general and human inventiveness in particular, causal influences that have obviously played an important role in human evolution. Indeed, the one lapse in Hodgson's otherwise impressive scholarship was his apparent unawareness (at least back in 1993) that many biologists have long since accepted the concept of "purposiveness" in evolutionary biology. (As noted earlier, many of them seem to prefer Colin Pittendrigh's term *teleonomy*, or evolved *internal* teleology, as distinct from an externally imposed purposiveness.) Ever since the late 1950s, leading evolutionists have appreciated the fact that purposive behavior plays an important causal role in initiating evolutionary changes. I have frequently quoted Ernst Mayr on this point, but similar statements can be found in the writings of Waddington, Dobzhansky, Simpson, Ayala, and Thorpe, among others. There are also detailed reviews of this issue in Corning (1983, 2003). In sum, this aspect of Lamarckian theory has long since been Darwinized and does not need to be imported to supplement an evolutionary paradigm.

So, how might one define economic evolution from a bioeconomic perspective? Bioeconomic evolution can be characterized as a consequential change in a society's (or a species') mode of adaptation—that is, in the "means of production" (in Marxian terms) of the requisites for biological survival and reproduction. It entails the differential selection of alternative adaptive modalities (instrumentalities of needs satisfaction). It is a multileveled selection process, and it is predominantly, though not exclusively, a teleonomic process. Although creativity and cumulative learning play an obviously important role, the process is also affected by entrenched cultural values, routines, customs, traditions (yes, including habits and institutions), and many other factors; the dynamics are synergistic.

One way of differentiating between evolutionary economics and what I am calling bioeconomic evolution is in terms of the theoretical question that is addressed. Evolutionary economics is, understandably, focused on causal explanations for how and why economic activities and institutions change over time. Johan Peter Murmann's elegant 2003 case study concerning the evolution of the synthetic dye industry in the nineteenth century is a model example of this approach. Murmann rightly characterizes it as a coevolutionary process that was shaped by a complex set of causes, from patent laws to the educational system, the cultural environment, and government actions. On the other hand, bioeconomic evolution is concerned with changes in the means of production for the satisfaction of our basic survival and reproductive needs, that is, changes in our "survival strategies." Some social theorists deploy Ernst Mayr's distinction between proximate and ultimate causation in relation to the formulation of evolutionary questions in economics. However, I believe such a categorical distinction has limited utility. It may blur the iterative and interactional relationship between the many different kinds of causal influences in a given context. Some may be immediate and contextual (proximate causes); others represent accumulated baggage from the past (ultimate causes). Indeed, in the special (though probably not unique) case of humankind, where future goals, foresight, and planning greatly influence current behavior, the imagined future, too, may be an ultimate form of causation.

Conclusion

Is bioeconomics (as I define it) "an impossible dream," as the old Broadway song would have it? In actuality, its essential elements are already well established in such disciplines as ecology, ethology, sociobiology, evolutionary psychology, behavioral ecology, and, not least, the health and nutritional sciences. (There is also relevant work in ecological and economic anthropology, biological anthropology, and human ecology.) Researchers in these fields routinely study animal (and human) behavior and social organization in terms of its adaptive consequences. These research endeavors share several focal questions: How does each species organize its survival enterprise? How does it allocate resources to meet its needs? How does it make choices among alternative strategies? And with what biological and adaptive consequences?

Since we are not exempted from the survival problem, there is no reason in theory why the same paradigm could not be applied more broadly to the human species. If economists can successfully penetrate the complexities and ambiguities of market prices to get to the bedrock of "real costs," it should also be possible to get to the bedrock of "real benefits." The problem, of course, is the leaden weight of vested interests—intellectual, academic, ideological, economic, even political. These saddle-weights may, in fact, constitute an insurmountable handicap. But perhaps economists will respond to Hodgson's plea for greater theoretical diversity.

However, experience suggests that bioeconomics—at least in the radical formulation proposed here—will not easily win converts. It is a discipline that will most likely have to be created in the interstices between economics, anthropology, and the life sciences. Nevertheless, I will hazard the prediction that, in the long run, the efforts of Nelson and Winter, Witt, Hodgson, Rothschild, Gowdy, and others who have been chipping away at the reigning dogma (and the priesthood) of the economics establishment will come to be viewed as major contributors to a more fundamental paradigm shift even than these pioneering thinkers have in mind. We will pursue this theme further in the next two chapters.

— ☙ —

The power of population is indefinitely greater than the power in the earth to produce subsistence for man. . . . Population, when unchecked, increases in a geometrical ratio. Subsistence increases only in an arithmetical ratio. . . . This implies a strong and constantly operating check on population from the difficulty of subsistence. . . . And the race of man cannot, by any efforts of reason, escape from it.

—Thomas Malthus

Anyone who thinks that growth can go on forever in a finite world is either a madman or an economist.

—Kenneth Boulding

SUMMARY: To reiterate, the ground-zero premise of the biological sciences is that survival and reproduction constitute the basic, continuing, inescapable problem for all living organisms. Life is, at bottom, a survival enterprise. Whatever may be our perceptions, aspirations, or illusions, this tap-root assumption is applicable to the human species as well. Survival is the paradigmatic problem for all human societies; it is a prerequisite for any other, more exalted objectives. Modern economic science, with its roots in the classical economists' simplistic assumptions about human nature and human motivation—that is, the rational, self-interested "economic man"—and with a single-minded focus on the flow of goods and services in expansive, self-equilibrating markets, has often been oblivious and sometimes cavalier about the survival problem. Indeed, over the years some economists have even denied the existence of objective human needs. Bioeconomics, which is grounded in the biological imperatives, therefore presents a fundamental challenge to the theoretical foundations of mainstream economics. It imposes an empirically based normative criterion on the performance of any given economy, as well as second-guessing various conventional economic measures. In effect, bioeconomics provides an alternative set of account books for assessing the performance of an economy, including especially the outcomes.

10

Bioeconomics as a Subversive Science

Introduction

Bioeconomics is one of several new interdisciplines that have sprung up in recent years at the interface between the life sciences and the social sciences, as pointed out in chapter 9. However, I believe that bioeconomics (at least as I define it) presents a more fundamental theoretical challenge to mainstream economics than is generally appreciated.

Some of the new interdisciplines—evolutionary economics, ecological economics, evolutionary anthropology, and the like—involve a borrowing of methodologies and models, or a search for biological analogies, as discussed in chapter 9. Others were inspired by a desire to reach out and embrace the many causal influences that disrespect traditional disciplinary lines. Bioeconomics also does this, but it involves something much more profound as well. It challenges the traditional (neoclassical) economists' assumptions about the nature and purpose of economic life; it disputes their model of human nature and human motivation; it internalizes and integrates what have often been treated as externalities by mainstream economists; it embraces nonmarket activity that may be vitally important to our well-being; and it hedges the traditional aspiration for a predictive "science" of economics. Indeed, it imposes a superordinate set of normative criteria on such conventional economic indicators as personal income, economic growth, and gross domestic product. These familiar measures may or may not serve what is viewed here as a more basic purpose—an objective, inescapable, ongoing problem that is the central organizing principle for most of our economic activity as a species, although sometimes it may be pursued unconsciously and indirectly. In any case, it has been largely ignored by conventional economists.

Biologist Edward O. Wilson got himself into trouble a quarter of a century ago by making similar claims for sociobiology, the new discipline that was launched with much fanfare in his massive tome *Sociobiology: The New Synthesis* (1975). Wilson asserted that sociobiology was destined to reformulate the foundations of the social sciences. However, things did not quite work out that way. Sociobiology was roundly attacked, even by some of Wilson's Harvard colleagues, and it became mired in an acrimonious controversy.

Despite this inauspicious beginning, sociobiology has been able to make some valuable contributions to our understanding of social behavior over the years. Yet its overall impact on the social sciences has been far less significant than Wilson had predicted. To many social scientists, sociobiology is forever tainted with the aura of biological determinism and even (unfairly) social Darwinism. It is probably fair to say that sociobiology has been rejected by the majority of practicing social scientists down to the present day.

Accordingly, I am acutely aware of the dangers associated with making overly optimistic claims for bioeconomics. I recognize the need to make the case to a skeptical audience, and to differentiate bioeconomics from sociobiology, evolutionary economics, evolutionary psychology, and the like. Indeed, bioeconomics (as I propose to define and use it) is emphatically not sociobiology in disguise. In the course of defining and unpacking this term, I will disagree with the premises of sociobiology in some very significant ways. It is clear that sociobiology as defined by Wilson got some things wrong, though some of it can be blamed on the theoretical state of the art in evolutionary biology at the time he wrote. In a nutshell, sociobiology (and its fraternal twin, evolutionary psychology) is focused on the genetics and the biology of social behavior and evolution—ultimate causation. In contrast, bioeconomics is focused on the strategies and actions of the phenotype (inclusive of its social, economic and ecological environments)—proximate causation. This casts a very different light on the evolutionary process, as I have tried to show in the first part of this book.

A (Very) Short History of Bioeconomics

Let us begin with some historical perspective on bioeconomics. The observation that economics and biology have a fundamental affinity can be traced back at least to Linnaeus (who is reputed to have coined the term *the economy of nature*), and perhaps even to Aristotle. Similar conceptualizations can be found in Adam Smith's *The Theory of Moral Sentiments* (1759) and *The*

Wealth of Nations (1964), in Darwin's *The Origin of Species* (1968), and in the writings of the founders of ecology—Charles Elton, A. G. Tansley, Raymond Lindeman, G. Evelyn Hutchinson, and others. (For a detailed history of ecological thinking, see Worster 1977.) As Hodgson noted (see chapter 9), some of the early economists—Alfred Marshall, Thorstein Veblen, and Joseph Schumpeter are three prominent examples—also acknowledged a kinship between the two disciplines, although relatively little productive use was made of this familial connection until the latter part of the twentieth century.

The term *bioeconomics* was actually coined by an obscure turn-of-the-century theorist, Hermann Reinheimer, one of whose provocatively titled books was *Evolution by Co-Operation: A Study in Bioeconomics* (1913). Reinheimer was intent on showing how cooperation produced benefits of various kinds, both in nature and in human societies. More recently, the term was employed by economists Kenneth Boulding (1978, 1981) and Nicholas Georgescu-Roegen (1976a, 1976b, 1977a, 1977b), although neither of these theorists, to my knowledge, ever provided a precise definition of the term. Boulding advanced a broad, multidisciplinary conceptualization of evolutionary dynamics and evinced over time a growing concern about environmental issues. Georgescu-Roegen was particularly interested in a thermodynamic approach to economic analysis, though his concerns were broader than that. Gowdy and Mesner (1998) point out that Georgescu-Roegen used the term *bioeconomics* to draw attention to the biological origin of economic processes. His overall focus was the problem of scarcity, especially the irreversible utilization of resources and the resulting environmental consequences. He was also a pioneering proponent of sustainability. (See also the usage in Adler-Karlsson 1977.) The term *bioeconomics* has also been used in recent years specifically with reference to the interface between the environment and human economic activities, such as fisheries management (e.g., Clark 1990 and references therein).

Another significant contribution to the emergence of bioeconomics has been the growing number of explicit theoretical and analytical linkages that have been forged between economics and biology over the past two decades or so, as Michael Ghiselin (1992) and Geoffrey Hodgson (1993) have documented. In the absence until quite recently of an organized bioeconomics movement, these interdisciplinary efforts were mostly uncoordinated and piecemeal, but they nevertheless established a firm theoretical bridge between the disciplines. Particularly notable was the work of Ghiselin (1969, 1974, 1978, 1986, 1987) related to what he has called "general economy," as well as the pioneering work of Gordon Tullock (1970, 1971a,

1971b, 1976, 1979, 1990), Gary Becker (1974, 1976), Becker and Tomes (1979), and Jack Hirshleifer (1977, 1978a, 1978b, 1982, 1985, 1986), and some of my own work (Corning 1983, 2000, 2002c). The burgeoning cross-disciplinary work in game theory, though not as a rule identified with bioeconomics, nevertheless fits within this general framework (see especially Maynard Smith 1982b, 1984; Axelrod and Hamilton 1981; Axelrod 1984; Nowak and Sigmund 1993; Sigmund 1993; Binmore 1994a, 1994b, 2004; Hirshleifer 1999). Three relatively new journals, *Ecological Economics, Evolutionary Economics,* and, most important, the *Journal of Bioeconomics,* have now become focal points for this work.

Defining Bioeconomics

Like many other loosely structured movements in the academic world, bioeconomics suffers from the "Humpty Dumpty effect"—that is, the anarchistic tendency (highlighted in Lewis Carroll's *Through the Looking Glass*) for a neologism to be defined in almost as many ways as there are theorists. The candidate definition for bioeconomics that I would like to propose here has the advantage, I believe, of being the most inclusive—and most fundamental—in its focus. But more to the point, the paradigm that I will utilize below and in chapter 11 is based on this definition.

My definition proceeds from what I have called the ground-zero premise of the biological sciences. The basic problem that we confront—along with all other living things—is the "struggle for existence" (in Darwin's pellucid phrase). Whatever may be our perceptions, aspirations, or illusions, biological survival and reproduction constitute the paradigmatic problem of the human species. Furthermore, the survival problem is ongoing, relentless, and inescapable; it will never be permanently solved. To repeat: Life is, at bottom, a survival enterprise.[1]

This taproot assumption about the human condition is not exactly news, although we very often deny it, or downgrade it, or simply lose touch with it. The survival imperative was clearly recognized by Aristotle in various writings (Nussbaum 1988, 1993). It was also the underlying assumption in Darwin's treatise, *The Descent of Man* (1874). Herbert Spencer and a slew of nineteenth-century social theorists also took the survival problem as a given. Today it figures prominently in some of our public policy debates, most notably those concerning poverty and various environmental problems. It has often been noted that the words *economics* and *ecology* are both related etymologically to the Greek word *oikos* (household), and various phenomena studied by economists and ecologists (competition, a division of labor, economies of scale, resource allocations, efficiency, productivity, etc.) are common to both disciplines. Some theorists have emphasized one or more of these as being the defining characteristic(s) of bioeconomics. While it is true that these phenomena are important, the problem is that they do not speak to the fundamental *purpose* of an economy. If the paradigmatic problem is biological survival and reproduction in an uncertain and changeable natural environment (which may also include various organized ecological, social, and political systems), it follows that the underlying (broad) purpose of any economy is related, first and foremost, to meeting the basic, continuing biological needs of the "household." This is not a normative statement. It is, rather, an empirical proposition (a truth claim) to the effect that earning a living is, quite literally, the organizing principle underlying most economic activity, whether we are conscious of this fact or not. In this light, bioeconomics may be defined as *the study of how living organisms (of all kinds) acquire and utilize various resources to meet biological needs.* This encompasses not only the choices that people make—their implicit survival strategies—but the consequences of those choices as well. (We will address the specifics, including the levels-of-analysis problem, in chapter 11.)

Obviously, this definition of bioeconomics requires an analytical frame-shift away from the conventional economic focus on tastes, preferences, rational choices, goods and services, gross domestic product, and the like. The phenomena encompassed by these well-established, even venerable concepts very often do not mirror our biological needs, or may go well beyond them. Or worse, they may even be detrimental to our biological needs. Indeed, the interface between the traditional, neoclassical paradigm and a biological perspective is a complex issue in itself, and a number of theorists in recent years have discoursed on this subject (see Corning 1983, and chapter 11; also Gowdy 1985, 1987, 1991, 1994, 2004a, 2004b; Ursprung 1988; Costanza 1989; Ehrlich 1989; Proops 1989; Witt 1991; Rosenberg 1992; Hodgson 1993, 2003b, 2004b; Vedeld 1994).

I will discuss this issue further below, but one major implication should be mentioned at this point. Many theorists stress the relative scarcity of, and competition for resources as a fundamental aspect of economics, or bioeconomics. Ecological competition is certainly a very important facet of the natural world, but it is less than universal; it is not a defining characteristic or a common denominator for living systems. Ecological competition is largely a function of the local supply and demand relationship for needed resources, which can vary greatly from one context to the next. Some species are adept at avoiding direct competition, and some resources are relatively plentiful.

For example, living systems tap on average less than 1 percent of the ambient solar flux (the basic energy source of the biosphere). Moreover, photosynthesis is a highly profitable biotechnology. The productivity of photosynthetic organisms, including even one-celled cyanobacteria, greatly exceeds what is required for meeting minimal energy needs (see chapter 13). Energetic profits, or "expansive energy" in Van Valen's (1976) formulation, have provided the fuel (literally) for the Malthusian population growth assumption that has been one of the pillars of evolutionary theory ever since Darwin's day. In many instances, the limiting factors in biological evolution (after Justus von Liebig's "law of the minimum") have been other needed resources—water, nitrogen, carbon dioxide, tolerable ambient temperature ranges, suitable nesting sites, and a host of other ecological factors. Competition for available energy is most frequent in heterotrophs, although many plant species must also compete for sunlight as a corollary of their competition for other resources. Nevertheless, available energy is not, overall, a scarce resource.

To finesse the problems associated with interpreting the term *scarcity* too literally, some theorists have redefined it to mean only that there is no free lunch in nature or society; there are costs associated with every benefit. Scarcity in this defanged definition refers only to the cost of extracting a needed resource and the necessity for making choices. In these terms, scarcity logically implies only the need for efficiency (the benefits must outweigh the costs) and for husbanding limited time and resources. However, this terminological segue also deflates the theoretical significance of scarcity. Ecological competition can no longer be characterized as an unavoidable imperative—as the basic organizing principle in nature. I happen to agree. I see the basic problem of survival and reproduction as the most important organizing principle and ecological competition as a secondary consequence.

Bio-Logic and Economic Science

In a very real sense, the basic premise of the social sciences during the course of the past century could be considered a null hypothesis. Several generations of our forebears in the social sciences have accepted without question (and many still do) the assertion that mere survival, and the provision of basic needs, is no longer a real problem for humankind, at least not in the so-called developed countries. This despite the fact that over the past century hundreds of millions of people have been left hungry, or in physical deprivation, or dead as a result of two world wars, the Russian and Chinese Revolutions, and the Great Depression, not to mention various lesser tragedies in more recent decades. Indeed, the influenza pandemic of 1918–19 alone is said to have killed more than 21 million people worldwide.

Furthermore, the Food and Agriculture Organization of the United Nations estimates that some 20 percent of the population in the developing and less-developed countries—about 800 million people—are chronically undernourished (Pimentel and Pimentel 1996; Ehrlich 1998). All told, about one-third of humankind suffers from the effects of undernutrition and/or malnutrition (WHO 1995, Combs et al. 1996), even though the world's total population is continuing to grow, if somewhat less rapidly than before (Bongaarts 1994; Smail 1997; Ehrlich 1998). More disturbing is the estimate by the well-known ecologists David and Marcia Pimentel that in the past 40 years almost one-third of the worldwide stock of arable land has been eroded—some of it irretrievably (see also Lal and Stewart 1990; Pimentel et al. 1995)—and that the per capita availability of fresh water (especially for irrigation) has begun to decline as well (Postel 1992; Gleick 1993; Pimentel et al. 1997).

Perhaps most ominous is the fact that increases in worldwide food production, following the boom years of the so-called Green Revolution, are no longer keeping pace with population increases. In 1997, the world food carryover (reserve stocks) was the lowest since 1960. World population has passed 6 billion and is now projected to reach 9.5 billion in 2050. Although reducing the large quantity of food wasted (mostly during storage and transport) offers hope for some significant short-term improvements in the developing and less-developed countries, there are currently no major opportunities available for dramatically increasing the world food supply over the long term (Pimentel and Pimentel 1996; Ehrlich 1998).

As the new millennium dawned, the situation looked ominous. World poverty threatened to overwhelm economic progress; the AIDS epidemic, involving much human suffering and devastating economic impacts, was a major challenge; the threat posed by weapons of mass destruction in the hands of terrorist fanatics emerged as an alarming new menace; and, finally, the problem of global warming loomed as an externality that threatens to disrupt the very basis of the human survival enterprise.

Nevertheless, in the social sciences the concepts of value relativism, cultural relativism, and cultural determinism—along with their co-conspirator, the behaviorist "reinforcement" learning paradigm in psychology—have long prevailed. Some social theorists (most notably the latter-day Marxists) blame human suffering largely on cultural factors, particularly capitalist economic and political institutions, and tend to discount the importance of basic needs per se. Then there are the phenomenologists, who deny that the concept of basic needs can have any external, objective meaning at all apart from the individual's subjective experience. Meanwhile, many other mainstream social scientists have proceeded from the assumption that our basic biological needs are only marginally relevant to social theory and that individual motivation can be treated as a "black box" into which various cultural influences are poured (as pointed out in chapter 9). Our social, economic, and political behaviors are therefore largely shaped by our wants, tastes, revealed preferences, subjective utility functions, and social norms—which are said to be infinitely variable and culturally determined. (For more detailed discussions and critical analyses of this paradigm, see especially Corning 1983; Doyal and Gough 1991; Edgerton 1992; Hodgson 1993; and Gowdy 2004a, 2004b.)

Modern economic science epitomizes this relativist position. Standard economic texts like John B. Taylor's *Economics* (1998) and Baumol and Blinder's *Economics: Principles and Policy* (1991) do not even list the terms *needs* or *basic needs* in their indexes. The closest approach to the subject, albeit obliquely, is the concept of inelastic demand. Economists have observed that the demand for some necessities may not be susceptible to finding ready substitutions, and therefore may not be very sensitive to price changes. Despite its pretensions, the science of economics remains largely descriptive, and its modest repertoire of low-level predictions is derived from correlations, not the underlying causal relationships. For instance, economists estimate that a 10 percent rise in the cost of a restaurant's meals will, on average, result in a 16 percent decline in volume (Baumol and Blinder 1991, p. 472). On average! In other words, it's a statistical artifact with much individual variation. But why? Even the vaunted "laws" of demand and supply are hedged with an all-other-things-being-equal qualifier, and in reality things seldom are equal. So the predictive power of these laws is uneven, and sometimes very poor.

Moreover, the so-called is-ought dichotomy in social theory, dating back to the philosopher David Hume, has long inhibited social scientists from passing value judgments on any given social practice or personal choice. We are told that we cannot deduce an ethical imperative from any empirical circumstance. Economist John C. Harsanyi's (1982) principle of "preference autonomy" (a.k.a. preference utilitarianism) epitomizes this posture: "In deciding what is good and what is bad for a given individual, the ultimate criterion can only be his own wants and his own preferences" (p. 55).[2] Similar assertions can also be found in the literature of anthropology, sociology, and psychology, not to mention social philosophy.

This tacit null hypothesis and its philosophical underpinnings are becoming increasingly untenable. Various developments in the life sciences and the social sciences alike over the past two decades—ranging from behavior genetics and the neurosciences to ecological anthropology and welfare economics—have, in effect, challenged the environmentalist/relativist paradigm and the *tabula rasa* model of human nature. "Economic man" is a caricature.

One important source of evidence for this proposition is the growing body of research on strong reciprocity (altruistic punishment), fairness, and group-serving norms that was noted in chapter 6. (On strong reciprocity, see Gintis 2000a, 2000b; Fehr and Gächter 2000a, 2000b, 2002; Sethi and Somanathan 2001; Fehr et al. 2002; Bowles and Gintis 2004; de Quervain et al. 2004. On fairness, see especially Corning 2002a; also see Kahneman et al. 1986a, 1986b; Rabin 1993; Fehr and Gächter 2000a, 2000b, 2002; Fehr and Schmidt 1999; Henrich and Boyd 2001; Henrich et al. 2001; Price et al. 2002. On norms, see especially Axelrod 1986; Sethi and Somanathan 1996, 2001; Ostrom 2000; Boyd and Richerson 2002; Young 2003; Gowdy 2004a, 2004b; Richerson and Boyd 2004.) Nevertheless, a broad theoretical framework based explicitly on what could be called a *biologic*—the core premise that the biological/survival imperatives mean just that—has lagged behind (but see Galtung 1980; Corning 1983, 2000; Doyal and Gough 1991; Gowdy 2004a, 2004b; and chapter 11).

The Collective Survival Enterprise—A Comparative Example

Conventional economists tend to view a society as, quintessentially, a marketplace for exchanges of goods and services among autonomous, self-interested individuals (and firms) with infinitely varying wants and preferences. In this model, a division of labor and patterns of mutualistic barter and exchange provide the means for individuals to satisfy their wants more efficiently. Indeed, efficiency has been the touchstone for conventional economic analysis. While this characterization, which traces back to Adam Smith's "invisible hand" metaphor (and even to Plato's writings), obviously has much merit, it is insufficient. As noted at the outset, the traditional economic performance measures are not isomorphic with the satisfaction of basic survival and reproductive needs for a population (again, see chapter 11). An organized society can more accurately be viewed as a "survival enterprise," a paradigm that casts a very different light on the relationship between economic performance and our bedrock survival needs.[3]

This metatheoretical shift can be illustrated with a classic anthropological study—a comparative analysis of the survival strategies pursued by four different groups inhabiting essentially the same ecological environment—a section of the Canadian Great Plains in the province of Saskatchewan. The data and analysis were the result of a field study by John W. Bennett (1969). (The study is now dated, but the general conclusions remain valid.)

The four groups that Bennett studied were (1) a small community of Cree Indians living on a reservation, (2) several small communes of Hutterite Brethren (a religious sect), and a number of (3) farmers and (4) ranchers of European stock. Each group brought a quite different set of cultural preadaptations into the so-called Jasper Region of Saskatchewan (a fictitious name), and each exploited the environment in quite different ways. Accordingly, Bennett was able to make systematic comparisons of the four different sets of strategies while, in effect, controlling for the natural environment. In addition, Bennett was able to establish linkages between the survival strategies adopted by each group and its relationships variously to the other Jasper groups, as well as to the external human environment and to different agencies of the Canadian government. In essence, Bennett found there was a subtle interplay and patterns of reciprocal causation in the relationships between the humans, the environment, and the larger economic and political systems.

The Cree Indians

Consider first the 120 Cree Indians, who had been herded by the Canadian Mounties onto 3,000 acres of marginal land. (There is some variation in soil quality, precipitation, mean temperature, etc., within the Jasper Region.) This area is much too small to support so large a human population, and it is not well-suited to farming, although some livestock grazing is possible. But the Cree brought no capital with them for buying livestock or developing the land, and they were outcasts who were permitted to hold only low-wage temporary jobs within the white persons' economic system. In other words, external politics set severe constraints on what resources were available to the Cree and what strategies they could pursue. To make matters worse, the Cree brought with them a set of cultural preadaptations that were unsuited to the new conditions in which they found themselves. Their ancestors had been nomadic bison hunters who maintained a very loose social structure. There was relatively little integration or cooperation, other than among immediate kinship groups.

Despite all of these disadvantages, the Cree followed a highly adaptive strategy—in effect a mixed strategy. At various times they sold aspen trees for fence posts and lumber; they leased their land to local cattlemen for grazing; they picked berries in season and hunted the small game that inhabited their land; they also filled the contract farm labor and service jobs in the Jasper Region when available; finally, they aggressively exploited an opportunity for outside subsidies. They received welfare payments from the Canadian government.

In general, these strategies served to support the Cree population at a minimally adequate level of living. The Cree were not self-sufficient, but they fully exploited the economic opportunities that were available to them. Politically, as of 1969 they had generally remained unorganized, a strategy that was consistent with their lack of political leverage and their dependency status. Whether or not this was the most effective strategy the Cree could pursue is, of course, debatable.

The Hutterites

In contrast with the Cree, the Hutterite communities have developed by far the most effective adaptation to the Jasper Region. In the first place, the Hutterites brought with them a highly integrated, managerial social/political organization large enough to permit economies of scale and a considerable division of labor, and their communities have lasted to the present day. Living in groups of about 150, they maintain very large holdings that permit diversification and optimal use of the somewhat variable soil and water resources. They raise and finish poultry, sheep, and cattle; grow grain, fruit, and vegetables; and supply themselves with various dairy products. They also preserve their fruit and vegetables; butcher meat; make tools, clothes, furniture, and soap; and construct their own dwellings and outbuildings. Thus, with the exception of scrap metals and small outlays for used machinery, fuel, and the like, the Hutterites are almost completely self-sufficient, both socially and economically. In addition, their strong religious and cultural system encourages hard work, frugality, cooperation, and group loyalty; there is almost no out-migration.

But if the Hutterites are aloof from Canadian society in some respects, they are aggressive participants in several other ways. They employ the most modern of agricultural techniques and technology (but at the least possible cost and with the most independence) and they sell cash crops on the Canadian grain and cattle markets to raise capital. The capital is used primarily for the purchase of more land for their growing population. In other words, they play the capitalist game without being fully integrated into Canadian society or dependent on its economy. At the same time, though, the Hutterites affect many economies by practicing egalitarian socialism with respect to their internal socioeconomic organization. There is also a strong internal political system that is actively involved in every major economic and strategic decision. In short, the Hutterites have a seamless, integrated economic, social, and political system.

Evidence of the Hutterites' success is the fact that (as of 1969) their communities had the highest rate of return on invested capital of any group in the Jasper area. They averaged 9 percent, compared to 6 percent for the ranchers and 3 percent for the farmers. They were also able to maintain the highest person-to-land ratio of any of the three self-supporting groups in the region, with about 120 persons per 6,000 acres. This compares with about 50 per 6,000 acres for the farmers and substantially less than that for the ranchers.

Consistent with this overall survival strategy, which was designed to maximize their independence from the Canadian economy, the Hutterites have generally kept a low political profile. They watch the national markets closely, but they are not in a position to influence them. Locally, on the other hand, there is little that they need in the way of services or facilities, so for the most part they do not participate in local politics. On the other hand, to minimize potential antagonism on the part of other Jasperites, the Hutterites have been generous in contributing volunteer labor to assist others, or for the benefit of the community as a whole. This has resulted in exactly the sort of live-and-let-live attitude that the Hutterites desire, and they generally avoid being discriminated against.

Farmers and Ranchers

In between the Cree Indians and the Hutterites, who represent the two extremes, the farmers and ranchers pursued strategies that were significantly divergent and that led to significant differences in lifestyle and political behavior. The farmers initially occupied 180- or 360-acre plots under the Canadian Homestead Act, primarily in family farm units. Not only was the acreage relatively small for so marginal a farming area, but the plots were laid out on a grid pattern without regard to microenvironmental differences of fertility or water resources. Thus, in the early days there were many failures. Over time, the more successful farmers succeeded by aggrandizing their holdings into larger units and diversifying their activities. In recent years, more farmers added cattle grazing and cattle finishing to their cash crops. In addition, they emulated the Hutterites by establishing irrigation cooperatives, grazing cooperatives, and so on. In other words, the farmers made substantial adjustments over time to improve their adaptive strategies.

Like the Hutterites, the farmers brought with them to the Jasper Region a set of cultural preadaptations   farming skills the Canadian market economy and monetary system (including the various goods and services that farmers need in return for their crops), plus the railroads, grain elevators, and the support of the Canadian government in the form of land, loans, irrigation projects, agricultural advice, and so forth. Furthermore, because farming on small family plots was only marginally profitable at best and because the farmers were impelled by the nature of their economic activities to engage in substantial manipulation of the land, there tended to be more emphasis on innovation, more sharing of information, and, over time, more collaboration.

In keeping with these economic imperatives, the farmers placed great emphasis on social life (in 1969 there were 15 churches and 32 clubs in the service town of Jasper, whose population was then about 2,500) and on family life (wives are an economic necessity on a family farm). By the same token, because of the farmers' relatively great dependence on the government, both for services in and around the town of Jasper and for aid with their individual enterprises, they were by far the most active group politically. At various times they engaged in collective action against the railroads, the grain elevator companies, and both provincial and national governments. Moreover, the strategic nature of the crops they produced gave them political leverage far beyond their numbers when it came to obtaining governmental assistance. Over the years they were extremely successful lobbyists.

The ranchers, on the other hand, were relatively more successful economically, and with less outside help. The Jasper Region land is generally more suited for cattle grazing. The ranchers' overhead costs were low (although initial capital inputs were high). They had to undertake only minimal changes to the land to make it productive, and they made a better living out of it, on balance, than the farmers, with less effort and less need to modify or improve their basic strategies through time. The number of ranches remained stable at about ninety for more than fifty years. Accordingly, the ranchers generally adhered to a norm of independence and individualism. They tended to be extremely conservative in their way of doing things, preferring to coast rather than innovate. They also emphasized material acquisitions more and social life less, and because many of them were comfortably well off, they enjoyed high prestige in the community.

Yet, if the ranchers' prestige was high, their political involvement and political clout were not commensurate. This was attributable partly to the fact that their numbers among the Jasperites were relatively smaller. But partly it was due to the fact that the ranchers had less need for governmental services and were preoccupied most intensely with taxes and with beef prices on the national markets. As a result, they, like the Hutterites, tended to be more aloof from politics. They were more active in local politics, however, particularly where projects involving tax money were concerned. On the other hand, the ranchers' involvement was not a function of wealth. If anything, it was the reverse.

The ultimate success of any group survival strategy, or strategies, from a biological/survival perspective can be measured in terms of population size over the longer run (see chapter 11). An appropriate question, therefore, is how well each of these four groups was able to do in maintaining or enhancing its numbers. The answer is revealing. The Jasper Cree Indian population fluctuated around a fairly stable number. The Hutterites, on the other hand, were growing rapidly through time, and their strategy for handling population growth was to divide and spin off new colonies. Parent communities sent out a vertical cross-section of the existing community, including a nucleus of trained people, and provided capital and support for the establishment of a new community, using the same time-tested formula.

The number of Jasper farmers, by contrast, declined over the years, partly due to failures and partly because the income to farmers was too low on the whole to permit them to subsidize their offspring when they came of age, either on the existing land or by establishing a new spread. Thus, in addition to substantial inputs from the national economy, the farmers also depended on the national economy to absorb the out-migration of their offspring. Individual family farming was not sufficiently productive in this region to sustain increases in the population. Of course, this was due partly to farm prices in national markets, as well as to the marginality of the land. But as the Hutterites demonstrated, farm prices and land quality were not the whole story.

Finally, the Jasper ranchers were able to maintain large enough spreads and had sufficient earnings to be able to support two and even three generations, in some cases. Thus, while the ranching population was not growing very rapidly, there was relatively little out-migration.

Although this brief summary cannot do full justice to a very detailed and complex book-length study, it should be evident that there were significant interrelationships between the natural environment, the survival requisites of the four distinct human populations, the cultural/economic and political preadaptations and resources that each brought with them, and the survival strategies—economic and political—adopted by each. Markets played a role, but they were only a part of the story. (For a classic study that provides a similar picture for hunter-gatherer societies, see anthropologist Marshall Sahlins's *Stone Age Economics*, 1972.)

Conclusion

In sum, it is the basic survival problem and its various "components," that defines the underlying purpose of a human society and shapes the strategies that are utilized to cope with the problem. A variety of both market and nonmarket instrumentalities and activities are utilized, as a rule, to satisfy our basic needs, along with various forms of political organization and action. It is an integrated enterprise—a collective survival enterprise—not a neatly compartmentalized set of specialized realms, much less an epiphenomenon of market dynamics.

In the next chapter, a limited effort will be made to operationalize this vision of society with an explicit analytical paradigm. In essence, this will involve a synthesis of three very different concepts and research traditions from three separate disciplines. From biology comes the concept of biological adaptation, which provides the theoretical foundation. From the social sciences, including welfare economics, comes the concept of basic needs, which provides an analytical framework with strong empirical grounding. And from the public policy field comes the methodology and research tools that are associated with the social indicators movement. Together, these three elements are synergistic; they provide a new way of viewing and analyzing economic and social phenomena.

— ℰℬ —

I do not think we have adequately determined the nature and number of the appetites, and until this is accomplished the inquiry will always be confused.

      —Socrates (quoted in Plato, the *Republic*)

SUMMARY: A key concept in biology is *adaptation*—commonly meaning both the functional requisites for survival and reproduction and the specific means that are employed for doing so by a given organism in a given environment. An organism is, quintessentially, a bundle of adaptations. Although the term *adaptation* is also familiar to social scientists, until recently it had been used only selectively, and often very imprecisely. Here a more rigorous and systematic approach to the concept of adaptation is proposed, in terms of basic needs. It is argued that much of our economic and social life (and the motivations behind our revealed preferences and subjective utility assessments), not to mention the actions of modern governments, are either directly or indirectly related to the meeting of our basic survival needs. Furthermore, these needs can be specified to a first approximation and supported empirically to varying degrees, with the obvious caveat that there are major individual and contextual variations in their application. Equally important, complex human societies generate an array of instrumental needs that, as the term implies, serve as intermediaries between our primary needs and the specific economic, cultural, and political contexts within which these needs must be satisfied. An explicit framework of Survival Indicators, including a profile of Personal Fitness and an aggregate index of Population Fitness, is briefly elucidated. Although this framework has been under development for some years, there is still much work to be done and much room for improvement. Finally, I suggest that a basic needs paradigm could provide an analytical tool (a "bio-logic") for examining more closely the relationship between our social, economic, and political behaviors and institutions and their survival consequences, as well as providing a predictive tool of some value.

11

Biological Adaptation in Human Societies:
A Basic Needs Approach

On the Concept of Adaptation

Theodosius Dobzhansky, one of the leading evolutionists of the twentieth century, was fond of characterizing the evolutionary process as a grand experiment in adaptation. And biologist Julian Huxley, in his landmark volume *Evolution: The Modern Synthesis* (1942), defined adaptation as "nothing else than arrangements subserving specialized functions, adjusted to the needs and the mode of life of the species or type . . . . Adaptation cannot but be universal among organisms, and every organism cannot be other than *a bundle of adaptations,* more or less detailed and efficient, coordinated in greater or lesser degree [italics added]" (p. 420).[1]

Adaptations are means to an end; they serve a purpose; they are teleonomic in nature. (As noted earlier, *teleonomy* is a term used in biology to connote evolved purposiveness, as distinct from an externally imposed teleology.) In George C. Williams's (1966) phrase, an adaptation is a "design for survival." Not everything in nature is adaptive, of course. Functional adaptation may be predominant in evolution, but it is not omnipotent. Darwin never took the position that everything in nature is useful, as Stephen Jay Gould and Richard Lewontin (1979) forcefully reminded us. There are also many fortuitous effects, some of which involve nothing more than the operation of the laws of nature. To use one of Williams's illustrations, when a flying fish leaps out of the water, that may well be the result of an adaptation, but its fall back into the water is not. On the other hand, what may be a fortuitous or random effect initially may well become an adaptation, if it persists and enhances the survival chances of the bearer and its progeny—in other words, if it is positively

253

selected. Indeed, the work on self-organizing phenomena, noted in chapter 4, underscores this point.

The assumption of a need for adaptation, then, is nothing more or less than a "bio-logical" deduction from the core premise stated in chapters 9 and 10, namely, that biological survival is an inescapable problem for all living organisms and that they must actively seek to survive if they wish to do so. Richard Lewontin (1978) has written: "The modern view of adaptation is that the external world sets certain 'problems' that organisms need to 'solve,' and that evolution by means of natural selection is the mechanism for creating these solutions." Of course, the evolved *internal* needs and characteristics of an organism also set problems that must be solved. More important, the very definition of what constitutes a problem often has a relational aspect. For example, most plants do not have the problem of locomotion or the need to obtain energy by consuming other plants and animals, although they share with all other species the need for energy. Likewise, fish and people have very different sorts of problems in and out of water.

Adaptation may also be a two-way street. An organism must adapt to its environments (living and nonliving), and in the process environments are often modified, perhaps in ways that in turn influence the organism. As noted earlier, Ehrlich and Raven (1964) coined the term *coevolution* to describe such dynamic interactions, citing as examples the stepwise directional evolution of predator and prey species via successive incremental adaptations to one another. And nowadays there is a growing literature in coevolution theory and niche-construction theory (see chapter 2).

Sloppy Theorizing

There has been much sloppy theorizing about adaptation over the years. Evolutionists often engage in *a priori* reasoning to the effect that there must be an adaptive (functional) explanation for every trait and, conversely, that natural selection can be invoked as an explanation for every biological phenomenon. Gould and Lewontin (1979), in a famous critique, called such reasoning "just so stories," after Rudyard Kipling's fanciful tales. However, John Maynard Smith (1975) points out that *a priori* reasoning is not necessarily wrong and may well be the most efficient way to proceed. Unless one is ready to set aside the core premise that survival and reproduction are the basic problem and to discount the necessity for adaptation (something a field-trained naturalist would view as ivory tower theorizing), then most traits probably evolved in relation to the problems of earning a living, even though they may not currently be optimal or in any way adaptive. For example, the number of known or presumed nonfunctional aspects of human morphology is exceedingly small.

Maynard Smith (1978) notes that it may not be necessary (and might even be considered foolish) to devise ways of testing the obvious—why animals have teeth, or why horses have legs. In such cases we can legitimately reason from a necessary function to be performed to appropriate structures for fulfilling that function, given the core premise. But when there is reason to be suspicious of the obvious explanation, when drift or allometry (nonfunctional correlated changes) might be plausible alternatives, or when the function of a trait or an organ is obscure to us and subject to debate, then experimental tests or evidence should be demanded and ad hoc explanations challenged. (For a discussion of the problems involved, see West-Eberhard 1992.)

Supporting evidence frequently can be found to buttress *a priori* functionalism. For instance, water bugs are normally dark-colored on top and have light-colored bellies, as camouflage against predation from above or below—according to the adaptationist explanation. The exceptions are those water bugs that swim on their backs; as an adaptationist would predict, their color patterns are reversed (Maynard Smith, 1975).

Another example, in human societies, involves some elegant field work described by anthropologist Andrew Vayda (1995). It happens that the Enga people of the New Guinea central highlands cultivate their staple sweet potato crops in large mulch mounds, typically more than half a meter high and three meters in diameter. Although the Enga, according to the researchers' informants, believe that sweet potatoes will not grow in unmounded bare ground, they do not themselves know exactly why the practice of mounding exists. One obvious explanation is that the mounds serve to enhance soil fertility and produce larger yields. However, the mounding practice is not universal in that region. In fact, the most plausible hypothesis is that the mounds serve to protect the sweet potatoes from radiation frost damage, a significant hazard at high altitudes. Careful studies, primarily by Waddell but also by Brookfield, have shown that the spatial distribution of mulch mounds corresponds with the distribution of the frost hazard in that region. Moreover, the size of the mounds and the minimum planting height increase at higher (colder) elevations—a finding that is consistent with the fact that the warmest temperatures are to be found in larger mounds, and at the tops of the mounds. In short, mulch mounding appears to be an unintentionally adaptive cultural practice.

Huxley (1942) suggested that there are three basic categories of adaptations: An organism must be adapted to the inorganic environment, to the organic environment, and to its own internal environment (so to speak). At the time Huxley wrote, no one seems to have objected to the fact that he did not include a fourth category (the sociocultural environment)—that is, socially constructed behavioral constraints, opportunities, tools, information, and other resources that are part of the adaptive environment for any organism that lives in a functionally interdependent social group. In the 1940s the consensus was that culture is a uniquely human invention that sets humankind apart absolutely from other species. However, this was an extreme, ideologically tinged reaction against the nineteenth-century social Darwinists and other advocates of biological determinism, not to mention the apologists for laissez-faire capitalism. Darwin did not accept either extreme separatism or extreme biologism, and he chided his co-discoverer, Alfred Russel Wallace, for exempting the evolution of the human brain from natural selection. Nevertheless, radical separatism came to dominate the social sciences in the twentieth century, as noted earlier.

Culture as an Adaptation

Today many contemporary theorists accept the views that were first developed in Roe and Simpson's *Behavior and Evolution* (1958) and Dobzhansky's *Mankind Evolving* (1962), which stressed the mutual interdependence of human nature and human behavior. It is obvious that there are unique aspects to human cultures. However, most theorists today seem to agree that the sociocultural category of adaptation is not unique to humankind; it is not independent of biological evolution; and it is not free from the biological imperatives. It should properly be added to Huxley's list as a class of biological adaptations. (See especially Wilson 1975; Barash 1977, 1986; Alexander 1979; Bonner 1980; Cavalli-Sforza and Feldman 1981; Corning 1983; Boyd and Richerson 1985; Durham 1991; Smith and Winterhalder 1992; Gowdy 1994, 2004a, 2004b; Whiten et al., 1999; deWaal 1999; Hammerstein 2003). First, many species have the rudiments of culture, at least according to Bonner's reasonable definition (the transfer of information by behavioral means, especially via social learning and teaching). Second, the functional products of culture—organized physical structures and social processes—have survival relevance and may therefore be instrumentalities of natural selection (properly understood). As Bonner writes, culture is "as biological as any other function of an organism, for instance respiration or locomotion" (1980, p. 11).

To be sure, many cultural adaptations in human societies do not involve a direct, conscious pursuit of biological/adaptive ends. These may be the farthest thing from our minds as we struggle with rush-hour traffic, income tax forms, final exams, or deadlines at work. In cultural adaptation, where most of our conscious efforts are focused, biological needs and purposes are often served in oblique and roundabout ways—and may even be ill-served. There is a very imperfect fit between what serves biological adaptation and the processes of sociocultural adaptation. In other words, there are many degrees of freedom and the potential for a disjunction to occur between our cultural practices and their biological/survival consequences. A great many factors—lack of information, bizarre social customs, destructive economic practices, malevolent political forces—may limit or constrain biological adaptation in humankind. If this were not the case, an adaptationist perspective and the traditional social science paradigm would be isomorphic—end of discussion.

Furthermore, human cultures often display a mirror image of biological adaptedness—traits or behaviors that are strictly speaking maladaptive and may significantly lower biological fitness. As noted in chapter 7, this was documented extensively by anthropologist Robert Edgerton in his important study *Sick Societies* (1992). As Edgerton puts it (paraphrasing George Orwell's famous line), "All societies are sick, but some are sicker than others" (p. 1). Even when a population/society as a whole may be reasonably well-adapted, Edgerton notes, there are likely to be some practices or behaviors that are harmful to health, well-being, and reproductive success. This is true both of the "folk societies" studied by anthropologists and of contemporary Western societies. In his extensive and detailed review of the evidence, Edgerton cites the following maladaptive practices, among others: infanticide, torture, wife-beatings, witchcraft, human sacrifice, lethal competition for women, patterns of feuding and revenge, female genital mutilation, female foot binding, rape, homicide, suicide, slavery, drugs, alcoholism, smoking, celibacy, and environmental pollution, not to mention many dysfunctional food and health care practices that increase infant mortality, reduce life-expectancy, and/or lower personal productivity. Some societies, in fact, seem to be systematically maladapted.[2]

Accordingly, biological adaptation (and its antipode, maladaptation) are variables for humankind, just as they are for any other species.[3] Adaptation involves much more than simply "filling our bellies," as one critic of an adaptationist paradigm charged, and even in affluent Western societies the provision of adequate food and shelter are problematical for a significant number of people (Riches 1997). But more to the point, the problem of meeting basic survival and reproductive needs is an imperative for every one of us, whether or not we are aware of it, or care about it. In fact, our biological needs routinely impose themselves on the daily rhythms of our lives. And if our basic needs are not met, there will be significant biological/adaptive consequences, not to mention psychological perturbations. What the value-relativists often overlook is the fact that survival and reproduction are inescapable daily imperatives for all of us. We must actively pursue the meeting of our survival and reproductive needs or we will fail to do so, with predictable consequences. In this light, an economic science that is focused exclusively on the psychology of human preferences and satisfactions, and is studiously indifferent to the bio-logic of adaptation, excludes by fiat a bedrock source of psychological motivation and causation in economic life.

The Problem of Measuring Adaptation

The core analytical challenge, then, is how do we measure adaptation? The ultimate biological criterion of adaptation is Darwinian "fitness". Traditionally, this has been defined as the ability of an individual to produce viable progeny, or of an interbreeding population to reproduce itself. However, in recent years the concept of *inclusive* fitness (the summed proportion of one's own genes shared by close relatives as well as progeny) has been increasingly favored as a more satisfactory measure. In population biology, which dominated evolutionary theorizing during the middle years of the past century, the primary tool used to measure adaptation was (and is) the selection coefficient, a quantitative measure of the *relative* reproductive efficacy of different genotypes in discrete breeding populations (demes). This rigorously analytical approach has been widely used in laboratory and field studies of microevolutionary change. However, the problems involved in applying this approach to the larger evolutionary process, including sociocultural evolution in humankind, are manifold. Only recently have biologists focused on the complex relationship between adaptation at the micro level (individuals) and at higher levels of organization (trait groups, social organizations, demes, species, ecological communities). Yet in dealing with complexly organized species such as humankind, nothing less than a multileveled approach will do. The most important unit of adaptation in humankind must often be defined in relation to units of economic and political organization—that is, units of functional interdependency—that go beyond anything in the rest of nature. By the same token, there has been a growing appreciation in recent years of the complex relationship in humankind between economic, social, psychological, and biological measures of well-being.

These and other limitations in the classical formulation have prompted calls for a less restrictive approach to measuring adaptation in *Homo sapiens* (e.g., Coelho et al. 1974; Hardesty 1977; Durham 1991; Smith and Winterhalder 1992). Various candidates have been proposed. There have been (1) efforts to develop criteria for defining and measuring the optimal population size, (2) attempts to specify in some concrete way the property of adaptability, or flexibility; (3) efforts to measure adaptive functions directly, and (4) applications of bioeconomic analyses, particularly benefit-cost analyses utilizing various proxy currencies (such as time or energy).

Energy-oriented analyses were especially popular in the 1960s and 1970s. Two different approaches were utilized. One, following the lead of anthropologists Leslie White, Marshall Sahlins, Elman Service, and others, stressed the *amount* of energy capture in various cultures. The other, which includes most of the empirical studies done to date, stresses the *efficiency* of energy capture (or the benefit-cost ratios). The shortcoming of this orientation is that energy capture is not the only important adaptive problem. Some of the constraints that have been encountered in energy-resource development, especially environmental constraints, testify to the multidimensional nature of the adaptation problem. From a biological perspective, energy throughputs are a means to the larger end of sustaining and enhancing the overall life process. As noted earlier, a relative scarcity of energy may be a limiting factor in societal development, in conformity with the "law of the minimum," but there are many other limiting factors—protein, for instance, and water, and the basic raw materials that are also absolute requisites for sustaining complex economies.

The "Struggle for Satisfactions"

Accordingly, many theorists believe that we need a more inclusive and multifaceted approach to measuring adaptation. The anthropologist Eugene Ruyle (1973) urged us to concentrate on the "struggle for satisfactions." The psychologist Robert W. White (1974), calling adaptation the master concept of the behavioral and social sciences, applied it to any means-ends, or goal-oriented, behavior (though surely he did not mean to include actions that are biologically maladaptive). Others, especially ecological anthropologists, have adopted an explicitly biological orientation. Donald Hardesty (1977), for example, defined adaptation as "any beneficial response to the environment," and it is clear from the context that he meant *biologically* beneficial. Anthropologist John Bennett (1976) conceptualized adaptation in terms of how human actors realize objectives, meet needs, and cope with conditions. Bennett wished to stress the cognitive/purposive elements in human behavior; he wished to treat adaptation as a goal-oriented process that is embedded in a cultural milieu. But he also made it clear that biological problems lie at the root (see the review of his book-length study in chapter 10). Vayda and McCay (1975) were also concerned with the "existential game" of survival and reproduction. In an article whose objective was to identify "new directions" in ecological anthropology, they argued for an emphasis on "health" and various "hazards" and "stresses."

More recently, the burgeoning new interdisciplines of evolutionary ecology, evolutionary psychology, and evolutionary anthropology have focused on attempting to explain human behaviors in terms of Darwinian adaptation. Thus, the anthropologists Eric Alden Smith and Bruce Winterhalder (1992) stress that adaptation in human cultures involves a "propensity" toward Darwinian fitness, even though it may not reflect a tight fit with the criteria of survival and reproductive success (see also Richerson and Boyd 1992). Meanwhile, the evolutionary psychologists John Tooby and Leda Cosmides (1990, p. 375) take the position that "present conditions and selection pressures are irrelevant to the present design of organisms and do not explain how and why organisms behave adaptively, when they do." Evolutionary psychologists seek to explain present behaviors in terms of postulated "ancestral environments." (The term *environment of evolutionary adaptation,* or EEA, is also frequently employed in this context.) Needless to say, neither of these movements seeks to measure adaptation per se. Rather, they aspire to account for various human behavior patterns in terms of their past or present contribution to adaptation.[4]

The Moral Order

To our knowledge, there have been at least three noteworthy attempts in anthropology to operationalize a broadly defined concept of adaptation. One is Raoul Naroll's *The Moral Order* (1983). Naroll, hoping to initiate a systematic science of cross-cultural evaluation (which he called "socionomics"), produced a data-rich comparative study of adaptation and maladaptation across all human societies. However, Naroll's purpose was not explicitly related to biological adaptation. His main concern was the cultural practices and core social values that support, or undermine, what he called the "moralnet"—the moral and ethical framework that he held to be the foundation of any society. Naroll's agenda was frankly normative. His objective was to develop a set of "indicators" that could monitor the ongoing condition of the moralnet. Although the United Nations, the World Bank, and other agencies publish data on the needs and adaptive problems of various countries, Naroll asserted that there was no "scoreboard" for the overall status of the global moralnet. His goal in developing such a scoreboard was to provide a policy/planning tool for "the creation of a stable human world order," which he called "the deepest historical task of our times" (p. 20). His proposed indicators for monitoring the moralnet included suicide, divorce, child abuse, mental illness, alcoholism, drug abuse and crime, among others. Naroll also developed a summary index of the quality of life in these terms that allowed him to rank the performance of various nations.

Although *The Moral Order* was an impressive effort and a useful source of comparative data on adaptation, from our perspective it ultimately amounts to only a partial view of the overall adaptation problem. It is a tool for assessing one important aspect of biological adaptation in human societies. Strictly from the viewpoint of biological adaptation, the moral and ethical framework of a society is a means (an instrumental need, in our terminology) that serves, or ill-serves, the broader adaptive needs of a society and its members. (For an expanded treatment of this subject, see David Sloan Wilson's *Darwin's Cathedral* 2002.)

Another noteworthy effort to utilize the concept of adaptation in anthropology was the theoretical program of Benjamin Colby and his coworkers, which was concerned with the concept of "adaptive potential" (see Colby et al. 1985; Colby 1987). Colby defined the term *adaptive potential* broadly (it included altruism and creativity, as well as what Colby called "adaptivity"), and it was seen by Colby as a basis for developing predictors of adaptation (he preferred the term well-being), including physical health, satisfaction, and happiness.

Sick Societies

More recently, the concept of adaptation was discussed in some detail by Edgerton in *Sick Societies* (1992), although his primary concern, as noted earlier, was with the antithesis of adaptation—maladaptation. Edgerton noted that the terms *adaptive* and *maladaptive* can have various meanings, depending upon which criteria are used and which level of cultural organization is involved—individuals, families, groups, or societies. By the same token, the causal dynamics of maladaptation are both multileveled and multifaceted. Some forms of maladaptation are the direct result of genetic influences that predispose an individual to poor physical or mental health, ranging from Parkinson's disease and Down syndrome to schizophrenia and manic-depressive psychosis. Other forms of maladaptation involve personal behavioral patterns with significant health or mortality implications—from smoking to extreme diets. Still other forms involve harmful cultural practices—say, unhealthy or highly stressful working conditions. As Jerome Barkow points out in his influential book *Darwin, Sex and Status* (1989), maladaptive cultural traits can also occur when there are environmental changes and the population fails to respond effectively, or when short-sighted ecological practices lead to environmental destruction, or when powerful elites serve their own interests in such a way as to harm others in the community.

Closely related in spirit to these anthropological writings, but very different in its disciplinary focus, is the literature in the field of welfare economics, and especially the work related to the concept of well-being. Although the term *welfare* has a long and distinguished history in economic theory, it has been used in widely varying ways over the years. One tradition is associated with the orthodox neoclassical formulation, which seeks to derive individual and collective well-being from the sum of individual utility functions or subjective satisfactions (see especially the discussions in Sen 1982; Elster and Roemer 1991; Hanley and Spash 1993). Others define welfare in terms of the preferences or goals of some collective entity—an organization, agency, or polity (e.g., Faber and Proops, 1990, who utilize a multilevel approach). Still others have advanced various external criteria, from GNP per capita to average life expectancy (e.g., Streeten 1981).

Jon Elster and John Roemer (1991), introducing the second volume of an important collection of conference papers concerned with interpersonal comparisons of well-being, point out that there are a number of complex issues associated with the concept: (1) How do you define it? (2) How do you validate it? (3) How do you measure it? And (4) how do the analyst's values or goals affect the answers to questions 1–3? (see also Elster and Hylland 1986). Thus, interpersonal comparisons of well-being might be used variously to achieve distributive justice; to establish some intersubjective standard for measuring well-being; or to explain economic behavior when interpersonal comparisons are among the factors that are influencing the actors themselves (i.e., when keeping up with the Joneses is an important motivator). Significantly, many of the participants in the well-being conference objected to the use of any purely subjective measure of psychological satisfaction as a standard, without regard for the objective situation. Two of the contributors to the conference, James Griffin and Thomas M. Scanlon, argued strongly for more "impersonal standards" that are based on widely shared values. Indeed, Scanlon observed that the very process of evaluating well-being is value-laden, no matter what standard is used. Scanlon's preferred alternative was to construct "a more concrete conception of welfare in terms of particular goods and conditions that are recognized as important to a good life even by people with divergent values" (Scanlon 1991, p. 39).[5]

Rawls's Framework

Perhaps the best-known attempt to construct such a framework is philosopher John Rawls's *A Theory of Justice* (1972), which has inspired an enormous critical literature (pro and con). Briefly, Rawls attacked relativistic notions of justice and equity and set out to develop a "universalistic" foundation. Using a highly contrived "thought experiment" that was reminiscent of the social contract theorists, Rawls posited a negotiation process that he claimed could be expected to produce a shared interest in the mutual provision of what he calls "primary goods"—that is, basic "rights and liberties, opportunities and powers, income and wealth" (1972, pp. 92–93). Rawls saw his primary goods as necessary prerequisites to being able to formulate any other life goals and to act upon them. Because all participants in this imaginary negotiation are required by Rawls to come to the bargaining table with a shared understanding about the world— but behind a "veil of ignorance" about their own pre-existing personal interests—the game is actually rigged. Everybody must start out equal in terms of perceived needs and presumed benefits. Rawls calls this the "original position," but it is obviously a very hypothetical construct, which various critics, on both the political left and the right, have attacked (more on Rawls in chapter 17).

Mention should also be made of the distinctly humanistic work on human needs by the Chilean economist Manfred Max-Neef and his coworkers (1989, 1991). Max-Neef postulated nine broad categories of human needs—from "subsistence" through "understanding," "leisure," and "identity" to "freedom"—and developed an analytical framework that involved a matrix of phenomena—"qualities," "things," "actions," and "settings"—that could be related to these needs.

Amartya Sen's Approach

The movement toward objectification of welfare economics was given further impetus by the prolific and important theoretical work of Amartya Sen and various colleagues over the past three decades (see especially Sen 1982, 1985, 1992; also Nussbaum and Sen 1993). In a series of writings that date back to the 1970s, Sen has mounted a major assault on the utilitarian, subjectivist model of well-being. To some extent paralleling and expanding the arguments of Rawls, Sen has challenged the adequacy of various "psychological" formulations of welfare that rest on desires, tastes, subjective utilities, or what have you. Sen charges neoclassical economics with circularity, vacuity, gross oversimplification and the use of psychological premises that are without foundation. Noting, for example, that sympathy and concern for others can also affect a person's welfare, or that individual welfare functions can be interdependent (as highlighted in game theory), or that social commitments may affect behavior, Sen argues that a narrow, materialistic concept of self-interest is not a sufficient definition of behavioral motivation, much less well-being. Furthermore, Sen points out, consistency in making choices is a pretty weak definition of rationality. In one famous passage from his Herbert Spencer Lecture at Oxford University in 1976, entitled "Rational Fools," Sen concludes, "The *purely* economic man is indeed close to being a social moron. Economic theory has been much preoccupied with this rational fool decked in the glory of his *one* all-purpose preference ordering. To make room for the different concepts related to his behavior, we need a more elaborate structure" (1982, p. 99).

Sen does not try to define what the end-state should look like for any given individual, but rather directs our attention to the *means* that are necessary for setting and pursuing personal goals. However, in contrast with Rawls, who was concerned about the "goods" (say food) that are needed to create various "opportunities," Sen focuses on the "capabilities to function"—the nutritional benefits of food versus food per se. Sen describes it as "a particular approach to well-being and advantage in terms of a person's ability to do valuable acts or reach valuable states of being" (Sen 1993, p. 30). In the current political jargon, Sen's focus is on "empowerment" rather than a person's subjective sense of satisfaction, which, as Sen notes, may or may not be concordant. Sen tells us that the functionings that may be relevant for well-being can vary from "elementary" ones, such as escaping mortality, morbidity, or hunger, to more "complex" and subtle conditions, such as achieving self-respect or enjoying social interactions. However, Sen demurs from proposing "just one list of functionings" (quoted in Nussbaum 1988, p. 152).

Sen also addresses the issue of poverty and basic needs in his framework. He speaks of a subset of capabilities that he calls "basic capabilities," and he defines these as "the ability to satisfy certain crucially important functionings up to certain minimally adequate levels" (1993, p. 41). Noting the extensive literature in recent years on the concept of basic needs (see below),

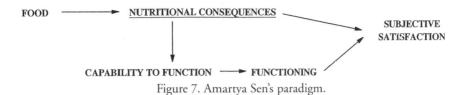

Figure 7. Amartya Sen's paradigm.

Sen argues that the basic capabilities approach is compatible with a basic needs approach and can greatly improve the use of income measures for defining poverty.

Sen's theoretical stance can perhaps be illustrated with the diagram shown in figure 7 (inspired by Doyal and Gough, 1991, but significantly modified).

A final point is that Sen clearly recognizes the concept of basic needs. Indeed, he and various colleagues have been much concerned about such pressing real-world problems as hunger and global poverty (e.g., see Drèze and Sen 1989; Drèze, Sen, and Hussain 1995). And yet, Sen's paradigm does not provide any explicit theoretical basis for his distinction between "capabilities" and "basic capabilities." Like so many other treatments of the concept of basic needs, its status in Sen's paradigm is at once intuitively obvious and theoretically murky. In short, what is missing in Sen's work is a way of grounding the concept of capabilities (requisites) that is both independent and directly measurable. Sen has demurred from elaborating his concepts in more specific detail, so they remain elusive as analytical tools for real-world situations. Sen leaves that task to others. How, then, can we apply and test Sen's concepts? As Scanlon (1993) argues, what is required is a "substantive list" of the elements that are needed to sustain life and make it valuable. Scanlon calls for an "objective index" of well-being that can pass two tests: (1) adequacy and (2) practicality.

Basic Needs and the Social Sciences

Actually, concerted efforts to measure the quality of life more objectively date back at least to the emergence of the so-called social indicators movement in the 1960s. Although the origins of this movement could perhaps be traced to the sociologist William F. Ogburn's *Social Trends* (1929), contemporary researchers generally identify Raymond Bauer's *Social Indicators* (1966) as the catalyst for the more recent and sustained efforts in this area. Following the publication of Bauer's path-breaking book, social indicators research enjoyed a period of rapid, well-funded growth.

Much of the impetus for the creation of a distinct body of data called social indicators arose out of a reaction against our heavy dependence on economic indicators as measuring rods for societal progress or well-being (especially the GDP and per capita income). The goal of the social indicators "idealists," as they were sometimes pejoratively called, was to develop a broad definition of the general welfare that subsumed economic growth and also accounted for various diseconomies, or economic externalities. Perhaps the most frequently quoted statement of this energizing vision (at least in the United States) can be found in *Toward a Social Report* (1970), a benchmark report sponsored by the (then) U.S. Department of Health, Education and Welfare and written principally by economist Mancur Olson.

> A social indicator may be defined to be a statistic of direct normative interest which facilitates concise, comprehensive and balanced judgments about the condition of major aspects of a society. It is in all cases a direct measure of welfare and is subject to the interpretation that, if it changes in the "right" direction, while other things remain equal, things have gotten better or people are "better off." (p. 97)

Just as economists had been able to articulate a theory of economic activity that established a framework for aggregating data on the production and consumption of goods and services, so the social indicators proponents aspired to develop a coherent system of objective measures of well-being. The goal was not just to augment and improve an ad hoc collection of social statistics that already were being gathered for sometimes obscure or narrow purposes. The hope was that it would also be possible to create a comprehensive statistical portrait, a tapestry by means of which we could view a society simultaneously as an integrated whole and in all of its major facets.

Social Indicators

Needless to say, the concept of basic needs has also played an important role in the social indicators movement. In addition to my early work on measuring basic needs (Corning 1970, 1975, 1978), which was little-noticed at the time, there was a study by the Stanford Research Institute (1975) for the U.S. Environmental Protection Agency concerning "Quality of Life Minimums" (QOLMs), which analyzed existing political standards in this area. Other studies include the important work sponsored by the Overseas Development Council on a "Physical Quality of Life Index" (PQLI)

(Morris 1979); the voluminous writings on a basic needs strategy for world development emanating from the World Bank (Streeten 1977, 1979, 1981, 1984; Streeten and Burki 1978; Hicks and Streeten 1979; Streeten et al. 1982); and the manifold efforts of various United Nations agencies since 1975 (see especially the so-called McHale and McHale Report, 1978).[6]

Unfortunately, none of these efforts was rigorously grounded theoretically. All rested on intuitive (albeit often compelling) pragmatic criteria. Although there was considerable overlap among the various attempts to formulate a shopping list of basic human needs, there were also significant differences among them, not surprisingly. Hicks and Streeten (1979), for instance, included nutrition, education, health, sanitation, water, and housing. Geist (1978) included among his basic "normative criteria" for human health the social milieu, education, nutrition, exercise, natural surroundings, and emotional security. Mazess (1975), a specialist in high-altitude peoples, had a physiologically oriented list of nine "adaptive domains" (see also Streeten et al. 1982; Streeten 1984; Miles 1985; Stewart 1985).

Attacks on the social indicators proponents came from many social scientists, who invoked value relativism and claimed that well-being is necessarily a personal and subjective affair. Included in their number were the many workers in the survey research field who, for obvious reasons, had a strong preference for perceptual indicators of well-being. The Survey Research Center's director, Angus Campbell, for instance, noted "the obvious fact" that "individual needs differ greatly from one person to another and that what will satisfy one will be totally unsatisfactory to the other. Indeed, the same individual may find the same circumstances thoroughly unsatisfactory at one stage of his life but quite acceptable at a later stage" (Campbell et al. 1976, p. 9).

Likewise, the sociologist Erik Allardt (1973, pp. 267, 272) asserted that: "A level of need satisfaction defined once and for all has hardly any specific meaning . . . . To a large extent, needs are both created by society and culturally defined, meaning that the satisfaction and frustration of needs have to be studied in a systematic context in which societal feedback processes are considered." Rist (1980, p. 241) was even more dogmatic: "Needs are constructed by the social structure and have no objective content."

In the same vein, the writers of a synthesis volume on the quality of life, published by the U.S. Environmental Protection Agency in the 1970s, claimed that "Quality of life means different things to different people. It can be stated that at the present no consensus exists as to what it is or what it means . . . QOL is viewed by many as not applying to the nation as a whole. In their view, the only way QOL could be applied at the macro-level would be by homogenizing the country and forcing everyone to accept the same value standards"(1973, pp. 1,11).

Finally, advocates for Third World countries attacked the very concept of social indicators as an imperialist tool that was meant to deflate the legitimate economic aspirations of the developing countries and/or deflect attention from the then-popular focus on redistributing wealth between the Northern and Southern Hemispheres (see Miles 1985; Wisner 1988). Still others accused the social indicators advocates of being politically naive. It was not realistic, they claimed, to think that the powers that be, especially in Third World countries, would allow the development and publication of such politically sensitive social outcome statistics.

In their important book on basic needs (see below), Len Doyal and Ian Gough (1991) conclude that "The movement for social indicators and human development appears to have run into the sand. . . . The decline and fall of the social indicators and human development movements was due first and foremost to the lack of a unifying conceptual framework" (p. 154). That was certainly true, but it was only part of the reason. As Nussbaum and Sen (1993, p. 4) point out, "The search for a universally applicable account of the quality of human life has, on its side, the promise of greater power to stand up for the lives of those whom tradition [read economic and political forces] has oppressed or marginalized. But it faces the epistemological difficulty of grounding such an account in an adequate way, saying where the norms come from and how they can be known to be the best." Doyal and Gough agree: "The earlier theoretical innovations . . . all suffer from one overriding defect. None of them demonstrates the universality of their theory, nor, the other side of the same coin, tackles the deeper philosophical questions raised by relativism" (ibid). In short, the search for a satisfactory metric, or measuring rod, for well-being and the quality of life has been severely hampered by the lack of a compelling theoretical foundation.

Basic Needs and Adaptation

We propose that the concept of basic needs can be defined in terms of the biological problem of survival and reproduction. To our knowledge, the first social scientist to espouse a basic needs approach to adaptation in significant detail was the anthropologist Bronislaw Malinowski (1944). For Malinowski, a society is preeminently an organized system of cooperatively pursued activities. It is purposive in nature, and its purposes relate to the satisfaction of basic needs—that is, "the system of conditions in the human organism, in the cultural setting, and in the relation of both to the natural environment, which are sufficient and necessary for the survival of the group and organism" (p. 90).

In contrast to the hyphenated structural-functionalism (so-called) of Comte, Durkheim, and their descendants, Malinowski's "pure functionalism," like Herbert Spencer's before him, was concerned with relating the complexities of cultural behavior to "organic processes in the human body and to those concomitant phases of behavior which in us all desire or drive, emotion or physiological disturbance, and which, for one reason or another, have to be regulated and coordinated by the apparatus of culture" (p. 74). The structure that Malinowski developed for his essentially biological functionalism is reproduced in table 4 in synoptic form.

Malinowski drafted the listing in table 4 only for the sake of simplicity; his textual discussion provides more detailed and more sophisticated treatment. For example, his "health" need has a dual significance. In a narrow sense it refers to the absence of physical impairment or sickness, but in a broader sense it is a condition that is affected by all the other categories (see below). Malinowski also went on to show that these primary needs give rise to a set of "derived" societal needs. (The concept of "instrumental needs," discussed below, is both similar and different.) Malinowski used the fork as an example. Can anyone doubt that the function performed by a fork (a "capability" in Sen's terminology) is a significant part of the explanation for the existence and the design of this commonplace cultural artifact? Yet the fork is not a cultural universal, so more information is needed to account for how the fork was invented and diffused and why it is used in some cultures and not in others.

In light of contemporary anthropological theory (not to mention the technical literature on social indicators), one finds many shortcomings in Malinowski's formulations (see especially the critique in Harris 1968). One

Table 4

Basic needs	Cultural responses
1. Metabolism	1. Commissariat
2. Reproduction	2. Kinship
3. Bodily comforts	3. Shelter
4. Safety	4. Protection
5. Movement	5. Activities
6. Growth	6. Training
7. Health	7. Hygiene

might take exception, for instance, to Malinowski's claim that his basic needs approach was the only valid set of external, or "etic," criteria for cross-cultural classification and comparisons (1944, p. 176). Nevertheless, we believe that his basic approach was sound, indeed essential to a view of human societies that is in touch with the biological fundamentals.

Another major progenitor of the basic needs approach is the humanistic psychologist Abraham Maslow (1954, 1962, 1967). Maslow's famous hierarchy of human needs involved nothing less than a theory of human nature and motivation. According to Maslow, the human being is neither a behavioral sponge (as the Behaviorists implied) nor a tormented neurotic (as some Freudians hold), but a natural innocent endowed with an array of biologically based needs. These needs ascend hierarchically through five categories, from "deficiency motivations" (which derive from such physiological needs as food, water, shelter, sleep, and sex) to "being motivations," to, at the apex, "self-actualization," a kind of beatific state in which one achieves the full use of one's talents and potentialities. Maslow's five categories are (1) physiological needs; (2) safety needs; (3) "belongingness" and love needs; (4) esteem needs; and (5) self-actualization, or "growth" needs (1954, pp. 80ff; cf. Max-Neef 1991).

Despite its popularity among various psychologically oriented social scientists, Maslow's hierarchy per se gained only marginal status among experimental psychologists because it did not have empirical support. Although it has been frequently invoked to justify a particular moral position, or to anchor some proposed model of social behavior, such uses are pseudoscientific. Fitzgerald concludes, "Most psychologists regard the purely empirical study and validation of a hierarchy of needs in Maslow's sense as presenting immense and (perhaps) insurmountable problems. It is clear that insofar as a potentially verifiable aspect can be abstracted from this ambiguous amalgam, Maslow's theory of human needs has not been empirically established to any significant extent" (1977, p. 46). Nevertheless, Maslow's more expansive vision of "human nature" had more than a kernel of truth, as I argue below.

Doyal and Gough's Theory

Another attempt to create a theoretical foundation for the concept of basic needs, and a major contribution to the debate, is Doyal and Gough's book, *A Theory of Human Need* (1991). As stated in their introduction, their goal was a "coherent, rigorous theory of human need. . . . We shall argue that basic needs can be shown to exist, that individuals have a right to the optimal satisfaction of these needs and that all human liberation should be measured by assessing the degree to which such satisfaction has occurred" (pp. 3–4).

Doyal and Gough's theory has a frankly normative aspiration—in their words, to undergird "the moral importance of the needs of individuals" and to support "the maximum development of the individual as a person" (p. 5). They also proclaim themselves to be strong advocates for a "political economy of needs-satisfaction as a constraint on the free play of market forces Although their theory is convergent (and to a degree compatible) with the Survival Indicators paradigm (see below), it also differs in some significant respects, most especially in its theoretical foundation and normative implications. It is important, therefore, to describe and discuss the Doyal and Gough theory very briefly, although we cannot do full justice here to their detailed explication and analyses.

Doyal and Gough begin with a full-dress rebuttal to the neoclassical/relativist attacks on the concept of basic needs. First, they point out that the relativist position is fatally compromised once it is acknowledged that there is such a thing as "perfect knowledge" (an objective external state that transcends the individual's subjective perceptions); or when it is recognized that wants can be manipulated externally and may not reflect a person's true wants; or if it is conceded that market forces may distort a person's real wants. The relativist claim to moral superiority (allegedly because it is the road to greater personal freedom) also leads to a *reductio ad absurdum* unless hedged with externally imposed limits, or constraints. Do our children (or worse, our teenagers) always know what is best for them? Should we indulge the strongly held preferences of rapists, bank robbers, swindlers, and other antisocial actors? In fact, the argument for a moral order as a necessary (objective) constraint and precondition for economic and political freedom goes back to Adam Smith (and to Plato and Aristotle before him).

Doyal and Gough also address the problem of defining basic needs. To be sure, the term is used in many different ways: from psychological motivations or "drives" (*sensu* Maslow), to strictly physiological requisites (food, water, sleep, waste elimination), to any conceivable want or preference whatsoever. Following the lead of philosopher Garrett Thomson (1987) in his thoughtful monograph on the concept of needs, Doyal and Gough argue that the bedrock implication of the term should be that some specific "harm" will occur if the posited need is unfulfilled, whether we are aware of it or not. (Galtung, 1980, advanced a similar idea under the term "disintegration.") Furthermore, some needs are universal. To quote Thomson (1987, p. 27): "Fundamental needs are inescapable; we cannot escape the fact that we must all ail and eventually die without [among other things] food, water, and air." Accordingly, Doyal and Gough focus on "goals which are instrumentally and universally linked to the avoidance of serious harm" (1991, p. 42).

The concept of objective and universal human needs is thus central to their theoretical task. Doyal and Gough argue that (a) our basic needs are equally needed by all (within a clearly bounded range of variation); (b) we are all equally harmed if these needs are not satisfied; (c) it constitutes an injustice if these needs are not fulfilled; (d) our needs take normative precedence over nonessential wants; and (e) most of us do desire the satisfaction of our basic needs. (The latter point is linked by Doyal and Gough to the recent rediscovery of human nature by the social sciences. However, they rightly stress that biological influences shape, but do not determine our choices and behaviors.)

In keeping with their normative agenda, a conspicuous feature of Doyal and Gough's argument is that "harm" (in their terms) refers to the broad concern for human fulfillment—most importantly participation in the life of the community—and not biological adaptation, strictly speaking. Thus, Doyal and Gough remain within the Western, humanistic moral tradition, which supports human aspirations as ends in themselves. A "basic need" in their terms refers to the preconditions for the fulfillment of our "being motivations" (in Maslow's terminology), in addition to bottom-line survival. Indeed, Doyal and Gough cite an array of theorists whose writings are supportive of this viewpoint, including Plato and Aristotle, Kant, Gewirth, Rawls, Habermas, Sen, Thompson, Braybrooke, Dworkin, and others. (Maslow could also be added to their list.) So, in the final analysis, their use of the term *basic needs* overlaps with, and embraces, Maslow's aspiration for human self-actualization and well-being. It is really a theory about well-being disguised as a theory of basic needs.

Accordingly, Doyal and Gough posit two global basic needs. One is "physical health," which encompasses physical survival but means much more to them than mere survival. (They cite the so-called "biomedical model" of health as a reference point, and they claim to be operationalizing the famous WHO definition of health as "a state of complete physical, mental and social well-being, not merely the absence of disease and infirmity.") The second basic need, Doyal and Gough claim, is "autonomy," by which they mean (a) a person's level of "understanding;" (b) his or her psychological capacity to make choices and act upon them; and (c) objective opportunities to act upon these choices, with emphasis on participation in social activities (see figure 8, below). Although they acknowledge wide personal

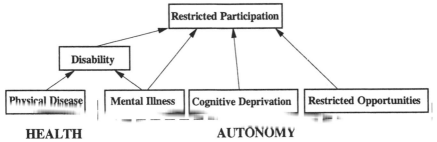

Figure 8. Doyal and Gough's basic needs framework.

and cultural differences, both in perceptions about autonomy and in the forms of expression that autonomy may take, Doyal and Gough insist that meaningful evaluations can be made both within a given culture and comparatively between cultures, in terms of the relative degree of need satisfaction. The concept of "optimum need satisfaction" is universally applicable, they claim, whatever the differences may be in specific cases.

In order to satisfy these two broadly defined basic needs in any given society, Doyal and Gough also posit a set of "intermediate needs," which they see as encompassing the range of specific "need-satisfiers" (a concept similar to Sen's notion of "capabilities"). "Basic needs, then, are always universal but the specific satisfiers are often relative" (1991, p. 155). These satisfiers generally refer to the goods and services provided by the economic, sociocultural, and political systems of a given society. However, embedded in each specific satisfier (say, a particular type of foodstuff) is what Doyal and Gough call their "universal satisfier characteristics" (i.e., the nutritional properties of the food). It is those universal satisfiers (what Sen refers to as "capability characteristics") that Doyal and Gough identify as the basis for their concept of "intermediate needs." To use a concrete example, the specific need-satisfier in the package of snack bars that I currently hold in my hand corresponds to its list of ingredients, but the universal satisfier is the percentage of various daily food values documented in the "Nutrition Facts" table printed on the side of the box.

Doyal and Gough note that there are many different lists of basic needs. As Braybrooke (1987) points out, a large "family of lists of needs" has resulted from the various social indicators projects sponsored by international agencies such as the International Labor Organization (1976), national governments such as Sweden's (Erikson 1993), and private organizations such as the OECD (1976). There is even a consolidated list produced by Braybrooke himself (1987, pp. 33–36). By contrast, Doyal and

Gough claim that their theory provides clarification because it dictates which intermediate needs (read universal satisfiers) are important in any culture for the satisfaction of their two overarching basic needs. "The only claim for inclusion . . . is whether or not any set of satisfier characteristics universally and positively contributes to physical health and autonomy" (p. 158). Their list of eleven intermediate needs includes:

1. Nutritional food and clean water,
2. A nonhazardous work environment,
3. Appropriate health care,
4. Significant primary relationships,
5. Economic security,
6. Safe birth control and childbearing,
7. Protective housing,
8. A nonhazardous physical environment,
9. Security in childhood,
10. Physical security,
11. Appropriate education.

Doyal and Gough then proceed to support their claims with a detailed, two-chapter review of the efforts that have been made by various workers to develop standards and measurement techniques relating both to their own postulated basic needs and to the array of supportive intermediate needs. Their conclusion: "One thing, we hope, is clear. Our theory of human need has a purchase, albeit a tenuous one, on existing evidence of need-satisfaction throughout the world" (p. 221).[7]

One other historical use of the concept of basic needs should also be noted in passing. Even though it has been regularly debunked by cultural relativists, the concept of basic needs has nonetheless played an important political role in the development of the so-called welfare state in Western societies over the past century. Beginning in 1883, when Chancellor Otto von Bismarck established the first "social insurance" program in the new German nation-state, an appeal to basic needs has figured in the development of a broad spectrum of social programs in Western countries. These programs include workers' compensation, public assistance, social security, health insurance, and the minimum wage, among others. The concept was also an explicit element of the New Deal philosophy of Franklin Roosevelt. As FDR put it in one of his famous fireside chats, "One of the duties of the State is that of caring for those of its citizens who find themselves victims of such adverse circumstances as makes them unable to obtain even the necessities of mere existence without the aid of others. That responsibility is recognized by every civilized nation . . ." (quoted in Corning 1969, p. 29). Thus, it seems paradoxical (to say the least) that the concept of basic needs has been regularly invoked in connection with social policy and regularly rejected in social theory.

Survival Indicators

The Survival Indicators approach is not grounded in any ethical concern or public policy objective. It is grounded in the empirical problem of biological survival and reproduction (adaptation) for the human species. It attempts to measure the current status of an individual, or a group, or a population as objectively as possible with respect to this transcendent human concern. It is not about what ought to be—well-being or happiness—but about basic needs, *sensu stricto*. It is addressed to the widespread criticism that the concept of social indicators, particularly in relation to basic needs, lacks a theoretical foundation. It does not seek to promote any desirable political objective, but rather seeks to specify and measure human adaptation as precisely as possible. It does not contradict various ad hoc, pragmatic, or normative approaches, nor is it antagonistic to the concept of well-being. But it does have a distinctive analytical focus that seeks to rationalize and give logical coherence to the effort to measure relevant aspects of the human condition in a systematic way. Finally, the Survival Indicators approach is not designed only for use with the many millions of people who daily experience the deprivation of their basic needs, or who have a genuine anxiety about the problem. It is also designed for use with those fewer among us whose basic needs are so well provided for that we may be complacent, or even oblivious to the problem. To repeat, biological adaptation is a problem that exists for all of us, whether we are aware of it, or care about it, or not.

The Survival Indicators paradigm has its roots in some empirical work that was done in the 1970s on the relationship between income and basic needs-satisfaction for welfare recipients in the State of California (Corning 1970, 1975, 1978). Initial attempts to develop a survival indicators framework and to formulate a master indicator of adaptation called the Population Health Index were presented in my 1983 book called *The Synergism Hypothesis: A Theory of Progressive Evolution.* (It should be stressed, again, that the term *progress* was used in that work with reference only to the evolution of functional complexity, and not in any normative sense.) The Survival Indicators framework is also briefly described in Corning (1996b) and (1997b). The present rendering includes an elaboration and refinement of the original formulation.

In the Survival Indicators paradigm, the term basic need is used in the strict biological/adaptive sense as a requisite for the continued functioning of an organism in a given environmental context; that is, denial of the posited need would significantly reduce the organism's ability to carry on productive activities and/or reduce the probability of its continued survival and successful reproduction. So defined, basic needs are not unique to humans alone; the term applies to all living things. Moreover, we agree with Thomson, Doyal and Gough, and others that the term need connotes a requisite for lack of which significant harm will occur, but we specify the nature of that harm in biological rather than moral terms—that is, in terms of normal functioning and productive activities related to meeting basic needs. This concept is further elaborated below.

Unpacking the Concept of Basic Needs

Several brief comments are in order with regard to this definition. One is that the concept of basic needs is not interpreted in a narrow, physiological sense. It is not just about food and water and waste elimination. Like Doyal and Gough (and Sen, Rawls, Edgerton, Naroll, Maslow, and Malinowski, as well as numerous researchers in the social indicators field), we recognize that human nature, and the very nature of the human survival enterprise, entails cognitive/psychological needs and a need for social relationships of various kinds. But these are not ends in themselves. Most of us are participants in a collective survival enterprise; our needs are satisfied through socially organized activities. More than that, the Survival Indicators paradigm recognizes that basic needs have a life cycle—a trajectory that includes growth and development, reproduction, child nurturance, and aging. The longitudinal dimension of the survival enterprise, often overlooked in other paradigms, is reflected in several of the basic needs domains listed below.

A second point is that the term *basic needs* is used here in both of the senses described earlier under the concept of adaptation, which we have traced back to Huxley (1942). Here adaptation refers both to the functional requisites for survival and reproduction in a given environmental context and to the specific means that may be required to achieve them. For example, we have a physiological need for a defined quantity of uncontaminated fresh water (a "primary need" in our terminology), as well as an "instrumental need" both for a source of fresh water *and* for appropriate technologies to obtain the water and satisfy the primary need—what Sen would call a "capability" and Doyal and Gough would call a "satisfier." We go beyond both of these important conceptualizations by attempting to specify in concrete terms the primary survival and reproductive needs that are served by various capabilities and satisfiers, as well as the linkages between them (the "substantive list" called for by Scanlon, 1993).

A related point is that the Survival Indicators paradigm involves a highly nuanced conceptualization of basic needs. In particular, we attempt to distinguish between (1) *primary needs,* (2) *instrumental needs,* (3) *perceived needs,* (4) *dependencies,* and (5) *wants* (or tastes and preferences). Basic needs refer only to the first two of these categories (primary and instrumental needs). Primary needs are irreducible and nonsubstitutable. One cannot substitute food for water, or sleep for air (small, but is a risk). Primary needs coincide with the broad functional requisites for adaptation. (They include a number of what Doyal and Gough define as "intermediate needs.") Instrumental needs, on the other hand, are the derived adaptive *means* (capabilities or satisfiers). Instrumental needs may be reducible to primary needs (may be subsumed), may be substitutable for various functional equivalents (e.g., beef as a protein source, rather than chicken or eggs), and may vary widely depending on the precise adaptive context (internal, external, or cultural).

Thermoregulation, for example, is a primary human need, but the instrumental needs for clothing, heating fuel, electric power, and/or thermally insulated shelter will vary from one climatic environment and culture to another. Similarly, mobility is an irreducible primary need, but within that category there may be instrumental needs for horses, bicycles, snowshoes, automobiles, or wheelchairs, depending upon the context. In fact, it could be argued that walking, though pan-cultural and biologically primordial in humankind, is not always essential—not a primary need. There are cultural contexts in which various substitutes for walking are available, among them prosthetic devices, the services of others (caretakers and carriers), and transportation and communications technologies that provide functional equivalents. Walking can be viewed as a biologically evolved capability that is derived from a primary need, just as a compact body build or a thick layering of subcutaneous fat may represent physiological adaptations for cold climates that have evolved in relation to our primary need for thermoregulation. By the same token, cultures may significantly alter the adaptive value of various physiological adaptations such as walking—or, for that matter, running. Human technologies may also compensate for various physiological deficiencies such as myopia, or diabetes, or even a defective organ.

It is also important to distinguish between needs and so-called drives, or internal sources of motivation. Needs are functional requisites; drives are psychobiological mechanisms (and various correlates) that we may perceive as needs. Human sexuality involves a drive that we sometimes colloquially call a need, but in reality it is an evolved instrumentality for serving our primary reproductive need. The empirical distinction between the two concepts (need versus drive) is evident both in the practice of birth control and in artificial insemination. By the same token, a person may eat either more or less than is nutritionally necessary in response to the promptings of hunger. Accordingly, in our paradigm the various motivational states (from whatever source), as distinct from basic needs, are categorized under (1) perceived needs, (2) dependencies, and (3) wants.

The litmus test for a *primary need*, according to this formulation, has nothing to do with whether or not the need is reflected in correlative psychobiological motivations (although most are). Nor does it matter that these primary needs vary—as they do in systematic ways that are more or less well understood (see below). Rather, it matters how much they vary, why they do so, and with what consequences. Primary needs (a) vary within a relatively narrow range, (b) are pan-cultural (universal), (c) cannot be substituted for one another or replaced by functional equivalents, (d) are largely independent of our "higher" motivations and the specific environmental and cultural context, and (e) may vary significantly as a result of biologically based individual differences (notably including age) and may vary somewhat less in relation to the environmental and cultural context. But most important, they are causally linked to the potential for doing harm in the strict biological/survival sense.

Instrumental needs serve the primary needs. Some instrumental needs are so pervasive as to be close to primary needs in their importance—for example, exogenous energy, protective shelter, basic utensils and tools, clothing, language skills, and walking. Such instrumental needs are in fact the focal concern of many recent efforts to develop basic needs indicators. In our view, many of the items in these paradigms are not primary needs at all but actually refer to instrumental needs. Our intention here is not to slight instrumental needs or diminish their importance but to categorize them properly with respect to their functional significance for our analytical objective.

It should also be emphasized that instrumental needs can vary widely, depending upon the context. For instance, the instrumental need for a means of waste removal can range from dug latrines to open sewers and the latest high-technology waste treatment plants. Likewise, telephones and automobiles may be of little use in a simple folk society but may constitute a need in the strict sense of the term for people who live in a complex developed society. Also, some instrumental needs take the form of economic goods and services, while others relate to features of the cultural environment. Naroll's "moralnet," for instance, could be viewed as an important instrumental need.

Finally, it is important to recognize that, over the long sweep of our cultural and technological evolution, our various primary needs have generated complex hierarchies of instrumental needs. Our need for mobility, for instance, has resulted not only in the invention of automobiles but in the creation of additional instrumental needs for auto mechanics, paved roads, stop signs, the oil industry, gas stations, highway patrols, traffic courts, and so on. In fact, many inventions have been catalysts for others. If necessity is the mother of invention, as the old saying goes, the reverse is also true: Inventions are the mother of necessity. Moreover, many of our instrumental technologies involve complex networks of economic interdependency. Taking away the tire industry, for example, would cripple the automobile and trucking industries, and very likely devastate our economy.

Dependencies are induced, often non-survival-related needs, some of which may even be destructive (such as an addiction to heroin, alcohol, or sugar; or compulsive gambling; or smoking). *Perceived needs* are those desired objects that the individual thinks he or she needs, regardless of the actual situation. And *wants* reflect the individual's less urgent motivations, goals, and aspirations, very possibly unrelated to any biological requisite. Of course, these categories often overlap. For example, a person's primary nutritional need for protein, carbohydrates, vegetables, and various vitamins and minerals may lead to the selection of a particular instrumental need, or satisfier (say, a Big Mac), which could also become a dependency if the person developed a strong psychological craving for Big Macs, or if Big Macs were the only food available. The person might also accurately perceive Big Macs as an instrumental means and, what's more, might actually enjoy them.

Income as a Surrogate Measure

A special word is in order here regarding the role of income as an instrumentality for the satisfaction of basic needs. Income is often used as a surrogate social indicator, but there are many problems associated with this approach (see Goldstein 1985; Ram 1985; Sen 1985; Doyal and Gough 1991). Sen argues strongly against the use of an income-based measure of well-being. On the other hand, income is also a necessary prerequisite (a means) for meeting basic needs in a great many human societies, as numerous social indicators theorists have recognized (e.g., Selowsky 1981; Ram 1985; Doyal and Gough 1991; Erikson 1993; Erikson et al. 1987). It is therefore highly relevant as an *instrumental need*, even though it is inadequate as a summary measure of primary needs-satisfaction, much less of well-being.

The distinction between our category of instrumental needs and Malinowski's "derived needs" should also be explained. Malinowski's concept refers to the cultural arrangements upon which humans have become dependent—that is, systems of economic cooperation, systems of rules and rule enforcement, educational systems, and political systems (1944, p. 125, *passim*). These derived needs include Radcliffe-Brown's structural functions and Talcott Parsons's functional requisites for social systems. In the Survival Indicators paradigm, by contrast, cultural modalities of various kinds are a subset of the much larger class of instrumental needs, only some of which are based on culture per se.

Likewise, it is important to draw a distinction between our instrumental needs and Doyal and Gough's "intermediate needs." Recall that Doyal and Gough defined intermediate needs in terms of their postulated status as inputs to their two basic needs—physical health and autonomy. The result, from our point of view, is a mixed bag that includes some primary needs (such as food, water, and physical security), some instrumental needs (such as education, health care services, shelter, and a nonhazardous work environment), and some items that we find questionable (such as access to cross-cultural knowledge). At bottom, the distinction between the two paradigms rests on how the concept of basic needs is defined.

Some Preliminary Points

There are several other preliminary points that should be mentioned briefly. First, primary needs vary, but not as much as the relativists imply. Nor are the variations a consequence of personal whim. The obvious case in point is nutritional needs, which are known to vary systematically (and to a substantial degree predictably) as a function of genetic and physical endowment, age, sex, reproductive status, and levels of physical activity. (For a sophisticated model that has been tested with various folk populations, see Leslie et al. 1984.) Indeed, our nutritional needs vary not just in terms of the number of calories but also in relation to a range of required nutrients. Nevertheless, adequate nutrition constitutes a universal primary need.

Second, a complex set of interrelationships exists among the various primary needs; all needs are not equally urgent at all times, and there is an implicit hierarchy. This circumstance greatly affects the organization of our behavioral systems and the patterning of our daily activity cycles. For example, if an individual's life or physical safety were suddenly threatened during a meal, it can be confidently predicted that the person would stop eating. Likewise, we routinely—and at times even mindlessly—interrupt other activities to respond to the promptings of hunger, thirst, fatigue, discomfort or pain, a physical threat, the need for waste elimination, and the like.

Third, there are many interactions among our primary needs; even though they cannot be "reduced" to one another, neither are they entirely independent. For example, communication (information flow) is at once an irreducible primary need and a prerequisite for the satisfaction of other primary needs—nutrition, physical safety, physical health, affective nurturance of the young—not to mention facilitating instrumental needs such as gainful employment. Likewise, waste elimination is a primary need that can also impact our physical health, just as a lack of proper nutrition, sleep, or satisfactory social relationships may affect a person's mental health, or physical health, or both.

Fourth, there are many potential conflicts among our needs. The obvious examples are situations in which physical safety or physical health might have to be jeopardized in order to obtain food or other necessities, or where personal nutrition, health, and safety might have to be sacrificed for the sake of one's offspring. Other things being equal, however, the individuals who are best able to satisfy the entire gamut of primary needs, including those of their progeny, will be better adapted (*sensu stricto*) and more likely to be successful in reproducing well-adapted offspring.

Fifth, harking back to Elster and Roemer's concerns about the validation and measurement of any concept of well-being, it is important to distinguish between the analytical and measurement problems that are associated with determining more precisely what our basic needs are (and their functional relationship to survival and reproduction) and the more "applied" problem of how best to measure needs-satisfaction for the purposes of social intelligence and social indicators. Our knowledge in many basic needs domains is still far from perfect, and we do not underestimate the problems involved in establishing more precise criteria for each need. In some cases, it may be that surrogate measures such as personal income or an individual's perceptual self-assessment might suffice as an indicator, as various researchers contend. But, in the end, any objective measure will be only as good as the state of the art in the biological, behavioral, and social sciences. Accordingly, we must view the Survival Indicators paradigm as a work in progress, not as the actualization of some Platonic ideal.

A further point is that the Survival Indicators paradigm is designed primarily to measure current adaptation. It is not explicitly future-oriented, even though it is certainly relevant to adaptability (*sensu* Colby and others). It is obvious, after all, that one factor in determining the future adaptability of an individual, or a population, is the current level of basic needs-satisfaction. However, the fact remains that the Survival Indicators framework is not

designed to make forecasts. It cannot anticipate such contingencies as automobile accidents, lightning bolts, earthquakes, tsunamis, plagues, wars, or
asteroid strikes—much less the effects of global warming. It can only enable
us to make various if-then predictions.

For instance, the Survival Indicators paradigm can provide some guide-
lines for assessing future survival challenges, helping us to calibrate the full
dimensionality of the problem of future adaptation. We noted earlier that
the world's population is expected to increase to perhaps 9.5 billion by
2050. This projection has served to focus our attention on the problem of
how to effect a major increase in food production (roughly 60 percent), a
formidable task. But this is only a part of the problem of providing for the
basic needs of 9.5 billion people. We must also collectively provide for
a comparable increase in the satisfiers for the entire spectrum of needs:
uncontaminated fresh water, adequate clothing, housing, fuel, waste disposal (and appropriate pollution-control facilities), public health services,
education, and so on, not to mention the vast quantities of such instrumental needs as capital and raw materials of various kinds.

A related point is that a basic needs approach to measuring adaptation
is not the same as an explanation of culture in terms of basic needs (as
Malinowski also insisted). Nor does it follow that every aspect of a cultural
system is adaptive (as Edgerton has shown). A particular item of culture
may be adaptive, neutral, or maladaptive in relation to basic needs. Some
items may be more or less directly related to a particular need. (Following
Maynard Smith's argument, would anyone doubt the adaptive function of
toilets—aside from some playful toddlers?) Other cultural items may be
only indirectly related to basic needs (how do we account for sidewalks, or
umbrellas?). Still other items may be apparently unrelated. (Can anyone
provide an adaptive explanation for television game shows, baseball, or
amusement parks?) Indeed, leisure activities are directly survival-relevant
only for people whose livelihoods depend upon them.

In addition, there are almost always some cultural practices that are
unequivocally maladaptive, as was documented in Edgerton's grim catalog.
Nevertheless, it is proposed that cultural systems do tend to track basic
needs-satisfaction over the course of time, however imperfectly, and that
concern for meeting basic needs (adaptation) is very often the implicit motivator for various individual and collective actions. To the extent that cultural practices are functionally related to the meeting of these needs, they
can be viewed as instrumentalities of human adaptation, whatever may be
our perceptions or, equally important, whatever the precise mix of causal
influences that may have produced such practices in the first place.[8]

For heuristic purposes, we also find it useful to distinguish between the levels of needs-satisfaction required for (1) minimal life support, (2) minimal ability to sustain transgenerational continuity (meaning successful reproduction and the nurturing of the young during the maturational process), and (3) optimal life support (meaning maximally efficient functioning and optimal reproductive output). In general, the analytical focus used in this paradigm is the second category. Minimal life support is well and in some circumstances (say, when there is a short-term crisis such as a drought, a war, an earthquake, or a blizzard), but over a prolonged period of time and for an entire population, minimal life support would be maladaptive in the strict sense; the population would be unable to reproduce itself. Conversely, the concept of optimal life support involves a much greater degree of uncertainty and normative judgments (as evidenced in Doyal and Gough's convoluted treatment). Optimal need-satisfaction, like the concept of well-being, involves criteria that are difficult even to define with any precision. Moreover, in strictly Darwinian/biological terms the notion of optimal success in leaving progeny is problematical. For the most part, evolutionary biologists rely on a relative standard—that is, *differential* reproductive success. Our approach strives to approximate the functional requisites for biological adaptation in human societies, but there are obviously some tradeoffs involved.

Finally, a word is in order regarding the recent surge of interest and research on happiness, both in psychology and in economics. From an economist's perspective, the challenge is to develop objective measures of happiness, independent of the traditional use as "revealed preference" choices in the marketplace, as a sufficient surrogate. As Bruno Frey and Alois Stutzer (2004) put it, happiness is "one of the most important issues in life—if not *the* most important issue" [their italics]. (Among the many recent publications on this subject, see especially Oswald 1997; Diener et al. 1999; Kahneman et al. 1999; Easterlin 2002; Frey and Stutzer 2002; and Camerer et al. 2003.) Although happiness is certainly a worthy subject, we believe that (like Abraham Maslow's concept of self-actualization) it belongs at the top of the biopsychological pyramid. Happiness surveys may well prove to be a useful indirect measure of basic needs-satisfaction, but our argument is that these needs should—and can—be measured directly.

The Survival Indicators Framework

Figure 9 shows our fourteen primary needs domains (so called because several of them have more than one element, or aspect). These represent what are postulated to be the irreducible functional requisites for biological adaptation

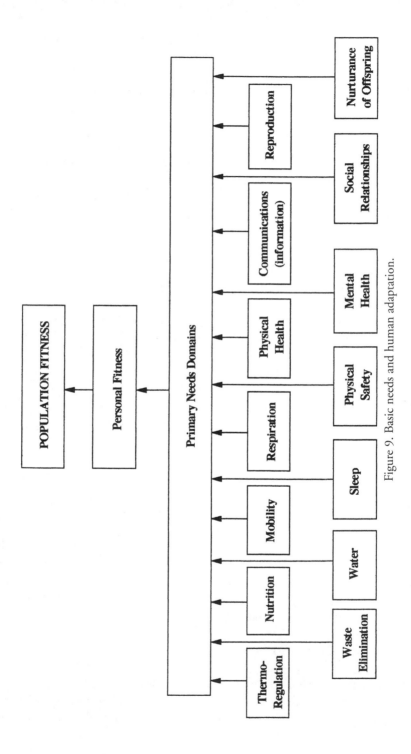

Figure 9. Basic needs and human adaptation.

in the human species. However, these domains are not entirely separate from one another; as noted earlier, there are many interrelationships among the primary needs. We will draw attention to some of these interrelationships in the course of our discussion. It should also be emphasized that our categories are not ad hoc or arbitrary, but neither do they have the status of Mosaic law. They were initially formulated more than a decade ago and, in retrospect, still appear to be valid. (Only one ~~population~~ has been added more recently). Nevertheless, our framework remains open to challenge and revision at any time if more, or fewer, or different categories can be justified. (The order of presentation and the distinction between the top and bottom rows on our diagram are somewhat arbitrary. All of these needs are viewed as being equally important in terms of their relationship to adaptive success, or to harm.)

It follows, therefore, that the outcome state—successful adaptation—is postulated to be a direct consequence of the meeting of these fourteen primary needs. Conversely, the failure to meet any one of these needs will result in varying degrees of harm—a decrement in the ability to engage in normal functioning and the pursuit of productive activities (as defined below). However, these criteria do not fully determine (predict) ultimate reproductive fitness. As noted earlier, the relationship is probabilistic because the satisfaction of these needs cannot guarantee future adaptation. In other words, the satisfaction of our basic needs is necessary, but not sufficient. Other things being equal, however, the chances of future survival and reproductive success should be vastly greater for those whose basic needs are fully satisfied.

Some of these primary needs domains may seem to be self-evident. Many of them can be found on other lists of basic needs. (We are not, after all, venturing into unexplored territory.) Others of our postulated needs may appear to be puzzling or vague (or controversial) and may call for some elaboration. In actuality, there are complications (and ramifications) associated with every one of these needs, some of which we view very differently from more conventional treatments. Accordingly, we will discuss each need in turn, although it will only be possible to provide a brief explanation here.

1. Thermoregulation

Maintenance of our body temperature within a very narrow range is at once a starting point and a prime example of the concept of a primary need. We often take this biological imperative for granted, or respond almost reflexively to various assaults on our internal thermostats. Yet thermoregulation is a critically important need—even though its demands on our time and resources are obviously context-dependent. Thermoregulation presents a very different kind of problem in the tropics or in the midday heat of Death Valley than it does in the arctic, or even between summer and winter or night and day in many locations. The thermoregulation problem also differs significantly for someone who is inactive or asleep versus a runner or cross-country skier in full stride. By the same token, there are well-documented physiological differences between individuals in terms of their comfort levels and their susceptibility to cold and heat and their comfort levels.

Nevertheless, this ongoing, inescapable need (even at times when we are for the moment relatively comfortable) illustrates the fact that our primary needs are objectively important and play a direct causal role in shaping our cultural patterns (and technologies), as well as our economic choices (and satisfactions). A portion of our daily activities typically involves the deployment of instrumental means for thermoregulation. Material adaptations can range from suitable clothing and shelter to the use of personal fans, shade trees, warm bedding, fossil-fuel heating systems, air conditioning, and so on. Relevant cultural practices can include huddling (sometimes even with animals), sharing "bundling beds" (very popular in colonial times), fire-building, going for a swim, or taking a midday siesta. Sometimes thermoregulation may involve taking off items of clothing, or turning off heaters. And when this need is not satisfied, for whatever reason, the consequences can range from a mild disruption of one's normal routine to a high-priority search for relief, or to health-threatening heat strokes, frostbite, or even death from exposure.

How can we measure need-satisfaction with respect to thermoregulation? At the individual level, one obvious way of assessing whether or not a person's need is satisfied is through observation or self-reports concerning the availability and use of appropriate instrumentalities—firewood, heating oil, warm clothing, shelter, blankets, and so forth. However, thermoregulation is also a need that is heavily dependent on instrumentalities that are provided collectively via our social, economic, and political systems. Accordingly, the range of instrumental needs related to thermoregulation in any society may include gainful employment and adequate personal income, the services of the coal or oil industry, public utilities, roads, railroads, merchant marine and trucking facilities, clothing manufacturers, the home construction industry, and much more. In short, much of our economic activity in a complex economy is in fact directly related to the meeting of primary needs such as thermoregulation. This becomes apparent when both the primary need and various indispensable instrumental needs are fully elucidated, and when the linkages between the two are clearly established. We will observe a similar pattern of instrumental needs dependency with respect to our other primary needs, and we will return to this critically important point later on.

2. Waste Elimination

This is another primary need that we often take for granted, even though it involves a non-trivial part of our economic activity. (Waste elimination here refers to the disposal of both bodily waste and the various liquid, solid, and gaseous wastes produced by our personal, social, and economic activities.) Like thermoregulation, this is also a need that is conspicuous for its potential to impact on other primary needs, ranging from physical health to the availability of uncontaminated air and water, sometimes sleep, and even thermoregulation (for anyone who has ever had to visit an outdoor privy in the middle of a cold night). Here we can see again that a primary need has inspired a large number of instrumental needs, especially in complex societies. And here, also, the problem of measurement refers to the ability to provide functionally adequate sanitation facilities at the individual level (by no means a given in many societies, even today), as well as an appropriate infrastructure of services and technologies—from sewer systems to garbage trucks to stack scrubbers. Anyone who has experienced a prolonged garbage strike can testify to the fact that waste removal is an indispensable basic need in a complex society. And this is trivial compared to the health consequences of pollution—as seen at Minimata, Bhopal, the Love Canal, Chernobyl, and the Yellow River (WHO 1992, 1995). As *The New York Times* headlined in a special series on pollution in the Far East: "Across Asia, a Pollution Disaster Hovers" (11/28/97:1). Dzerzhinsk, the heavily polluted site of Russia's chemical weapons production facilities (among other activities), bears witness to the potential consequences. Life expectancies among the inhabitants there have fallen to 42 years for men and 47 for women (Greene 1998). In other words, complex industrial systems have greatly expanded the scope of our primary need for waste elimination, and with it the range of instrumental means that are needed to satisfy it.

3. Nutrition

It is a safe bet that nutrition is included on virtually every social indicators shopping list. Yet, as noted earlier, even an obvious primary need like nutrition has many components, many variables, and even perhaps some remaining unknowns. Appropriate quantities of calories are not enough to satisfy this need, no matter how many may be available to us—a point recently underscored in a report that the Chinese diet is seriously deficient in iodine, which has resulted in a very high incidence of mental retardation in that country. (And this is only the latest example in an age-old litany of nutritional ignorance and its maladaptive consequences. Remember scurvy?) Indeed, malnutrition of one kind or another remains a serious problem in many parts of the world, despite our much better understanding today of what constitutes an adequate diet. Conversely, it is possible to consume too much of a good thing—sugars, fats, and overdoses of certain vitamins being especially notable problems in some developed societies.

Not only does a simple term like *nutrition* mask the complexities involved in providing for this primary need, but it does not even begin to account for the vast human enterprise, and the enormous range of absolutely essential instrumental needs, upon which our nutritional needs also depend. The list includes, among other things: fertile soil, a suitable climate, water, irrigation systems, fertilizers, seeds, tools in great profusion, farm machinery of great complexity, pesticides, animal husbandry, processing and packaging industries and personnel, and transportation and distribution systems. In addition, exogenous energy inputs of various kinds are required to power farm equipment, move water, provide fuel for transportation systems, make fertilizers, process foodstuffs, and, not least, to cook the many foods that would be toxic or infectious if eaten raw. With the exception, perhaps, of the few remaining hunter-gatherers, pastoralists, and subsistence horticulturalists (and even they depend on primitive technologies), the rest of the world's economies depend on a formidable array of food production technologies, and this says nothing about the enormous quantity of information and human skill that is also involved. If there is food on your table (or in your local restaurant) tonight, it is only because a vast human food chain performed its job with only minor glitches.

In this light, it is a bit fatuous to claim, as some theorists do, that the basic need for nutrition is not an important consideration in a developed economy. It is an adaptive modality (in the strict sense) in which vast numbers of people participate worldwide, and it is a relentless daily challenge everywhere (as we are reminded when there is a drought, a freeze, a flood, a hurricane, a famine, a trucking strike, or a contamination of the food supply). Whether or not this vast system can be sustained, much less augmented, for the long term remains to be seen.

4. Water

Any list of social indicators that lumps food and water together can be charged with being a bit cavalier about the distinct challenges associated with providing fresh water resources and the multifaceted use that humankind makes of fresh water (and sometimes also salt water). Our need for uncontaminated drinking water is incessant and often urgent, and so are the water needs of the plants and animals that sustain us. These needs are obvious. But water also serves other primary and instrumental needs, particularly those related to personal hygiene, food processing and preparation, waste disposal, and fire-fighting, not to mention a plethora of important manufacturing processes. Accordingly, the instrumental problems related to water acquisition, storage, transportation, and pollution have played an important role in the evolution of human societies ever since the dawn of civilization, and probably earlier. Our need for water has been responsible for many instrumental technologies over the centuries, including catchment basins, wells, irrigation systems, water containers of various kinds, pumps, hoses, viaducts, baths, sewer systems, dams, desalinization technologies, the soft-drink industry, and more. And again, the problem of meeting this primary need on a daily basis is an ongoing challenge. Fresh water resources are currently diminishing worldwide, and some experts in this area consider fresh water supplies to be the most critical limiting factor for the continued growth of human populations (see Postel 1992; Gleick 1993; Pimentel et al. 1997).

5. Mobility

The ability to purposely change locations is a universal primary need in virtually all species of animals. (To a limited extent, this is also technically true of plants with regard to the problem of dispersing pollen and seeds and the spreading of roots, and sometimes of propagules.) Moreover, mobility is a universal primary need in humankind and at the same time an instrumentality that is critically important to our ability to satisfy other primary needs. It is a need that can easily be taken for granted (it figures in relatively few lists of social indicators), and yet it should not be overlooked. We are reminded of this fact when we observe people who are immobilized—those who are hampered by a variety of congenital defects, or the victims of land mines, or paraplegics who have suffered war injuries or debilitating accidents, or elderly persons with various maladies.

Equally important, mobility is a primary need whose scope has greatly expanded in complex modern societies. Our panoply of transportation technologies—horses, bicycles, automobiles, mass-transit systems, wheelchairs, maybe even sidewalks—are instrumental needs in the strict sense. Many of us would be unable to meet our basic needs without daily access to these technologies. And once again, this fact is driven home to us when the particular technology upon which we may depend fails us for whatever reason—a transportation strike, a snowstorm, a fuel shortage, a malfunctioning automobile, a low battery, a fare increase, and so forth. Although substitutions are frequently available, or changes in lifestyle can be made in order to mitigate the need, the need for mobility itself is, in fact, an incessant human preoccupation. We never solve it; we are compelled to cope with it every day.

6. Sleep

If mobility is often overlooked as a primary need, sleep may seem to be even more problematic—except for the fact that human societies, and human behaviors, are significantly shaped by this need. As noted earlier, we spend approximately one-third of our lives sleeping, and a great deal of economic activity (and inactivity) worldwide is oriented to providing for this need—from work and production schedules to beds and bedding, bedrooms and sleepwear, alarm clocks and sleeping pills, and, of course, hotels, hostels, campgrounds, and sleeping accommodations on planes, trains and ships. Indeed, even wars are fought with due respect for this basic need. Nor is sleep a need whose satisfaction can be taken for granted. A host of factors may interfere with our sleep requirements: work or academic pressures, family stresses, illness, insomnia, jet lag, sleep apnea (snoring), noisy neighbors, night shifts, late-night television, and so on. The consequences of insufficient sleep can range from mild fatigue and loss of efficiency to life-threatening impacts on personal health, or even a fatal accident. And at the risk of belaboring the obvious, it is a biological/adaptive need that can never be solved. In fact, a recent report on sleep deprivation in the United States characterized the problem as "epidemic" in scope. Two national polls and a Congressional study concur that an estimated 70 million Americans have problems getting adequate sleep. More than 20 million suffer from potentially serious sleep apnea (often exacerbated by obesity and a lack of exercise), and another 50 million Americans are afflicted by one (or more) of 80 other recognized sleep disorder syndromes. The estimated loss in national productivity, in dollar terms, may be as high as $70 billion. Not surprisingly, there are now at least 3,000 sleep clinics in the United States. It is a growth industry (Sullivan 1998).

7. Respiration

We added this primary need to our original list after it was pointed out that adequate respiration cannot, after all, be taken for granted, and that instrumental means are often required to satisfy it, especially in modern societies. Respiration, and the provision of an adequate supply of clean, oxygenated air can be a serious problem at high altitudes, in enclosed spaces (mines, submersibles, and modern habitations such as high-rise buildings), during a fire, or when swimming. Suffocation, for various reasons, is a significant cause of accidental deaths each year, and air pollution has become a serious health threat in various localities. Moreover, there is no known substitute available for oxygenated fresh air.

8. Physical Safety

Avoidance of physical injury or death is included on most social indicators lists; its relationship to any definition of basic needs, well-being, or happiness is self-evident. It is also an example of a need that is greatly affected both by internal, personal influences (and behaviors) and by a host of external influences. Personal factors include such things as fatigue, alcohol or drug use, forgetfulness, distractions, errors in judgment, self-inflicted accidents, failure to utilize available information (e.g., not reading the instructions) and, not least, a broad range of deliberate behavioral choices, from engaging in risky sports to flying military jets or participating in criminal activity. External influences on personal safety are equally wide-ranging. To name a few: faulty systems, bad weather, earthquakes, fires, the mistakes of others, personal conflicts, feuds, deliberate acts of criminal violence, government violence, terrorism, and wars.

Clearly, physical safety is an ongoing primary need in any society, and the instrumental means for mitigating threats are almost endless. They range from personal factors such as lifestyle choices, proper education, and corrective lenses to subtleties such as handrails and properly lighted stairs, automobile seat belts, effective police protection, low unemployment rates, a strong "moralnet" (after Naroll), a strong military, and even tranquil foreign relations. Although we often do not connect such far-reaching aspects of our economic and political systems with this primary need, they are nevertheless highly relevant.

9. Physical Health

As noted earlier, this primary need is so obvious that many theorists, including a number of workers in the social indicators field, consider it to be virtually equivalent to a definition of basic needs-satisfaction (and even well-being), or at least a good surrogate indicator (see especially WHO

1980; Caplan et al. 1981; OECD 1985; Doyal and Gough 1991; Brock 1993.) To be sure, physical health is a state that can be affected by many other variables: genetic defects, prenatal assaults, postnatal diet, immunizations, housing conditions, public health measures, health education, lifestyle, family and social relationships, job stresses, and many more. It is certainly highly pertinent to biological adaptation and even to well-being. (Indeed, health researcher Richard Wilkinson, 1996, 2001, has documented extensively the role of chronic stress in health/disease patterns and has linked it to the many economic, social, and political conditions that can contribute to personal stress.) On the other hand, physical health is also a highly labile concept; much depends on how the term *health* is defined and measured. Some theorists define it very broadly, in the spirit of the well-known WHO definition quoted earlier—"a state of complete physical, mental, and social well-being . . ." Consequently, these theorists tend to be expansive in their choice of health indicators. Doyal and Gough (1991), for instance, define physical health in such a way that all eleven of their "intermediate needs" and associated indicators are treated as "inputs". (Doyal and Gough use the same indicators as inputs into their proposed companion need for autonomy and claim that their two basic needs together embrace the WHO definition.)

We prefer to define physical health much more narrowly (and conventionally) as the absence of inborn errors of metabolism (genetic or ontogenetic); the absence of disease, parasites, and other physically debilitating conditions (diarrhea, for example, is a major problem in Third World countries); and such directly health-related variables as muscle tone, cardiovascular conditioning, and personal hygiene. To repeat, we fully appreciate the many other factors that can influence physical health (nutrition, pollution, public order, working conditions, sleep, etc.). However, we have assigned these health-related factors to our other categories of primary needs, and we think properly so. We seek to adhere closely to the principle that each primary need category should contain irreducible, nonredundant elements of the overall adaptation problem. Later on we will develop a more inclusive summary measure of adaptation (originally termed Personal Health but here renamed Personal Fitness to avoid confusion with our concept of physical health) that is designed to serve as a conceptual umbrella for these interrelationships. As suggested in our diagram, personal fitness represents a synthesis of all of our primary needs domains, and it will serve as the basis for our proposed master indicator of population adaptation: the Population Fitness Index.

10. Mental Health

The inclusion of mental health as a primary need might seem questionable to some readers, except for the fact that biological adaptation also implies the capacity of an organism to engage in productive, life-sustaining activity. Mental health is not used here in relation to personal fulfillment, happiness, or a carefree existence. Rather it refers to a state of mind that allows an individual to carry on normal functioning and self-care without significant impairment (harm). There is a very large research literature on various cognitive, mental, even emotional dysfunctions in animals and humans alike (reviewed in depth in Corning 1983). Furthermore, in the case of a complex social animal such as *Homo sapiens*, the concept of mental dysfunction extends to more subtle aspects of individual psychology, such as self-esteem, emotional stability, and social integration and status (or its antipode, social isolation).

In the past, our perspective on this important aspect of human behavior was wracked by polarized attitudes and bitterly competing schools of thought (and methodologies). On the one hand, Sigmund Freud, the founder of psychoanalysis, argued that neuroticism is inherent in society due to the misfit between human nature and the unnatural demands and constraints of civilization (Freud 1961). At the other extreme, skeptics like Thomas Szasz (1961) have argued that mental illness is a "myth"—a syndrome fabricated by therapists and supported by the tendency of a society to label as "sick" any behaviors that are deviant or eccentric (see also Hirst and Woolley 1982). However, in the past twenty years or so a new consensus seems to have emerged to the effect that mental illness (a) is a very real phenomenon, (b) is cross-cultural in nature, (c) takes many different forms, and (d) is affected by a great many different causal factors, both biological and environmental. (Among the many references, see Beck 1967; Foster and Anderson 1978; Clare 1980; Murphy 1982; Naroll 1983; WHO 1983; Gilbert 1984; Busfield 1986; Cohen 1988; Helman 1990; Edgerton 1992). Moreover, the consequences for mental functioning can range from mild anxiety or minor cognitive dysfunctions to total incapacitation and even death.

The list of mental disorders for which biological causes are directly implicated includes, among others, schizophrenia and the schizoid spectrum, manic depressive psychosis, various depressive syndromes, Alzheimer's disease, Parkinson's disease, porphyria, Down syndrome, autism, aphasia (various speech disabilities), alexia (various reading disabilities), some forms of mental retardation, and possibly certain psychopathic behavioral disorders. Of course, many environmental and cultural influences can also greatly affect cognitive, emotional, and behavioral functioning, from childhood experiences (see below) to the many stresses and traumas to which adults are exposed. For instance, there is much evidence that the lack of productive employment can be a cause of serious depression (Beck 1967; Gilbert 1984; Warr 1987), and so can the loss of a loved one, or a divorce, or the loss of a child, or the failure to achieve some desired goal.

Needless to say, the instrumental needs related to mental health are very broad. Some are very personal and derive from the individual's genetic inheritance or family and community relationships (i.e., Naroll's moralnet). Others involve economic or political forces that can have an impact on mental health. But in any case, considerable progress has been made over the past two decades in developing a broad taxonomy of mental disorders, along with better diagnostic tools and better indicators of mental health (Naroll 1983; Beiser 1985; Warr 1987; Helman 1990). Particularly relevant is the American Psychiatric Association's *Diagnostic and Statistical Manual of Mental Disorders.*

11. Communications

The ability to transmit and receive information from the environment, including feedback in the strict sense of the term, is a fundamental property of all complex organisms and an absolutely essential tool in ontogeny, adaptation, and reproduction alike. Moreover, the scope of this need is even greater in a social species, where an individual's behaviors must very often be coordinated with others to accomplish many of the tasks associated with the survival enterprise. Unfortunately, the term *information* is used in many different ways. Potentially it could refer to an indiscriminately vast domain—an unmanageable, and unmeasurable, diversity of processes and technologies. Here we confine our definition to what I call *control information*—that is, *the capacity (know how) to control the acquisition, disposition, and utilization of matter/energy in purposive (teleonomic) processes.* (For a detailed explication of this concept, see chapter 14.) In the present context, control information refers to what an individual needs to know and/or communicate to others in order to effectuate the process of adaptation. It includes the informational requisites for being able to control the activities and events that are related to meeting instrumental adaptive needs in a given situation. These informational needs are also highly context-specific. They may or may not include literacy or higher education, cooking or computer skills, work skills or social skills, depending on the context.

How do we know when harm has occurred with respect to this need? The answer is, when the lack of information or communications skills significantly affects a person's ability to meet his or her other primary needs (to carry on normal functioning). An obvious example is a lack of knowledge about basic hygiene, with the result that the person may be much more susceptible to debilitating diseases or parasites. Likewise, a lack of knowledge about some physical threat (say, poisonous mushrooms) can have fatal consequences. (Another, perhaps familiar example is the experience of being in a foreign country without knowing the language; the pronounced feeling of helplessness may only be relieved when an interlocutor is found who can communicate with you in your own language.)

12. Social Relationships

This is another primary need that could be interpreted so broadly as to encompass virtually all of our interactions with others. To complicate matters, there is increasing evidence that our social needs are critical to the ability to engage in normal functioning and are at the same time an intrinsic psychological motivation, an evolved aspect of human nature. Deficiencies in this domain may also have serious psychobiological consequences for our mental health, our physical health, or both. In addition, social relationships are a primary means of gaining access to information in any social context, as well as providing the cooperative social structure within which many of our survival-related activities are pursued—from production to reproduction. Indeed, our social needs begin at birth. They play a key role in child development, and they remain critically important throughout the maturation process (see below). The challenge associated with evaluating and measuring this primary need, then, is how to narrow its scope in such a way as to focus on its role in adaptation and its potential for doing harm to our capacity for survival and reproductive success. The precise forms that these relationships may take differ significantly from one society to another, but there are certain common elements. These include stable, supportive, caring relationships with parents and/or other closely associated adult caretakers; acceptance by and supportive interactions with peers; and a positive social environment, meaning that the operative social values and goals are not alienating, destructive, or exploitative—that is, harmful.

13. Reproduction

We move now to the first of two primary needs domains that may seem to be problematic and debatable. They are not usually found on lists of basic needs, with the notable exception of Doyal and Gough's treatment. (We have broken the process down into two distinct needs, because they entail distinct challenges and very different kinds of instrumental needs.) Under the heading of reproduction, the focus is on conception and the status of the mother and fetus, along with attendant information, nutrition, health services, and the like. Of course, the birthing process also involves a distinct set of health risks, services, and skills. However, this need also begs the question: In a world where excess population growth may in fact seem to be a threat to our survival—a part of the problem—how can we justify including reproduction and child nurturance among our primary needs? One reason is that it is absolutely essential from a long-term perspective; it happens to be an inescapable part of the struggle for existence. Nature has made reproduction an integral part of the adaptation problem for all living species, like it or not, and a significant portion of our collective activity as a species is devoted to reproduction and its aftermath. Indeed, the world population problem is not a result of reproduction per se but of *too much* reproduction—an excess over what is needed to sustain ourselves, at least at the population and species levels. (We will address the "levels of analysis" problem below.)

A second reason for including reproduction on our list of needs is that it is a very strong felt need and a conscious lifetime preference for vast numbers of us. According to various surveys on this subject over the years, the number of people who actively do not want children at all is rather small, although many more males than females seem to be somewhat indifferent on the subject (Wright 1994). There are also some people who make conscious sacrifices of their reproductive potential, while many more are doomed to be disappointed, or will reproduce as an unpremeditated consequence of following their biopsychological urges. Nevertheless, reproduction is a strongly held human value. However, we are also aware that in recent times, at least, there has been a very imperfect fit between material abundance or wealth and reproductive success (see especially Coale and Watkins 1986; Knauft 1987). In fact, the research in life-history theory suggests that human reproductive strategies are complex and highly variable (MacDonald 1997).

Two major questions are raised by the inclusion of reproduction and child nurturance as primary needs. One is, how do we interpret a failure to reproduce? Does this mean that the individual—male or female—is maladapted?

In a strictly biological sense, the answer is yes (at the individual level), and many people, sad to say, are acutely disappointed if they are unable to produce children. However, this is preeminently a primary need that can also be viewed from an aggregate, population-level perspective. Even if many individuals in a human population do not reproduce, the population as a whole may be well adapted if it is able to reproduce itself over successive generations. Indeed, many nonreproducing individuals may nevertheless contribute in various ways to the successful reproduction of a population. (We will return to this point in relation to our Population Fitness Profile.)

The second question arises out of the fact that a conflict may occur between the primary needs of the parents and those of their offspring. The sociobiological term *parental investment* can involve a zero-sum relationship in which reproduction and child nurturance require a sacrifice of parental needs, and of parental adaptation. How are these tradeoffs to be made commensurate? The answer in strict biological/adaptation terms is that parental sacrifices are likely to be more adaptive if they can be minimized. Indeed, the prolonged period of childhood dependency and the complex nurturing needs of human children put a premium on long-term parental health, competence, and support. In short, the adaptation of parents and their offspring is not easily decoupled. However, if a choice must be made, the biological/adaptive paradigm favors the children—especially if it ensures the "magic number" of 2.1 offspring, on average (the replacement rate).

14. Nurturance of Offspring

From a biological perspective, the last of our primary needs is not merely an afterthought or an easily compartmentalized subtask. It is the culmination of the process of adaptation. In a very real sense, it has to do with the investment we make in our biological future. We are not, of course, referring to our conscious, culturally shaped goals and values, but to the ultimate goal, and the associated imperatives, of the survival enterprise. Accordingly, child nurturance is a primary need that is "more equal" than any of the others (to paraphrase Orwell again). It entails every one of the other primary needs—both for the children and their caretakers—during a period of dependency that may be confined only to the first few years or may persist well into adulthood, depending upon the particular culture and the economic niche that is being occupied within that culture. Not only do young children have the same array of physical and material needs that apply to adults—thermoregulation, food, water, mobility, and so on—but they also have very specific developmental needs—cognitive, social, and emotional needs (and skills) that are increasingly well-understood (see Kellmer-Pringle 1980; Corning 1983; Naroll 1983; Doyal and Gough 1991).

Some of these nurturing needs may be provided for through shared goods (heat and shelter are common examples), and others may be provided for at the margin (investments of time in child-care activities). But other nurturance needs unavoidably require the provision of economic surpluses, or trade-offs. Indeed, outright conflicts may arise between the needs of the next generation and the needs (and wants) of the present generation, and any society (and its political system) that makes *de facto* zero-sum choices in favor of the older generation is maladaptive in the strict sense. Unfortunately, our record (historically) as a species in meeting the need for child nurturance has been uneven. Witness the worldwide data on infanticide, avoidable infant mortality, child neglect and abandonment, and even child exploitation and slavery (see Naroll 1983; MacPherson 1987; UNICEF 1987; Edgerton 1992; World Bank 1996). And this is only the most obvious list of failures. When an affluent society collectively subsidizes services for its elderly population (via Social Security, Medicare, senior housing, etc.) yet fails to provide adequate nutrition, shelter, health care services, and education for many of its children, that is maladaptive in the strict sense. By the same token, when satisfaction of the wants of the present generation involves the overexploitation of a renewable resource, or the production of health-threatening externalities that will impose both economic and public health burdens on future generations, that too is maladaptive in the strict sense.

Indicators of Biological Adaptation

We believe that these fourteen primary needs, and the instrumental means that are required to satisfy them in a given context, provide a solid foundation for evaluating adaptation in human societies in a manner that accords (albeit imperfectly) with Darwinian criteria.[9] Equally important, we believe this framework provides a logical and solidly grounded definition of basic needs for the purpose of social monitoring and welfare economics. We postulate that the denial, or serious deprivation, of any one of these needs will cause significant harm in relation to an individual's chances of continued survival and successful reproduction. Although this perspective is to varying degrees compatible with other paradigms, it has the advantage of providing a rigorous external criterion for defining and measuring basic needs and their satisfaction in any given society.

So how does one go about measuring basic needs-satisfaction? If the problem of defining our basic needs is anything but simple and straightforward, measuring them presents an even greater challenge. In fact, it is not possible to develop an all-purpose set of Survival Indicators. There are a number of different ways of measuring needs-satisfaction, depending on the analytical focus, the analyst's objectives, and various practical data-gathering considerations (recall Elster and Roemer's cautionary remarks earlier in the chapter). In addition, there are many complex measurement and validation issues, especially where instrumental needs are involved. What constitutes adequate shelter, for instance? Or an appropriate level of education and training? Or sufficient income? And who should make the call on these issues—the individual, or some outside "expert" using bureaucratic or technical criteria?

However, none of these problems are *terra incognita*. As indicated earlier, there already exists a very large body of social intelligence—the fruit of many years of research and development by many researchers. A great many ongoing data-collection activities are already in place, and there is currently a broad array of useful social indicators. Particularly notable are the poverty indicators that are published annually by the World Bank (e.g., 1996), which currently cover 191 economies worldwide.[10] Other important sources of data include the Food and Agriculture Organization of the U.N., the World Health Organization (WHO), the United Nations Development Program (UNDP), and a number of programs at the national level, especially in the Scandinavian countries. In the United States, the Departments of Agriculture, Commerce, Education, Health and Human Services, Housing and Urban Development, and Justice, as well as a wide range of nongovernmental agencies, collect data that are relevant to basic needs-satisfaction as we have defined the term here.

The United Nations Development Program is especially noteworthy. The UNDP annually publishes a series of global human development measures, including a Human Development Index (HDI) and a mirror image of it called the Human Poverty Index (HPI). These indexes are severely constrained by the difficulties associated with obtaining relevant data in many countries. Nevertheless, they represent quite useful outcome measures. The HDI measures longevity (life expectancy at birth), knowledge (adult literacy), and the standard of living (GDP per capita). (In 2002 Norway ranked first, while the United States was an embarrassing sixth.) The HPI, by contrast, represents a measure of overall progress (or lack of it) in achieving human development. It measures rates of premature deaths, levels of illiteracy, lack of improved water sources, and the percentage of children under five who are underweight (see the UNDP *Human Development Report*, 2002).

Also important is the so-called community health movement in the United States, along with the parallel work on healthy cities—sponsored by the WHO and supported by a large constituency of city governments around the world—which began in the mid-1980s (Stoto 1992; Hancock 1993). A culmination of sorts occurred with the publication by the U.S. Department of Health and Human Services of the massive "Healthy People 2000" report in 1990. This landmark report was the product of a broad national effort that involved a consortium of nearly 300 national and state agencies and organizations, including the U.S. Institute of Medicine, the National Academy of Sciences, and the U.S. Public Health Service. Among other things, there were inputs from 22 different expert groups, as well as from some 10,000 participants in various hearings and reviews.

The focus of this prodigious effort was to develop an array of national objectives for improving community health, along with the establishment of better measures for monitoring health outcomes. The report identified more than 300 national health objectives, some with multiple parts, and called for the ongoing maintenance of some 400 statistical series, about one-quarter of which did not then exist. (Many other data series required significant improvements.) When this report was published, there already existed some twenty-five different national surveys or data reporting systems that were deemed to be relevant for community health, ranging from the Annual Survey of Occupational Injuries to the Continuing Survey of Food Intake, the National Health and Nutrition Examination, the National Crime Survey, the National Household Survey of Drug Abuse, and the National Nursing Home Survey.

Consequently, many survival-relevant indicators already exist, encompassing many of the basic needs. These statistics include, among others: calorie consumption levels, the incidence of malnutrition, access to safe drinking water, availability of sanitation facilities, poverty levels (using various standards), unemployment, work-related injuries and illnesses, access to health services, immunizations, the incidence of violent crimes, schooling, and such sensitive health-related statistics as infant and maternal mortality and life expectancy. (Unfortunately, the implementation of these measures in the less developed countries has been spotty at best.)

What can the Survival Indicators paradigm add to this ongoing collective effort? First, it provides a framework for ordering and rationalizing various existing indicators in terms of the underlying biological survival problem. Second, it expands the horizon of the existing body of social and health indicators to include some additional areas of concern that are often slighted (e.g., thermoregulation, mobility, and mental health), or that are typically defined and measured in rather narrow terms (e.g., fresh water supplies and sanitation). Indeed, the Survival Indicators approach is concerned also about the health of the economy and the environment—in other words, all of the systems that are survival-relevant. But most important, the Survival Indicators paradigm has a broader and deeper objective than simply to provide better statistics for monitoring the social correlates of economic development, or social well-being, or optimum personal development, or even community health—however important these objectives may be.

The Survival Indicators paradigm addresses a question that many social theorists are not even asking, even though they should be—namely, how are we doing in terms of the basic survival problem? This question, in turn, implies a multileveled, multifaceted approach to measurement. Adaptation is a phenomenon that can be addressed at the individual level or at the population level. It can be directed to the primary needs level or to the provision of instrumental needs, and it can focus either positively on documenting needs-satisfaction or negatively on the evidence of harm—that is, decrements (or failures) in terms of meeting basic needs. At the individual level, we refer to the use of a Personal Fitness Profile and a Personal Fitness Index; at the population level, we use the terms Population Fitness Profile and Population Fitness Index. We briefly describe these alternative approaches below.

A Personal Fitness Profile

This involves direct assessments of an individual's status in relation to the fulfillment of each of his or her primary needs in a given context. The term *Personal Fitness Profile* does not refer to physical fitness, needless to say, but to fitness in the Darwinian sense. It is not equivalent to health, or well-being, or the absence of relative deprivation but focuses pointedly on a person's functional capacities and the resources of various kinds that are needed to support them. We define *Personal Fitness* here as *the capacity to function effectively in relation to the activities that are instrumental to survival, reproduction, and the nurturance of offspring in a given environment; it involves the ability to carry on normal functioning and to engage in productive activity.*

Several points are in order here regarding the concepts of normal functioning and the ability to engage in productive activities. We are not here referring to self-actualization, or optimum human development, or the like. We are referring to the more limited capacity to provide for one's own basic needs—self-care and the ability to engage in whatever daily activities are required for adaptation in a given environment. Personal fitness is not unrelated to a broad definition of personal health, but it is not as expansive as the WHO definition. Also, it should be stressed that the term *normally* is not as vague and imprecise as it may seem. *Normally* means *in accordance with performance norms that can be specified in various ways that are not mutually exclusive*: societal work and productivity standards, medical assessment standards (such as those used by WHO), specific physiological and mental tests (such as those that are routinely used by military and business recruiters and law enforcement officials), and (especially) self-evaluations. In fact the National Center for Health Statistics utilizes similar concepts in its routine surveys of a large sample (well over 100,000) of the U.S. population. Among other things, the survey reports on the number of people whose activities of daily living and instrumental activities (such as employment) have been restricted during the reporting year due to acute illnesses or various chronic conditions. The NCHS defines physical disability as a "reduced ability to perform tasks one would normally do at a given stage of life." However, we use a somewhat expanded definition of normality to encompass the avoidance of restrictions caused by a decrement to any one of the basic needs—lack of sleep, physical danger, serious family conflicts, a lack of gainful employment, and so forth.

How can this outcome state be measured? We believe that it is necessary to couple aggregate data of various kinds, as well as evaluations made by outside observers who are informed by technical knowledge (e.g., nutritional standards and the nutritional content of various diets, or the objective safety risk in a particular environment), with survey protocols that permit self-evaluation in terms of needs-satisfaction. In fact, many surveys of this nature already exist, especially in the health field. In the United States, for example, there are the various Medical Outcomes Studies (MOS), the Sickness Impact Profile, the Duke Health Profile, the McMaster Health Index Questionnaire, and the Quality of Well-Being Scale, among others (Ware 1993).

A major deterrent to the further development of a more comprehensive Personal Fitness Profile is the fact that it would be very expensive to develop and administer to a large population on a continuing basis, especially in Third World countries. And this says nothing about various political constraints. It is probably not realistic for the foreseeable future. Furthermore, many of the data that are collected, including various health surveys, provide only macro-level statistics—the number of doctors per 100,000 persons, or the average number of calories consumed per person in a large population. However, the objective of the Personal Fitness Profile is to make individual assessments. An appropriate analogy might be the distinction between the individual health questionnaires that are administered by insurance agents, military recruiters, or personal physicians and the data that are collected by public health agencies for the purpose of monitoring specific categories of health-related problems.

There are certainly possibilities for doing more fine-grained analyses of basic needs-satisfaction using macro-level data. For instance, inferences are often made about basic needs-satisfaction from the relationship between personal income and the costs for various instrumental goods and services (food, water, energy, shelter, etc.) in a given context. The various poverty-line income measures that have been developed exemplify this approach. (On this issue, see especially Goldstein 1985; Ram 1985; Doyal and Gough 1991.) By the same token, more could be done to evaluate whether or not people utilize their resources efficiently in providing for their basic needs (see especially Streeten 1984). (It has been pointed out that the pawnshops of Reno and Las Vegas are filled with evidence that people do not always use their financial resources wisely.) What the Personal Fitness Profile strategy can add to this process is a more coherent focus, a more comprehensive shopping list (our primary needs framework), and an emphasis on measuring the wherewithal to be able to engage in productive activity.

A Personal Fitness Index

Our proposed Personal Fitness Index utilizes a negative approach. In other words, it is designed to measure degrees of harm, or decrements to normal functioning by a given individual in a given context as a result of deficits in satisfying one or more of the fourteen primary needs—that is, in obtaining the relevant instrumental means. Over the years, many researchers have fixed upon some version of personal health as a surrogate indicator for well-being. One notable example is the so-called Olson Indicator, "Expectation of a Healthy Life," which appeared in *Toward a Social Report* (1970). Another is the State of Health index developed by A. J. Culyer et al. (1972). A third example is the Health Status Index developed by Milton Chen and his coworkers (1975), which utilizes three scales to measure physical activity, mobility, and social activity. Other well-known examples include the Overseas Development Council's Physical Quality of Life (PQLI) index, a composite of indices for infant mortality, life expectancy at age one, and literacy (Morris 1979); and the United Nations Development Program's Human Development Index (HDI), mentioned earlier. (See also the Sickness Impact Profile of Bergner et al. 1976, and the Quality of Life Index of Spitzer et al., 1981.) However, all of these indexes have been criticized. Not only are there numerous problems of definition and measurement, but these constructs lack a coherent theoretical foundation. What do they measure (and not measure), after all?

Doyal and Gough are skeptical about the possibility of developing any summary index of needs-satisfaction. They write: "Though we should not foreclose the search for summary measures of human well-being, the idea of a single indicator (like GNP per head) will probably remain a search for the Holy Grail" (1991, pp. 167–68). Ironically, only three pages later in their text, Doyal and Gough themselves point the way to a possible solution to the problem. Although we disagree with the normative focus of their approach, we share their view that some form of restriction (or harm) can serve as a summary measure of needs-satisfaction. For Doyal and Gough, the focus is on restrictions in the ability of an individual to participate in the life of the community and attain optimum personal development. For us, the concern is with restrictions in an individual's ability to engage in life-sustaining activities, whether alone or interacting with others. Compare Doyal and Gough's graphic representation of their framework (figure 8) with our Survival Indicators framework in figure 9.

Doyal and Gough even suggest an appropriate measuring rod. They reference the WHO *International Classification of Impairments, Disabilities and Handicaps* (1980), but they see this construct as providing only one indirect indicator for their normative objectives. By contrast, we see various forms of physical impairment, from whatever cause, as a direct measure of a shortfall in the desired outcome state. Restrictions in the ability to engage in productive activities can arise from a deficiency in any one of our postulated primary needs domains (although reproduction and child nurturance are obviously special cases). Such impairments could be due to malnutrition, a congenital disease (say rheumatoid arthritis), a disabling accident, a paralyzing mental illness, a dysfunctional family environment, a lack of relevant education or skills, racial or gender discrimination, or even unemployment (especially in the many countries that do not provide a safety net of social insurance). Note also that instrumental needs are fully accounted for in this formulation; they are an integral part of the process that produces either a full satisfaction of the primary needs or some level of deprivation. To illustrate, when a severe ice storm during the El Niño winter of 1997–98 knocked out electrical power to a substantial part of Eastern Canada for several weeks, many thousands of people who depended upon electrical heaters, stoves, and the like, were severely affected and, in many cases, were forced to move to emergency shelters. In other words, there was a significant shortfall for those people in relation to their primary need for thermoregulation, due to a failure in the system of instrumentalities upon which they depended, and this imposed a significant restriction on their normal activities.

Accordingly, we posit that an appropriate personal *fitness unit* could be defined as: *one day free of restriction in the ability to function normally.* We call our fitness unit a "Darwin" (tentatively), and we posit that each person has a theoretical maximum of 365 Darwins per year (366 in leap years)—assuming no functional restrictions of any kind. Our Personal Fitness Index number, then, can be derived from how closely a given individual approximates the theoretical maximum. The simplest method is to multiply the percentage attained by 100 or 1,000. (Of course, partial restrictions of various kinds introduce a number of complications and require various forms of estimation, as described below.) Most of us fall short to varying degrees. Even a person in perfect health who is gainfully employed may suffer from jet lag, a hangover, a bout of the flu, a crushing disappointment at work, grief over the loss of a loved one, bad weather, or a variety of other negative influences. Conversely, even an unhappy slave might come close to achieving the maximum index number; as noted earlier, the Survival Indicators paradigm is not (directly) concerned with personal freedom, or fulfillment, or well-being.

A health assessment tool such as the MOS SF-36 survey (Ware 1993) suggests the possibility of developing and implementing such an index. Each of the eight SF-36 scales measures decrements from normal functioning (although three of the scales—for "general health," "vitality", and "mental health"—are bipolar and, in effect, measure optimum levels as well). Likewise, the U.S. National Center for Health Statistics routinely collects data on disabilities as a part of its annual National Health Interview Survey. (Some of the health indexes cited earlier might also be useful.) Accordingly, it may be possible to develop a scale that would permit more precise, quantitative self-assessments of restrictions to normal functioning, from whatever cause. (Ware, 1993, notes that various attempts have already been made to develop summary indexes using the MOS survey instruments. Apart from the serious methodological problems involved in trying to combine the eight SF-36 scales, Ware points out that they lack a theoretical basis. Although our proposed metric would, we believe, address the theoretical problem, it would introduce new measurement and scaling problems.)

A Population Fitness Profile

This concept provides a framework for the use of aggregate indicators of adaptation, ranging from public health statistics to economic measures relating to the per capita provision of various instrumental resources. Measures of distributive equity would also be highly relevant. Among the obvious candidate measures at the primary needs level are life expectancy, infant and child mortality, accidental deaths and injuries, suicides, violent crimes, diseases, substance abuse data, and pollution levels. Relevant instrumental needs would include such indicators as employment and income data, access to health services, immunizations, shelter, transportation, schooling, public health measures, and many more.

One primary need that was not addressed (above) at the individual level of adaptation is reproduction. Given the fact that individual reproductive output varies widely in any given population, even when the population as a whole may be growing, we believe that this aspect of human adaptation is most appropriately measured at the population level. For a very small population with abundant resources, overall population growth is obviously adaptive. But for large human populations, especially those that are pressing against the limits of their resources, population stability over time is arguably a more adaptive strategy in strict Darwinian terms. This criterion, in turn, implies a bipolar measuring rod; reproduction at the replacement level would be viewed as optimal, and anything either above or below that rate would be less adaptive. (The analogy with Pareto optimality is often invoked in this regard.)

A Population Fitness Index

It may be that the most inclusive and practicable measure of fitness for any given human population will be found at the aggregate level, where statistical sampling techniques and routine bureaucratic reporting procedures provide a more economical means of acquiring the necessary database. In brief, our Population Fitness Index is based on the degree to which a given population falls short of its collective capacity to function normally and engage in productive activity during a given unit of time. Thus, over the period of one year, the maximum number of Darwins available to an entire population would be equal to the size of the population multiplied by 365. (Births during the year would add units to the total stock, just as deaths would deplete it.) Of course, no population ever realizes its maximum potential productivity. Decrements or losses occur through mortality, morbidity, and a plethora of other restrictions to normal daily activity. The Population Fitness Index, then, represents a population-wide summary measure of the actual degree of harm to a given population—the decrement to its aggregate functional capabilities.

Culyer and his colleagues (1972) used a somewhat similar measure in their composite State of Health index. However, they proposed an arbitrary ten-point scale ranging from 0 for "normal" to 10 for "dead." (See also the Population Health Index of Chen, 1975.) Our approach, based on the use of a common measuring unit, allows each death to be quantified in terms of the number of days of productive activity lost during a given period. Similarly, days in fractions of days lost through morbidity and restricted activity can also be quantified. These can then be summed and subtracted from the maximum number of fitness units potentially available to the population as a whole. When this total is divided by the theoretical maximum number, the result represents an overall measure of Population Fitness. We hasten to add that we do not underestimate the many difficulties involved in measuring losses to functional capabilities. Sometimes the effects of some negative influence—insufficient sleep, jet lag, a hangover—may be very subtle and difficult to quantify. Likewise, someone with a severe physical handicap may, with the help of various prostheses and other accommodations, lead a highly productive life with only limited impairment. Nevertheless, as Abraham Maslow put it, "What needs doing is worth doing, even though not very well."

An Illustration

For trial purposes, we offer the following tentative illustration, using the United States as our model population. As of July 1, 1994, according to the 1997 edition of the *Statistical Abstract of the United States,* there were approximately 260,682,000 people living in this country, which can be treated for our purpose as the average (resident) population for the year as a whole. This implies a potential total stock of some 95.15 billion potential person-days of normal activity (or Darwins) for that year. However, on any given day during 1994, a total of about 5.47 million Americans were incarcerated in federal and state prisons (1.05 million), local jails (486,000), mental hospitals of various kinds (516,000), homes for the elderly (1.38 million), nursing homes (1.55 million), and acute-care hospitals (481,000). The total loss of normal activity was therefore about 1.9 billion Darwins. On any given day in that year there were also 11.35 million Americans (about 4.4 percent of the population) who were reported to be restricted by illnesses or other disabilities, representing a loss of about 4.14 billion Darwins for the year. In addition, there were some 4.3 million more Americans each day who required home health care (1.9 million), hospice care (61,000), visits to outpatient or emergency room facilities (438,000), or visits to physicians' offices (1.9 million). Assuming (for estimation purposes) that each reported case represents a total loss of productive activity for that day, the total cost in productivity for the year was about 1.57 billion Darwins.

Much more difficult to estimate, but nonetheless very important, were the productivity losses due to unreported ambulatory illnesses, lack of sleep, stress, obesity, the influence of alcohol or drugs, emotional difficulties, untreated chronic conditions, learning disabilities, injuries, and a variety of other personal disruptions. This number could be nearly half the population by some estimates, but let us conservatively put the figure at about 52 million (20 percent of the population) and assign them an average 20-percent loss of functional efficiency on any given day—a total decrement of 3.8 billion Darwins. Adding up our estimates, we get a total functional loss of about 11.41 billion person-days, or 11.99 percent of the potential stock of Darwins. Thus, the overall Population Fitness Index number for the United States for the year 1994 is estimated to have been 880 out of a possible 1,000. Another way of putting it is that the U.S. population had a loss of about 12 percent of its potential for productive activity in 1994. (Whether or not the remaining 88 percent was in fact used productively is another matter.)

Compared to a sophisticated economic measure like GNP, our Population Fitness Index is admittedly a very crude indicator. Some obvious shortcomings include the fact that much loss of productivity goes unreported, whereas some reported losses are bureaucratic artifacts and are not empirically valid. Estimates of functional decrements based on various categories of disability can be very misleading, especially where compensatory prostheses are available. Some productivity losses are also implicit in the statistics on unemployment, but they were not included because the exact relationship is very uncertain. The functional losses suffered by our homeless population are also difficult to gauge. Conversely, the losses associated with the prison population may be overstated, since many prisoners do engage in various personal and/or prison-related activities. Likewise, visits to doctors' offices may or may not be associated with a loss of functional capacities. Some of these visits involve routine medical examinations. Others may involve a variety of conditions that cause pain or discomfort but do not materially affect a person's functioning. Still others may involve conditions that are already reflected in our other categories of statistics. On the other side of the ledger, many patients endure much more than a single day of lost productivity in connection with a visit to the doctor, whereas those who visit non-physicians (say, chiropractors) or various unorthodox healers are not even included in the statistics.

Nevertheless, we believe that our Population Fitness Index is theoretically sound. It attempts to synthesize and summarize the various sources of harm or interference with our ability to engage in self-care and productive activity. It represents a useful, if imperfect, index of adaptive success for a human population, as well as (in theory) permitting comparisons to be made either between populations or of the same population over the course of time. Furthermore, it is a measure that is susceptible to improvement. As noted earlier, this is a work in progress.

Some Implications

Implicit in the Survival Indicators framework is a major shift in the way economic, social and political phenomena are viewed and analyzed. As suggested above, the ongoing survival/reproduction problem, and the basic needs that are associated with this problem, apply to all societies at all times. Moreover, much of our economic activity is devoted to meeting these needs, even sometimes when we label the instrumental consumer products luxury items. Fur coats, after all, do serve a primary human need; they also keep the wearer warm. (Of course, many substitutes for fur coats are available today, but for some of our remote ancestors living in high latitudes, or high altitudes, fur coats were nonsubstitutable instrumental needs.) In a similar vein, king-size beds enable us to satisfy our primary need for sleep, even though less imposing accommodations may serve just as well.

From a biological perspective, our primary needs provide the inner logic (the bio-logic) of economic life. They represent the agenda that implicitly guides our economies, and it is possible to view all of economic, social, and political life in terms of their relationship (if any) to the survival imperatives. As we have suggested, much of our economic activity is in fact instrumental to our survival; it is either directly or indirectly related to the satisfaction of our biological needs. To be sure, some economic activity is very tangential or not at all related. In fact, some activities are destructive to our adaptive needs, to reiterate Edgerton's thesis. Smoking and hard drugs are obvious examples, but so is almost any other activity that is carried to extremes—for the simple reason that our survival and reproductive needs are manifold. If we satisfy any one of these needs to excess, we may well jeopardize other needs. (For a book-length treatment of this issue, see Frances Ashcroft's *Life at the Extremes*, 2002.)

Many insights about economic, social and political life may be gained by viewing them from an adaptation perspective. For instance, it might shed further light on such traditional economic concepts as discretionary income, demand elasticity, and the logic of substitutability. But more important, our biological needs create economic imperatives that allow us to formulate many if-then predictions about our economic choices and behaviors. Many of these predictions already make intuitive sense to us. For instance, we can predict in general (but not in every detail) what would happen if the water supply for a major metropolitan area (say the reservoirs that serve San Francisco) were to be suddenly, irreversibly contaminated. Likewise, we can make predictions at the individual level about how a person's priorities will change as a consequence of the prolonged deprivation of any one of their primary needs (excepting, possibly, reproduction and child nurturance).

As a thought experiment, imagine how difficult it would be to continue working, or studying, in the context of an extended denial of such primary needs as sleep, food, water, waste elimination, or heat (on a very cold day). Similarly, an immediate physical threat is likely to interrupt whatever else we are doing. These things happen often enough, and they produce predictable consequences. Moreover, most of the world's people spend the vast majority of their available time and energy engaged in activities that are directly or indirectly related to satisfying basic needs. (A small-scale survey of time use by Americans some years ago suggested that the same is true in the developed countries as well. See Corning 1979.) To deny the relevance of our primary biological needs is to deny reality.

One of the major challenges for bioeconomics, then, is to utilize the biological/adaptation perspective as an analytical paradigm. This, in turn, implies a need to revise our basic assumptions about the underlying purpose of human societies and the consequences of economic life. To repeat, an organized society may be viewed as, quintessentially, a "collective survival enterprise." The bulk of our economic activities and processes are related to meeting our basic survival needs. Moreover, the functional interdependencies that exist within any complex economy are both profound and inescapable—and contingent. Such a paradigm shift presents an important theoretical opportunity. But more important, at this critical juncture in our evolution as a species it is also an increasingly urgent moral imperative.[11] If bioeconomics as I define it is a subversive science, it is also a science that, I believe, grounds economics more firmly in the biological realities of the human condition.

PART III

From Thermodynamics and Information
Theory to Thermoeconomics and Control
Information

— ∾ —

When I use a word, it means what I choose it to mean—neither more nor less.
—Humpty Dumpty (in Lewis Carroll, *Through the Looking Glass*)

SUMMARY: Our whimsical title reflects our dismay over the rampant confusion regarding the use of key concepts from thermodynamics and information theory in various disciplines, but especially in relation to theories of biological evolution. After a brief introduction to this challenging literature, we begin by drawing a critically important distinction between *order* and the *informed functional organization* that characterizes living systems. We then outline what we believe is the appropriate paradigm for theorizing about the role of energy and information in biological processes; in essence, our paradigm is cybernetic. This is followed by a brief discussion of thermodynamics, with particular reference to its application to living systems. Two concepts that are well developed in the engineering literature but not commonly used elsewhere provide an approach that we believe is both more rigorous and more readily understood, namely the control volume frame of reference and the concept of available energy. In this chapter we define both of these concepts in precise mathematical terms. We also critique some of the misuses of thermodynamic concepts. (This chapter is based on a paper co-authored by the late Stephen Jay Kline, Woodard Professor of Science, Technology and Society and Professor of Mechanical Engineering, Emeritus, at Stanford University.)

12

To Be or Entropy: Thermodynamics, Information, and Life Revisited

Introduction

Entropy may fairly be called one of the great buzzwords of twentieth-century science. The very abstractness and obscurity of the term evokes in laymen an aura of mystery and arcane knowledge. But more important, the scientific law that is associated with the concept (the second law of thermodynamics) has long been treated with special reverence as one of the fundamental principles of the natural world. Indeed, entropy has often been portrayed as a dark force that somehow governs the fate of our species and dooms our progeny to oblivion—in the eventual "heat death" of the universe.

The practice of making such cosmic claims for entropy dates back to Rudolph Clausius, the physicist who originally coined the term. (He derived it from the ancient Greek word for "transformation.") In *Abhandlungen über die Mechanische Wärmetheorie* (1864), Clausius wrote: "The energy of the universe is constant; the entropy of the universe tends towards a maximum" (quoted in Harold 1986). Clausius also coined the term *heat death* (*Wärmetod*).

This dour vision has long since become the conventional wisdom of the western scientific establishment. Over the course of the past 130-odd years, it has been echoed by countless other theorists (see, for instance, Lotka 1922; Bridgman 1941; Schrödinger 1945; Shannon and Weaver 1949; von Bertalanffy 1952; Koestler 1967; Morowitz 1968; Lehninger 1971; Georgescu-Roegen 1971; Miller 1995; Riedl 1978; Wicken 1987; Weber et al. 1988.) Even the so-called far-from-equilibrium "open" thermodynamic systems—often associated with living organisms—are said by some to be only local exceptions to the cosmic primacy of the second law.

313

Unfortunately, many people are confused about what entropy is. In a major article on thermodynamics published in the journal *Science* back in 1972, chemist Witold Brostow bemoaned the many loose and inaccurate renderings of the second law that he had found in the scientific literature, including even in college-level textbooks on the subject. However, Brostow's article had no detectable influence. In the intervening years the problem has become much worse. (Mario Bunge, 1986, claimed to have counted at least twenty different renderings of the second law, and we have encountered at least a dozen definitions of entropy.) Entropy is now a household word for any kind of disorder, disorganization, uncertainty, waste, confusion, inefficiency and, most flagrantly, willful sabotage. Entire best-selling books have been devoted to exploring the (supposed) philosophical, ideological, economic, even social and psychological implications of the second law (see especially Rifkin 1980; see also Georgescu-Roegen 1971; Chase 1985; Swenson and Turvey 1991; Schneider and Sagan 2005).

Inevitably, the confusion that has plagued various interpretations of the second law over the years has also infected the disciplines of information theory, economics, and biology, where terms such as *entropy, negentropy,* and even the concept of information are used in a bewildering variety of ways. Worse yet, the various attempts to meld thermodynamics and information theory, not to mention the recent efforts to apply these paradigms to the explanation of biological evolution, have only served to thicken the already dense theoretical fog. The story told by Tribus and McIrvine (1971, p. 122) about how the pioneer information theorist Claude Shannon came to call his formal theoretical function "entropy" is revealing in this regard. Shannon was thinking of calling it simply "uncertainty," but mathematician John von Neumann suggested that he use the term *entropy* instead, first because it was being used for a similar function in statistical mechanics, but "more important, no one knows what entropy really is, so in a debate you will always have the advantage."

Rampant Confusion

Such imprecision has resulted in rampant confusion in various disciplines, but especially in biology. To cite a representative example: Biophysicist Rupert Riedl, in a major theoretical work on *Order in Living Organisms* (1978), noted that information has been defined as being "equivalent" both to entropy and its antipode, negentropy (see also Brillouin 1962; Wilson 1968). Riedl's view of this contradiction was that both versions may be correct. His proposed solution to this conundrum was that the appropriate definition should depend on how the terms are being used and what kind of information is being referred to. The problem with this ecumenical approach, however, is that information cannot be equivalent to two terms that are diametrically opposed to one another without corrupting the concept of information.

In the same vein, biologist Jeffrey Wicken in his book *Evolution, Thermodynamics and Information* (1987) initially adopts Shannon's concept of information, a formulation that refers to certain statistical and quantitative properties associated with the messages that are transmitted in formal communications systems. Then, in an acknowledged theoretical segue, Wicken proceeds to deploy the concept of information as a *causal agency* in biological evolution. In order to do so, however, Wicken must shift to using a *functional* definition of information as an evolved, purposive artifact, a definition that more nearly accords with our common-sense understanding of the term. Wicken advances the notion that organisms are "informed thermodynamic systems," although he demurs from addressing the unresolved challenge of how to measure functional information empirically. He characterizes it as "a very perilous enterprise. . . . We aren't even close to knowing how to quantify it" (pp. 27–28).

The work of biologists Daniel Brooks and E. O. Wiley, especially *Evolution as Entropy* (1988), should also be noted. Here evolution is characterized as, quintessentially, "a thermodynamic process" that takes place in "phase spaces" (an abstract mathematical construct) where the vast complexities of the real world are reduced to a limited "constraint." What Brooks and Wiley propose, in essence, is another in the current genre of anti-Darwinian, structuralist theories of evolution, with the singular feature that the postulated autocatalytic trend toward increased complexity (that is, unaided by any external catalyst or other agency) is a result of entropy. Indeed, the authors speak of having identified a "natural law of history" (p. xiii). In this vision, the process of biological complexification is said to be associated with an expansion of the informational phase space, which in turn expands the theoretical domain of statistical uncertainty (entropy) in biological information. In other words, what Brooks and Wiley have done is to focus on the statistical properties of information (Shannon's entropy) rather than thermodynamic (energetic) entropy, and they have blurred the distinction between the two.

In Winston Churchill's phrase, the daunting "terminological inexactitude" in this domain—and its theoretical consequences—have become a serious impediment, the cause of much imprecision and misinterpretation. As the Roman statesman/philosopher Seneca observed almost two thousand

years ago (rendered here in compressed form): "When the words are confused, the mind is also." Accordingly, we propose to take arms against this
sea of troubles (to borrow a line from Hamlet's soliloquy). In this chapter,
we begin by drawing a critically important distinction between physical
order of various kinds and the functional organization that characterizes living systems. We then outline what we believe is the appropriate paradigm
for theorizing about the role of energy and information in biological processes. In essence, our paradigm is cybernetic. This is followed by a brief discussion of thermodynamics, with particular reference to its application in
biological processes. The discussion includes a critical review of how the
concepts of entropy and negentropy have been utilized in the life sciences.

In the next two chapters, we discuss what we refer to as the thermoeconomics of living systems—that is, an economic and cybernetic approach to
analyzing the role of energy in biological evolution—and we relate this paradigm to a distinction between various statistical, structural, and logical
properties of "Shannon information" and what we call cybernetic control
information. Finally, we suggest how control information can be measured
empirically; we propose a methodology for linking thermodynamics and
information theory that contrasts sharply with the existing approaches to
this problem. Some implications are also explored.

Order versus Organization in Biological Processes

In no small measure, the confusion that pervades the theoretical literature
on thermodynamics, information and evolution can be traced to a more
fundamental metatheoretical issue, namely: What do various theorists mean
by "order/disorder" and how do they relate these concepts to biological
organization, which is characterized by the property of evolved functional
design (teleonomy)? In fact, order has been defined in many different ways,
although some theorists do so only by implication. Often the problem of
accounting for biological organization in nature is equated with the problem of explaining order—which is not really kosher.

The point of entry for dealing with this preliminary issue is the association, historically, between the concept of order and the concepts of
entropy/negentropy in thermodynamics. Clausius defined order and disorder in strictly phenomenological terms. In an ordered state, energy is aggregated in such a way that it has the potential for doing useful work. Hence,
the concept of entropy (or thermodynamic disorder) was proposed by
Clausius as a measure of the degree of energetic *dispersal* or *dissipation* and,
consequently, its unavailability to do work.[1] In this formulation, a state of

maximum entropy corresponds to a complete state of energetic disorder which, paradoxically, also represents an equilibrium condition. Although this version of the entropy concept has had many practical applications over the years, it also suffers from a serious limitation. As noted earlier, the material world is in fact organized in a multileveled and hierarchical manner, but Clausius's concept was focused only on energetic order at the macroscopic level.

This shortcoming was rectified when physicists Ludwig Boltzmann (1909) and J. Willard Gibbs (1906), inspired by the pioneering work of James Clerk Maxwell in kinetic theory, independently addressed the relationship between the energetics of the macroscopic and microscopic levels—say, a body of gas in a container versus the dynamics of its constituent atoms. The dilemma is that the behavior of a thermodynamic microstate cannot be precisely predicted, for two reasons: first, because there is an inherent degree of stochastic fluctuation (indeterminacy) at that level, and second, because a human observer cannot know precisely the initial microstate of the system or make the necessary observations.

Accordingly, Boltzmann and Gibbs deployed statistical techniques resembling those that were developed for games of chance to describe in probabilistic terms the degree of energetic order/disorder at the microstate level, with the presumption that, within certain important constraints and limitations, these microstate statistics would correlate with the properties of the macrostate as well (see below).[2] Gibbs pointedly called his statistical formalizations "entropy analogues" to emphasize the fact that they were mathematical approximations only, not direct measures of the real thing. Today, physicists and engineers often distinguish between classical entropy and statistical entropy for much the same reason. (Later formulations, reflecting the development of quantum theory, added yet another level of microstate indeterminacy to the measurement of thermodynamic order and disorder.)

In any case, thermodynamic entropy as defined by these pioneers is a state function, comparable to temperature or pressure. Entropy in this sense is not a thing or a force. It is a property of matter with the peculiar attribute that it is designed to measure the relative absence of something, namely, energetic order. When the entropy of a medium increases, its work potential decreases—which is why, somewhat confusingly, entropy equations relating to work potential typically carry a negative sign. Conversely, what physicist Erwin Schrödinger called "negative entropy" (and Leon Brillouin subsequently called "negentropy"), was conceived to be a thermodynamic measure of order in the sense of a highly improbable concentration of work potential.

Schrödinger's Paradigm

The problems associated with trying to link these thermodynamic concepts of energetic order/disorder—which were designed for use in strictly bounded and constrained physical systems—to living systems, and especially to biological evolution, did not begin with Schrödinger. Others who had attempted to link energy and evolution included Lamarck, Spencer, Boltzmann, and Lotka, among others. But Schrödinger was the first to define living systems specifically in terms of the second law, and subsequent formulations in this vein have built upon his unique vision. In his legendary book, *What is Life?* (1945), Schrödinger characterized living systems as being, in essence, embodiments of thermodynamic order (negative entropy) that have managed to circumvent the inherent tendency of real-world processes to dissipate energy and become ever more entropic over time. Organisms do so, Schrödinger explained, by "extracting order" from their environments. He also spoke of organisms "sucking orderliness" from their environments, and he identified as their most distinctive feature their "well-ordered" state. Although it is often said that organisms feed upon energy, Schrödinger declared that this is "absurd . . . what an organism feeds upon is negative entropy" (p. 72).

In other words, what matters most in living systems is their ability to resist the cosmic determinism of the second law and to create local conditions of increased thermodynamic order. Schrödinger then proceeded to define negentropy, not in any independent, phenomenological way, but in mathematical terms as the reciprocal of Boltzmann's expression for entropy.[3] A crucial corollary of this formulation, which has echoed down through the years as received wisdom, is the proposition that living systems do not thereby violate the second law because they must pay for their increased order (negentropy) by producing an equivalent amount of entropy in the environment to compensate.

Schrödinger's poetic metaphor is seductive. It has been quoted on innumerable occasions over the years. But in fact, it too could be called absurd. In the first place, it reduces the complexities of living systems to a monolithic thermodynamic process and conflates thermodynamic order with functional organization—purposive designs for adaptation in a great variety of specific environments, including a number of different energy regimes and levels of organization. (To repeat, Schrödinger was not the first to blur the distinction between thermodynamic order and physical order, a source of confusion that has since become endemic.) Metabolism is only one aspect of the many-sided problem of earning a living in the natural world. In effect, Schrödinger truncated the challenges associated with physical organization, self-development, self-maintenance, and reproduction into a single parameter, thus distorting the very nature of the evolutionary process (see below).

Schrödinger's vision also caricatures the energetics of living systems, which have developed ingenious and highly efficient (i.e., profitable) mechanisms for capturing or harvesting available energy in various forms and then using it for various purposes, from doing useful work to building live mass (more on this below). Contrary to Schrödinger's assertion, it is more accurate to say that organisms feed on available energy and create thermodynamic, structural, and functional order than to say that they feed on order (cf. Morowitz 1968, p.19; Perutz 1987).

But the most serious issue is that Schrödinger's basic hypothesis is untestable, since his definitions of entropy and negative entropy are circular and have no empirical referents. In this context, we have no idea how to go about measuring either one. It is not at all like measuring the temperature gradient of the gas molecules in a defined system. Indeed, as we will explain, there is reason to question Schrödinger's assertion that the process of biological evolution has been accompanied by an equivalent increase in entropy "production" [sic] in the environment.

Prigogine's Vision

Physicist Ilya Prigogine's vision is similar to Schrödinger's in that it, too, characterizes living systems in thermodynamic terms as far-from-equilibrium "dissipative structures" that feed on energy (Prigogine 1978; Prigogine et al. 1972a, 1972b, 1977; Nicolis and Prigogine 1977, 1989). What Prigogine adds, most importantly, is an evolutionary mechanism that he characterizes as a "universal law of evolution" (1972a). According to Prigogine, "order"—which he does not explicitly define but which he uses both in thermodynamic/process and structural terms (a disturbing ambiguity)—evolves spontaneously in "open" systems via continuous energetic inputs. These may produce structural instabilities, which may in turn produce perturbations, or "fluctuations" in the direction of greater "complexity" (also not precisely defined). Prigogine refers to this causal dynamic as the principle of "order through fluctuations" (1972a), and he characterizes a living system as "a giant fluctuation stabilized by exchanges of matter and energy" (Prigogine et al. 1977, p. 18). As he says, this is an autocatalytic theory of evolution. He also agrees with Schrödinger that "during evolution a system may reach a state where entropy is smaller than at the start" (1972a).

One problem is that Prigogine makes no distinction between "order" and functional "organization." In fact, he uses the terms interchangeably. Thus, he sees no difficulty in applying the same explanatory principle both to the formation of convection cells (Bénard cells) in a pan of heated water and to the complex control mechanisms associated with glycolysis, or the highly coordinated functional transformations that occur over time in a colony of the cellular slime mold *Dictyostelium discoidium* (Prigogine et al. 1972a, pp. 27–28; 1977, pp. 32, 34). This is a theory that seriously overreaches (see below).[4]

Enter Information Theory

The problem of how to define order and organization in the natural world became vastly more complicated with the introduction of information theory in the late 1940s. As noted above, Shannon's (1948) concept of information was concerned with the problem of measuring uncertainty in the communication of messages between a sender and a receiver. As Shannon intended the term *entropy* to be used, it referred only to the degree of statistical uncertainty (disorder) in a given context before the fact, whereas *information* referred to the capacity to reduce that statistical uncertainty. If one uses the binary bit as a basic unit of measure, the degree of informational uncertainty (entropy) can therefore be defined empirically as a function of the number of bits required for its elimination.[5]

Attracted by the mathematical isomorphism between Shannon's entropy and the Boltzmann/Gibbs formalizations for statistical entropy in thermodynamics, many other theorists since the 1940s have tried to apply information theory directly to thermodynamics, an enterprise Shannon himself is said to have discouraged (see, for instance, Tribus 1961; Brillouin 1962; Tribus and McIrvine 1971; Gatlin 1972; Jaynes 1983; Brooks and Wiley 1988). In general, these efforts share a tendency to lose sight of the original purpose of statistical entropy measures. A fascination with the general properties of uncertainty and/or unpredictability has led succeeding generations of theorists away from the roots of statistical mechanics, which was concerned with the creation of analogues for Clausius's energetic entropy. One consequence of this broadened quest to explore uncertainty as a widespread property of the phenomenal world is that the concepts of entropy and information have been stretched far beyond their original meanings, and have been given several inconsistent meanings. (A more serious objection is that the two forms of entropy are like apples and oranges; thermodynamic entropy has dimensions, whereas informational entropy does not. Quantities with different dimensions, in this sense, cannot properly be equated.)

Wicken (1987) makes a convincing case against the notion that Shannon's information/entropy concepts can be treated as generalized measures of order/disorder in nature. As Wicken emphasizes, Shannon's entropy bears no relationship to the state of the phenomenal world; it relates to the efficiency or effectiveness with which a message is communicated from a sender to a receiver and the degree of uncertainty reduction. Because Shannon's entropy represents the pre-informational context of a communications interchange, it can only be measured (it at all) before the fact. Information in action (at least as Shannon used the term) is designed to reduce or eliminate informational uncertainty. Equally important, information is also—inherently—a functional phenomenon; it controls the work that is done via cybernetic control processes. (We will have more to say on this matter in chapter 14.) Thus, it is an ontological (or at least semantic) error to use the same concept both as a measuring rod for uncertainty/predictability and as a causal agency in the production of order/organization in the real world. One of biologist André Lwoff's (1962) famous paradoxes can be used to illustrate the problem. If a virus is introduced into a cell-virus system, its Shannon information content (negentropy) is thereby increased, but if the cell dies as a result of this increase, both its information and its thermodynamic order will be destroyed (maximum entropy).

These difficulties notwithstanding, there has been a proliferation of thermodynamic/informational theories of evolution over the past three decades, accompanied by multifarious definitions of order, organization, and information.[6] One notable exception is the work of Bernd-Olaf Küppers (1990), which provides an approach to the relationship between order, organization, and information—based on the work of Manfred Eigen and his colleagues (1981)—that is very close to, indeed a springboard for the paradigm that will be developed here. Küppers presents an extended argument for the assertion that purposiveness in the sense of functional (means-ends) organization is a fundamental, irreducible property of life. Drawing on the functional parallels between the camera and the human eye (evidence that has been invoked in many debates about evolution), Küppers points out that it would be absurd to recognize the purposiveness of the former but not of the latter. Furthermore, this purposive aspect can be traced to the earliest prebiotic RNA molecules, whose capacities for storing, transporting, and using energy and information played a critical functional role in the early stages of the evolutionary process. They served as building blocks for more complex functional assemblages. Evolved internal purposiveness, or *teleonomy* (to reiterate Colin Pittendrigh's enduring term), is thus a defining characteristic of living systems (see also Mayr 1974b, 1988 and Dobzhansky et al. 1977).[7]

The Teleonomy Problem

A second key assertion, in our view, is that teleonomy cannot be derived from any of the laws of physics and chemistry; it cannot be derived from the second law, or dissipative structures, or entropy in any form (see below; also Kline 1995, 1996; Corning 1983; Polanyi 1968; Anderson 1972). Available energy is necessary and self-organizing or even autocatalytic processes (e.g., hypercycles) might play a creative role, but every complex functional process requires a highly specific pattern of relationships and interactions among various constituent parts if the combined emergent outcome is to occur. Recall the argument in earlier chapters. I referred to it as "synergy minus one." If any major part is missing, the system may not work. As Yockey (1977) has shown, the probability that such functionally related configurations of parts could arise by chance alone is infinitesimally small. A process of differential selection among replicable functional variants (which can be called random only in a very restricted sense) was necessary, even at the prebiotic stage of evolution. Equally significant, the capacity to exercise *control* over these functional relationships and interactions was an indispensable concomitant. (This, in essence, is our definition of biological *control information*—see chapter 14). Accordingly, the evolution of functional information coincided with the evolution of life itself (see also Eigen et al. 1981).

A much-used but appropriate example is the hemoglobin molecule. (It was mentioned earlier, but I will elaborate here.) Hemoglobin is tetrameric protein. Its four protein monomers form a quaternary structure that allows for allosteric (cooperative) interactions possessing both functional and informational properties. Equally important, each monomer is folded around a ring-shaped heme group carrying an iron atom that is capable of reversibly binding and transporting atoms of oxygen, carbon dioxide and, we have recently learned, nitric oxide, all of which play a vital role in the functioning of complex living systems. However, the highly complex, highly specific structural and functional properties of hemoglobin molecules (we have left out many of the details) can be destroyed by the replacement of even a single amino acid. There is, in other words, an interdependent functional synergy in hemoglobin molecules, the combined properties of which are irreducible. (The fact that there are a number of functionally equivalent variants of the hemoglobin molecule does not negate the basic point.)

Theodosius Dobzhansky, one of the leading evolutionary biologists of the past century, was fond of characterizing organisms as "designs for survival" (see especially his *Mankind Evolving*, 1962). Just as a refrigerator, a computer, or an automobile represents a specific arrangement of parts—a unique teleological structure—that is capable of performing a limited range of functions, so organisms have been designed, via the causal dynamics associated with the evolutionary process, to have a specific "vocation." As Dobzhansky put it (quoted in part in chapter 4; here is the full quote):

> The origin of organic adaptedness, or internal teleology, is a fundamental, if not the most fundamental problem of biology. There are essentially two alternative approaches to this problem. One is explicitly or implicitly vitalistic. Organic adaptedness, internal teleology, is considered an intrinsic, immanent, constitutive property of all life. However, like all vitalism, this is a pseudo-explanation; it simply takes for granted what is to be explained. The alternative approach is to regard internal teleology as a product of evolution by natural selection. (Dobzhansky et al. 1977, pp. 95–96)

Survival Is the Problem

An appropriate way to conceptualize the nature of biological purposiveness is in relation to the fundamental, inescapable, and always context-specific problem of biological survival and reproduction. The framework, or paradigm, within which the survival process occurs is a highly varied, uncertain, and changeable natural environment. Therefore, the parameters of the survival problem for each organism will differ, depending on its particular characteristics, the nature of its specific environment and, most important, the precise organism-environment *relationship*. Moreover, the survival problem is a multifaceted affair. Living systems must cope with an array of vital concerns, and the means that have evolved over the course of time for meeting survival and reproductive needs are correspondingly varied and complex.

Accordingly, available energy and, equally important, access to relevant information about how to capture and utilize it are two universal requisites for survival and reproduction. However, the capture of available energy alone provides an insufficient definition of the survival problem, or of the phenomenon of adaptation. Even the problem of acquiring available energy is not a simple matter, and living organisms have evolved a variety of means for doing so. In this perspective, monolithic thermodynamic theories of evolution are fundamentally flawed, because they are based on an implicit definition of the problem of survival that is grossly oversimplified. Another way of putting it is that monistic thermodynamic theories make only a narrow cut through the hyperspace of parameters and state variables that define the evolutionary process. Theoretical simplifications are a legitimate and time-tested way of gaining insight and understanding in science. However, it is also incumbent upon a theorist to acknowledge the attendant limitations and, more important, to avoid claiming too much for the results. Unfortunately, this caveat has been widely abused in relation to various thermodynamic theories of evolution. (The subject of overclaims is discussed at length in Kline 1995.)

How should the process of adaptation be conceptualized? The appropriate model, we believe, is cybernetic. As noted in chapters 4 and 6, the term *cybernetics* derives from the Greek word *kybernetes*, or "steersman," and it is the root for such English words as governor and government. A cybernetic system is by definition a dynamic, purposive system; it is designed to pursue or maintain one or more goals or end-states. The key to understanding a cybernetic process—say, a smart bomb as distinct from a ballistic missile—is the concept of feedback. Technically, feedback denotes information that a cybernetic system uses to monitor and adjust its behavior in order to attain or maintain a desired goal-state. Thus, cybernetic systems are controlled by the *relationship* between endogenous goals and the internal or external environment as experienced via informational processes (see Wiener 1948; Buckley 1968; von Bertalanffy 1968; Powers 1973; Miller 1995; Corning 1983).

As noted in chapter 4, the systems theorist William T. Powers (1973) has shown that the behavior of a cybernetic system can be described mathematically in terms of its tendency to oppose an environmental disturbance of an internally controlled quantity. It should also be reiterated that cybernetic control processes may produce results that resemble the so-called dynamical attractors of chaos theory, but they are achieved in a very different way. Without some internal reference signal (teleonomy), there can be no feedback control, although there can certainly be self-ordered processes of reciprocal causation at work, or perhaps Darwinian processes of co-evolution and stabilizing selection. Indeed, the existence of systemic purposiveness (teleonomy) distinguishes organisms (and superorganisms) from ecosystems and, of course, inert, naturally occurring objects. The mere fact of functional interdependence is insufficient to justify the use of an organismic/cybernetic analogy. Although cybernetic systems must operate within the constraints of the laws of physics, chemistry, and so forth, cybernetic causation, by definition, introduces *unique* historical and configural (that is, situation-specific) influences into the degrees of freedom that exist in the natural world. To reiterate the argument in chapter 4, living systems are characterized by cybernetic (teleonomic) properties, which transcend the laws of physics. Another way of putting it is that organisms are distinguishable from, say, crystals or geysers in that their cybernetic properties introduce an emergent, partially independent source of causation into the natural world. This model of living systems will figure importantly in the paradigm that we develop in chapters 13 and 14.

Thermodynamics and Living Systems Revisited

With this perspective in mind, we turn now to a reconsideration of the relationship between thermodynamics and biology. A starting point is the often overlooked metatheoretical observation that science involves the construction of mental representations—more or less imperfect schemata that necessarily simplify the properties of the real world. Thus, an analytical system can be anything that a particular theorist wants it to be. But unless the system is precisely defined beforehand (for everyone concerned, including the theorist), much confusion will almost inevitably result. To illustrate: In the classic Gibbs equation for a simple one-level thermodynamic system, its truth claims involve no less than eleven specific restrictions, which makes it applicable only to a very limited set of contexts. Nearly all of the examples relating to thermodynamics in physics and chemistry textbooks are confined to this very simple class of systems. However, in the secondary literature on thermodynamics (i.e., the work of theorists who have not had formal training in the subject), it has typically been the case that such limitations are not appreciated. This accounts for many of the theoretical overclaims noted above.

Another metatheoretical problem that must be kept in mind is that the biological realm is multileveled and hierarchical in nature. We noted earlier that even a simple thermodynamic system such as a bottle filled with gas molecules is ordered at more than one level. Nevertheless, for most practical purposes such systems can be analyzed at a single level. (We refer to them as one-level analytical systems.). However, in complex, functionally organized living organisms this is not possible. The basic properties, and the dynamics, are distinctly different at different levels of organization. The energetics of muscle action cannot be treated within the same analytical system as photosynthesis, or glycolysis, and the complete organism requires yet another framework. Thus, analyses that might be appropriate at some levels (or perhaps for understanding certain microevolutionary processes) will not be appropriate at the level of whole organisms, populations, species, and ecosystems—or for analyzing macroevolutionary processes. Equally important, the levels of biological organization are not entirely separable from one another even for analytical purposes, for there are complex patterns of mutual constraint and functional interaction between levels. Thus, one needs to employ a multilevel analytical paradigm, along the lines laid out in chapter 6. (A more extensive treatment of this and related issues can be found in Kline 1995.)

The metatheoretical problem of choosing the appropriate analytical framework matters a lot, because different paradigms often lead to different results and different conclusions. A famous example is the so-called irreversibility paradox, which has haunted thermodynamics theory for most of the past century. At the micro level, the formal equations that physicists have developed for the second law over the years allow for reversibility in a thermodynamic process (under carefully specified model conditions). Yet at the macro level, these processes are always irreversible, both in theory and in our empirical observations. So which version of the second law is correct? Some theorists have maintained that the paradox is real, a perplexing and deeply mysterious property of nature itself. However, Kline (1995) shows that the paradox is the result of faulty reasoning; there are serious logical errors involved. Our mathematical representations are not more trustworthy in this case than our direct experience of the real world. We need to keep in mind Alfred Korzybski's (1933, p. 38) dictum that "the map is not the territory."

"Control Volumes" and "Available Energy"

With this as a preface, we can now consider two key concepts that are not well-known in physics, chemistry, or biophysics. In these disciplines it is almost always the case that conventional analyses of thermodynamic processes deploy a framework that involves as a core assumption a fixed, unchanging mass. It is often referred to as a "control mass" analysis. Within this framework, there are three subcategories, depending on the assumptions that are made about allowable interactions across the system's boundaries. In an *isolated* system, there are presumed to be no interactions; in a *closed* system, only energy can cross the boundary; and in an *open* system, both energy and batches of mass can cross the system boundary. It is the latter type of system that physicists such as Prigogine, as well as various biophysicists, have utilized to theorize about evolution (e.g., Morowitz 1968; Harold 1986).

However, we maintain that the entire category of control mass systems is in fact inappropriate for the analyses of whole organisms, ecosystems, and macroevolutionary processes, because living systems at these levels are not systems of fixed mass. The flow of matter and energy through these systems more nearly resembles a jet engine than a bottle containing a fixed quantity of gas molecules. Accordingly, control mass analyses, which are appropriate for some purposes and some levels of living systems, should be supplemented at higher levels of biological organization by a different type of analytical framework, which is often referred to as a *control volume*. A control volume is defined as *a volume in space through which mass, and often energy, freely flows*. In addition to jet engines, some examples include heat exchangers, chemical plants, and turbomachinery of all kinds. Control volume analyses (and the equations used to describe them) were first deployed by Leonhard Euler in the eighteenth century, in the form of differential equations. Algebraic forms were perfected over half a century ago by an MIT group (see especially Keenan 1941; Shapiro 1953; Reynolds 1965). However, for a number of reasons this methodology remains little appreciated to this day, outside of engineering. The lack of multidisciplinary communication is unfortunate, because the application of control mass equations to systems involving regular (and variable) throughputs of mass/energy will generally give misleading or even erroneous results.

A second major concept relates to how energy is characterized and treated within a given analytical framework. Energy in conventional thermodynamics takes one of three broad forms (somewhat awkward, but time-honored): work, embodied or stored internal energy, and heat. Each form is convertible into the other, but their sum is always constant for any process in a finite system (in accordance with the First Law of Thermodynamics). As we noted above, Clausius's formulation of the second law, and the related concept of thermodynamic entropy (which was inspired by the earlier work of Carnot on the efficiency of heat engines), asserts that there is an upper limit to the conversion efficiency of internal thermal energy or heat flows into work. Entropy, in fact, meters the *inefficiency* and energetic degradation in a thermodynamic process. However, the problems associated with utilizing the entropy concept in bioenergetic and evolutionary analyses are manifold. Entropy is difficult to comprehend; it cannot be measured directly; and it omits other fundamental constraints on thermodynamic processes that are covered by the First Law and the equation of state.[8]

Fortunately, there is an alternative formulation—first suggested by Gibbs in an obscure footnote in the 1870s and later developed by Keenan (1941, 1951)—that is called "available energy." Broadly defined, available energy meters the amount of energy in a given system and its surroundings that can be converted to work. It has the advantages of being easy to understand and use, and it incorporates all three principles of thermodynamics (including entropy) into a single function. The available energy function — customarily represented as "A" in thermodynamic analyses—is precisely defined and widely used by engineers (in Europe it is often referred to as "exergy"), but for reasons that are unclear to us, it has been little-used in bioenergetic and evolutionary analyses (but see Schneider and Kay 1995; also Schneider and Sagan 2005). Instead, biologists have used either the Helmholtz free energy function or the Gibbs free energy function, both of which are, in fact, special cases of the more general availability function. These free energy functions are important, but they cannot cover the full range of energetic processes in living systems.

Perhaps the most important advantage of an available energy approach is that it shifts the focus of bioenergetic (and evolutionary) analyses away from the second law and the arcane obscurities of the entropy concept without having to deny or ignore its effects. Instead, it directs our attention to energy and its ability—within the constraints of the system, the environment, *and* various entropic effects—to do work, and thus to hold the system out of equilibrium and/or build biomass over the course of time. Needless to say, the concepts of available energy and control volumes are also utilized in the paradigm that we develop in chapters 13 and 14.

A Critique of the Biology and Thermodynamics Literature

Space does not permit us to include a detailed critique of the many misuses of thermodynamic concepts that we alluded to above, but we highlight some of the more blatant inexactitudes that we have found in the literature and offer a few brief comments on each.

- **Entropy always increases in the phenomenal world.** Not so. Within finite, bounded systems entropy often decreases. Living organisms are routinely able to increase their available energy content and their capacity to do work. Likewise, various processes inside organisms involve entropy decreases. And this is not unique to living systems. If we close the door of a refrigerator that is not running and plug it into a power source, the temperature of the air inside will go down. While the entropy of the refrigerator and the room taken together increases, the entropy inside the refrigerator decreases. This illustrates the many ways in which we can use available energy to decrease entropy in finite systems.
- **The second law does not apply to living organisms (open systems).** This idea seems to be a result of applying the application of the control mass form of the second law to systems that require a control volume analysis.

When we examine the control volume equation for the second law (see Corning and Kline 1998a, Appendix A), it is apparent that the law is also applicable to living systems. Nor do life forms violate the second law, although they may be able to postpone its effects, sometimes indefinitely (see below). In short, one must do the entropy bookkeeping in the proper frame of reference.

- **Entropy drives the evolutionary process.** Entropy is a state function—a property—and cannot be equated to a drive or a force, any more than temperature can be equated to energy. Entropy is a constraint on thermodynamic processes, not a cause of them; it measures the waste associated with any real-world dynamic process. A focus on entropy as a way of trying to understand living systems is analogous to trying to understand a horse by studying horse manure (to paraphrase Steve Kline).

- **Negentropy drives the evolutionary process.** We know what available energy is. We can measure its effects, and we can give it a precise mathematical definition. However, we don't have a clue about what (thermodynamic) negentropy is, except in formal theory. In truth, when it is defined (as Schrödinger does) as a decrease in thermodynamic entropy, negentropy is really another term for an increase in available energy. We prefer to say that the glass is half full, rather than speak of a lack of emptiness. Of course, entropy and negative entropy are now also used (rather loosely) with reference to the general property of physical disorder/order in the universe (e.g., Georgescu-Roegen 1971; Hawking 1988; Penrose 1989). Thus, Faber (1985, p. 317) informs us that "thermodynamics is that branch of physics which deals with systems of great numbers of particles." In this vulgarized form, the term negentropy is in reality a convoluted synonym for order—literally, an absence of an absence of order.

- **Living systems must pay for their thermodynamic order with an equivalent amount of entropy.** There is not a one-to-one correspondence between the creation of order and an increase in entropy. In a perfect (i.e., reversible) process, there would be no increase in entropy at all. But more relevant for our purpose are the many cases in which efficiencies can be achieved that result in per-unit entropy reductions. We provide here two examples, one in technology and one in biology. Power plants in the year 1900 required about eight times as much coal per kilowatt-hour of electricity output as do the best power plants operating today. In living systems, similarly, Schmidt-Nielsen (1972) showed that the energy consumption associated with locomotion is far lower per pound for large animals than for smaller animals (the regression line is -0.4). In fact, some energetic transformations within living systems have remarkably high efficiencies (very low entropy production). The point is that it is the inefficiencies—that is, the wastes or irreversibilities—and not the ordering processes per se that create entropy.

- **Irreversible thermodynamic processes drive evolution by creating self-organizing dissipative structures via fluctuations arising from thermodynamic instabilities.** All the equations of physics taken together cannot describe living systems, much less explain them. Indeed, the laws of physics do not even contain any hints regarding cybernetic processes or feedback control. Thus, the term dissipative structures does not adequately describe the informed, purposive organization of living systems. It is comparable to characterizing jet engines—which are painstakingly designed and manufactured with extremely precise dimensional properties and tolerances—as dissipative structures. They are not self-designed, nor are their dissipative properties among their most salient features. (It might also be worth noting—admittedly a reductio ad absurdum—that a maximally dissipative structure would be incapable of doing any work.) This is also an example of the fallacy of using one-level analytical systems to explain multilevel processes (again, discussed in Kline 1995). Finally, and most serious, the domain of irreversible thermodynamics is restricted to states where the ordinary temperature and pressure can still be defined locally, but many biological processes do not satisfy this criterion. An illustration is the process that initiates the capture of available energy and the miracle of photosynthesis in living systems—namely, the activation of high-energy states within atoms via irradiation by sunlight. This fundamental life-creating process cannot even be described by irreversible thermodynamics (on this point, see also Morowitz 1968).

- **Steady-state systems operating at near-equilibrium conditions obey a law of minimum entropy dissipation.** Often called Prigogine's principle, this supposed law is sometimes portrayed as an inherent economizing influence in thermodynamic processes (for example, Proops 1985; Wicken 1987). However, this principle is true only in some special cases. Obvious counterexamples are turbulent flows, say, in various liquids and gases. Equally important, Gage et al. (1966) provided a definitive disproof that any variational principle of this kind could exist in a dynamic system. Although Gibbs showed in the 1870s that such a principle is applicable to static systems, it is not applicable to living systems operating in an approximate steady state. Indeed, there is no monolithic, general variational principle of any kind based on a single variable that applies to living systems.

- **Macrolevel entropy in the physical world can be derived from the microlevel statistical equations developed by Boltzmann and Gibbs.** Some physicists have claimed that this relationship has long ago been validated, but that is not true. Consistency has been shown to exist only for a few very special cases. Moreover, we believe that this micro-to-macro derivation cannot be done. So confident are we of this conclusion that a formal $1,000 bet, at odds of ten to one, was offered by Kline (1995) to any physicist (or anyone else) who could develop a complete, logically correct statement of the second law from statistical mechanics. (The rules were very specific, but nobody has even attempted to claim the prize.) We highlighted the problem above in relation to the so-called irreversibility paradox. More than that, there are at least five logical difficulties (detailed in Kline 1995) that make any such effort extremely dubious.

It should also be noted that the micro–macro conflict with respect to entropy is in fact a special case of a larger claim, associated with some physicists and biophysicists, that macrolevel phenomena can be described and explained better with more powerful microlevel equations, or methods. This assertion has even led some theorists to suggest that quantum mechanics might somehow nullify the second law, and that molecular machines could generate the anti-entropic dynamics of the evolutionary process. Evolution, then, might be characterized as quantum coherence writ large. The problem, very simply, is that multiple microlevel exceptions will add up to a macrolevel contradiction that violates the oft-neglected "correspondence principle" of Niels Bohr—a founding father of quantum theory. To put it bluntly, micro and macro formulations of the laws of physics cannot contradict each other. If they appear to do so, the fault lies not with nature but with the analyst's reasoning. Accordingly, theories that purport to explain macrolevel thermodynamic phenomena with inconsistent microlevel equations are demons that must be exorcised (see chapter 13). A somewhat expanded discussion of classical and statistical thermodynamics, and the relationship between them, is contained in endnote 8.

- **Entropy is relevant to an understanding of complex human economies and societies, but its applicability is strictly limited.** One example is sociologist (and systems scientist) Kenneth Bailey's (1990) important Social Entropy Theory (SET). Bailey advanced a "holistic" model of society as a complex "dissipative structure" (after Prigogine), an elaboration on the holism of such sociologist pioneers as Durkheim, Merton, and Parsons, as well as general systems pioneers such as Ludwig von Bertalanffy (1968) and James Grier Miller (1995). Unfortunately, Bailey's effort was marred by the conflation of energetic and physical entropy, a complete absence of teleonomy and cybernetics from the model, and confusion about the nature of biologically relevant information (more on this in chapters 13 and 14).

A Conclusion and a Preface

To anticipate the basic elements of the paradigm described in chapters 13 and 14, we use as a starting point the vision developed by biophysicist Alfred Lotka in a pioneering essay entitled "Contribution to the Energetics of Evolution" (1922). Lotka offered us only a sketch, but it was conceptually sound. He argued that, given the obvious importance of energy in biological processes, natural selection would tend to favor organisms whose "energy-capturing devices" were more efficient at directing "available energy" (Lotka specifically employed this term) into uses that would be favorable to the preservation of a species. Furthermore, Lotka noted, "If sources are presented, capable of supplying available energy in excess of that actually being tapped by the entire system of living organisms, then an opportunity is furnished for suitably constituted systems to enlarge the total energy flux through the system."

In pursuing our own version of this paradigm, we disagree with the thermodynamic/entropic school that energy flows (much less entropy) are the driving force or principal causal agency in evolution. We see available energy as a variable free good (subject only to extraction costs) that has interacted with matter and purposive information to facilitate the broader process of selection and evolutionary change. Living systems harvest available energy in various forms. Some of it becomes embodied in biomass; some is utilized for work; and some becomes waste heat and is ultimately returned to the environment with an entropy level that is no higher than would otherwise have been the case. (The latter point is often overlooked. The dissipated energy from the solar flux is, in any event, finally re-radiated back into space; living systems do not add anything to the ultimate entropy of the universe.)

Furthermore, in the absence (to our knowledge) of any concrete evidence, we dispute the statements of Schrödinger and many other theorists to the effect that the creation of biological organization is necessarily accompanied by an equivalent compensatory increase in thermodynamic entropy in the universe, thereby preserving the inviolability of the second law. (We argue that the second law can be satisfied perfectly well if various deferred payments, with an indefinite repayment schedule, are taken into account; the second law contains no timetable.) As noted earlier, we believe Schrödinger's assertion is unverifiable. But more important, we believe this conceptualization is basically unhelpful. Even if Schrödinger's formulation were correct, the question is begged: So what? Of what relevance is entropy to an explanation of biological evolution? It forms a constraint; it is not a cause. We believe the keys lie in a thermoeconomics framework, and especially the concepts of (a) control volumes; (b) available energy; (c) cybernetic, purposeful organization; and (d) control information—which, as we attempt to show in the next two chapters, lend themselves to quantitative, empirical analyses.

— ∽ —

Horse manure does not explain a horse.

—Stephen Jay Kline

SUMMARY: "Thermoeconomics" is based on the proposition that the role of energy in biological evolution should be defined and understood, not in terms of the second law of thermodynamics, but in terms of such economic criteria as productivity, efficiency, and especially the costs and benefits (or profitability) of the various mechanisms for capturing and utilizing available energy to build biomass and do work. Thus, thermoeconomics is fully consistent with the Darwinian paradigm. Economic criteria provide a better account of the advances (and recessions) in bioenergetic technologies than does any formulation derived from the second law.

13

Thermoeconomics: Beyond the Second Law

Introduction

The second law of thermodynamics is one of the pillars of the physical sciences, and rightly so. It has withstood the test of time, including numerous, often ingenious efforts to find exceptions or dispute its hegemony.

In the life sciences, however, the so-called entropy law has had a more checkered history, as discussed in chapter 12. The fact that energy plays a central role in living systems, and in evolution, has long been appreciated. The so-called "power of life" was a centerpiece of Jean Baptiste de Lamarck's nineteenth-century evolutionary theory. Herbert Spencer elaborated on this theme in the latter part of the nineteenth century with his grandiose "universal law of evolution." According to Spencer, energy was the driver of an inherent evolutionary trend toward increased complexity, both in nature and in human societies. As noted earlier, physicists Ludwig Boltzmann (1909) and Alfred Lotka (1922, 1945) also defined evolutionary progress in energetic terms, and many latter-day theorists have followed suit. But, as noted earlier, it was the physicist Erwin Schrödinger, in his legendary book *What is Life?* (1945), who catalyzed the modern approach to thermodynamics and evolution. Schrödinger characterized a living system as, quintessentially, an embodiment of thermodynamic order—what he termed "negative entropy."

Though there are serious problems with Schrödinger's paradigm, it has nevertheless enjoyed an immense influence over the years. Among other things, Schrödinger inspired many subsequent efforts to explain the evolutionary process, and especially the evolution of complexity, in terms of various interpretations of the laws of physics, including the second law. There are, needless to say, major differences among these theories, but the common theme is the claim that biological evolution has been driven by forces, or propensities, or tendencies that are inherent in nature, as opposed to the workings of natural selection (which one member of this school characterized as an "uninvited guest"). Sometimes energy, or some form of information, or both, are said to be the keys to how living systems are able to transcend the entropy law, but at other times the entropy law itself is identified as the primary causal agency.[1]

A Plethora of Laws

Recall physicist Ilya Prigogine's claim to have discovered a "universal law of evolution" (Prigogine et al. 1972a, 1972b), and biologists Daniel Brooks and E. O. Wiley (1988), who laid claim to a "natural law of history." Their core hypothesis was that "biological evolution is an entropic process. . . . Increasing complexity and self-organization [arise] as a result of, not at the expense of, increasing entropy."[2] Likewise, Rod Swenson (1989, p. 187) touts what he calls the "law of maximum entropy production," which he says forms "the cornerstone to a theory of general evolution within which biological and cultural evolution are special cases." Biologist Jeffrey Wicken (1987, 1988, pp. 152–53) characterizes entropy as a "teleomatic drive" toward disorder that underlies biological variation and gives direction to evolutionary change. "Speciation is driven by the randomizing directives of the second law," he tells us (p. 144). On the other hand, Wicken also claims that free energy fueled the prebiotic phase of evolution with "an inexorable determinism." (Wicken, as noted earlier, also made a problematic attempt to incorporate information into his paradigm.)[3]

In a similar vein, biophysicist Harold Morowitz, in one of his early works (1968, p. 146), proposed that the evolutionary process was the necessary result of "the constant pumping" of energy, mainly from the sun. ". . . the flow of energy through the system acts to organize that system. . . . Biological phenomena are ultimately consequences of the laws of physics" (p. 2). More recently, Eric Schneider and James Kay (1994, 1995, p. 171), citing Morowitz as a progenitor, advance what they describe as a "Unified Principle of Thermodynamics." They tell us that "life emerges because thermodynamics mandates order from disorder whenever sufficient thermodynamic gradients and environmental conditions exist."[4]

This theory is further elaborated in a new popularization coauthored with science writer Dorion Sagan, (Schneider and Sagan 2005). Schneider and Sagan claim (in their draft manuscript) that energy flows "generate, perpetuate, elaborate," biological complexity. "Life is organized by energy flows." It was "sired by energy flow." It is "ruled" by energy and its transformations. Indeed, the second law of thermodynamics defines the very purpose of life. "Nature abhors a gradient," they claim, and life arose in order to reduce energy gradients—in much the same way that tornadoes serve to dissipate the pent-up energy in the gradient between high and low pressure air masses. The emergence of life is "causally connected to the second law," they say. Indeed, the second law is variously characterized by Schneider and Sagan as a force that "governs," "organizes," "selects," "generates," "determines," "mandates," "pushes," and "leads to" biological structure and organization. The second law is the "source" for the overall directionality observed in the evolutionary process, they say (cf. the social entropy theory of Kenneth Bailey 1990, mentioned in chapter 12).

There have also been several different claims to the discovery of a new "fourth law of thermodynamics." Morowitz's suggestion that energy flows are autocatalytic and serve to organize a system is often referred to as a new law of physics—although Morowitz himself demurs from this view and advances a more complex paradigm (see below). Economist Nicholas Georgescu-Roegen (1977a, 1977b, 1977c, 1979) also formulated a "fourth law of thermodynamics," which he asserted governs economic life. Calling the entropy law the "taproot" of economic scarcity (1979, p. 1041), Georgescu-Roegen (1977a, p. 269) posited that, in a closed system such as a human society, "material entropy must ultimately reach a maximum." There is no way to escape it, he argued, and economies must work within this cosmic constraint.

There is much of great value in Georgescu-Roegen's pioneering work, but he too was (sadly) burdened by the conceptual muddle in thermodynamics and information theory (see also chapters 12 and 14). Though he struggled gamely to utilize these concepts in economics, in the end he came to recognize that the entropy law had, at best, a metaphorical role to play in explaining economic life. On the other hand, unlike many mainstream economists, he was always clear about the basic purpose of economic life: "Apt though we are to lose sight of the fact, the primary objective of economic activity is the self-preservation of the human species" (1971, p. 277). Thus, Georgescu-Roegen's use of the term *bioeconomics* was ultimately consistent with how we have defined it here, and he could properly be considered a "godfather."

Finally, there is Stuart Kauffman's "fourth law of thermodynamics" (mentioned earlier),which he describes as an inherent tendency for the biosphere to become increasingly diverse and complex (Kauffman 2000, p. xi).

Kauffman and others also regularly invoke self-organization and autocatal-ysis as inherent ordering influences in evolution.

Problems with the Thermodynamics Paradigm

There can be no doubt that many autocatalytic and self-ordering processes do exist in nature (see chapter 4), but there are serious—no, fatal—prob-lems associated with elevating these local influences into a general law (or laws) that govern the overall trajectory of the evolutionary process.

To reiterate, some of the flaws in what could loosely be called the "ther-modynamics paradigm" include the following: Many of these second law theorists seriously misinterpret and thus misuse the concept of entropy; oth-ers utilize deficient concepts of "information" that cannot be operational-ized; many blur the crucial distinction between statistical or structural forms of order, on the one hand, and evolved, goal-directed, functional organiza-tion; not least, they have been misled by some of the very "gods" of physics into conflating energetic order/disorder and physical order, which in many cases is not correct (see below). But most serious of all, these theorists for the most part discount what I have been calling the ground-zero premise of the biological sciences, namely, that life is a contingent phenomenon and that survival and reproduction combine to define the paradigmatic problem of all living organisms. Life is quintessentially a survival enterprise (and a reproduction enterprise, of course), the parameters of which are locally defined by the nature of the organism and its specific environment. The pre-cise organism-environment relationship is a key determining factor in the ongoing evolutionary process.

As noted in chapter 12, both Erwin Schrödinger and Ilya Prigogine helped to promote an expansive definition of the entropy law that, we main-tain, is both unwarranted and significantly overstates the role of entropy in the natural world. To reiterate, some of the confusion associated with the use of thermodynamics in evolutionary theory is the result of a major theo-retical segue that occurred with the development of statistical mechanics in the latter part of the nineteenth century. The problem arose when some leading theorists of that era assumed that there is an isomorphism between statistical order, energetic order, and physical order. As a consequence, sub-sequent generations of physicists and laymen alike have often uncritically accepted the claim that the entropy law applies to everything in the uni-verse. Thus, biologist Ludwig von Bertalanffy (1952) wrote: ". . . according to the Second Law of Thermodynamics, the general direction of physical events is toward decrease of order and organization." Likewise, biologists

Brooks and Wiley (1988, p. 36) speak of a general physical law that "predicts that entropy will increase during any real series of processes." Georgescu-Roegen (1979, p. 1039) assured us that "matter matters too"— the material world is also subject to the second law. (See also Georgescu-Roegen 1971.) Physicist David Layzer (1988, p. 23) asserts that "all natural processes generate entropy." Economist Malte Faber (1985, p. 317) tells us that "thermodynamics is that branch of physics which deals with systems of great numbers of particles."

More surprising, physicist Stephen Hawking (1988, p. 102) refers to "a physical quantity called entropy, which measures the degree of disorder of a system. It is a matter of common experience that disorder will tend to increase if things are left to themselves. (One has only to stop making repairs around the house to see that!)" Similarly, physicist Roger Penrose (1989, p. 308) informs us that "the entropy of a system is a measure of its manifest *disorder* [his italics]. . . . Thus, [a] smashed glass and spilled water on the floor is in a higher entropy state than an assembled and filled glass on the table; the scrambled egg has a higher entropy than the fresh unbroken egg; the sweetened coffee has a higher entropy than the undissolved sugar lump in unsweetened coffee." It follows, then, that "the second law of thermodynamics asserts that the entropy of an isolated system increases with time" (p. 309). Penrose goes on to associate the second law specifically with the "relentless and universal principle" that organization is continually breaking down.

Is the Earth Dissipating?

One problem with this formulation is that we know of no evidence for the assertion that the material world has an inherent tendency to dissipate. If this were the case, presumably somebody by now would have calculated the depreciation rate for the earth as it progressively deteriorated. Though stars burn out and aggregates of individual gas molecules may readily dissipate, the stable molecular bonds that hold solid chunks of matter together do not spontaneously break down. (The unstable exceptions do not disprove the rule.)

Another problem is that energetic order and physical order are not always isomorphic. A case in point is a volume of water molecules that becomes increasingly disordered as energy inputs convert ice crystals to liquid water and then steam. This is a case where energy inputs result in a progressively increasing physical *disorder*! (This crucial point can also be illustrated with a thought experiment. Two equally heated crystals, one in lattice form and the other in a disordered pile of shards, nevertheless could in theory produce exactly the same work output under appropriate conditions.)

In fact, much of the physical disorder we experience is energy-driven! Take Hawking's decaying house metaphor. This is not an example of an inherent entropic trend but of the effects of gravity, earth movement, wind, weather, solar radiation, oxidation, human use, and termites, among other things. Likewise, in Penrose's examples, it is the joint action of gravity and a solid surface, not entropy, that are responsible for breaking the water glass. Energy inputs are also needed to scramble the egg, and well-understood physical processes (including the stirring actions of the coffee drinker) are responsible for dispersing sugar cubes. (I can testify to the lack of entropy when I fail to stir my coffee!)

Equally dubious is the claim that the general trend in the universe is toward increased entropy. Indeed, entropy has often been portrayed as a dark force that somehow governs the fate of our species and dooms our progeny to oblivion in the eventual "heat death" of the universe, a portrayal that dates back to Clausius (see chapter 12). His dour vision has long since become the conventional wisdom of the western scientific establishment. However, there is reason to doubt the conventional wisdom. In a nutshell, the heat death scenario overlooks the role of gravity. Alongside the well-documented trend toward increased entropy in the universe, new free energy is being aggregated as we speak in the ongoing process of star formation and stellar nucleosynthesis. These energy-ordering processes are driven by the non-entropic influence of gravity, in utter contradiction to the second law!

As physicist Freeman Dyson (1971, p. 20) explained it, ". . . in the universe the predominant form of energy is gravitational . . . gravitational energy is not only predominant in quantity but also in quality; gravitation carries no entropy . . . [Moreover] in the universe as a whole the main theme of energy flow is the gravitational contraction of massive objects, the gravitational energy released in contraction being converted to energy in the form of motion, light and heat." In other words, even as the existing stock of available energy in the universe is being dissipated, more is being created by that great engine of negentropy in the universe, gravity. Physicist F. A. Hopf (1988, p. 265) observed that the conventional wisdom about entropy in cosmic evolution might be "an artifact of our ignorance about how to handle thermodynamics when gravity is important."

It should also be pointed out that a portion of the available energy that is mobilized by gravity and emitted from our sun does work of various kinds on earth and ends up being trapped and embodied in matter and living systems. Some of it also gets recycled and reused in various ways, so it is not entirely lost to entropy. To be sure, the vast majority of the energy that bombards the earth and the many billions of other celestial objects is ultimately dissipated. But this would have happened in any case; living systems do not in any way increase the overall energetic entropy of the universe. Indeed, some of that entropic energy is positively beneficial; it warms our planet and in other ways makes our environment hospitable to life.[5]

Forget Entropy

A corollary assumption of the heat death scenario, and one of the pillars of modern physics, is that dissipated available energy ultimately goes to equilibrium (i.e., maximum entropy) in the vacuum of space and forms part of the residue of background radiation that is suffused throughout the universe. The problem with this scenario, it seems increasingly evident, is that the vacuum is not a vacuum. Rather, we simply cannot detect and measure what is going on out there. It has been a major embarrassment to cosmology for some years that perhaps 95 percent of the predicted mass of the universe is missing and unaccounted for. Various theorists have struggled with this and other important paradoxes (such as quantum entanglement and quantum nonlocality). For instance, Haisch et al. (1994) and others have developed what they call the "zero-point field" theory, which posits an undetected omnidirectional field that can account in a new way for how inertia and gravity work—they are viewed as effects produced by the field. In quantum mechanics, similar claims are made for the hypothetical Higgs field.

More recently, in light of the growing evidence that the universe is expanding at an accelerating rate, some cosmologists have revived Albert Einstein's postulate of a "cosmological constant" in the form of undetected "dark energy" that may be driving the dynamics of the universe in ways that are not yet understood. In either case, the available energy that is being created and dissipated in the part of the universe we can detect may be vastly outweighed by the energy we cannot detect. Though it is pure speculation at this point, it could be that the energy we define as entropic is not being dissipated at all. Instead, it is being absorbed back into the vast energy pool in which we are embedded. In any case, we are far less certain than we were only a few years ago about either the dynamics of the universe or its ultimate fate. (Dyson, quoted in Overbye 2002, concludes that "all bets are off"). But of one thing we can be reasonably certain: Entropy will have little to do with it.

More to the point, it is evident that entropy has had relatively little to do with biological evolution. To repeat, entropy is a state function like temperature or pressure; it cannot be equated to a drive or a force any more than temperature can be equated to energy. Entropy represents a constraint on thermodynamic processes, not a cause of them; it measures the energetic wastes associated with any real-world dynamic process. It's a cost of doing business in the biosphere.[6]

Contrary to Schrödinger's formulation (to repeat), we believe that it is more accurate to say that living organisms feed upon available energy to create thermodynamic (energetic) order, as well as biological organization, rather than saying that they feed upon a statistical measure called "order." Furthermore, we believe that energetic order, physical order, and biological organization are not equivalent to one another. But most important, we believe that the role of energy in evolution can best be defined and understood in economic terms. By this we mean that living systems do not simply absorb and utilize available energy without cost. They must capture the energy required to build biomass and do work, and they must invest that energy in development, maintenance, reproduction, and further evolution. To put it baldly, life is a contingent and labor-intensive activity, and the energetic benefits must outweigh the costs (inclusive of entropy) if the system is to survive. Indeed, energetic profitability is essential to growth and reproduction. This could be called the first law of thermoeconomics.

Accordingly, there are three core assumptions that provide the conceptual framework for thermoeconomics: (1) life is a contingent phenomenon, and adaptation to specific, varying environmental conditions and constraints is an ongoing challenge for all living systems; (2) functional variation is endemic in nature and any form of biological order (or organization) is always subject to stringent testing and editing by natural selection; and (3) living systems are by their very nature purposive (cybernetic) in character, and their adaptation and evolution over time have been shaped in part by functional control information (see chapter 14).

This paradigm sharply contrasts with the thermodynamics paradigm, which allows (and even invites) externally driven models of living organisms, and with attendant laws of evolution. Many of these theorists (by no means all of them) assume that available energy is a free good that can simply be poured into a living system, and that the environment presents at most only limited constraints. In contrast, the thermoeconomic perspective is fundamentally Darwinian, assuming that the struggle for existence is a process in which living systems must unfailingly earn a living in the economy of nature. In this paradigm, there is no "order for free," as Stuart Kauffman would have it; all forms of order must also have a Darwinian seal of approval.

An Illustration: Maxwell's Demon

One way of illustrating this paradigm shift is by revisiting perhaps the most famous of all thought experiments—namely, Maxwell's demon. In his classic text, *Theory of Heat* (1871), physicist James Clerk Maxwell proposed a means by which, supposedly, the second law might be violated. Maxwell conjured up a fanciful being that would be stationed at a wall between two enclosed volumes of gases at equal temperatures. (The term *demon* was actually coined by a contemporary colleague, William Thomson.) The demon would then selectively open and close a microscopic trapdoor in the wall, in such a way as to be able to sort out the mixture of fast and slow gas molecules between the two chambers. In this manner, Maxwell suggested, a temperature differential would be created that could be used to do work, thereby reversing the otherwise irreversible thermodynamic entropy.

We suspect that Maxwell never thought his successors would take his demon very seriously, but many have. That is why, in the late 1920s, physicist Leo Szilard (1929) was compelled to argue, in a professional journal, that the energetic costs associated with the demon's efforts (he focused on the gathering of information) would cancel out any gains from the sorting process; the demon had to be part of the thermodynamic accounting.[7] Then, in 1949, Leon Brillouin added the argument that in order to be able to see the molecules, the demon would also need illumination. Following Szilard's lead, Brillouin (1949, 1968) stressed that the information required to do the sorting involved an offsetting (entropic) cost.

Many other theorists since the 1940s have made similar arguments (see especially the papers collected by Leff and Rex 1990), but the demon refuses to die. For instance, physicist David Layzer (1988, p. 37) revived the issue with a proposal that the demon could be replaced by "a tiny robot" that would be programmed with information about the positions and velocities of all the gas molecules after an "initial moment." This would allow the trapdoor to be opened and closed automatically. Of course, Layzer conceded, "such a calculation would need to be based on an immense quantity of data . . . but that is all right in a thought experiment." No, it is not all right. One cannot arbitrarily set aside the constraints of the real world and then claim to have found a way to violate the second law. Layzer's argument fails if the vast energetic cost of designing, building, and operating the robot, and of acquiring the necessary information, is included. Furthermore, as shown in Kline (1997), the very notion that it could ever become possible to track and sort individual molecules in a volume of gas is scientifically and technically "wildly unfeasible."[8]

Another problem with Maxwell's paradigm, mostly overlooked, is that the demon would be attempting to derive work from a thermal gradient in a control mass with a fixed energy content (an isolated system). If, for example, the two volumes were hooked up to a heat engine that was coupled to a means for recapturing the energy from the work output, the objective would be thwarted by the Kelvin-Planck dictum. This principle states, in effect, that one cannot create a perpetual motion machine; the output would not be completely reversible. So, Maxwell's classic model, even with the assistance of modern technology, is not a paradigm for progress.

A Cyanobacterium in Sunlight

The fundamental problem with Maxwell's demon is that it was not really an experiment in thermodynamics but a surreptitious—unacknowledged—experiment in biology, cybernetics, and thermoeconomics. Maxwell himself can be blamed in part for creating this muddle. In the famous and much-quoted passage from his book about his imaginary creature, Maxwell wrote that the second law is true "as long as we can deal with bodies only in mass, and have no power of perceiving or handling the separate molecules of which they are made up. But if we conceive a being whose faculties are so sharpened that he can follow every molecule in its course, such a being . . . would be able to do what is at present impossible to us" (quoted in Leff and Rex 1990, p. 4). Setting aside the egregious implication that such a perceptual feat—tracking every molecule in a volume of gas—might *ever* become possible, Maxwell then proceeded to make a serious conceptual error. He claimed that his hypothetical creature could "without expenditure of work" create an energetic differential in a divided vessel. What, no work? At a stroke, this assertion effectively removed the demon from the realm of realism. Of course, Maxwell was only using his metaphor as an illustration of the fact that statistical methods are important to microlevel thermodynamic analyses. He did not pose it as a serious theoretical problem. Unfortunately, many of his successors *have* taken it seriously. Leff and Rex (1990) provide an annotated bibliography with some 250 references, many of which are concerned either with exorcising or resurrecting the demon.

Beginning with Leo Szilard's famous 1929 paper, Maxwell's thought experiment was redefined in such a way that it forced physicists to include the costs of the demon, especially the informational costs, in the thermodynamic bookkeeping, rather than treating them as externalities. This in itself was a major contribution, whatever may have been the ultimate flaws in Szilard's argument (see the critique in Corning and Kline 1998b, appendix A). In addition, there were the (usually) overlooked economic costs associated with designing, building, and operating the demon. (A recent example can be found in C. H. Bennett 1988.)[9] As an increasing degree of realism was introduced into the debate, along with various doomed attempts to add technological improvements to the demon, the physics community ultimately converted the experiment into a problem in information theory and, lately, into a pedagogical tool in introductory physics courses.

The ultimate failure of physicists to design a feasible Maxwell's demon highlights the fundamental problem associated with defining the evolutionary process in purely thermodynamic terms. The demon paradigm shows us, inadvertently, why it cannot be done. In a nutshell, there is no way to operate the demon at a profit. Contrary to the claims of many physicists and biophysicists over the years, the evolution of living systems can best be explained not in terms of the laws of physics (or the concepts of entropy and negentropy), but in terms of thermoeconomics. The laws of thermodynamics describe underlying physical conditions and constraints with which bioenergetic and human-made technological systems must cope, but they do not encompass or explain the informed, purposive actions of a cybernetic control system such as Maxwell's demon. In living systems (and, by extension, in human technology), the locus of causation is not confined to the energetics; it is crucially dependent also on the information-based actions of purposeful biophysical structures and processes. In order for living systems to function, work must be done to acquire and make use of available energy, which necessarily entails extraction or production costs and cybernetic control activity.

In effect, the structures and activities associated with the capture and utilization of energy for purposeful work introduce a new set of bioeconomic and cybernetic criteria into thermodynamic processes. This suggests the need for such familiar economic concepts as capital investments, operating costs, efficiency, and even amortization. (Consider, for example, the annual "retooling" by deciduous trees.) A good model for the role of energy in living systems is a cyanobacterium in sunlight. Nature has vastly improved on Maxwell's demon by developing a highly efficient energy-capturing system that regularly operates at a profit. It is time to give bacteria the credit they deserve, and to give Maxwell's demon a decent burial—or perhaps a cremation.

The Thermoeconomics Paradigm

Harold Morowitz, one of the leading figures in biophysics and a major contributor to our collective effort to understand more fully the origins of life, inadvertently provided an illustration of the need for a broad, thermoeconomics paradigm in his path-breaking (and still valuable) volume on *Energy Flow in Biology* (1968). Recall how he proposed that the evolutionary process has been driven by the self-organizing influence of energy flows, mainly from the sun. "The flow of energy through a system acts to organize that system. . . . Biological phenomena are ultimately consequences of the laws of physics" (p. 2).

This, unfortunately, was an overstatement. If energy flows were all that mattered in the evolution story, then we should expect to find complex living systems everywhere on earth and, indeed, everywhere else in our solar system (we assume that the laws of physics are also applicable there). So there must be something more involved—some other ingredient—and in fact there is, as Morowitz himself acknowledged later on in his book. In the penultimate chapter, where he explored ecological aspects of energy flows, Morowitz wrote "at this point, our analysis of ecology as well as evolution appears to be missing a principle" (p. 120). His conclusion: Although the flow of energy may be a necessary condition to induce molecular organization, "contrary to the usual situation in thermodynamics . . . the presence or absence of phosphorus would totally and completely alter the entire character of the biosphere" (p. 121). And, we might add, so would the absence of water, or carbon dioxide, or oxygen (for aerobes).

Furthermore, as Morowitz noted earlier in his text, the lowest trophic level in the food chain is dependent on exogenous sources of free nitrogen, which would otherwise be a limiting condition (Liebig's Limit) for the entire biosphere (as opposed to the abundant supply of energy that actually exists). Finally, and most significant, Morowitz acknowledged that the functionally organized cyclical flow of matter and energy in nature requires a cybernetic explanation. In his words, ". . . the existence of cycles implies that feedback must be operative in the system. Therefore, the general notions of control theory [cybernetics] and the general properties of servo networks must be characteristic of biological systems at the most fundamental level of operation" (p. 120). Exactly so. Biological evolution takes place within a situation-specific array of constraints and needed resources, and its course is also greatly affected by various kinds of control information—from enzymes to genes, to nerve impulses, to cultural information (memes). Thermodynamics, and especially entropy, has little to say about such matters.

Equally important, natural selection has played a major role in shaping the process, perhaps from the very outset. Some theorists (e.g., Wicken 1987; Depew and Weber 1995; Kauffman 2000) hold the view that biogenesis (the origin of life) was shaped by the laws of physics and that a historical process (natural selection) came later. Setting aside the growing suspicion that the laws of physics may themselves be artifacts of cosmic evolution, the assumption that the process of biogenesis was somehow indifferent to the specific historical environment and followed an autocatalytic, self-organizing course that got things right on the very first try is a dubious proposition, I would argue. Given the ubiquity of variation in nature—plus the high frequency of failures and the evidence that functionally important evolutionary inventions and improvements do not as a rule follow a smooth, predestined course, but instead emerge from a messy process of "trial-and-success" (to use Julian Huxley's felicitous term once again)—it is more likely that history, and particularly economic criteria, were codeterminants from the outset. Indeed, by its very nature the process of biogenesis created dependencies—the need for a suitable environment and the need for access to a variety of material resources (namely, carbon, hydrogen, nitrogen, oxygen, phosphorous, and sulfur), in addition to an abundant supply of available energy.

Accordingly (to repeat), a number of familiar economic criteria were likely to have been important from a very early point—capital costs, amortization, operating costs, and, most especially, economic profitability. The returns had to outweigh the costs, especially with regard to energy capture. This historical aspect in turn provided an opportunity for synergistic functional innovations and improvements that were differentially favored by natural selection. To reiterate: The Synergism Hypothesis asserts that the synergistic effects produced by various combinations of elements, parts, or individuals are themselves an important causal agency in evolution. Functional effects are also causes—they are important determinants of natural selection. Moreover, many of these synergistic innovations have involved new methods of energy capture.

The Thermoeconomics of Biogenesis

This important evolutionary trend can perhaps be illuminated by reviewing a few highlights. We begin near the beginning. Among the many useful sources, see especially Morowitz (1968, 1978a, 1992); Lehninger (1971); Broda (1975); Harold (1986); Curtis and Barnes (1989); and Nicholls and Ferguson (1992). Until recently, it was widely believed that photosynthesis—the ability to feed upon direct energy inputs from the solar flux—was preceded by fermentation—the consumption of energy-rich organic compounds, such as the simple sugars that formed spontaneously in the prebiotic environment, in the presence of solar radiation (Broda 1975; Curtis and Barnes 1989). However, it was not a free lunch, for there would also have been significant acquisition costs.

Another problem with fermentation as a biotechnology was that it was based on exploiting a strictly limited resource in a relatively inefficient manner. For instance, when yeast cells are placed in a barrel of sugar solution, they can recover (in the form of ATP) only about 35 percent of the energy content during alcoholic fermentation; the rest is lost as entropy (mostly in the form of waste heat). But more important, as Broda noted, this was ultimately a dead-end strategy. A growing population of living organisms would have been dependent upon a limited and ultimately shrinking resource base. Absent the invention of a means for directly tapping the abundant renewable energy resources of the sun, the evolutionary process might have come to an early end.

However, in recent years a radically different scenario for the origins of life has emerged from the work of a number of theorists, including David Deamer, Harold Morowitz, and several others (see especially Deamer 1978; Deamer and Oro 1980; Deamer and Pashley 1989; Morowitz 1978b, 1981, 1992; Morowitz et al. 1988). This new scenario focuses on the role of amphiphiles—elongated fatty molecules that are like the lipids in modern cells. Amphiphiles, which evidently were present in the prebiotic environment, have the unique property that they are hydrophobic at one end and hydrophilic at the other end; they will align themselves with respect to a water medium. Thus, these molecules can self-assemble into vesicles, which are envelopes that might have provided a protected enclosure within which various forms of protochemistry could arise with the aid of raw material resources and an energy source such as free-floating protons. These resources could have been selectively transported across the amphiphile membrane from the surrounding aqueous medium. This development, in turn, might have set the stage for development of a primitive precursor of photosynthesis, using chromophores that contain photosensitive chlorophyll and retinal molecules.

It is an elegant concept, and the case for a spontaneous, autocatalytic process of this kind is quite plausible; much evidence has been marshaled to support it. However, the developers of this scenario also recognize that each step would have involved new energy and resource dependencies and many opportunities for functional improvements. As Morowitz (1992, pp. 93, 175–76) puts it, "the necessity of persistence in [this] non-equilibrium domain leads to a Darwinian-like struggle for survival [and differential selection] long before there are organisms in the conventional sense." This is consistent with the earlier argument of Eigen and Schuster (1977) regarding the likely role of Darwinian selection in the emergence of autocatalytic "hypercycles". Even Depew and Weber (1995, pp. 469–70), who prefer to label it "thermodynamic or chemical selection," nonetheless embrace the underlying principle: "In a world in which autocatalytic cycles compete for efficiency in finding, utilizing, and dissipating energy sources, however, there would have been keen selection pressure for any entity that could increase these efficiencies by storing the information needed for autocatalysis and for expanding autocatalytic prowess . . ." In short, natural selection was probably primordial in the evolutionary process.[10]

Energy Progress in Evolution

In a nutshell, the story of energy in evolution has little to do with entropy; it has more to do with progressive improvements in bioenergetic technologies. This can be seen clearly in the development of photosynthesis, a highly sophisticated nanotechnology for exploiting a virtually unlimited energy resource with fantastic profit potential. Even photosynthetic bacteria are able to capture much more available energy than is required for their own immediate maintenance needs. However, the ability of the so-called prokaryotes to exploit atmospheric sources of carbon (CO_2) for building biomass was only marginally more efficient than anaerobic fermentation (Lehninger 1971; Harold 1986). Its principal virtue was that it provided access to an abundant new source of raw materials.

The next significant technological improvement represented a major breakthrough and was highly synergistic. According to the serial endosymbiosis theory (SET) of Lynn Margulis and Mark McMenamin (1993), when primitive eukaryotic protists—one-celled organisms with an enclosed nucleus and various specialized functional units called organelles—developed (or more likely, enveloped) the ancestors of modern plant chloroplasts, they acquired potent new energy-capturing capabilities. Each chloroplast is a specialist (at least in modern land plants) that contains several thousand photosynthetic systems, consisting of a reaction center and 250 to 400 chlorophyll and carotenoid molecules—perhaps as many as one million antenna pigments altogether. Moreover, each eukaryotic cell may contain forty to fifty chloroplasts (Curtis and Barnes 1989). In other words, eukaryotes can capture many orders of magnitude more energy than their prokaryote ancestors.[11]

A crucial corollary, however, is that the specialization and increased productivity achieved by chloroplasts in turn depends upon a combination of labor in which these specialists are supported by a larger collaborative enterprise, including the metabolic functions provided by the mitochondria and an array of other life-sustaining activities. The result is an interdependent system that is vastly more productive—one that, among other things, is capable of producing some fifteen to twenty times as much available energy (net of entropy) as do prokaryotes (Margulis and Sagan 1995; Ridley 2001).

The next major development in the energy story is associated with the evolution of metazoans, complex multicellular organisms that developed new ways of exploiting the synergy principle. Now each eukaryotic cell, with its forty to fifty million antenna pigments, became a contributor to a vastly larger enterprise in which many photosynthetic cells combined forces and developed entire energy-capturing surfaces, each square millimeter of which might contain half a million chloroplasts. And this already huge number (perhaps 2.5×10^{12} pigment molecules) could in turn be multiplied by the total light-capturing surface area of a given plant. For a single deciduous tree, the total number of pigment molecules might be astronomically large—perhaps 5×10^{22}.

Freeloading—better known as predation—may also be a (relatively) low-cost way to obtain available energy, and this alternative strategy is also likely to have developed early on in the evolution of the prokaryotes. However, a major evolutionary breakthrough occurred when a new class of predators (heterotrophs) developed the ability to utilize an accumulating biological waste product (oxygen) to bypass the rigors of photosynthesis and extract energy directly from the biomass of the so-called autotrophs (such as plants and grasses), using oxidative combustion. This represented a significantly more economical biotechnology. Equally important, it freed the heterotrophs from the need to sit in the sun all day and remain connected to an array of solar panels. However, as Fenchel and Finlay (1994) point out, these increasingly complex forms of energy capture and metabolism were the result of synergistic functional developments that provided adaptive economic advantages. They were not the result of thermodynamic instabilities, fluctuations, or bifurcations.

Finally, various organisms have developed the ability to capture and exploit exogenous energy subsidies to enhance their survival-related activities and reduce internal energy costs. These subsidies range from solar radiation to tidal currents, alluvial flooding, prevailing winds, and even gravity. In humans, needless to say, they have had a major effect in shaping not only the destiny of our species but the course of evolution itself. For example, modern agricultural practices (as of the mid-1980s) required about ten calories of subsidy for every calorie of output. However, the total output per agricultural worker went up proportionately. To repeat, an American farmer then could raise enough food to support him or herself and forty-five to fifty other people (and the number is probably higher today), but a New Guinea horticulturalist could support only four to five people (E. P. Odum 1983).

In sum, the development of novel bioenergetic technologies in the evolutionary process has had little to do with entropy or dissipative structures and much more to do with engineering—design improvements in the ability of living systems to capture and utilize available energy. It is the organized use of available energy in evolved, informed (cybernetic) structures that has been the key, as noted earlier. And the explanation for these changes lies in their economic advantages, as Lotka (1922) long ago suggested. No detailed cost-benefit analysis of this progressive trend has yet been undertaken, to our knowledge, but it is unlikely that we would be surprised by the findings. In fact, this trend supports one of the axioms of evolutionary theory, tracing back to Malthus and echoed by Darwin, which holds that living organisms have evolved the capacity for unchecked multiplication in the absence of various environmental constraints. However, to reiterate: It is not a free lunch.

Does Entropy Pay the Bill?

One of the most striking trends in the evolution of bioenergetic technologies has been the improvement in productivity and efficiency over time—in other words, entropy reduction. Before we consider some aspects of this trend, however, it is necessary to confront two items of conventional wisdom about thermodynamics that are directly related. One is the claim, going back to Schrödinger, that living systems must pay for their thermodynamic order with an equivalent amount of entropy. To repeat, there is no one-to-one correspondence between the creation of order and an increase in entropy. In a perfect (i.e., reversible) process, there would be no increase in entropy at all. But more relevant for our purpose are the many cases in which efficiencies have been achieved that result in per-unit entropy reductions. Here we provide just two examples, one in technology and one in biology. Power plants in the year 1900 required about eight times as much coal per kilowatt-hour of electricity output as do the best power plants operating today. Similarly, in living systems Schmidt-Nielsen (1972) showed that the energy consumption associated with locomotion is far lower per pound for large animals than for smaller animals (the regression line is −0.4). The point is that the inefficiencies—i.e., the wastes or irreversibilities—create entropy, not the ordering processes per se.

Furthermore, many bioenergetic processes are remarkably efficient and entail very little entropy. Internal conversion of chemical energy (ATP) to mechanical work within animal muscles, for instance, ranges from about 66 to 98 percent efficient (Kushmerick et al. 1969; Blake 1991). Likewise, there is almost no entropy associated with the light-dependent reactions in photosynthesis. McClare (1971, 1972) has suggested that there may be a time function associated with the thermalization of energy (and the creation of entropy) in living systems. Very rapid photochemical and biochemical energy conversion processes may, in effect, be more efficient and may reduce energy wastage.

Finally, from a broader, cosmic perspective, what difference does a little more entropy in the universe make? For the sake of argument, let us say that 1 percent of the free energy that living systems are able to capture from the solar flux to do work and build biomass ends up being permanently stored or reused elsewhere in nature. This is better than if all of it were dissipated into deep space, which might otherwise have been the case. So, contrary to Schrödinger's assertion, living systems may actually reduce (very slightly) the total entropy of the universe, at least for the lifetime of this planet.

The other major misapprehension about thermodynamics and evolution has to do with the notion that there is some inherent economizing influence embedded in the laws of physics. This was mentioned in chapter 12. It bears repeating. Often called Prigogine's principle, the claim is made that as thermodynamic processes approach an equilibrium condition, they obey a law of "minimum entropy dissipation." However, this principle is true only in some special cases. Obvious counterexamples are turbulent flows, say, in various liquids and gases. We also noted that Gage et al. (1966) provided a disproof that any variational principle of this kind could exist in a *dynamic* system. Although Gibbs showed in the 1870s that such a principle is applicable to *static* systems, it is not applicable to living systems in general. As discussed in Corning and Kline (1998a), there is no monolithic, general variational principle of any kind in physics, based on a single variable, that applies to living systems. In other words, we must look instead to thermoeconomics (and natural selection) to explain the progressive improvements in energetic efficiency that can be observed in the evolutionary process.

Thermoeconomic Trends in Evolution

There are two distinct thermoeconomic trends in the overall evolutionary process that can be viewed as a reflection of progressive improvements in the capacity of living organisms to acquire and utilize available energy. One such trend, mentioned in the discussion above, relates to the total quantity of energy throughputs. For instance, Karasov and Diamond (1985) have shown that small mammals can process food up to ten times faster than lizards of similar size, with the same or greater extraction efficiencies, because of a greater intestinal surface area. A second trend, identified by Lotka (1922, 1945), has involved an increase in the total energy flux of the biosphere. Ecology textbooks refer to this quantity as the global gross primary production. Indirect evidence of this trend can be found in correlated environmental changes, most notably the reduction in atmospheric carbon dioxide and the increase in atmospheric oxygen over time (see E. P. Odum 1983).

Although evolutionists remain uncertain about many of the details, a related trend has to do with an increase in the earth's total biomass. Wesley (1989), following Ehrenvärd, estimates that there has been a twentyfold increase in biomass from the Cambrian era to the present day. The energetic significance of this increase can be likened to capital/asset accumulation, and we have adopted the term *structural energy* to label this phenomenon. The term refers to energy that is stored in various forms, much of it temporary (like ATP) but some as permanent as the inorganic matter that is aggregated or even manufactured by living organisms. (Thus we differ from Leigh Van Valen 1976, who originally coined the term. He excluded all forms of energy, such as ATP, that might be used for maintenance, growth, and reproduction.)

Structural energy in our usage includes not only the biomass tied up in currently living organisms but also the vast quantities of organic detritus contained in fossil fuels—coal, oil, tars, oil shale—as well as limestone, reef corals, petrified wood, and other inorganic products of organic activity. M. King Hubbert (1971) estimated that the total (initial) quantity of coal (before human consumption began in earnest) amounted to some 15.28 trillion metric tons, half of which can be commercially mined. The remaining oil reserves have recently been estimated to be about 10 trillion barrels (Davis 1990). These estimates represent an enormous accumulation of structural energy (to say nothing of atomic or chemical energy).

Efficiency is also an important concept in thermoeconomics. But as Blake (1991) has pointed out, it is a multifaceted concept that can refer variously to energy capture, chemical conversions, biomechanical work, locomotion/propulsion costs, thermoregulation, and so on. Natural selection sometimes maximizes for one or more forms of energetic efficiency, but more often it produces compromises among various survival-related criteria.

One example concerns the energetic costs of reproduction. The costs vary enormously from one species to another, and the reasons are always multi-factored and complex; they do not correlate closely with an obvious variable such as body-weight (Harvey 1986). Another example concerns human energetics. The cost of transport for a running human (in oxygen con-sumption per unit of body mass per unit of distance traveled) is higher than for many other mammals and birds. Yet humans also excel in endurance, and these two quite distinct adaptations reflect different selection criteria (Carrier 1984).

Improvements in efficiency can be achieved in at least three different ways. One has to do with a decrease in entropy, or the degree to which avail-able energy is fully utilized (often called first law thermodynamic effi-ciency). As we noted earlier, energetic evolution has not always resulted in increases in this type of efficiency. Photosynthetic plants waste a lot of energy in evapotranspiration, and so do many animals when there is no externally imposed need to economize. Likewise, human technologies are notoriously inefficient. For example, it requires two joules of energy from coal to produce one joule of electrical power, and automobiles have maxi-mum energetic efficiencies in the neighborhood of 35 to 40 percent. Overall, only about half of the exogenous energy inputs for human tech-nology are used productively.

Second law thermodynamic efficiency, on the other hand, refers to the fraction of (net) available energy that is utilized to do work in an energetic process. Thus, to use an example provided by Ayres and Nair (1984), a space heater may operate at 70 percent first law efficiency, meaning that only 30 percent of the energy inputs go up the chimney, whereas its second law efficiency may be only 4 percent. Only a small fraction of the heat is turned into mechanical work, whereas the rest may briefly serve to warm our house but is ultimately dissipated.

However, the natural world also provides many examples of a third type of energetic efficiency, namely, adaptations to minimize the absolute quantity of energy used in meeting various biological needs. These adap-tations range from shelter building to hibernation, heat sharing, nest sharing, physiological adaptations (such as fur, feathers, subcutaneous fat layers, etc.), and many others. For instance, recall the study by Le Maho (1977), which documented that huddling behaviors among emperor penguins enabled these animals to reduce their individual energy expendi-tures by 20 to 50 percent during the winter months. It is also important to note that one organism's waste may become the food supply for others.

Consider the many decomposers and scavengers that utilize otherwise wasted energy, or the recycling of oxygen and carbon dioxide between aerobic heterotrophs and photosynthetic organisms. Such interactions require us to do our energy bookkeeping at the ecosystem level, as well as at lower levels in the biosphere (see especially Ulanowicz 1980, 1983, 1986).

Issues in the Bioenergetics of Evolution

Two other issues concerning the bioenergetic aspect of evolution should also be mentioned briefly. One is related to a broader question in evolutionary biology, namely, does natural selection tend to maximize for any particular value or objective? Is there a discernable overall trend or general direction to the process? Some theorists have suggested that in light of their necessary role in biological systems, energy-capturing capabilities are likely to be a major target of selection. This idea was first suggested by Lotka (1922, 1945), who formulated a "law" of maximum energy flux. Van Valen (1976) refined this idea further with his so-called "third law of natural selection." Van Valen posited that natural selection would be likely to maximize, not for energy flows per se, but for what he called "expansive energy"—that is, energetic surpluses or profits that, over time, would enhance the capacity of the biosphere to expand the total quantity of biomass. The progressive improvements in bioenergetic technology cited above would seem to lend some support to this hypothesis, and culturally evolved energy-capturing technologies have, manifestly, played a major role in the emergence of complex human societies. The problem with this line of reasoning is that natural selection cannot maximize for any one parameter over the long run, because complex organisms have a package of important functional requisites. Energetic improvements are not likely to occur at the expense of other survival imperatives. In other words, no trend is an island, to borrow a metaphor.

Finally, there is the vexing issue of complexity in evolution. It is generally agreed that there have been significant increases in biological complexity over the course of evolutionary history, but there is also widespread disagreement about how best to measure complexity, and about its evolutionary significance (see the discussion of this issue in chapter 4).[12] Few, if any, Darwinian theorists think that natural selection would maximize for any form of complexity per se; complexity is likely to be an artifact of various functional advantages (see Corning 1983, 1995, 1996a; Bonner 1988;

Maynard Smith and Szathmáry 1995; Ridley 2001; Szathmáry et al. 2001). On the other hand, many anti-Darwinian theorists seem to think that evolution might do just that.

We believe that an unbiased reading of the fossil record and the diversity of currently living systems will not support the non-Darwinian hypothesis. Complexity—thermodynamic or otherwise—is a contingent survival strategy that is continuously subject to testing and revision in light of fundamentally economic criteria. (Consider the fate of such energy-intensive creatures as large dinosaurs.) From this perspective, it is the functional consequences of various kinds of complexity that have been responsible for differential survival and reproduction of complex systems over the course of evolutionary history. The explanation lies in the economic costs and benefits in a given set of environmental conditions, not in some inherent trend.

Thermoeconomics and Economics

Finally, a word is in order about the long-standing but uneasy relationship between energy and the discipline of economics. As noted earlier, the roots of this relationship can be traced back to Jean Baptiste de Lamarck, Herbert Spencer, Ludwig Boltzmann, and others in the nineteenth century who drew attention to the central role of energy in living systems. However, the first serious attempts to link energy and the science of economics date back to the 1880s. What economist Philip Mirowski (1988a, 1988b) called the "neo-energetics movement" (which he characterized as an "underworld of unorthodox economics") included an assortment of theorists who embraced the idea that energy could be equated with economic value, an approach that bore a resemblance to the better-known labor theory of value. The Belgian physical chemist and manufacturer, Ernest Solvay, was a prominent early advocate of this view.

In the twentieth century, the demographer and physicist Alfred Lotka (1922, 1945) was the first major theorist to view the role of energy within an economic context, and he spoke of using an energetic perspective to illuminate the "biophysical foundations of economics." However, it was physical chemist and Nobel laureate Frederick Soddy who, in the 1920s and 30s, became the most vigorous proponent of an energy theory of economic value. Soddy wrote (1933, p. 56): "If we have available energy, we may maintain life and produce every material requisite necessary. That is why the flow of energy should be the primary concern of economics."

Meanwhile, a contemporary of Soddy's, Frederick Taylor (the father of scientific management), developed a similar but more narrowly conceived labor-energy theory of value that was subsequently espoused by other theorists.

In the post-World War Two era, a number of anthropologists and ecologists also embraced energy-centered theories of cultural evolution, notably Leslie White (1943, 1949, 1959), Richard Adams (1975), Fred Cottrell (1953, 1972), E. P. Odum (1971), H. T. Odum (1971), and H. T. Odum and E. P. Odum (1982), among others. As Mirowski (1988b) observed, although there was much advocacy for the idea, the theory was never developed in a convincing way. Beginning in the 1970s, however, a "new breed" of economists and ecologists arose that began to make serious attempts to quantify energy and apply it to economic analyses. In addition to the pioneering work of Nicholas Georgescu-Roegen (1971, 1976a, 1977a, 1977b, 1977c, 1979) and the assessment of Georgescu-Roegen by Dragan and Demetrescu (1986), see especially Hannon (1973), Slesser (1975), Gilliland (1975), Huettner (1976), Berndt (1978), Berry et al. (1978), Costanza (1980), Boulding (1981), Parsons and Harrison (1981), Bryant (1982), Roberts (1982), Ayres and Nair (1984), Proops (1983, 1985, 1987), Van Gool and Bruggink (1985), H. T. Odum (1983, 1988), and Giampietro et al. (1993). A detailed history and critique of energy-economics can be found in Mirowski (1988a, 1988b). More recently, the work of Howard Odum and his colleagues and students on the concept of "emergy" (a conversion metric for various kinds of energy) has significantly advanced the cause of measuring the energy costs associated with economic activity (see especially H. T. Odum 1986, 1996). Also relevant is systems scientist James Simms's (1999) quantitative approach to the behavior of living systems in terms of energy expenditures and the capacity of an organism to direct the use of energy to do work.

Energy Economics and Bioeconomics

Despite these advances, there is yet no economic science that links energy expenditures to their bioeconomic benefits—the servicing of our basic needs (see chapter 11). To my knowledge, this remains terra incognita. Moreover, from a bioeconomic perspective, the enduring goal of using energy as a universal metric for measuring economic value is ultimately a search for fool's gold. The reason, simply, is that energy must be viewed as only one input to the survival enterprise. No work can be done without it, but the benefits can only be weighed in terms of externally defined human values. Moreover, and this point is crucial, all energy is not created equal in terms of its acquisition and utilization costs and benefits; it is not a free good. Therefore, it cannot be treated as an economic common denominator, any more than a market basket of currencies with fluctuating relationships to each other—say, marks, yen, pounds, and dollars—could serve as a universal monetary measure.

Then there is the problem of entropy—not as a causal agency of some sort but as a constraint and a cost of doing business in the biosphere. Unfortunately, the theorists in energy economics have sometimes been ill-served by their sources in the physical sciences. As the history of Maxwell's demon illustrates, the laws of thermodynamics are blind to the economic and cybernetic (control) aspects of the energy flows in living systems. Furthermore, as noted earlier, the conflation of energetic entropy and physical disorder has seriously misled some economists. To repeat, horse manure does not explain a horse.

Conclusion

Thermoeconomics adds to both evolutionary biology and economics a perspective in which the energetic costs and benefits in relation to meeting survival and reproductive needs are the keys to understanding the energetics of living systems. We believe that an economic (and cybernetic) paradigm provides a better predictor of the advances and recessions in biological complexity than does any formulation derived from the second law. Indeed, living systems may complexify, or simplify, for reasons that are unrelated to the gross energy throughputs. As we noted earlier, there are many cases in nature where reductions in energy use may reflect greater efficiency and even increased complexity (by some criteria). An obvious example is the collective (per capita) energy economies achieved by socially organized species such as honeybees, army ants, emperor penguins, and, of course, humans.

We believe that the entire strategy associated with various attempts to reduce biological evolution and the dynamics of living systems to the principles either of classical, irreversible thermodynamics or statistical mechanics—that is to say, to manifestations of simple, one-level physical systems—is a theoretical cul-de-sac. Physics is highly relevant to biology, but its explanatory arsenal can deal with only a part of the multileveled, multifaceted causal hierarchy that is found in living systems. We believe that we have outlined a potentially fruitful alternative approach, one that is capable of shedding new light on the relationship between energy and the evolutionary process. In so doing, we believe that we have also brought this aspect of evolution more firmly into the Darwinian paradigm. We see thermoeconomics as being fully consistent with Darwinian evolutionary principles, and we believe that this alternative approach will bear much fruit.

— ❧ —

"Information" is a relational concept that assumes meaning only when related to the cognitive structure of the observer.

—Heinz von Foerster

SUMMARY: Norbert Wiener's cybernetic paradigm represents one of the seminal ideas of the twentieth century. It has provided a general framework for analyzing communications and control processes in purposeful systems, from genomes to empires. Especially notable are the many important applications in control engineering. Nevertheless, its full potential has yet to be realized. For instance, cybernetics is used relatively little as an analytical tool in the social sciences. One reason, we argue, is that Wiener's framework lacks a crucial element—a functional definition of information. The functional (content and meaning) role of information in cybernetic processes cannot be directly measured with Claude Shannon's statistical approach, which Wiener also adopted. Although so-called Shannon information has made many valuable contributions and has many important uses, it is blind to the functional properties of information. Here, a radically different approach to information theory is described. After briefly critiquing the literature in information theory, a new kind of cybernetic information is proposed, which we call "control information." Control information is not a thing but an attribute of the relationships between things. It is defined as *the capacity (know-how) to control the acquisition, disposition, and utilization of matter/energy in purposive (cybernetic) processes.* We briefly elucidate this concept and propose a formalization in terms of a common unit of measurement, namely the quantity of available energy that can be controlled by a given unit of information in a given context. However, other metrics are also feasible, from money to allocations of human labor. Some illustrations are provided, and we also briefly discuss some of the implications. (The paper upon which this chapter is based was the winner of the U. K. Cybernetics Society's 30th Anniversary Prize Competition in 1999.)

14

Control Information: The Missing Element in Norbert Wiener's Cybernetic Paradigm?

Introduction

Norbert Wiener's *Cybernetics: Or Control and Communication in the Animal and the Machine* (1948), can truly be called one of the seminal scientific contributions of the twentieth century. Thanks to Wiener's inspired vision, cybernetic control processes are now routinely described and analyzed at virtually every level of living systems, including social, political, and technological systems.[1] Cybernetic processes, including especially feedback processes, are observable in morphogenesis (the translation of genetic instructions into a mature organism), in cellular activity, in plants (see Gilroy and Trewavas 2001), in the workings of multicellular organisms with differentiated organ systems, in the behavioral dynamics of socially organized species (such as *Apis mellifera*, the true honey bee), in the operation of household thermostats, in robotics, in aerospace engineering, and much more. Cybernetics has given us a framework for understanding one of the most fundamental aspects of living systems—their dynamic purposiveness, or goal-directedness. Much productive research has flowed from this paradigm, in fields as disparate as control engineering, molecular biology, plant physiology, neurobiology, psychology, and economics.

And yet, cybernetics is still far from realizing its full potential. For instance, it has been relatively little-utilized in the social sciences, despite the efforts of such theorists as Karl Deutsch (1963), David Easton (1965), William Powers (1973), James Grier Miller (1995), and the present author (1983), among others (see also François 2004). One reason for this shortfall, we believe, is that an important element is missing from Wiener's paradigm, and this omission has diminished its utility as an analytical tool.

Actually, Wiener's oversight involved more than an omission. To be precise, Wiener pointed his followers down a false trail, and this has had unfortunate consequences over the years, not only for the development of cybernetics but also for the related fields of semiotics and information theory. The problem, in essence, has to do with how information is defined and measured. Wiener failed to develop a functional definition of information, which is essential to an understanding of the role and dynamics of communication and control in cybernetic systems. Instead, he adopted an engineering approach similar to that of his colleague Claude Shannon, the "father" of information theory.[2]

Information Theory

Shannon, in his classic 1948 article and his 1949 book with Warren Weaver, confined his formulation of "communications theory" (as he initially called it) to the problem of measuring uncertainty/predictability in the transmission of "messages" between a sender and a receiver. As Shannon and his co-author wrote, "The fundamental problem of communication is that of reproducing at one point either exactly or approximately the message selected at another point. Frequently the messages have *meaning*. . . . [But] these semantic aspects of communication are irrelevant to the engineering problem" (p. 3).

As noted earlier, in Shannon's usage information refers to the capacity to reduce statistical uncertainty. If one were to use the binary bit as a unit of measurement, the degree of informational uncertainty would be a function of the number of bits required to eliminate it. Shannon also adopted the thermodynamics term *entropy* to characterize the degree of statistical uncertainty in a given communications context before the fact. More formally, Shannon's information can be represented by the equation:

$$Ix = \log_2 1/P_x \qquad\qquad (1)$$

where the information content I of an event x in bits is the logarithm to the base 2 of the reciprocal of its probability. Shannon's expression for entropy, then, was:

$$H = -K\sum_i P_i \log_2 P_i \qquad\qquad (2)$$

where K refers to Ludwig Boltzmann's famous constant (1.38×10^{-16} erg/°C) and P_i refers to the number of equiprobable states.

The justification for calling this quantity entropy came from its similarity to Ludwig Boltzmann's and Willard Gibbs's statistical equations for thermodynamic entropy. As noted in chapter 12, this conflation of terms and meanings served only to exacerbate an already serious muddle. It blurred the distinction between thermodynamic (energetic) entropy and physical (structural) order/disorder. The former usage refers to the availability of energy to do work, whereas the latter usage may be quite unrelated to any work potential. (There is more on this matter below.) Shannon was careful to differentiate between informational entropy and thermodynamic entropy, but other information theorists have not been so punctilious. Some of Shannon's followers have even suggested that there is an isomorphy, or equivalence, between statistical, energetic, and physical order/disorder, but that is not correct.[3] One consequence of this conceptual and theoretical conflation was that Shannon's form of information came to be viewed by many theorists as having more potency as an instrumentality for creating order/organization in the natural world than any purely statistical measure can properly support. It imputes causal efficacy to the statistical properties of the messages themselves, without regard to their content.

Unfortunately, Wiener followed the same approach. In his landmark book, published in the same year that Shannon's classic article appeared (1948), Wiener did discuss the functional aspect of information in various places (e.g., Chapter 7, "Information, Language and Society"), but his formal definition and mathematical treatment involved what he called "a statistical theory of the amount of information" (p. 10). Thus, "the transmission of information is impossible save as a transmission of alternatives . . . . Just as the amount of information in a system is a measure of its degree of organization, so the entropy of a system is a measure of its disorganization" (pp. 10, 11). Later on Wiener described enzymes, animals, and other cybernetic processes as "metastable Maxwell's Demons, decreasing entropy. . . . Information represents a negative entropy" (p. 58). (In fact, Wiener did not provide an explicit formalization in his long, discursive, and mathematically challenging chapter on the subject; instead, he focused on how to measure the "amount" of information.)

The suggestion that information is somehow equivalent to negative entropy (i.e., Schrödinger's neologism either for available energy or for statistical/structural order, depending upon which version of the term *entropy* is being referenced) has also encouraged a tendency to reify the concept of information. Biologist Tom Stonier (1990) is perhaps the most emphatic proponent of this view. He argues that information is "real." He writes, "*Information exists*. It does not need to be *perceived* to exist. It requires no intelligence to interpret it. It does not have to have *meaning* to exist. It exists [his emphasis]" (p. 21). Information is an embedded property of all physical order, Stonier says.

Some Pitfalls

What we refer to as statistical and structural (that is, order-related) formulations of information theory have made many important contributions to communications technology, computer science, and related fields. However, these approaches cannot lead to a unifying theory of information for the simple reason that they are blind to the functional (teleonomic) basis of information in living (and human) systems, as Shannon himself acknowledged. Indeed, objections to various overclaims for information theory began almost immediately after Shannon published his path-breaking formulation. As early as 1956, Anatol Rapoport published an important rebuttal article entitled "The Promise and Pitfalls of Information Theory." Rapoport noted that "it is misleading in a crucial way to view 'information' as something that can be poured into an empty vessel, like a fluid or even energy." In what might, in retrospect, be considered a major understatement, Rapoport commented that "the transition from the concept of information in the technical (communication engineering sense) to the semantic (theory of meaning) sense" will be "difficult."

In a similar vein, Heinz von Foerster (1966, 1980, *inter alia*) stressed the functional importance of information for living systems. The nonsense sentences "Socrates is identical" or "4 + 4 = purple" differ profoundly from sentences that have meaning. Likewise, the aggregate number of light photons that might be processed by the retina of a human eye is less relevant from a functional point of view than the analytical and interpretive processes that go on in the brain (the uses that are made of those photons). To repeat the quote from von Foerster in this chapter's epigraph, "'Information' is a relational concept that assumes meaning only when related to the cognitive structure of the observer."

MacKay (1968) also pointed out that Shannon's information, and similar formulations, are crucially dependent upon the existence of a sender and a receiver; otherwise, one is only describing a physical process—a flow of electrons, photons, and the like. For instance, a television screen may display 10^7 bits of statistical information per second. If one were to transmit an entirely new pattern once each second, the number of bits involved (the amount of information) would soon become astronomical, but it would have absolutely no meaning to a viewer. Similar arguments can be found in Ackoff (1958), von Bertalanffy (1968), G. Bateson (1972, 1979), Cherry

(1978), Krippendorff (1979), Maturana and Varela (1980), Eco (1986), Brier (1992), and Qvortrup (1993), among many others.

Nevertheless, the literature associated with statistical and structural information theory has continued to grow over the years, while the problem of meaning—and more broadly, the functional aspect of information—has been ignored, skirted, or acknowledged but largely passed over by the workers in information theory, with some recent exceptions. Other theorists have finessed the problem by working within the framework of a particular information coding system, whether it be DNA codons or phonemes. Yet the fundamental problem remains unresolved. If information is said by some to do work, how can it be differentiated from energy? If information is equated with thermodynamic order, how does it differ from available energy, or physical order (depending upon which version of the term is being referenced)?

But more important, from a functional perspective information is not equivalent either to thermodynamic entropy or negative entropy (order). If it were, why confuse matters by using different terms for the same thing? In fact, this conflation of different phenomena involves a fundamental dimensional error. Information (properly defined) has no dimensions, whereas thermodynamic entropy has the dimensions of energy divided by temperature. It is comparable to equating voltage with length, or mass with velocity. Indeed, physicist Rolf Landauer (1996) has devised a thought experiment that illustrates his argument that there is no minimum energy expenditure that is necessarily associated with information flows; in theory, the information flow could be made reversible (see also C. Bennett 1988).

Also, information (unlike energy) can be endlessly reused; there is no law of informational entropy. Nor is information conserved; it can be multiplied indefinitely. It has also been observed that, in some communications systems, information may flow in the opposite direction from the energy flow (as in the old-fashioned Morse code telegraph, for example). Also, highly organized biological systems tend to be relatively more efficient users of energy. They use information to economize on energy consumption and, in so doing, validate the distinctions between information, energy, and biological organization.

Other Problems

A further objection is that information by itself cannot do anything. It cannot control a thermodynamic process without the presence of a "user" that can do purposeful work. In other words, information must be distinguished functionally from the process of exercising control, yet many theorists simply take this operation for granted, as James Clerk Maxwell did with his demon (and as many other physicists have done since). It is this overlooked aspect—this "free ride"—that has allowed physical scientists to theorize about informational processes without acknowledging the necessary role of cybernetic control processes. To repeat, cybernetic processes cannot even be described by the laws of physics (see Corning and Kline 1998a).

Another theoretical problem with traditional information theory concerns the contexts in which information has no statistical aspect. This can be illustrated by embellishing an example used by Wicken (1987) to show how Shannon information depends upon the existence of alternatives. Flipping a coin repeatedly is said to produce information—a unique sequence among many possible alternatives. But if the coin is two-headed, the outcome of each flip is pre-determined, and so no statistical information is generated. Now suppose that there are two bettors, one of whom does not know that the coin is two-headed (at least initially). Consequently, some money might change hands, even though no statistical information is produced. Furthermore, after a few flips of the coin the "sucker" might get suspicious and challenge the process, precisely because of the absence of statistical properties. Clearly, some other kind of information—what we call "control information"—was also involved in this situation.

Defining information as a manifestation, or embedded property, of physical order (e.g., Tribus, Riedl, Brooks and Wiley, Stonier, Wicken, and others) presents similar difficulties. First, there is the problem of defining order in any empirically consistent, measurable way. We do not gain anything by conflating certain properties of the physical-biological world with a concept that has an inescapably functional connotation for living systems. To the contrary, we obscure the many properties of information that cannot be associated with physical order per se, such as the feedback in cybernetic processes that can even produce disordering effects. (Feedback is highly sensitive to phase relationships in periodic systems; in a poorly tuned system, it can produce all manner of destructive consequences.)[4]

In fact, whole categories of information in living systems are excluded altogether by equating information with order. For many organisms, physical phenomena of various kinds (gravity, the earth's magnetic field, thermal or chemical gradients, moisture, even the ambient flow of solar photons) provide useful information. Living organisms are constantly sensing, filtering, storing and deleting data on a real-time basis, but only some of it is used. This information is not so much ordered as it is sensed or detected, and then utilized in purposeful ways—only a portion of which can be said to be order-creating. One example is the role of facial expressions in shaping the interactions among humans (and other animals), as Paul Ekman (1973, 1982) has demonstrated, following Darwin's lead in *The Expression of the Emotions in Man and Animals* (1965). Facial expressions, with or without intent, can convey important information, but only to another animal that can properly interpret their meaning.

But perhaps most important, any definition of information that equates physical/statistical order with functional organization commits a fundamental typological error. Biological organization has properties that are not reducible to physical order. (On this point, see chapter 12 and Corning and Kline 1998a.) To repeat, cybernetic processes have the perverse property of being relational—they are always dependent upon the relationship between a given system (inclusive of its goals) and its specific environment.

Control Information

Accordingly, we propose that a categorical distinction should be made between what we have called statistical and structural definitions of information (which have important uses) and control information—which we designate "I_C" and attempt to formalize below. We define control information as *the capacity (know-how) to control the acquisition, disposition, and utilization of matter/energy in purposive (teleonomic) processes.*

Control information has a number of distinctive properties. First and foremost, it does not have any independent existence. It is not a concrete thing, or a mechanism. It is defined (and specified) by the *relationship* between a particular cybernetic system (a user) and its environment(s). In this paradigm, the physical environment contains latent or potential control information, but this potential does not differ in any way from the physical properties of the environment. Moreover, this potential is actualized only when a purposeful system makes use of it. In other words, the very existence and functional effects produced by control information are always context-dependent and user-specific. A few examples may help to clarify this seemingly paradoxical, even counterintuitive notion.

First, imagine a traffic intersection with a stoplight that has just turned red. The information conveyed by the photons of light emitted by the stoplight, and the behavioral consequences that ensue, will depend completely upon the circumstances. A motorist who does not see the light may drive right through the intersection. Another motorist, in a hurry late at night, might observe the light and then deliberately decide to ignore it. However, to the inhabitant of a remote, hunter-gatherer society—say, a Yanomamö tribesman—the red stoplight may represent only a puzzling apparition, while it may be only a bright-colored light to an infant. Thus, the user and the informational source together determine the informational value and the degree of behavioral control that results.

In the second example, imagine that a large boulder straddles a hiking trail in a mountainous area. The physical properties of the boulder are invariant, but the information extracted by four different hikers, and the functional consequences, may vary considerably. One hiker may see the boulder merely as an obstacle and will take action to walk around it. A second one, very tired, may see it as a place to sit down and rest. A third hiker may recognize it as the landmark for a diverging trail that he was instructed to take. Now imagine a fourth hiker who is a gold prospector. Observing a small vein of gold, he proceeds to demolish the boulder to remove the gold and, in the process, destroys forever the boulder's informational potential. Again, the informational process involves an interaction—a specific system-environment relationship.

A final example involves the properties of language. Linguists have long insisted that the functional properties of language (or meaning) cannot be reduced to an invariant, quantitative unit, such as a binary bit. Thus, the letters in RAT, TAR, ART, and TRA have energetic and statistical properties that are equivalent. Yet the meaning (if any) depends on the configuration—the "gestalt." In fact, written language involves an essentially arbitrary relationship between configurations of two-dimensional physical patterns and the associations, if any, that are produced in the specific reader's mind. This explains why the same configuration of letters can have very different meanings in different languages. We all know what the word *gift* means in English. In German it means poison.

The key point here is that control information causes purposeful work to be done in or by cybernetic systems. In accordance with the classic definition, if energy is "the capacity to do work," *control information is the capacity to control the capacity to do work.*

Virtually everything in the universe might, potentially, have informational value (i.e., be used by cybernetic systems for some purpose), but control information is not located in the physical objects alone. To reiterate, it is defined by the relationship between a given object and a given observer/user. Indeed, biological systems vary tremendously in their ability even to detect different aspects of the external world. Thus, the pheromone signals that control the behavior of army ants will go unnoticed and ignored by humans. Elephants can detect very low sound frequencies and dogs can detect very high frequencies that humans cannot even hear. Hawks have some eight times as many photoreceptors per square millimeter of retina as have humans, so there is a definite physical basis for the old expression about being "hawk-eyed."

As the foregoing discussion indicates, control information has a number of distinctive properties. First, control information is always relational and context-dependent, and has no independent material existence; it cannot be identified or measured independently of a specific cybernetic process. However, it *can* be measured (see below). Moreover, there may or may not be a sender, or a formal communications channel, or a message for that matter, but there must always be a user—a living system or a human-designed system. For instance, if you disassemble an automobile into its 15,000 or so component parts, it will no longer be able to utilize instructions from a driver.

Second, control information does not exist until it is actually used. An unread book, an unread genome, or an undetected pheromone represents only potential or latent information. Accordingly, the various mechanisms that exist in nature and human societies for coding, storing, and transmitting potential information are reducible to their underlying physical processes. Their informational properties arise only from the variety of ways in which these physical media may actually be used for informational purposes. To be sure, one can always make estimates or predictions about it, but control information cannot actually be measured except *in vivo* and *in situ*.

Control information, therefore, has no fixed structure or value. It is not equivalent to any specific quantity of energy, or order, or entropy, or the like. To illustrate, a single binary bit may (in theory) control an energy flow as small as a single electron or it could be the signal to launch a nuclear war; its power can vary tremendously, depending upon the context. (Another way of stating it is that all bits are not created equal.) Control information is analogous to money; its value is not intrinsic (despite our perceptions) but is defined in terms of specific transactions—what it can "buy."

Potential control information is very often embodied in various information-storage and transmission media—from DNA templates to the sound patterns in spoken language—but the vehicle must not be confused with its driver. Control information is equally prevalent in the state properties of physical objects—temperature, mass, velocity, viscosity, and so forth. There is no fundamental physical distinction between the two types of information; there is only a functional distinction.

Very often control information has synergistic properties, emerging from an ensemble of informational components or fragments that may be combined in many different ways. Language provides an obvious example.

A change in the arrangement of an identical set of letters (or words) converts the declaration "I shall go" into the question "shall I go?" Similar informational synergies are commonplace also with physical phenomena. Thus, the sight of a swarm of bees coming at you conveys an aggregate informational effect that is lacking if there is only a single bee.

By the same token, much of the information used by (and within) organisms involves processes that might be characterized as inferential—that is, they derive from the weight of the evidence rather than from a deterministic message. To illustrate: You may hear a fire alarm; you smell smoke; you see people running out of your building; you assess the context and your experiential database and may infer that there is a fire and that it would be advisable to vacate the building. In a similar vein, it could be said that the testimony presented at a trial consists of informational components, but only the verdict represents control information (i.e., produces definitive action).

Lies, myths, misinformation, or disinformation of various kinds may also serve as control information, insofar as they affect a user's behavior. It is not the veracity that counts in the control information paradigm, but the functional effects that are produced. (Recall the two-headed coin example above.) There is, in fact, a large literature in biology on the evolution and use of deception as a strategy for achieving various functional outcomes. (For a very different approach to analyzing the role of information in living systems, see Simms 1999.)

Formalizing Control Information

The term *control information* may be novel, but the concept itself is not idiosyncratic or alien. Many other theorists over the years have articulated similar ideas. To cite a few examples: Raymond (1950) pointed out that information controls the expenditure of energy. Rapoport (1956) characterized information as a means for resisting the second law and reducing entropy. MacKay (1968) noted that information "does logical work"—it has "an organizing function" (well, some of the time at least). Biologist Paul Weiss (1971) insisted that information and biological functions are inseparable. As noted earlier, Wicken (1987) differentiates between statistical information and what he calls "functional information," which he associates with the creation of biological "structures." Similarly, Küppers (1990), following Manfred Eigen, takes the argument to the level of nucleic acids and the very origins of life and speaks of the functional role of template-based information in creating living structures.

The problem, of course, is how to convert this perspective into an analytical framework. Specifically, the question is, how can you measure something that does not exist as a concrete physical entity? Our proposal, in essence, is that it can be measured in relation to what it does—in relation to its power to control and utilize available energy and matter in or by a purposeful system. One can measure its qualitative effects, or its meaning, in terms of the results that are produced—the cybernetic work that is accomplished. Potentially, there are many different ways of measuring these results. However, we have chosen to confine our measuring-rod (initially) to the thermoeconomic realm—that is, the capacity to control purposeful work. Accordingly, our basic formalization utilizes available energy. Our definition is as follows:

$$ICf = \ln Au - \ln Ai \qquad (3)$$

where A = available energy as defined by Keenan (1941, 1951), or the energy available to do work net of the entropy of a system and its surroundings, namely,

$$A = E + PoV - ToSC \qquad (4)$$

where E is the total stored energy, V is the volume, SC is the (Clausius) entropy of the system, Po is the pressure, and To is the absolute temperature of the surroundings. Accordingly, in our formalization Au = the total quantity of available energy potentially accessible for cybernetic control in a given situation by a given cybernetic system; Ai = the total available energy cost associated with bringing the available energy under control and exercising control over its use—inclusive of the cost of reducing/eliminating Shannon entropy (SS) or the cost of Shannon information (IS)—and f represents a multiplier for the quantity of a given type of informational unit that may be present in a given context. Use of the ln form allows one to handle a large range of numbers while expressing both the magnitude and efficacy (or power) of a given unit or ensemble of information. Also, if we take the exponential we get the amplification ratio, a measure of the relative efficiency of a given informational unit or ensemble. Thus,

$$\exp ICf = [Au/Ai] \qquad (5)$$

This formalization, it should be noted, deals only in the currency of energy. Yet cybernetic processes utilize many different kinds of currencies—from electron flows to biochemical interactions, animal and human "signals," manufacturing processes, and even monetary transactions. We believe that the utility of our formalization can be broadened by making appropriate conversions from these units into energetic equivalencies—a well-established technique in energetic analyses dating back to various efforts to develop energy theories of economic value in the 1930s. Here we propose instead to use energetic equivalencies as a common currency for measuring control information. A similar approach can be found in the efforts of H. T. Odum (1988) to develop an energetic measuring-rod for the cost of various kinds of embodied information in human societies. Specifically, Odum uses an energy-scaling factor (solar emjoules per joule) of energy inputs, which he calls "emergy." However, we use the more conventional available energy measure, and we focus instead on the benefits (or outputs) produced by information.

Some Illustrations

We can illustrate this formulation by revisiting examples provided earlier. In the red stoplight example, the signal produces a clearly observable change in the behavior of any motorist who responds by stopping, and this can readily be converted to a quantity of purposeful work output. (A proper accounting should also include the work of the automobile.) But what about the motorist who runs the stoplight? Here the analysis becomes more subtle and difficult. The potential information very likely causes a change in the driver's degree of alertness, heart rate, blood pressure, and so on, and may also result in a slowing down or speeding up (or both) of the automobile. The energetic consequences would be much smaller, but they would still be significant; the information would exercise some influence over the behavior of the driver (and the car). Conversely, in accordance with our definition, no control information would exist for the motorist who did not see the light, or for the Yanomamö tribesman, or for the infant, and there would be no measurable energetic consequences.

Similar energetic analyses could be done for the hiker example. In each of the four hypothetical cases, the boulder generated different quantities of control information by virtue of its influence on the behavior of each hiker. Likewise in the language example, it is axiomatic that words have the power to influence human behavior. A time-honored example is the proscription against shouting "Fire!" in a crowded theater. This venerable legal dictum illustrates both the potential power and the context-dependent nature of control information. Indeed, advertisers and their agencies spend untold billions of dollars or pounds each year trying to find just the right words to induce a desired behavior in the recipient.

Let us also consider a comparative example—operating an automobile versus pedaling a bicycle. The costs in monetary terms for operating a given automobile in a given setting are already quite well-known and could be converted to energetic equivalents. However, we must be careful to separate the costs associated with actually performing the work from the control costs for the process. From this perspective, the control information costs (A_i) turn out to be relatively low compared to the work that an automobile can perform (A_u). To simplify the analysis, the control information cost (A_i) could be equated with the labor (time/energy) consumed by the controller—the driver. So, the quantity (power) of the control information associated with driving a car could be calculated in terms of the available energy consumed by the car in doing work, minus the labor cost for the operator $(A_u - A_i)$. Now compare this with pedaling a bicycle. The control costs (A_i) are approximately the same, whereas the available energy that can be controlled (A_u) is reduced to the muscle work performed by the rider/controller in propelling the bike. Obviously, driving a car greatly amplifies the power of a given quantity of neuronal activity (control information).

The economic aspect of our approach should also be mentioned. As noted above, our basic equation for control information is designed to measure not the total available energy involved in a particular context but the profits, net of the entropy and the informational costs associated with the exercise of control. This approach, we maintain, brings our equation out of the realm of theory and locates it in the real world of economic analyses, where the relationship between costs and benefits plays an important, even decisive, role in determining whether or not potential information becomes actualized. If the efficiency (benefit-cost ratio) is very low, the likelihood that a given form of information may actually be utilized to exercise control will be reduced commensurately. It is likely to remain in the realm of latency. Indeed, our equation (5) above expresses precisely the reason why we will never see a real-world Maxwell's demon, even if it were technically feasible. There is no way that we know of for the demon to achieve an energetic profit. Maxwell's demon has unwittingly identified a law-like principle of control information theory: If the energetic costs of a particular type of control information exceed the potential energetic returns, there will be selection against its emergence and perpetuation.

What about the relationship between control information and organization (biological structures)? Many theorists have pointed to the key role of information in building and maintaining biological systems. It is also a truism that much biological information is encoded, stored, and transmitted in various ways. Indeed, information is an integral part of all biological processes. To some theorists, therefore, it has seemed logical to seek a concrete informational measuring-rod for biological organization. We believe that no such structural measuring-rod will be found. We believe that it is important to maintain a clear distinction between (a) the properties of the various physical media that may serve informational purposes and (b) their precise functional dynamics. By insisting that structural information, like any other kind, is latent (like an unread book) and of no direct functional significance until it is actually used in some way, we do not then have to explain such paradoxes as the fact that a significant portion of the DNA in the genome of any given species may not code for anything—in other words, it may not have any informational value. (The question of *why* so much "junk" DNA exists is another matter; it appears that it may not be superfluous after all.) In our scheme, latent structural information becomes control information if and when it is utilized, and its power is a function of its organizing ability—the organizing work that it can do with the available energy at hand, in relation to a given system.

It should also be noted that we have made no provision in our paradigm for developmental or capital costs—say, the energetic investment in designing and building a demon, or an automobile. Aside from the formidable analytical challenges and the problem of infinite regress (how far back do you go with the bookkeeping process?), ignoring those costs would be likely to produce some highly skewed results. A more logical approach is to follow the lead of economists and accountants, who utilize various cost-allocation and amortization procedures to apportion the developmental costs for various economic processes. Thus, in our automobile-versus-bicycle example, the (external) information costs associated with learning to drive or ride a bicycle, as well as the cost of providing traffic control systems (stoplights, road signs, etc.), might add a very small increment to the total information costs if allocated over the number of uses and users.

Control Information and Semiotics

To anyone who is familiar with the large and productive field of semiotics (the doctrine of signs), the concept of control information may seem to be quite similar. In fact, these two formulations are convergent, but they have different purposes and foci. As articulated by Thomas A. Sebeok (1986), one of the leading figures in modern semiotics, the doctrine of signs and their meanings traces its roots to ancient Greece (see also Nöth 1990).

Indeed, it has been an important theme in the entire tradition of philosophical discourse, from Plato and Aristotle to St. Augustine, Leibniz, Locke, Berkeley, and Charles Sanders Pierce.

A key element of the semiotics paradigm in its contemporary form is the requirement for a *source,* or a producer of *messages* that are communicated via some channel to a *receiver,* or a destination. In other words, it envisions a highly structured process rather like the basic paradigm in information theory. However, semiotics embraces all elements of that process. Equally important, semiotics focuses on the functional properties and meanings of the messages. It is concerned with the content, and not with the physical or statistical properties per se, as in traditional information theory. Although the semiotics paradigm rather obviously applies to human language and communications systems, it has also been applied by semioticians to communications processes in other living systems. There is even a nascent new interdiscipline called biosemiotics (Hoffmeyer 1997).

The control information paradigm is distinctive in three ways. First, it does not presuppose a discrete source of messages or structured channels. In our paradigm, every aspect of the phenomenal world represents latent information that may be detected and used in a myriad of different ways in cybernetic processes, and its role may be entirely passive. Indeed, even the absence of something may be of informational significance to a cybernetic system.

Second, our focus is on the user—a cybernetic system and his or her (or its) goals and capabilities. Control information is always defined in terms of the functional relationship between the source and the user. But most important, our paradigm provides a way of measuring the meaning of various signs in terms of one or more quantitative metrics. We have proposed a way of measuring the relative power and efficacy of semiotic processes in cybernetic systems, and we believe that semiotics as a science can benefit from the use of our control information concept.

Conclusion

We believe that the concept of control information provides a new tool for analyzing cybernetic processes, and informational processes, both in nature and in human systems. It provides both a qualitative and a quantitative measure of information in terms of the functional consequences that are produced by a given informational unit in a given context. Moreover, it has many practical applications. Indeed, it is used implicitly in many different fields, from advertising to politics and education. As noted above, it also lends itself well to economic analyses. Thus, control information enriches Wiener's original vision by providing a new and more fruitful way of measuring the relationship between goals, communication processes, and control functions. We believe that control information does indeed provide the missing element in Norbert Wiener's cybernetic paradigm.

PART IV

Evolution and Ethics

— ☙ —

Charlie Allnut: What ya bein so mean for, miss? Man takes a drop too much once in a while, it's only human nature.

Rose Sayer: Nature, Mr. Allnut, is what we are put in this world to rise above.
— *The African Queen* (C. S. Forester)

SUMMARY: Evolutionary ethics is a subject that has been debated ever since Darwin's day. The basic issue is whether or not human ethical systems can be explained—and justified—in terms of evolutionary principles. In recent years there has been an upsurge of publications devoted to this issue, including many new books (as well as a number of books on Darwinism and religion) and countless journal articles. Indeed, an Internet search using the term *evolutionary ethics* yielded 65,400 citations of various kinds. As this outpouring of publications suggests, there has been a great diversity of views on the subject over the years. However, the sea changes in evolutionary theory that were described in earlier chapters also have major implications both for our ethics and our understanding of the moral impulses that shape our lives and societies. In this chapter I critique the history of evolutionary ethics, concluding with an argument favoring the proposition that our ethical systems are products of human evolution and are genetically grounded; they are more than simply cultural inventions, or the actualized ideas of ancient philosophers.

15

Evolutionary Ethics: An Idea Whose Time Has Come?

Introduction

Can there be any doubt that ethics is a cutting-edge issue? We are daily assaulted by routine private acts of violence, chicanery, and deception, as men and women (and children) make choices or act out compulsions with ethical ramifications. We are also daily witnesses to ethically abhorrent political acts—the Oklahoma City bombing, the gas attacks in the Tokyo subways, the assassinations of political leaders in Mexico and elsewhere, the brutal civil war in Bosnia, the tribal bloodbath in Rwanda, the ruthless destruction of Chechnya, the insurgency in Iraq, and, not least, the terrorist attacks on the World Trade Center and the Pentagon.

Can evolutionary ethics play a part in addressing this age-old problem? Although evolutionary ethics traces its roots back to the nineteenth century and the "Synthetic Philosophy" of Herbert Spencer, its role in the ethical discourse of the past century has been checkered, to say the least. It played a prominent part in the ethical and political dialogue of the late nineteenth and early twentieth centuries—most visibly in connection with Social Darwinism and the eugenics movement—but from the 1930s to the 1970s it was totally eclipsed by cultural determinism and value relativism. During this minor dark age, a number of prominent biologists continued to write *ex cathedra* on biology and ethics (Julian Huxley, Warder C. Allee, Theodosius Dobzhansky and C. H. Waddington come to mind), but it was not until sociobiology forced its way through the previously barred doorway into the social sciences that evolutionary ethics regained legitimacy. Several book-length monographs and numerous articles on the subject have appeared over the past few years. However, the sea changes in evolutionary theory generally, and in our view of human evolution in particular, that were documented in part 1 of this volume provide a significant opportunity to reassess the status of the field and to address the broader question: Is evolutionary ethics an idea whose time has finally come?

Some Historical Perspective

First some historical perspective—admittedly glimpsed through a very small peephole. The use of nature and/or human nature as a grounding for ethics can be traced at least to Periclean Athens. To Plato and others of the so-called idealist school, human communities have their origins in the ability of individuals to meet their basic physical needs (including self-protection) through collaborative efforts; mutual aid, reciprocity, and the division of labor are the root causes. (I quoted Plato in chapter 8.) But if utilitarian ends are the basic incentives for social life, a community can also become the instrument for human development—specifically, for the realization of "the good life" and for taming the darker side of human nature. In the *Republic*, his utopian masterpiece, Plato proposed to vest such a perfecting role in specially trained philosopher-kings. But in later works, specifically the *Statesman* and the *Laws*, Plato opted for the second-best alternative of using government and law as instruments for societal improvement.

Aristotle, Plato's most famous student, had less confidence in human nature: "Man when perfected is the best of animals, but when separated from law and justice he is the worst of all" (1946, book 1, part 2). Accordingly, governments and legal systems exercise a vitally necessary constraining influence on human behavior. To paraphrase a line from the poet Robert Frost, good laws make good neighbors. However, Aristotle like Plato endowed the political community (or *polis*) with an overarching ethical purpose, namely, that of molding the raw material of its members into a "self-sufficient" and "harmonious" whole. In Aristotle's view, the true nature of a person, or a *polis*, involves what he/she/it is capable ultimately of becoming. To put it in modern terms, it is not the genes but the phenotype that defines human nature and human potentialities. Here is the model for a variety of progressive modern visions—socialism, the New Deal, the Great Society, and so forth.

A very different view of human nature and the social order was advocated by the Greek philosophers of the Sophist, Skeptic, Epicurean and Cynic persuasions (the very terms give the game away). The Sophist Antiphon, who actually predated Plato's school, preached the shocking idea (to his contemporaries) that all laws are merely conventions and that what is natural is the pursuit of self-interest. Human nature, in other words, is grounded in egoism. Morality, law and justice are at best the embodiment of enlightened self-interest. Thus, civilization is a product of artifice and expediency; it is not a moral Jell-O mold. Many years later, when the Greek city-states were in decline, the Epicureans revitalized and embellished these ideas by advancing a materialistic pain-pleasure principle and an early incarnation of the Benthamite slogan "the greatest happiness for the greatest number."

Here, then, were the philosophical roots of social contract theory, of eighteenth and nineteenth century liberalism, and of late twentieth century neoconservatism. Now, fast-forward through more than two millennia of philosophical writings, during the course of which these and other assumptions about human nature and society were utilized to anchor various systems of ethics and political theory. The list of theorists includes, among others, Cicero, St. Augustine, St. Thomas Aquinas, Marsilio of Padua, Machiavelli, Grotius, Hobbes, Locke, Montesquieu, Hume, Rousseau, Comte, Burke, Bentham, Mill, Kant, Hegel, Marx, and Charles Darwin. Yes, Charles Darwin.

Darwin's Darwinism

In more ways than are generally appreciated, Darwin himself laid the theoretical foundation for what later came to be called evolutionary ethics. One of his contributions was a more sophisticated understanding of natural selection and its behavioral implications than is found in many of the social Darwinist's (and Neo-Darwinist's) caricatures. His most famous slogan "the struggle for existence" was, as Darwin himself pointed out, somewhat hyperbolic. The problem of survival and reproduction in fact encompasses a great variety of specific circumstances, from plentiful resources and easy living to extreme scarcity, from mutualistic symbioses to literal cases of "nature, red in tooth and claw." Mutual aid, moreover, is commonplace, as Darwin wrote in *The Descent of Man*:

> Animals of many kinds are social; we find even distinct species living together; for example, some American monkeys; and united flocks of rooks, jackdaws and starlings. . . . The most common mutual service in the higher animals is to warn one another of danger by the united senses of all. . . . Social animals perform many little services for each other; horses nibble and cows lick each other for external parasites. . . . Animals also render more important services to one another; thus wolves and some other beasts of prey hunt in packs, and aid one another in attacking their victims. Pelicans fish in concert. The

Hamadryas baboons turn over stones to find insects, etc.; and when they come to a large one, as many as can stand around, turn it over together and share the booty. Social animals mutually defend each other. Bull bisons in North America, when there is danger, drive the cows and calves into the middle of the herd, while they defend the outside. . . . (1874, pp. 115–17)

Some nineteenth and early twentieth century ideologues, who wrote as though the existence of cooperation falsified Darwin's theory, seem not to have read his work. Nor, one suspects, did some of his more carnivorous defenders (see below). In fact, in *The Origin of Species* (1859) Darwin explicitly theorized that cooperative behaviors, including the division of labor and even altruism, could well have evolved via natural selection.

Darwin's views about human nature and the evolution of human societies were developed in the first half of *The Descent of Man* (1874), and in *The Expression of the Emotions in Man and Animals* (1965). The principal thesis of these works was that the human species had arisen through the same materialistic processes that governed the rest of the natural world and that *Homo sapiens* shares a common descent with all other animals, including a very close relationship with the higher non-human primates. But more important for our purpose, Darwin attributed our dominant position in nature and our remarkable cultural attainments to evolved social, moral, and mental faculties, in combination with our language abilities.

Though often portrayed as an apologist for cutthroat competition, Darwin in reality placed our sociality and our moral faculties highest among those qualities that have contributed to our success as a species. Following a discussion in *The Descent of Man* devoted to the role of social behavior and sympathy (what contemporary sociobiologists would call altruism) in various animal species, Darwin dealt at length—though in a speculative fashion—with "man as a social animal." In essence, he proposed that our moral systems should henceforth be studied as a branch of natural history—that is, within an evolutionary framework. Darwin's take was that morality is indeed a product of the evolutionary process. He believed that our "social instincts," including even our capacity for sympathy, kindness, and the desire for social approbation, are rooted in human nature. The rudiments of these behaviors, he pointed out, can be found in other social species as well.

Darwin also stressed the role of social cooperation, reciprocity, and mutual aid in human evolution, especially in food-getting but also in conflicts with other groups and other species:

In the first place, as the reasoning powers and foresight of the members became improved, each man would soon learn that if he aided his fellow-men, he would commonly receive aid in return. From this low motive he might acquire the habit of aiding his fellows. And the habit of performing benevolent actions certainly strengthens the feelings of sympathy which gives first impulse to benevolent actions. . . . But another and much more powerful stimulus to the development of the social virtues is afforded by the praise and blame of our fellow-men . . . and this instinct no doubt was originally acquired, like all other social instincts, through natural selection. (1874, pp. 146–47)

Darwin was well aware of the fact that these social instincts might seem at first glance to contradict the imperatives of natural selection and his own admonition, in *The Origin of Species,* that no organism can evolve an adaptation for the "exclusive good" of some other, unrelated organism. This would falsify his theory, he acknowledged. Accordingly, Darwin proposed a solution to this puzzle that now goes under the heading of "multilevel selection" theory. In modern terminology, Darwin suggested that natural selection operated at three different levels—between individuals, between families of close kin, and between social groups—and that it was possible for the three forms of selection to be aligned with one another rather than being in conflict; there are many forms of mutualistic cooperation that may simultaneously benefit individuals, families, and groups. Indeed, Darwin believed that competition between various "tribes" (group selection) also played a major role in shaping the course of human evolution. "Natural selection, arising from the competition of tribe with tribe . . . would, under favourable conditions, have sufficed to raise man to his high position" (1874, p. 148). The tribes that were the most highly endowed with intelligence, courage, discipline, sympathy, and fidelity would have had a competitive advantage, he argued. Alluding directly to the inherent tension in human societies between competition and cooperation, Darwin made the following observation:

Selfish and contentious people will not cohere, and without coherence nothing can be effected. A tribe rich in the above qualities would spread and be victorious over other tribes; but in the course of time it would, judging from all past history, be in its turn overcome by some other tribe still more highly endowed. Thus the social and moral qualities would slowly tend to advance and be diffused throughout the world. (1874, p. 148)

In sum, Darwin believed that socially organized groups—and the moral systems that gave them coherence—played a key role in human evolution.

Organized human societies are not simply cultural artifacts; they are products of our evolution as a species and have played a vital role in the success of our ancestors over many thousands of generations.

The Founding Father

Darwin himself did not venture explicitly into the realm of evolutionary ethics, but one of his important contemporaries did. It was the polymath and social theorist Herbert Spencer—considered by many in the nineteenth century to be the preeminent thinker of his age—who was in fact the founding father of evolutionary ethics. Spencer has been caricatured and libeled so relentlessly over most of the twentieth century that it is difficult to climb the wall of prejudice that has been built up around him (but see Corning 1982). Briefly, there are two Herbert Spencers—the young ideologue and polemicist of the *Social Statics* (1850) and various public policy debates (this is the Herbert Spencer who inspired the social Darwinist writers, though technically he was not one of them) and the mature theorist whose monumental, ten-volume *Synthetic Philosophy* (published between 1862 and 1893) placed him among the great intellects of the nineteenth century.

Like Plato and Aristotle, Spencer viewed society as a utilitarian instrumentality—a system of exchanges and mutual benefits that arose out of the struggle for existence: "Cooperation . . . is at once that which cannot exist without a society, and that for which society exists. . . . The motive for acting together, originally the dominant one, may be defense against enemies; or it may be the easier obtainment of food, by the chase or otherwise; or it may, and commonly is, both of these," Spencer wrote in *The Principles of Sociology* (1874–82, p. 244). Moreover, the progressive evolution of human societies has been the product of an *interaction* between what would now be called ecological, psychological and socioeconomic forces, including both cooperative and competitive or antagonistic forces. In concluding his overview chapter in *The Principles of Sociology*, Spencer penned a statement that is, to my mind, an underappreciated classic:

> Recognizing the primary truth that social phenomena depend in part on the natures of the individuals and in part on the forces the individuals are subject to, we see that these two fundamentally distinct sets of factors, with which social changes commence, give origin to other sets as social changes advance. The pre-established environing influences, inorganic and organic, which are at first almost unalterable, become more and more altered by the actions of

the evolving society. Simple growth of population brings into play fresh causes of transformation that are increasingly important. The influences which the society exerts on the nature of its units, and those which the units exert on the nature of the society, incessantly co-operate in creating new elements. As societies progress in size and structure, they work on one another, now by their war-struggles and now by their industrial intercourse, profound metamorphoses. And the ever-accumulating, ever-complicating super-organic products [it was Spencer, not Emerson, who coined the term *superorganism*], material and mental, constitute a further set of factors which become more and more influential causes of change. . . . (1874–82, pp. 435–36)

One aspect of Spencer's formulation should be stressed, namely, that he is here clearly suggesting a basis for resolving one of the more vexing problems in the social sciences—the nature of the relationship between the individual and society and the causal potency of each in social behavior and social change. Spencer's views, which were derived from both his psychology and his sociology, were similar to but also differed somewhat from those of Plato and Aristotle. To Spencer, human nature (man's psychological propensities and mental faculties) and society are involved in a coevolutionary process: "The phenomena of social evolution are determined partly by the external actions to which the social aggregate is exposed and partly by the nature of its units . . . observing that these two sets of factors are themselves progressively changed as society changes."

Though Spencer is often portrayed as a conflict theorist who sought to account for societal evolution through a competitive struggle for the "survival of the fittest" (another term coined by Spencer, not Darwin), actually he was a pacifist who abhorred war and held a dualistic view. He suggested that societies can be ranged along two ideal types (to borrow Max Weber's term), "militant" and "industrial" (economic). Whereas the former type had predominated in the past, it was Spencer's view that the latter would do so in the future, and that the overall direction of societal evolution was toward material affluence, peaceful integration, personal freedom and the withering away of the state—a vision of the future that he shared with, of all people, Karl Marx. (Remember that Spencer died at the apogee of the Victorian era, more than a decade before the paradigm-shattering struggle of World War I.)

Spencer's "science of ethics," which provided a foundation for what became known as evolutionary ethics, was derived from his vision of society. As articulated in *The Principles of Ethics* (1898), the final two-volume unit of his encyclopedic opus, the "science of right living" as he called it consisted of an application of the scientific method to the problem of determining which ethical principles and moral precepts would best be able to harmonize a given society at its particular stage of evolution. The criteria for evaluating ethical issues should be their consequences both for the super-organism and its members, recognizing their interdependence:

> So that from the biological point of view, ethical science becomes a specifica-tion of the conduct of associated men who are severally so constituted that the various self-preserving activities, the activities required for rearing offspring, and that which social welfare demands, are fulfilled in the spontaneous exer-cise of duly proportioned faculties, each yielding when in action its quantum of pleasure; and who are, by consequence, so constituted that excess or defect in any one of these actions brings its quantum of pain, immediate and remote. (vol. I, p. 100)

In other words, ethical prescriptions must be tailored to the results that they are likely to produce in specific contexts with regard to the ultimate purpose of society (as Spencer saw it)—the greatest happiness (broadly interpreted) of the greatest number, but with an appreciation also for the fact that individual satisfactions in complex societies are both biologically based and very often interdependent. Here, then, are the philosophical roots of evolutionary ethics—a unique amalgam of Aristotle, Benthamite Liberalism, Darwinism properly understood (although Spencer was also a dogged Lamarckian), nineteenth century psychology, and Marxist idealism (minus the dialectic).

A "Gladiators' Show"

Spencer's newborn science of evolutionary ethics was almost immediately disputed when the well known biologist of that era, Thomas Henry Huxley (dubbed "Darwin's bulldog" for his vociferous public defenses of Darwin's theory), spoke out on this subject in his famous (some say infamous) Romanes lecture of 1893 (reprinted in Nitecki and Nitecki eds., 1993). Huxley shocked his listeners, and subsequent readers, by disavowing Dar-winism as a basis for ethics. The "cosmic process," as Huxley called it, is nothing but "relentless combat"—a war of every man against every man in Hobbes's dour image. Huxley also characterized the natural world a "gladi-ators' show" in which the losers go to the wall. Nature is, indeed, "red, in tooth and claw" (in poet Alfred Lord Tennyson's famous phrase).

So how can one build a social ethics on this model of evolution? How indeed? Huxley had painted himself into a corner in which he could not find any ethical corollaries. The only way to avoid this trap was to promote the human capacity to transcend nature: "Social progress means a checking of the cosmic process at every step and the substitution for it of another, which may be called the ethical process . . ." (J. S. Huxley and T. H. Huxley 1947, vol. 1, pp. 435–36). By substituting the "State of Art" for the "State of Nature," Huxley claimed, human societies could ensure the survival of those who are ethically the best. (This begs the question, of course—why bother?) Huxley likened the process of cultural improvement to that of a gardener who transforms nature into an ordered regime.

Not surprisingly, Huxley's dim view of evolutionary ethics prompted a number of public rebuttals, including one by the famed philosopher and educator John Dewey:

> I have discussed this particular case [Huxley's garden metaphor] in the hope of enlarging somewhat our conception of what is meant by the term *fit*; to suggest that we are in the habit of interpreting it with reference to an environment which long ago ceased to be. That which was fit among animals is not fit among human beings . . . because the conditions of life have changed, and because there is no way to define the term *fit* excepting through these conditions. The environment is now a distinctly social one, and the content of the term *fit* has to be made with reference to social adaptation. . . . We have then no reason here to oppose the ethical process to the natural process. (reprinted in Nitecki and Nitecki 1993, p. 100)

Another critic, Leslie Stephen, expanded on Dewey's argument by pointing out that morality can be based on purely prudential grounds. Following Spencer's reasoning, men may find that peace is preferable to war, that the division of labor and reciprocity can be mutually advantageous and that a personal morality can be derived from our dependence on others for the meeting of our needs. A set of ethical rules—and a system of enforcement designed to prevent anyone from cheating—are in our own best interest. Stephen concluded:

> An individualism which regards the cosmic process as equivalent simply to an internecine struggle of each against all must fail to construct a satisfactory morality, and I will add that any individualism which fails to recognize fully the social factor, which regards society [merely] as an aggregate instead of an organism [i.e., Spencer's superorganism], will, in my opinion, find itself in difficulties. (reprinted in Nitecki and Nitecki 1993, p. 88)

In Robert Bolt's award-winning morality play, *A Man for All Seasons* (1960), there is some dialogue between Thomas More and his son-in-law, William Roper, that speaks forcefully to this point (to which we will return below):

> Roper: So now you'd give the Devil benefit of law!
>
> More: Yes. What would you do? Cut a good road through the law to get after the Devil?
>
> Roper: I'd cut down every law in England to do that!
>
> More: Oh? And when the last law was down, and the Devil turned round on you—where would you hide, Roper, the laws all being flat? This country's planted thick with laws from coast to coast—man's laws, not God's—and if you cut them down—and you're just the man to do it—d'you really think you could stand upright in the winds that would blow then? Yes, I'd give the Devil benefit of law, for my own safety's sake. (pp. 37–38)

The "Law of Competition"

If Huxley at least held out the hope that we could rise above our beastly nature, the social Darwinist writers extolled its virtues. Social theorists including William Graham Sumner, E. B. Tylor, Albert Keller, Gustav Ratzenhoffer, and others took their inspiration from Spencer's "survival of the fittest" image and emphasized raw competition. Fairly typical was this *pronunciamento* by Tylor (1889, p. 7): "The institutions which can best hold their own in the world gradually supersede the less fit ones, and . . . this incessant conflict determines the general resultant course of culture." Likewise, business magnate John D. Rockefeller, in a Sunday school address, assured his audience that "The growth of large business is merely a survival of the fittest. . . . This is not an evil tendency in business. It is merely the working out of a law of nature and a law of God" (quoted in Lux 1990, p. 148). However, it was the steel baron and philanthropist, Andrew Carnegie—never a man to mince words—who penned the most inflammatory expression of the social Darwinist credo in an 1889 essay known as "The Gospel of Wealth." "While the law [of competition] may be sometimes hard for the individual, it is best for the race, because it ensures the survival of the fittest in every department. We accept and welcome, therefore . . . great inequality of environment, the concentration of business, industrial and commercial, in the hands of the few, and the law of competition between these, as being not only beneficial, but essential for the future progress of the race" (1992, p. 132).

The opposing side in this increasingly harsh debate was perhaps most eloquently represented by the Russian émigré anarchist and naturalist, Prince Pyotr Kropotkin. In his famous polemic, *Mutual Aid: A Factor of Evolution* (1902), Kropotkin specifically refuted Huxley's tooth-and-claw image of the natural world. Among other things, Kropotkin argued that there was abundant evidence of cooperation in nature that falsified Huxley's one-sided interpretation of Darwin's theory. "During the journeys which I made in my youth in Eastern Siberia and Northern Manchuria . . . I failed to find—although I was eagerly looking for it—that bitter struggle for the means of subsistence, *among animals belonging to the same species* [emphasis in original], which was considered by most Darwinists (though not always by Darwin himself) as the dominant characteristic of the struggle for life . . ." (p. vi). Kropotkin claimed that cooperation is more important than competition in nature and is the key to progressive evolution. Kropotkin also insisted that social groups were important units of evolution.

The Modern Synthesis

During the period of the so-called modern synthesis in evolutionary biology, from the 1930s to 1960s, biologists generally seemed comfortable with the idea of evolutionary ethics. For instance, both Julian Huxley (grandson of T. H. Huxley) and Theodosius Dobzhansky wrote approvingly about this subject (see Huxley and Huxley 1947; Dobzhansky 1967). The modern synthesis was also deemed to be compatible with group selection of various kinds, just as Darwin had proposed. For instance, Sewall Wright (1968–78) at the University of Chicago coined the term *interdemic selection*—that is, selection between discrete breeding populations, or demes—and he developed what he called a "shifting balance" model, which he believed was of the utmost importance in producing evolutionary changes. Ernst Mayr, likewise, characterized evolutionary change as a population-level phenomenon, meaning that populations and species are the ultimate units of evolution, not individuals. Mayr (1963, 1976) also developed what he called the "founder principle," which envisioned small, reproductively isolated groups as a significant source of evolutionary innovation. Meanwhile, various students of animal behavior, such as William Morton Wheeler (1927) and Warder C. Allee (1931), stressed the cooperative aspect of animal behavior and social life. As noted earlier, Wheeler also promoted the idea of "emergent evolution," and he borrowed from Spencer the idea that a socially organized group can be likened to a superorganism.

One of the chief casualties of the Neo-Darwinian revolution (described in earlier chapters) was Darwin's explanation for human evolution, along with his reasoning about our social and moral faculties. Thus, for example, the philosopher Helena Cronin, in a popular 1991 book, *The Ant and the Peacock,* came to the conclusion that Darwin "lets us down" (p. 327). Why so? Because he relied on the supposedly flawed concept of group selection in explaining human evolution. Likewise, the science writer Robert Wright, in a provocative popularization about the new field of evolutionary psychology called *The Moral Animal* (1994), wrote off Darwin's explanation of humankind altogether: "The more you think about it, the less likely it seems"(p. 186). Despite his book's affirming title, Wright concluded that we are not moral animals after all but only "potentially" so; what passes for morality is "ruthlessly" subordinated to our self-interests, he declared. (A more detailed critique of Wright's book follows below.)

Reviving Darwin's Darwinism

The rejection of Darwin's Darwinism, and of Spencer's evolutionary ethics, was largely the result of two serious, interrelated misconceptions. The first (discussed in chapter 2) was that cooperation and sociality depend on altruism and are therefore severely constrained phenomena. To the first generation of Neo-Darwinians and sociobiologists, it seemed that only kin selection (and maybe reciprocal altruism) might be able to circumvent this obstacle, since group selection was widely viewed as being impotent. The second misconception was that true ethics necessarily requires altruism; enlightened self-interest does not count. In short, it was a classic double bind.

With regard to the first issue, the tendency to equate altruism and cooperation was clearly misguided, and the current revival of group selection theory in evolutionary biology can be attributed, in considerable measure, to a growing recognition that cooperation can often be a win-win process; cooperating groups might provide mutual advantages to their members. To be sure, cooperation may impose costs on the cooperators, but these may be offset by equivalent or greater benefits. In other words, cooperation can also involve the relatively straightforward economic calculus of costs and benefits, and the main constraint may be how these costs and benefits are toted up and distributed among the cooperators and whether or not the tendency to cheat (defect) is policed. Cooperation does not by definition require genes for altruism; there can be "egoistic cooperation" as well as "altruistic cooperation."

Indeed, one of the most important forms of cooperation in nature involves interactions that produce combined effects (synergies) that are largely self-policing because they are interdependent. This is frequently the case with symbiotic relationships, as well as in socially organized species (like humans) that exhibit a division of labor, or teamwork. Recall Maynard Smith and Szathmáry's rowing metaphor, in which two oarsmen, each with only one opposing oar, depend on each other for the performance of the boat; if one of the oarsmen slacks off, the boat will go in circles and neither oarsman will reach his or her goal.

As the evidence for cooperation as a widespread phenomenon in nature has continued to mount in recent years (see chapters 2 and 3), it has become increasingly clear that it is not a minor theme, or a phenomenon that depends on altruism or some hypothetical cooperative gene. Cooperation is a common response to the problems of living—a major survival strategy that is co-equal with competition in its importance. Indeed, competition via cooperation is a common strategy in the natural world. Even some of the most vociferous Neo-Darwinians have conceded as much, though their declamations are sometimes buried deep inside their texts. Consider again these quotes from Richard Dawkins, cited in chapter 2:

> In natural selection, genes are always selected for their capacity to flourish in the environment in which they find themselves. . . . But from each gene's point of view, perhaps the most important part of its environment *is all the other genes that it encounters* [emphasis in original]. . . . Doing well in such environments will turn out to be equivalent to "collaborating" with these other genes. (1987 pp. 170, 171)
>
> In a sense, the whole process of embryonic development can be looked upon as a cooperative venture, jointly run by thousands of genes together. Embryos are put together by all the working genes in the developing organism, in collaboration with one another. . . . We have a picture of teams of genes all evolving toward cooperative solutions to problems. . . . It is the "team" that evolves. (1987 p. 171)

Recall also that, in his signature book *The Selfish Gene,* Dawkins conceded that genes are not really free and independent agents. "They collaborate and interact in inextricably complex ways. . . . Building a leg is a multi-gene co-operative enterprise." Recall also his metaphor from rowing. "One oarsman on his own cannot win the Oxford and Cambridge boat race. He needs eight colleagues. . . . Rowing the boat is a co-operative venture." Furthermore, Dawkins noted: "One of the qualities of a good oarsman is teamwork, the ability to fit in and co-operate with the rest of the crew" (1989 pp. 39, 41).

What Dawkins is talking about here, of course, is group selection. Moreover, this form of group selection does not depend on altruism; it involves win-win relationships. Moreover, the cooperative, group selection model also fits the most likely scenario for human evolution, namely that we evolved from closely cooperating social groups; the context of human evolution most likely required close cooperation to offset the otherwise fatal vulnerabilities of these "Miocene midgets," as Milford Wolpoff has dubbed them. (What I have called the "synergistic ape scenario" is discussed in detail in Corning 2003.)

Reviving Spencer's Ethics

The other misconception, namely that evolutionary ethics must be based on altruism, can be addressed by returning to Spencer's vision. In effect, Spencer argued that an ethical system can be based on "prudential" grounds, or enlightened self-interest. Spencer's argument was hardly new, of course. It can be traced back at least to the Greek Stoics. Spencer's contribution was to relate enlightened self-interest to the biological problem of survival and reproduction and to assert that individual interests and the public interest were not necessarily incompatible or opposed to one another; they can be harmonized.

The crucial conceptual issue here was identified by Leslie Stephen in his rebuttal to Huxley's paradigm. If a society is viewed merely as an aggregate of individuals who have no common interests, and no stake in the social order, then why should they care? But if a society is viewed—more realistically in my view—as an interdependent "collective survival enterprise," then each of us has a vital, life-and-death stake in its viability and effective functioning, whether we recognize it or not. Another way of putting it is that much of our public ethics, and the cultural institutions that we have evolved for encouraging—and enforcing—our ethical principles and rules, are also an expression of evolutionary ethics. The two are not radically different spheres. David Sloan Wilson's recent book, *Darwin's Cathedral* (2002), develops the case for this argument in depth.

Yet, in hindsight, Spencer's evolutionary ethics, while necessary, was insufficient. Who, after the tragedy of 9/11, can doubt the reality of altruism as a significant aspect of human societies. We have grieved even for those we did not know and have donated billions of dollars to help their families. Moreover, there is much research showing that we do indeed seem to have an innate moral sense. For instance, a sense of fairness seems to have a strong, if imperfect, pull on our preferences and our conduct, to the point that we may even be willing to make sacrifices on its behalf. (This will be the subject of chapter 17.)

How can this be? What adaptive advantage could a sense of fairness have bestowed on our remote Pleistocene ancestors, such that it was "blessed" by natural selection and incorporated into the undergarments of our evolving human nature? The most likely explanation, in a nutshell, is that the principle of fairness came to play a central role in reconciling conflicting interests within our ancestors' groups, bands, and tribes. Darwinian group selection was most likely a powerful supplement to Spencer's prudential economic calculus. To quote again Darwin's observation in *The Descent of Man*: "A selfish and contentious people cannot cohere, and without coherence nothing can be effected." Competition may be an engine for enterprise, and economic progress, but mutually beneficial cooperation is the fundamental organizing principle underlying all human societies. Indeed, there is mounting evidence that our sociality and readiness to cooperate far exceeds that of any other primate.

The Moral Animal

Two recent best-selling volumes on evolutionary ethics, Robert Wright's *The Moral Animal* (1994) and Frans de Waal's *Good Natured* (1996) highlight both the theoretical problem and its potential solution. It is worth taking the time to consider their arguments.

The Moral Animal is, first of all, a work of formidable intelligence, wit and skill (and is deeply researched to boot). It was lavishly reviewed, highly praised and deserves its reputation as a somewhat provocative synthesis of evolutionary psychology. Wright calls his volume a "sales pitch" and, indeed, it is a highly partisan rendering of the subject with an admittedly cynical edge. Fair enough. Ambrose Bierce in his *Devil's Dictionary* defined a cynic as "a blackguard whose faulty vision sees things as they really are, not as they ought to be" (1967, p. 55). Wright provides a very creditable review and analysis of the literature on such topics as male versus female reproductive strategies (including the highly charged issue of monogamy versus polygyny), parental investment, friendship, and the nettlesome problems of deception and treachery in human relations, not to mention the convoluted relationship between our ethical preachments and our practices.

In many respects, Wright got it right. As he points out, evolutionary psychology supports the view that there is indeed a biologically based

human nature but that it is also highly flexible and adaptable (not genetically determined). Wright invokes a metaphor from electronics—the idea of
knobs and tunings—to characterize the relationship between our genes and
our environment. We have many biologically based biases and urges, but we
are also capable of learning from experience and controlling these urges—as
Darwin, and Huxley, and Rose Sayer (each from a different perspective)
suggested. Wright also reviews the case for gene selfishness as the guiding
principle behind our behavior; often it is the invisible anchor for our psychological proclivities. The good news is that our moral impulses may also
have a biological basis. The bad news, Wright claims, is that these impulses
are highly selective, inconsistent, and "ruthlessly" subordinated to our self-
interest; they do not seem well attuned to the good of the species. In short,
says Wright, we delude ourselves in thinking that our morality is not really
self-serving.

While it is hard to find fault with Wright's reportage, the spin he puts
on it, his interpretation, is deeply disappointing. At the very end of this
"feast of great thinking and writing" (as one reviewer put it), Wright finds
himself in an ethical cul de sac. Evolutionary psychology can give us no
more ethical guidance than can the musty musings of an Antiphon or a
Plato (we can't escape from the naturalistic fallacy). Nevertheless, in what
amounts to a non sequitur, Wright resurrects a dubious argument for utilitarianism and the greatest happiness of the greatest number. "It's just about
all we have left," he tells us (p. 336). Indeed, having adopted the Darwinian
definition of true morality as the conscious control over our urges (which
sets the bar pretty high), he concludes that we are *not* moral animals (his
title be damned); we are only "potentially" moral. The final result is a logical tangle—a whole that is far less than the sum of its impressive parts.
Because the issue of selfishness is crucial to whatever future evolutionary
ethics may have, it is worth taking the trouble to try to disentangle it.

Self-Interest Revisited

We begin with a simple question: how does Wright (or Antiphon, or Jeremy
Bentham, or even Adam Smith for that matter) define "self-interest"? Often
the term is used to connote a zero-sum relationship in which the "self" gains
at the expense of some "other." Indeed, Neo-Darwinians seem to relish the
idea that, where selfish genes are concerned, the self is ruthless (a flagrant
anthropomorphism). Of course, there is a large philosophical literature on
the concept of "enlightened self-interest"—a form of selfishness that may
overlap with the interests of others (say collective goods like safe drinking

water or the common defense), which goes under the heading of "mutualism" in evolutionary biology. In theory, there are at least three distinct kinds of self-interest: (1) those that are also consistent with—or supportive of—the interests of others, (2) those that are neutral in their effects on others, and (3) those that conflict with the interests of others.

Unfortunately, self-interest often gets linked only to the competitive, zero-sum relationship (number 3 in the list), and Wright's treatment reflects this bias. Although Wright speaks at several places in his text (mostly in passing) about the phenomenon of "non-zero-sumness" in human societies, his clumsy euphemism obscures the vastly important role in humankind of synergy (and mutualism); it is the equivalent of referring to white as "non-black."

The root of the problem, I believe, is a flawed vision of a human society as no more than a vast set of dyadic transactions between individuals in an essentially competitive arena, when in fact it is *also* a complex system of ongoing relationships and interdependencies, many of which are mutually beneficial, from the division of labor in production and reproduction to shared public roads and public order. Competition and cooperation are the "warp and woof" of human societies, to resurrect that old but still useful weaving metaphor. Evolutionary psychology (and its now politically incorrect predecessor sociobiology) have tended to underrate the cooperative dimension of human societies. Our species is sui generis—vastly more dependent than any other social species on economic "niches" that are created by the needs, wants and activities of others, and on joint efforts that produce both collective goods and divisible "corporate goods" (as I call them). To cite one hypothetical example: suppose that two hunter-gatherers each are able to collect enough firewood at dusk to feed a campfire for half the night. If the two of them pool their hoards and share a fire, they will both have enough firewood to stay warm through the entire night—and, equally important, to ward off potential predators. That's synergy, and human societies are rife with it.

Not only is non-zero-sumness (synergy) very important in human societies but it casts our moral impulses in a very different light. Human nature (as best we can discern at this stage) is highly adapted to exploiting the human potential for socially produced synergies, which necessitates fitting ourselves into a social group. We (mostly) enjoy associating and working with others and are highly attuned to the opinions, and influences and "approbation" of others (in Darwin's term), precisely because it is most often in our self-interest to be so motivated; our imperfect social and moral propensities are thus not opposed to our self-interests but are more or less aligned with them; often, in fact, our social needs become ends in themselves. To be sure, moral actions usually require some sacrifice, but, with some notable exceptions, the costs can be viewed as trade-offs for compensating benefits of various kinds. From an evolutionary perspective, we would expect moral impulses to be motivated and supported at the psychological level if they are in fact instrumental to positive synergies and, ultimately, to our reproductive success. Moreover, we would expect to find that these propensities are also influenced by the specific cultural context. Wright himself illustrates this point with the irreverent, and somewhat scatological, exploitation of Charles Darwin's life as a "test case." Darwin, and many other Victorians, demonstrated that a society can after all be a vehicle for moral betterment—for an increase in mutually beneficial civility and domestic tranquility (for the "good life" *sensu* Plato). In other words, we can make cultural choices that will encourage or discourage moral conduct—that is, conduct that is responsive to the needs and attitudes of others.

And yet, we also remain hard-core egoists. The point is that we are not one thing or the other; the idealists and the cynics are both partly correct. The great, inescapable paradox of the human condition is that both the market/exchange metaphor and the superorganism metaphor are partially valid. Human societies (in all but some pathological cases) represent a unique blend, in evolutionary terms, of all three kinds of self-interest, and the endemic conflicts within every society—indeed, within each individual— are a reflection of the interplay between them.

Wright himself unwittingly provides a possible illustration. Why, he asks, has the cultural practice of monogamy arisen in the face of the presumed reproductive advantage of male polygyny? This would seem utterly to contradict the bedrock premise of evolutionary psychology. Wright's explanation is that the advantages must have outweighed the disadvantages. It is a dangerous and destabilizing state of affairs to have relatively few males controlling the reproductive resources of the females in any given society while many more males are denied access to reproduction. So, in a kind of biological Magna Carta, some of our male ancestors made a bargain among themselves to share the females more or less equally, Wright says. But why should peaceful coexistence and reproductive cooperation matter more for humans than for, say, chimpanzees or elephant seals? Precisely because human economies, by and large, involve much more intense, ongoing cooperation and economic interdependency; the benefit side of the social order is typically much greater, as is the potential cost of internal conflict (just read the daily newspapers or watch CNN).

But what about the naturalistic fallacy—the prohibition against deriving ethical "oughts" from any empirical "is"? Even if our ethical impulses make evolutionary (adaptive) sense, so what? Here it may be fair to accuse Wright of letting us down. As noted earlier, Wright could not discern any basis for ethics in Darwinism or evolutionary psychology. "Can morality have no meaning for the thinking person in a post-Darwinian world? This is a deep and murky question that (the reader may be relieved to hear) will not be rigorously addressed in this book" (p.329). Nevertheless, a few pages later Wright presents an argument for utilitarianism as a basis for morality. He claims that the "happiness" criterion is "unscathed" by the naturalistic fallacy because happiness is in fact a value that "we all share" (pp. 334–35). Furthermore, happiness has a non-zero-sum property; everyone's happiness can go up if everyone treats everyone else nicely (synergy). In other words, we can derive an ethical system from a shared and/or interdependent set of social values. Ethics are not ends in themselves but instrumental means; *if* we all prefer happiness, *then* an ethical system can promote our common objective.

Why, in the name of Darwin, can't the same logic be applied to the biological problem of survival and reproduction? Forget happiness. Let's focus on *evolutionary* ethics. If we all (or almost all) seek to survive and reproduce, and if our survival and reproductive success—not to mention the longer-term reproductive success of our progeny (call it "posterity")—is largely dependent, ultimately, upon the collective survival enterprise—the tacit *raison d'être* of a complex human society—why can't we use our shared Darwinian "interests" as the basis for an evolutionary ethics? If we take the long view, and the large view, any ethical system that is conducive to "the survival and reproductive success of the greatest number" would, on balance, also be likely to be conducive to our own survival and reproductive interests. That, I submit, is a logical (and sturdy) foundation for an evolutionary ethics, although I am also well aware that there are some pitfalls to be avoided.

A useful analogue for parental investment within the framework of an enlightened, post–Neo-Darwinian evolutionary ethics might be community investment. After all, this is what the Goodwill, the Salvation Army, Habitat for Humanity, and the like, not to mention our many charitable foundations, are implicitly all about. So we are not just talking theory here. We are talking about what people actually do in the real world. And the good news is that the public interest, or general welfare, is not the chimerical fantasy of incurable romantics whose genes are headed straight for the evolutionary dustbin. Rather, these quaint old-fashioned terms—whose roots trace far back in the tradition of discourse—are conceptual container-ships for the non-zero-sumness (synergy) in society.

Good Natured

The case for an ESI model of evolutionary ethics is buttressed, compellingly, by Frans de Waal's book on the origins of moral behaviors. *Good Natured* (a clever title) is a rare treat, a work that combines good primatology, good social science, good moral philosophy, good writing and, not least, good will. De Waal sets out to show via the research literature in other social species, especially our primate cousins, that morality is not opposed to our animal instincts. "We are moral beings to the core" (p. 2). The mixture of good and evil that can be observed in human societies reflects a duality that can be seen also in other socially organized species. Our close relatives are very often selfish in the zero-sum sense of the term, yet they also exhibit such enlightened behaviors as sharing, succorance, empathy, attachment, reconciliation, tolerance, even concern for the community and for social harmony. (De Waal defines "community concern" somewhat stiffly as "the stake each individual has in promoting those characteristics of the community or group that increase the benefits derived from living in it by that individual and its kin" p. 207.)

Furthermore, there is evidence that other-regarding behaviors are encouraged and supported in primates and humans alike by a substrate of psychological and emotional rewards and punishments. "The fact that the human moral sense goes so far back in evolutionary history that other species show signs of it plants morality firmly near the center of our much-maligned nature," de Waal concludes. "It is neither a recent innovation nor a thin layer that covers a beastly and selfish makeup" (p. 218). De Waal also nicely deflates the conceit that true morality requires conscious deliberation: "Animals are no moral philosophers," he concedes, "but then, how many *people* are?" (p. 209). Animals occupy a number of floors of the "tower of morality," as de Waal puts it, but it is a bit gratuitous to claim that only the very top of the tower can be labeled "moral." De Waal likens our moral propensities to language acquisition, a human trait that exhibits both a specific biological predisposition (and associated machinery) and extensive learning.

Some of de Waal's earlier work, especially his book *Chimpanzee Politics* (1982), has been criticized for excessive anthropomorphism (for the record, by some of the very same people who freely employ the blatant anthropomorphism of selfish genes). Indeed, even Robert Wright fires a barb or two at de Waal (he calls de Waal's primate stories "almost soap-operatic") before appropriating some of them to use as reruns for the entertainment of his own readers. In *Good Natured,* de Waal responds to his critics with a review of the formal primate research literature and an argument for parsimony in theorizing about the striking comparisons between primates and humans. (De Waal also repays Wright with a sharp critique of Wright's charge that we are all self-deceiving hypocrites.)

One aspect of de Waal's book should be highlighted. A centerpiece of human relationships, and human morality, is our apparently universal (albeit imperfect) sense of "equity." It is deeply embedded in the human psyche, and it provides a somewhat erratic moral compass for human relationships—especially for our concepts of justice and fair play.

Consider the classic story "The Little Red Hen"—one of the all-time best sellers among children's books. The Red Hen works hard and is frugal. One day she finds some grains of wheat and decides to plant them. She asks her friends (a dog, a cat, and a pig, in one version of the story), "who will help me plant these seeds?" Well, her friends all have more important things to do, so she plants them herself. And so it goes at each successive stage in the production process—tending and weeding the garden, harvesting the wheat, threshing the grain, grinding the flour and baking the bread. At each step the Red Hen asks for help, but her friends are always too busy. Yet, when it finally comes time to eat the bread, her friends are more than willing to help; they're eager to do so. By then, of course, it's too late. Now, a Marxian scholar, or a well-trained defense lawyer, might object that the Red Hen should not have eaten all the bread herself, but many generations of children, unburdened by the teachings of our moral philosophers and legal scholars, seem to have gotten the point.

De Waal finds suggestive evidence that this sense of justice has its roots in the finely tuned "economy of sharing and exchange" in chimpanzees. If so, this has enormous theoretical significance; it links one of our most fundamental ethical principles to, yes, natural selection. Our own acutely developed sense of justice (by and large) may be without precedent in nature, but so are our language skills, our manual dexterity and a variety of other evolved traits. Indeed, as noted earlier (chapter 6), there is now a substantial body of supportive scientific evidence regarding this distinctive human trait under the headings of "fairness" and "strong reciprocity." In addition to the earlier references cited in chapter 6, see especially the review article in *Nature* by Fehr and Fischbacher 2003. Also, see the elaboration in chapter 17.)

One significant shortcoming in both Wright's and de Waal's presentations—a criticism that can be applied to most recent discussions of human nature—is that human nature is not a fixed, cookie-cutter set of traits. Individual differences, both biological and culturally induced, are as important with respect to personality characteristics and social behavior (including the moral dimension) as with any other evolved trait. Both of these authors make some generalizations about our behavioral propensities that are not always true, and the exceptions matter a lot. Thus, Wright asserts that the males of our species may be biologically predisposed to promiscuity and that monogamy is somewhat at odds with a Darwinian perspective. How, then, do we account for the wide variations in male behavior in our society? While some males cheat regularly and marry often, others make lifelong commitments or become celibates. Consider the couple in Sacramento, California, that, in the spring of 1997, celebrated their eighty-first wedding anniversary. (He was 101 years old and she was 97.) In strictly Darwinian terms, these centenarians have done fairly well: they have fourteen children, seventy-five grandchildren and, so far, forty-three great-grandchildren. Real-world ethical systems must take account of the variations that exist in any society, for whatever reason. That is why we have both formal and informal systems of rewards and sanctions to back up and reinforce our ethical norms. What some of us may be inclined to do, or not do, spontaneously, others may need to be "persuaded" to do for the sake of the "general welfare."

Maybe, after all, Herbert Spencer was right. An "ethical science," he asserted (1898, vol. 1 p. 100), should strive to harmonize "self-preserving activities," the "activities required for rearing offspring" and the "social welfare," so that individual self-interests will mesh with the interests of others (including the superorganisms through which our various needs and wants are met). A tall order, of course, but at least this ideal is grounded in the biological fundamentals and is consistent with Darwinian principles. It can perhaps provide a general framework within which to address specific ethical issues.

Tending to Huxley's Garden

If a guiding metaphor for evolutionary ethics might be useful, we probably can do no better than an improved version of the image that was introduced by T. H. Huxley in his Romanes lecture. Recall how Huxley suggested that a society can be likened to a domestic garden, where the task of the gardener (i.e., an ethical system) is to struggle with the hostile forces of nature to achieve an ordered regime and achieve the gardener's goals. John Dewey, in his rebuttal to Huxley, proposed a more benign image of the garden plot as a place where the gardener works *with* nature to make improvements and create conditions for abundant growth. From our vantage point, it seems likely that both versions of the garden metaphor are partly correct. Our ethical systems must, at one and the same time, weed out the dandelions and fight the aphids and snails while simultaneously planting, fertilizing, watering, pruning, and harvesting the plants upon which we have come to depend for our very sustenance.

So my answer to the leading question posed in this chapter is that, yes, the time is indeed ripe for evolutionary ethics—and has been, ever since Darwin. A better question is, are *we,* finally, ripe for evolutionary ethics?

— ✌ —

Democracy is the worst form of government—except for all the others.
—Winston Churchill

SUMMARY: This essay was inspired by a book-length treatment of this vitally important subject by two political scientists. Here I critique their views and advance an alternative view that, I contend, more accurately represents the nature of human nature, the reality of the human condition and the future prospects for democracy.

16

The Sociobiology of Democracy:
Is Authoritarianism in Our Genes?

Introduction

In the famous (some would say infamous) final chapter of his discipline-defining volume, *Sociobiology: The New Synthesis* (1975), biologist Edward O. Wilson invited us to consider humankind as if we were zoologists from another planet. In this light, Wilson said, "the humanities and social sciences shrink to specialized branches of biology" (p. 547). One of the functions of the new discipline of sociobiology, Wilson suggested, was "to reformulate the foundations of the social sciences..." (p. 4). Wilson cautioned, however, that it "remains to be seen" whether or not the social sciences can be "truly biologized" in this fashion.

More than a quarter of a century later, it still remains to be seen. In his more recent book, *Consilience* (1998), Wilson keeps the faith. But he concedes that we still know relatively little for certain about the biological foundations of human nature. Of course, there are some important exceptions. We do know that many mental and physical diseases (over 1200 to date) have genetic bases, and many of these are tied to the effects of a single gene. We also know that individual differences in cognitive abilities and personality traits are significantly influenced by genetic differences (although we still do not know precisely how). We know that there are many behavioral universals in the human species that strongly suggest a biological basis, from facial expressions to parent-infant bonding and language acquisition. And yet, we are still in the dark about the precise genetic bases of normal human behaviors. Wilson notes that the science of human behavior genetics is "still in its infancy." In time, "a clearer picture of human nature will emerge" (p. 147).

403

One implication is that we should be extremely cautious about biologizing human behaviors, especially when it involves something as complex and overlain with cultural influences as democracy. So it is disconcerting, to say the least, to see two of the leaders in the so-called biopolitics movement, political scientists Albert Somit and Steven A. Peterson, boldly go where Wilson fears to tread. In their recent book, *Darwinism, Dominance and Democracy: The Biological Bases of Authoritarianism* (1997), Somit and Peterson provide us with an important cautionary tale. Their book is significant both for what it says and for what it fails to say about the nature and nurture of democracy.

An Unpopular Thesis

Somit and Peterson's "predictably unpopular thesis"—as they disarmingly put it—is that "the most important reason for the rarity of democracy is that evolution has endowed our species, as it has other primates, with a predisposition for hierarchically structured social and political systems." Humankind, they assert, has "a genetic bias toward hierarchy, dominance, and submission" (p. 3). In other words, authoritarianism is in our genes.

However, Somit and Peterson assure us that we need not view this as "a counsel of despair" or a prophecy of ultimate doom for the democracies. Our hope for the future, they claim, lies in the fact that humankind is also unique among the primates, or any other species for that matter, in being able to adopt beliefs and practices that run counter to our genetic self-interests—from celibacy to monogamy to, yes, even democracy. The key to this countervailing influence is another biologically based propensity in humankind—our "indoctrinability." (Somit and Peterson acknowledge that this term is "lamentably awkward." To me, it is also lamentably constricted and insufficient; I'll come back to this point.)

Accordingly, Somit and Peterson see themselves as engaging in an act of political "consciousness raising," to borrow a phrase from their implicit nemesis, Karl Marx. Once we acknowledge our innate authoritarian tendencies, then we can take appropriate steps to resist them. In actuality, this is a line of reasoning among conservative Darwinians that dates back at least to "Darwin's bulldog," Thomas Henry Huxley, in the 19th century (see chapter 15), and it is echoed in the writings of various contemporary biologists, including George Williams, Richard Dawkins, and Richard Alexander. Recall that in his legendary Romanes Lecture of 1893 (reprinted in Nitecki and Nitecki 1993), Huxley claimed that nature is indeed "red, in tooth and claw," as poet Alfred Lord Tennyson so nicely put it, but that humankind can transcend "the cosmic process" and substitute the "state of art" for the "state of nature." Huxley's *Deus ex machina* (to borrow a phrase from an altogether different realm) has always begged the question, then and now: why bother? If hierarchy accords with nature and, after all, best serves the genetic interests of our species, why not just go with the evolutionary flow? Why fight the cosmic process?

Somit and Peterson express a personal preference for democracy, though they don't spell out exactly why. Indeed, they're even somewhat cynical and disparaging about what the term *democracy* means, since it has been used, and misused, in so many different ways. Yet they also take pains in their penultimate chapter to propose various policies and actions that might bolster the prospects for democracy. Unfortunately, their prescriptions are not very compelling since they cannot also tell us why it is important to foster democracy. Why should we resist the nepotistic depredations of the Suharto family, or the heavily armed dominance behavior of the drug lords, or the "ethnic cleansing" in Bosnia and Kosovo for that matter? Where is the ethical anchor for the ship of state? (This is another point that we will return to below.)

Nevertheless, it would be dangerous—a wanton act of denial—to reject Somit and Peterson's thesis just because we don't like it, or because we find their defense of democracy inadequate. In opposition to those who would like to believe that democracy is the more natural state of affairs and that authoritarianism is a culturally evolved aberration, Somit and Peterson marshall an array of contrary evidence. It includes: (1) the research literature on social behavior in other species, especially our primate relatives, in which dominance hierarchies are ubiquitous; (2) the large body of supportive research in social psychology, including Stanley Milgrim's famous and much-replicated experiments on obedience to authority; (3) the history of the past two thousand years or more, which has been heavily weighted in favor of authoritarian regimes; (4) the writings of the great philosophers, who have seldom come out in favor of full-fledged democracy and have frequently denigrated it; (5) the fragility of most democratic experiments in recent decades (it seems that an array of favorable cultural and economic circumstances may be ineluctable prerequisites); (6) the relative infrequency even today of "true" democracies (yes, there is a difference it seems), despite the fact that democracy is currently the reigning ideology in global politics and is paid lip service even by tyrants.

What's Missing?

So what's wrong with Somit and Peterson's argument? If their thesis about human nature and hierarchies may have some merit, then what is missing from this picture?

In a nutshell, what is missing are (1) some other, countervailing elements of human nature, and (2) a more adequate conception of the role of nurture in political life. Somit and Peterson acknowledge the importance of an interaction between nature and nurture; it's the party-line among sociobiologists these days. But they reduce nurture to the truncated concept of indoctrinability—which implies, perhaps unwittingly, a biological susceptibility to external manipulation by political operatives (or perhaps political science professors). In reality, nurture is a large and complicated domain that has many dimensions and plays a far more potent role in shaping human societies than Somit and Peterson suggest. Moreover, nurture is also a part of human nature—a part of our evolved biological heritage—although its precise content is obviously highly variable. But perhaps most important, the concept of social dominance is murky; it turns out to be a complex phenomenon that may have many different causes and consequences. It is not so simple as originally conceived by ethologist T. Schjelderupp-Ebbe (1922)—nor is it unambiguously tied to reproductive success. (For more on this issue, see the recent discussions by Bercovitch 1991; Capitanio 1993; Drews 1993; Moore 1993; Pusey et al. 1997; Wrangham 1997. See also the discussions in chapters 6, 8, and 15.)

To begin at the beginning, our hominid ancestors diverged from the rest of the primate line more than five million years ago, and we have undergone a radical psychological makeover since then. Our primate instincts are overlain with a large, calculating (Machiavellian) neocortex, as well as a greatly intensified degree of sociality—propensities for social cooperation, sensitivity to social "approbation" (to repeat Darwin's term) and even ethical sensibilities—that are also biologically grounded. (Again, see the discussion of these points in chapter 15.) As a rule, humans are neither exclusively competitive and hierarchical nor egalitarian and cooperative but an inextricable admixture of both. Moreover, the precise mix depends upon the context, including the influence of biologically based personality differences—which makes any gross generalizations about human nature extremely slippery.

Thus, our behaviors are also greatly affected by peer pressures, social pressures, and the expectations of others (including the norms of groups and organizations), not to mention personal calculations of potential costs and benefits. We may kowtow to the boss, or the President, not because we are following our irresistible primate instinct for submission but because the person in authority will fire us—or kill us—if we don't. Furthermore, the boss, or President, may be inspired to assert dominance behaviors not by virtue of his or her innate biological superiority but because he or she is empowered by a socially recognized set of behavioral expectations that are backed by social, economic, even physical sanctions (or at least by well-armed henchmen). As Somerset Maugham observed in the opening lines of his classic novel, *The Moon and Sixpence* (1941, p. 19), the "greatness" achieved by a leader may be "a quality which belongs to the place he occupies rather than to the man. . . . The Prime Minister out of office is seen, too often, as a pompous rhetorician. . . ." (For more on this point, see Rubin 2002; Corning 2004a.)

Conversely, we may follow a leader not because of an innate urge to be subservient but because it is in our self-interest to do so; we may recognize that somebody else can be more effective than we are in mobilizing and leading our group/organization/polity. (Why, after all, does anyone work for somebody else's election?) Or maybe we're just too busy; we're glad to let somebody else take on the burden of running the PTA or serving on the City Council. In fact, the economics of political dominance and submission—the potential costs and benefits to all concerned—are of crucial importance in making sense out of political hierarchies in modern societies. This often implicit calculus is, after all, the basis for such important terminological distinctions as "consent," "coercion," "legitimacy," "subjugation," "despotism," and "tyranny"—distinctions that are the very meat and potatoes of political science.

Is Democracy in Our Genes?

In addition, a case can be made for the assertion that democracy (properly defined) is also consistent with human nature and Darwinian principles; hierarchy and democracy are not incompatible opposites. Here the argument gets more subtle and complicated, so I ask for the reader's patience.

We need first to address an important preliminary issue. What is democracy and why bother to defend it? Why should we care? Broadly speaking, there are two polar views on the subject. The more liberal view is based on the notion of individual equality and civil rights. In ancient Greece, where the term *democracy* was coined, it meant direct participation in government and a share in the decisions and actions of the community, or the *polis*. (Of course, in Aristotle's day, political equality did not extend to women, or slaves, or to landless peasants. Universal suffrage came much, much later.)

The conservative view of democracy, on the other hand, is based on the assumption that people are not equal, either in terms of their talents and abilities or their station in life. Societies will always have elites and hierarchical relationships, and government will always be disproportionately influenced by the interests of the powerful few. Accordingly, for many theorists, democracy should be defined more narrowly in terms of participation by those who are qualified to do so. Indeed, the illiterate masses have often been distrusted by elitist democrats. Moreover, as societies have grown in size and complexity beyond the small, intimate city-states of ancient Greece, which numbered only a few thousands or tens of thousands, the notions of indirect democracy and representative democracy have evolved apace; government has become an increasingly elaborate component of ever larger and more complex division of labor in modern societies.

Despite the differences between them, the common denominator in both liberal and conservative theories of democracy is the idea that the citizenry, or at least the "weightier part" (as Marsilio of Padua put it), must possess the collective ability ultimately to determine the course of government, to constrain abuses by the rulers and, perhaps, to peaceably remove and replace the rulers if they fail to meet their fiduciary responsibilities.

Another way of looking at the matter is in terms of the cybernetic model of politics, as discussed in chapter 6. To repeat, a unique feature of a cybernetic system is that its behavior is directly influenced by inputs of various kinds and by feedbacks from the environment. Feedback is critical. It involves the ability to guide and ultimately to control the behavior of the system in order to keep it on course toward the realization of its goal(s). The classic example is the way a household thermostat works, but most cybernetic systems are more elaborate.

The point here is that government is quintessentially a complex cybernetic system. To oversimplify a bit, modern mass democracy can be viewed as an array of mechanisms and practices that are explicitly designed to exert or facilitate feedback controls. (To be sure, there may also be goal-changing, or system-changing feedbacks, as well as feedforward and a variety of other cybernetic elements, but these are not essential to our argument.) Many of the specific institutions of modern democracies, from universal suffrage and free competitive elections to the rule of law and civil rights are culturally evolved "tools" that have been created to undergird the basic objective of a democratic system. These tools give the citizenry the means to control the controllers; call it the cybernetic model of democracy.

From a cybernetic perspective, therefore, the concept of a hierarchy is not opposed to democracy (read feedback controls). Nor is hierarchy equivalent to authoritarian rule. Hierarchical organizations—from football teams to automobile manufacturers, political parties and armies—are not simply reflections of dominance competition and, implicitly, of reproductive competition. Hierarchical organization is also a functional imperative—a cybernetic requisite for any organization with collective goals and a division (or combination) of labor. (Somit and Peterson gloss over the distinction between dominance and leadership, which are different from one another both functionally and psychologically.)

Somit and Peterson also discount the argument that the establishment and spread of democracy in the past few millennia reflects a trajectory of cultural learning and political evolution. They do not give much weight to the invention of law and legal systems, the rejection of hereditary rule and the "divine right" of kings, the subordination of rulers to the rule of law, the rise of an independent judiciary, empowered legislatures, secret ballots, fixed terms of office, freedom of speech, checks and balances and other political inventions that represent important instrumentalities of cybernetic control. But perhaps this is because their theory is monolithic; in their view, the hierarchies that are everywhere to be found in human societies are primarily a reflection of dominance competition. The contrary view suggested here is that there has been a dualistic causal dynamic at work; there may be both personal dominance relationships (and power struggles) *and* functional reasons for the hierarchies that are ubiquitous in complex human societies. In fact, there is often an interplay between these two great shaping influences.

By the same token, Somit and Peterson tacitly discount the role of coercive power in establishing and maintaining authoritarian rule. Forget the "whisperings within," as the biopsychologist David Barash so elegantly put it. The masses of humankind often submit to tyranny, not because they are following their primordial primate instincts but, more important, because they are making the best of a bad lot: better to survive in poverty and subjugation than to be tortured and killed by El Grande's thugs. However, if you give them a choice. . . .

A Different Spin

Indeed, it is possible to put an entirely different spin on Somit and Peterson's historical evidence for an authoritarian bias in human history. Authoritarianism often succeeds because coercive power of some sort is an essential tool for maintaining internal order and (most often) for external defense, in virtually every society. It does not take a very great leap of imagination for a leader to recognize that military or police power can also be used to pursue personal, family, ethnic or factional advantages—or simply to suppress political opposition. In other words, authoritarian regimes often have both the incentives and the means for perpetuating themselves in power and exploiting it for their own purposes, and democratic institutions are often "murdered" because those who control coercive power naturally do not want to be constrained by cybernetic feedback controls, much less removed from office. So authoritarianism may also arise from the pure calculus of self-interest. (It should also be pointed out that the philosophers of antiquity who so often disparaged democracy were mostly writing for the literate and powerful few, not the powerless and illiterate masses; their intended audience was not likely to have been very open-minded on this subject.)

But why, after all, should democracy matter? If authoritarianism is in any case hard to resist and democracy is more fragile and precarious, why bother trying to shore it up—much less trying to improve its chances of success? Here I believe we can get some help from Darwin himself, and from some supportive research literature in sociobiology, primatology, anthropology, psychology, and even the management sciences.

First, recall Darwin's views on human evolution (discussed in chapter 15). Far from seeing it as an anarchic process that was governed by competition, Darwin postulated that our social and moral faculties—our "social instincts"—were of primary importance. Recall Darwin's argument: "A selfish and contentious people will not cohere, and without coherence nothing can be effected. A tribe rich in the above qualities would spread and be victorious over other tribes . . ." (1874 p. 148).

The accumulating evidence supports Darwin's scenario. The precise patterning and sequence of human evolution remains a matter of continuing debate. Nevertheless, Darwin correctly identified one crucially important factor—the role of functional interdependency. (This subject will be discussed further in chapter 17.) In so doing, Darwin also pinpointed the bottom-line answer to the "why bother" question. To repeat, an organized society may be characterized as a "collective survival enterprise"; we depend upon one another in a myriad of different ways for the meeting of our various survival and reproductive needs. Moreover, the collective performance of the group may also greatly affect the fate of any individual member, especially in the context of competition with other groups. Indeed, we very often do things together as teams (or "tribes") and have even evolved a psychological propensity to enjoy cooperation (not coincidentally). In other words, rulers and followers generally need each other. Both natural selection and various cultural mechanisms may work together to constrain dominance behaviors and dominance competition (both ultimate and proximate causes, in evolutionary jargon). And, as societies have become ever more complex, the cultural constraints have become increasingly important; a prime corollary of system complexity is a high degree of functional interdependency among the parts.

Consider the Evidence

This is not just a theory. There is considerable evidence in the research literature on social animals (especially the primates and social carnivores) that such constraints and cybernetic controls do in fact exist. Significantly, these constraints are most notable among species with a high level of functional interdependence (savanna-living baboons, wild dogs, wolves and the like)—the very species whose adaptive strategies most likely resemble the pattern in evolving hominids. (Among the many references on this point, see especially Kummer 1968, 1971; Wilson 1975; Lopez 1978; Strum 1987; Dunbar 1988; Harcourt and de Waal 1992; Wrangham et al.1994; de Waal 1996, 2001.)

Support for the notion that social constraints on dominance behavior are an important part of our evolutionary heritage can also be found in anthropology, a discipline that Somit and Peterson apparently overlooked. What Christopher Boehm (1997, 1999), director of the Jane Goodall Research Center at USC, calls the "egalitarian syndrome" is fairly common among the small hunter gatherer societies that are found in remote areas of the world—the very societies whose lifestyle most nearly resembles that of our Pleistocene ancestors. Boehm's term refers to a pattern of social pressures that serve to dampen status rivalries, encourage cooperation, facilitate consensual decision making and police free-riders. (See also James Woodburn's classic 1982 study of "Egalitarian Societies.")

To cite one example, the !Kung San (Bushmen) hunter-gatherers of Africa's Kalahari Desert have elaborate means for exerting social pressure on any member who gets too pushy, or arrogant, or selfish. The group will act collectively to admonish and restrain the individual, with the ultimate (painful) sanction being banishment from the group.

The work of primatologist Frans de Waal and various colleagues on coalitions and alliances in chimpanzees and other primates also reinforces this argument (Harcourt and de Waal 1992). De Waal (1996) maintains that egalitarianism in general and coalition behavior in particular are important counterweights to dominance behaviors, in animal and human societies alike. Submission to a dominant animal is primordial, de Waal points out. But, "we need to know what's in it for the subordinate." De Waal's answer amounts to a seconding of Darwin's motion: "The advantages of group life can be manifold . . . increased chances to find food, defense against predators, and strength in numbers against competitors . . . . Each member contributes to and benefits from the group, although not necessarily equally or at the same time . . . . Each society is more than the sum of its parts" (pp. 9, 102). Indeed, there is even some evidence for participatory group decision making in other social animals as well (see Conradt 2003; Corning 2004b).

This paradigm is also consistent with a large and important body of research both in psychology and in the mental health field on what is variously called "personal autonomy," "self-determination," "competency," "self-efficacy," and "personal empowerment."[1] Although there are some differences of emphasis associated with each of these terms, the common core has to do with a basic psychological need in humankind that is antithetical to the proposition that supine submission to authority is more consistent with human nature and thus requires an offsetting cultural indoctrination. Just ask the parents of a teenager about dominance and submissiveness among the offspring that carry their genes! Moreover, there is much evidence that the suppression of personal autonomy, self-control, efficacy or whatever may have significant functional, health and even reproductive consequences. (There is, in fact, a large research literature on the subject of "learned helplessness.") Submission to authority, I contend, has a very large component of nurture associated with it—from parental sanctions to the menace of a police-state's gestapo.

Finally, there is a large and compelling literature in the management sciences on what has been called the "new paradigm" of organizational effectiveness, much of which traces its roots to the pioneering work of management guru W. Edwards Deming on "quality management" back in the 1950s and 1960s. (Deming's influence was particularly important in the development of the Japanese management model, which emphasizes consensus building and quality improvement teams). Again, there are a number of variations on this theme: "Total Quality Management" (TQM), "servant leadership," "self-managed teams," "flattened organization," "decentralized decision-making," "worker empowerment," "team leadership," "consensus building," and so on.[2] In practice, there may be discrepancies between what managers say they will do and what they actually do. But the overall results of this movement have been impressive. Over the past two decades, many industries have made major improvements in productivity and product quality under the general heading of "process redesign." And this could not have been achieved without teamwork.

What Can We Conclude?

In sum, modern-day authoritarian regimes bear no resemblance to the small, egalitarian and mostly kin-based groups that characterized our evolving hominid ancestors, and the relationship between contemporary authoritarian governments and the Darwinian criterion of reproductive success is problematical to say the least. Even in private industry—a traditional bastion of authoritarianism—the hierarchical model of organization has often proved to be less effective than something that is more democratic and participatory. Therefore, it could be argued that modern authoritarianism involves a distortion of human nature, and that the machinery of modern democracy amounts to a set of evolved cultural analogues for the informal, personal feedback controls that existed among our Pleistocene progenitors.

So the other half of Somit and Peterson's half-truth amounts to this: Although we may have a deep evolutionary legacy of behavioral proclivities and biases, these are complex in nature and are ultimately of less importance in understanding the interplay of authoritarianism and democracy in today's world than are the many cultural influences—from child-rearing practices to socialization, peer pressures, social customs, the political culture, the mass media, economic conditions, institutional protections, and, not least, the power resources and decision calculus of both the rulers and the ruled.

Thus "indoctrination"—Somit and Peterson's prescription for preserving democracy—will not suffice to ensure its future. (The sad lesson here is that Somit and Peterson became the captives of their categories.) We are still left with Plato's dilemma: *quis custodiet ipsos custodes?*—who will guard the guardians? More specifically, how can we build and maintain political institutions and practices—and public respect for them—that are strong enough to constrain the rulers and protect the ruled, sometimes under very difficult economic and social conditions and very often at the expense of the rulers and/or some powerful economic or political faction? This is the primary challenge for democracy, as it always has been.

Perhaps we should also keep it in mind that democracy is not simply an old family recipe that can be mixed up on the spot, whenever the right ingredients happen to be present. It is an evolved and evolving cultural institution with many facets—an unfinished work in progress. Moreover, each new instantiation involves a learning curve as new institutions, rules, norms, and practices are developed, or mimicked (or both) in what had previously been an alien environment. We should also keep in mind Winston Churchill's famous judgment: "Democracy is the worst form of government—except for all the others." Somit and Peterson themselves quote this quintessential bit of Churchillian wordplay in their volume, but it is buried in one of their footnotes. They might have served us better if they had made it the frontispiece for their book.

Postscript

After my review of *Darwinism, Dominance and Democracy* was published, the authors responded with a "reaffirmation" of their thesis (Somit and Peterson 2001). To summarize, they continue to adhere to the view that humankind has a genetic bias toward authoritarian dominance and submission hierarchies. It is a part of our "primate nature," they say, and it is sufficient to explain the "rarity" of democracy. Democracy "runs counter to our basic inclinations," they claim. Indeed they see the basic requisites for democracy (economic, social, educational, and political) as "continuing to deteriorate" (at least in this country) and they believe that it "may be too late" to save democracy here. The authors even chide me for my supposed "optimism" and conclude that "the data run otherwise."

In my "reply" (Corning 2001b), I reiterate that there is broad evidence across several disciplines—from ethnography to organizational psychology and even primatology—that we are also equipped by our nature for more democratic practices and that there is good reason to believe that democracy also has ancient roots in humankind. Modern authoritarianism may in fact be a recent, pathological condition that can be associated with the rise of large, complex societies during the past 10,000 years. Accordingly, authoritarianism may be explained more parsimoniously in terms of proximate cultural, economic and political factors rather than in terms of a genetic predisposition. A reductionist explanation may be gratuitous and unnecessary. I also cite the many scholarly references that reflect a more balanced gene-culture coevolution paradigm. (They include, among others, Corning 1983, 2003a; Boyd and Richerson 1985; Durham 1991; Weingart et al. 1997; Sober and Wilson 1998; Dunbar et al. 1999; Ehrlich 2000; Hammerstein 2002; D.S. Wilson 2002; Ridley 2003; Richerson and Boyd 2004.)

Although I am hardly a starry-eyed optimist about the future of democracy, as Somit and Peterson charge, neither am I a dour pessimist whose forebodings could become self-fulfilling prophesies. My own prognosis is based on what I call the "learning curve" of political evolution. In the end, I am hopeful that more would-be authoritarians will come to accept Churchill's wry verdict that the alternatives to democracy are much worse.

— ❧ —

Life is unfair.

—John F. Kennedy

Funny, I always believed that the world was what we make of it.
—(Ellie Arroway: from the movie *Contact*)

The color of truth is grey.

—attributed to André Gide

SUMMARY: Ever since Darwin, social theorists have freely—and sometimes promiscuously—exploited evolutionary theory to advance various ideological agendas. Here I exercise my right to preempt any ideological interpretation of Holistic Darwinism. I believe that this paradigm—inclusive of the accumulating evidence about human evolution (the state of nature) and the complexities of human nature—provides the basis for a new bioeconomic ideology that more accurately represents the reality of the human condition. A biologically grounded approach to social justice enables us to define a middle ground between capitalism and socialism and to finesse the ideological standoff between them. I call this new ideology "fair shares," and I propose a normative framework that includes three complementary principles: (1) goods and services should be distributed to each according to his/her "basic needs" (see chapter 11); (2) "surpluses" beyond the provision for our basic needs should be distributed according to "merit" (which I will attempt to clarify); and (3) in return, each of us is obliged (under the well-established evolutionary principle of reciprocity) to contribute to the collective survival enterprise in accordance with his/her ability. Though none of these principles is new, in combination they define a biologically grounded "third way." Needless to say, many issues are raised, and questions begged, by this formulation. Some are addressed here. Philosopher John Rawls, in his celebrated—and much debated—theory of justice as fairness, utilized a variation on the hypothetical "state of nature" in social contract theory to advance a claim for the least advantaged in society. While Rawls's search for an ideological middle ground was commendable, such artificial constructs are no longer sufficient or necessary.

17

Fair Shares: A Biological Approach
to Social Justice

Introduction

Charles Darwin changed the ground rules for the ideological debate, and the time has come to replace our one-sided, often self-serving, and sometimes destructive grab bag of political ideologies with an empirically grounded vision that better reflects the way human beings and human societies actually work. An unbiased reading of Darwin's *The Descent of Man* (1874)—what could be called Darwin's Darwinism—is a starting point. But more compelling is the steadily accumulating body of scientific evidence on human evolution coupled with our growing understanding of the biological (and psychological) underpinnings of human nature and, not least, the evidence directly in front of us in our day-to-day experiences. These issues were discussed in earlier chapters. Here I will reiterate, and elaborate on, some key points to make the case for a new, middle-ground ideology that I call "fair shares."

Let us begin with Darwin. Recall his view that socially organized groups—superorganisms—played a leading role in human evolution. To repeat Darwin's bottom line: "Selfish and contentious people will not cohere, and without coherence nothing can be effected. A tribe rich in [intelligence, courage, discipline, sympathy, and fidelity] would spread and be victorious over other tribes . . . Thus the social and moral qualities would slowly tend to advance and be diffused throughout the world" (Darwin 1874, p. 148). In other words, human societies—and the moral systems that give them coherence—are not simply cultural artifacts; they are also products of our biological evolution and have played a vital part in our success as a species.

What the Evidence Shows

Consider first what the political philosophers refer to as the "state of nature" (prior to civilization). Much of the evidence that has been assembled on the evolution of humankind in recent decades is generally concordant with Darwin's scenario (see Klein 1999; Wolpoff 1999a; P. R. Ehrlich 2000; Corning 2003). As noted earlier, it now seems clear that the five-million-year (plus) span of human evolution involved at least three distinct transitions. In each of these transitions, sociality and social organization were the keys to our competitive success; human nature and evolving human societies have been indelibly shaped by a collective survival strategy. I will briefly summarize the case for this scenario.

The first and perhaps most crucial of these "major transitions" (to borrow a term from Maynard Smith and Szathmáry) was related to a shift from an arboreal lifestyle to terrestrial living, just as Darwin had supposed. However, this transition most likely did not happen all at once. For one thing, it involved substantial costs and risks. As their foraging ranges expanded, so did the time and energy required to exploit this new niche, and the earliest known australopithecines were imperfect bipeds—competent but not as efficient as the much later *Homo erectus/Homo ergaster*. More important, the exploitation of a terrestrial environment introduced new life-and-death dangers, both from predators and competitor species as well as rival protohominid groups (see Brain 1981, 1985; Foley 1995; Isbell 1995). To steal a line from anthropologist Charles K. Brain, this scenario could be called "Man the Hunted" (and woman the hunted, of course).

So, the question is, how did a diminutive ape (less than three feet tall) with constrained mobility on the ground and no natural defensive weapons solve the problem of shifting to a terrestrial habitat, broadening its resource base, and, over time, greatly expanding its range? It is likely that social organization was a key factor. In a patchy but abundant environment that was also replete with predators, competitor species, and sometimes hostile groups of conspecifics, group foraging and collective defense/offense was the most cost-effective strategy since it significantly increased the odds of survival. In other words, there were immediate payoffs for collective action that did not have to await the plodding pace of natural selection. (A number of other theorists over the years have also endorsed the group-defense model of human evolution, including George Schaller, Alexander Kortlandt, John Pfeiffer, Richard Alexander, Richard Wrangham, and others.)

Group selection may very well have been involved in this process, but it was not exclusively, or even primarily, based on altruism. To reiterate, it also involved "collective goods" that everyone shared. It did not require a cooperative gene. It required a degree of sociality (a common characteristic among the primates) and a degree of intelligence about means and ends as well as costs and benefits. Furthermore, these nascent superorganisms were (most likely) formed around a nucleus of closely related males. So individual selection, kin selection, and group selection would have been aligned and mutually reinforcing—as Darwin himself suggested.

Why would these protohominid males have bothered to defend the females and infants? For one thing, the males might not have been sure of their paternity if the females followed a reproductive strategy involving multiple matings and, perhaps, disguised ovulation (as our close relatives the bonobos evidently do). Another factor was that all of the infants would have been closely related—nephews, cousins, or even younger siblings. A third point is that, in a species with a long reproductive cycle and a short life span, each offspring was relatively more valuable. Finally, in an organized, interdependent superorganism, the defense of other members of the group was most often not a matter of altruism, or reciprocal altruism (mutual sacrifices), but of teamwork in a win-win (or lose-lose) situation. It could be called a synergy of scale.

Was there also a division of labor? Contemporary human societies of all kinds typically have a division of labor along sexual lines, and it is possible that a rudimentary version of this pattern also existed among the early australopithecines, with the females caring for the infants and juveniles and the males being mainly responsible for providing protection. (Modern-day savanna baboon troops offer a frequently cited analogy.) It is also likely that these creatures deployed tools that could double as digging sticks and weapons. This would have more than compensated for any lack of natural armament, such as the outsized canine teeth of baboons (see Kingdon 1993; Wrangham and Peterson 1996).

We may never know for certain about this and many other details relating to human evolution, but group living, group foraging, and a cooperative division of labor allowing for greater access to a more dangerous but abundant terrestrial environment is likely to have been primordial in the hominid line. This in turn set the stage for a second transition to the more humanlike *Homo erectus/Homo ergaster*, which, most likely, became systematic hunters and gatherers. And this important shift set the stage for the final emergence of the larger-brained, technologically more sophisticated, loquacious *Homo sapiens*. Thus, the so-called "social contract" was, in fact, a multi-million-year biological/survival contract based on mutualism and close cooperation, not an arm's-length exchange of goods and services. The

actual "state of nature," despite the gratuitous assumptions of the social contract theorists, involved socially organized groups.

This is not to say that individual competition, status rivalries, and internal social conflicts somehow disappeared. Then, as now, it is likely that
there existed a somewhat precarious interplay between competition and
cooperation, between the self-interests of individuals and the interests of the
group. Indeed, a dynamic tension between individual and group interests is
a common phenomenon in social mammals generally. The key to the evolution of sociality in our hominid ancestors lay in the costs and benefits to
each individual for cooperation or non-cooperation. Reciprocity and reciprocal altruism may have played a role, but the benefits associated with being
included in a group—and the high cost of ostracism and isolation—must
also have been a major factor. The superorganism was a vitally important
survival unit; it produced collective survival advantages that were defined in
terms of life or death, and each individual had a stake in its preservation and
enhancement.

In other words, the "public interest" was rooted in the group's potential
for producing public goods. For instance, a larger group was more likely—
all other things being equal—to benefit from synergies of scale in confrontations with predators or competitors and, later on, with potential prey.
Likewise, more effective leadership and group decision making could have
been selectively important, as anthropologist Christopher Boehm (1996,
1997, 1999) argues. These collective benefits provided an overarching
incentive for containing conflict, enhancing cooperation, and punishing
cheaters and free riders.

The theoretical implications of this rendering of the "state of nature"
are, briefly, as follows. The Neo-Darwinian "selfish gene" model (like the
Hobbesian model in political theory) is fundamentally flawed. It is unlikely
that we evolved as isolated individuals pitted in relentless competition with
one another. Nor did we inhabit a Lockean world of autonomous individuals. The state of nature, literally for millions of years, was characterized by
an overriding need, with commensurate rewards, for mutualism, reciprocity, and even some altruism, all of which served to constrain, limit, and mitigate (though it hardly eliminated) reproductive competition.

In accordance with this scenario of human evolution, claims for individual rights—for reproductive advantages, for freedom, or even for private
property—are not inconsequential, but they are ultimately subordinate to
the needs of the rest of the community as an interdependent, collective
survival enterprise. In other words, individual rights must ultimately be
subsumed by the public interest. Moreover, this is not simply a normative

statement—a gratuitous "ought." As we shall see, it also represents an empirical reality—a decision rule with predictable consequences that can be ignored only at great peril.

The Science of Human Nature

In light of this (very brief) account of human evolution, what can we infer about human nature and the nature of the biological/survival contract that holds human societies together? The answer is that we don't have to infer anything; there is an increasingly compelling body of evidence on the subject that is consistent with the scenario rendered above. Political theorist Andrew Heywood (1999), early on in an otherwise commendable survey of the philosophical literature, asserts the following:

> It is important to remember that in no sense is human nature a descriptive or scientific concept. Even though theories of human nature may claim an empirical or scientific basis, no experiment or surgical investigation is able to uncover the human "essence." All models of human nature are therefore normative: they are constructed out of philosophical and moral assumptions, and are therefore in principle untestable. (pp. 7–18)

On the contrary, an evolutionary biological approach provides a well-validated empirical or scientific basis for defining the essential characteristics of human nature and for understanding our normative priorities. Moreover, the nascent science of human nature, which spans many scientific disciplines, is gradually fleshing out more of the specific details. As the distinguished biopsychologist Melvin Konner (2002) noted in a recent issue of *Nature,* "In the era of genomics and brain imaging, hypotheses about human nature are more testable than ever" (p. 121).

It is important, therefore, to reiterate the argument developed in previous chapters. The "ground-zero" premise of the biological sciences is that survival and reproduction are the basic, continuing, inescapable problems for all living organisms. To repeat, life is at bottom a "survival enterprise." Furthermore, the problems of survival and reproduction are multifaceted and relentless; they can never be permanently solved. Biological survival and reproduction remains the "paradigmatic problem" of our species.[1]

Accordingly, the conceit that society is merely a facultative arrangement, or a marketplace, or a vehicle for material or moral improvement diminishes or even denies its true nature and purpose. To reiterate, an organized, interdependent society is quintessentially a collective survival enterprise—a superorganism. Its fundamental purpose is to provide for the basic survival and reproductive needs of the population—past, present, and future. It can accurately be called a biological contract. Although the great eighteenth century English conservative theorist Edmund Burke had in mind a somewhat different point (and a different cosmology), he captured the essence of this idea in this famous, much-quoted passage:

> Society is indeed a contract . . . [But] the state ought not to be considered as nothing better than a partnership in trade of pepper and coffee, calico or tobacco, or some other such low concern, to be taken up for a little temporary interest, and to be dissolved by the fancy of the parties . . . As the ends of such a partnership cannot be obtained by many generations, it becomes a partnership not only between those who are living, but between those who are living, those who are dead, and those who are to be born. Each contract of each particular state is but a clause in the great primeval contract of eternal society. (1999, p. 368)

This biological contract, along with the imperatives associated with pursuing it, encompasses the bulk of human activity and human choices world wide. To be sure, survival per se may be the farthest thing from our (conscious) minds as we go about our daily lives. Nevertheless, our mundane behaviors are mostly instrumental to the underlying survival challenge. They reflect the particular survival strategy—the package of economic, political, and cultural tools—by which each society organizes and pursues the ongoing survival enterprise. (Recall the discussion in chapter 11.)

Although most modern theorists, following the philosopher David Hume, admonish us not to breach the "is-ought dichotomy," the fact is that numerous survival-related "oughts" have been programmed into our genes over the long history of our species and the much longer history of life on Earth. We are endowed with an array of existential, biologically based human values that are virtually universal, and we mostly choose to follow their dictates. Moreover, all preferences are not created equal. This allows us to seek regularities, make predictions, and link human nature to human behavior in comprehensible ways.

The first and most important generalization about human nature is that each of us is defined, in considerable measure, by the array of basic needs elucidated in chapter 11. These needs are essential to our survival and reproductive success, and we come into the world being oriented toward satisfying them. As we noted earlier, the concept of basic needs is not new. Its roots go back to Plato and Aristotle, and it has been used in a variety of ways over

the years, ranging from a narrowly focused preoccupation with food, clothing, and shelter to psychologist Abraham Maslow's expansive claims for
"self-actualization." On the other hand, the very concept of basic needs has
also come under severe attack in recent years. Andrew Heywood (1999)
summarized the argument as follows:

> Needs are notoriously difficult to define. Conservative and sometimes liberal
> thinkers have tended to criticise the concept of "needs" on the ground that it
> is an abstract and almost metaphysical category, divorced from the desires and
> behavior of actual people . . . It is also pointed out that if needs exist they are
> in fact conditioned by the historical, social and cultural context within which
> they arise. If this is true, the notion of universal "human" needs, as with the
> idea of universal "human" rights, is simply nonsense. (p. 298)

Not only is this conclusion incorrect but it ignores the large and growing body of empirical research, most notably under the sponsorship of the
United Nations, the National Academy of Sciences, and other agencies, not
to mention the work related to the "survival indicators" program (see chapter 11) that gives scientific credence and lends considerable precision to the
concept of basic needs.

As noted in chapters 10 and 11, the survival indicators framework
implies a fundamental shift in the way economic, social, and political phenomena are viewed. The performance of an organized society can be evalu-
ated in terms of how it relates to, or impacts upon, the package of basic
needs that defines the parameters of the ongoing problems of survival and
reproduction. The overwhelming majority of economic activity world wide
is devoted to the satisfaction of these basic needs. Moreover, and this point
is crucial, vast numbers of people the world over, even in the advanced
industrial societies where basic needs deprivations are supposedly no longer
a problem, come up short and are at serious risk.

The Political Animal

If earning a living, in a very broad sense, is the fundamental vocation of
the human species, then the psychological substrate of human nature—
the perceptual, mental, and emotional tools that we deploy to pursue the
survival enterprise—is also a product of our long evolutionary heritage.
And that heritage, most likely, is profoundly social in nature. We pursue
our survival and reproductive needs for the most part within a nested set
of goal-oriented (cybernetic) social units, from families to work-groups

and tribes, clubs and churches, small villages and towns, elaborate trade and exchange networks, large scale business enterprises, densely populated cities, and, not least, governments. This is hardly news, but the point is that it also reflects a deep-rooted part of human nature. "Nurture"—the culture that envelops us—molds, shapes, and differentially rewards and punishes the precise patterns of social cooperation in any given society, but our "nature" also potentiates it and participates in making this happen. Indeed, effective social cooperation critically depends upon evolved, exquisitely engineered psychological facilitators—including our superlative communications skills, our capacity for forming emotional attachments (ranging from parent-infant bonding to pair bonds to patriotism), our susceptibility to social pressures (the "praise and blame" of others, as Darwin put it), our receptiveness to participation in cooperative social hierarchies, and our willingness to follow leaders. In fact, recent work in paleoanthropology by Robin Dunbar and others strongly suggests that the evolution of our outsized brain was related to the increased mental demands imposed by living in larger, more complex social groups (superorganisms). For a recent summary of this work, see Dunbar (2001). Others have found a correlation between brain size and various measures of "cognitive skills" (Reader and Laland 2002). Thus, when Aristotle characterized *Homo sapiens* as a distinctively "political animal" (meaning that we are designed for life in an organized political community), he identified one of the most fundamental characteristics of our species. And its roots may go back millions of years.

How do we know? The evidence is all around us: in the complex social organization and behavior of our closest primate relatives, the chimpanzees and bonobos; in the accumulating evidence (summarized above) related to human evolution; in the many studies of other societies by anthropologists; and, most compellingly, in the exponentially growing research literature on human nature across a broad spectrum of scientific disciplines. These disciplines include, among others, molecular biology, human behavior genetics, neurobiology, evolutionary psychology, sociobiology, human ethology, anthropology, developmental and social psychology, sociology, and even behavioral economics where the hypothetical "economic man" is being challenged by research on how we actually behave in the marketplace. Among the many references on this point, see especially Masters and Gruter (1992); de Waal (1996, 2001); Ridley (1997); Weingart et al. (1997); Sober and Wilson (1998); Dugatkin (1999); P. R. Ehrlich (2000); and Gowdy (2004a, 2004b). Also, see the discussion and references in chapter 6 and chapters 9 through 11. Finally, there is overwhelming evidence of purposeful sociality in our everyday experience.[2]

But if we are highly social—even to the point of being altruistic on many occasions (witness the outpouring of nameless contributions for the victims of the 9/11 terrorist attacks and the recent tsunami)—we are also (quite obviously) self-interested, acquisitive, and highly competitive. More than that, we are often (not always) motivated to strive for personal achievement and personal influence and power. We invented capitalism but not the motivations that energize it. Likewise, we invented political democracy but not the political competition that invigorates and sometimes corrupts it. Indeed, competition and the aggressive pursuit of self-interest is a ubiquitous feature of the natural world and human societies alike.

Many theorists have claimed that competition is the primary driver of evolution—the very heart and soul of natural selection. And Darwin himself stressed its importance. But—to reiterate—Darwin also had a broader view. He understood the role of cooperation and symbiosis, and he was also well aware of the fact that natural selection is a metaphor and not a mechanism, much less the judge in some kind of Olympics competition. As noted earlier, natural selection refers to differential success (or failure) among differing individuals (or groups) in the multifaceted business of earning a living and reproducing in the economy of nature. Sometimes survival and reproduction can be a cakewalk, especially when a new niche is being exploited and rapid population growth is possible. At other times, differential survival is a result simply of being in the right place (or wrong place) at the right time. Sometimes differential survival is the result of direct competition between predators and their potential prey, or of a head-to-head ecological scramble for scarce resources within or between species. At still other times, though, differential survival and reproduction may be the result of cooperation. Call it competition via cooperation.

As discussed at length in earlier chapters, the key to the emergence and continuity of cooperation in nature (and in human societies as well) is the functional synergy that cooperation produces—the economic payoffs, broadly defined—with respect to one or more aspects of the business of earning a living. In accordance with the Synergism Hypothesis, the synergies produced by cooperation can take many different forms and serve various survival needs. Among other things, there may be synergies of scale, functional complementarities, joint environmental conditioning, cost and risk sharing, resource sharing, information sharing, and division of labor ("combination of labor").

The Collective Survival Enterprise

Human societies are based on synergy—cooperative effects that are not otherwise attainable. And the public interest or common good is not about the pursuit of happiness, or the "greatest happiness for the greatest number." It is first and foremost concerned with meeting the basic survival and reproductive needs of the population as a whole. This is the common denominator—the universally shared individual interest in (or at least a shared prerequisite for) every organized society, whether we are conscious of it or not, and it represents the very foundation of political "legitimacy"—the willing consent of the citizenry.

Competition to secure our basic needs—and much more when we can—is endemic in human societies, just as it is in other socially organized animal societies. But so is cooperation and interdependency. The more complex the society, the more deeply dependent we are upon the skills, efforts, and support of others, not to mention the accumulated stock of cultural tools and resources that have been passed down to us over many generations. Indeed, most of us are far more completely dependent on the services of others than we recognize (until we get into trouble or something breaks down). A vivid appreciation of this deep interdependency was articulated by, of all people, Adam Smith in *The Wealth of Nations* (1776/1964):

> In civilized society . . . [man] stands at all times in need of the cooperation and assistance of great multitudes. . . . [M]an has almost constant occasion for the help of his brethren. . . . Observe the accommodation of the most common artificer or day-labourer in a civilized and thriving country, and you will perceive that the number of people whose industry a part, though but a small part, has been employed in procuring him this accommodation, exceeds all computation. The woolen coat, for example, which covers the day-labourer, as coarse and rough as it may appear, is the product of the joint labour of a great multitude of workmen. The shepherd, the sorter of the wool, the wool-comber or carder, the dyer, the scribbler, the spinner, the weaver, the fuller, the dresser, with many others, must all join their different arts in order to complete even this homely production. How many merchants and carriers, besides, must have been employed in transporting the materials from some of those workmen to others who often live in a very distant part of the country! How much commerce and navigation in particular, how many ship-builders, sailors, sail-makers, rope-makers, must have been employed in order to bring together the different drugs made use of by the dyer, which often come from the remotest corners of the world! What a variety of labour too is necessary in

order to produce the tools of the meanest of these workmen! To say nothing of such complicated machines as the ship of the sailor, the mill of the fuller, or even the loom of the weaver. Let us consider only what a variety of labour is requisite in order to form that very simple machine, the shears with which the shepherd clips the wool. The miner, the builder of the furnace for smelting the ore, the teller of the timber, the burner of the charcoal to be made use of in the smelting house, the brick-maker, the bricklayer, the workmen who attended the furnace, the millwright, the forger, the smith, must all of them join their different arts to produce them. . . . If we examine, I say, all of these things, and consider what a variety of labour is employed about each of them, we shall be sensible that without the assistance and cooperation of many thousands, the very meanest person in a civilized country could not be provided, even according to, what we very falsely imagine, the easy and simple manner in which he is commonly accommodated. (pp. 10–11, 12)

As economists from Adam Smith to the present day will attest, exchange, trade, and organized markets play a vital role in facilitating the collective survival enterprise in almost every society, and may well have done so for hundreds of thousands of years (or much longer). However, it is also wise to remember that these instrumentalities in turn depend upon various social underpinnings like honest dealing and trust (what economist Arthur Okun called the "invisible handshake") as well as explicit rules and policing. Adam Smith himself emphasized the moral underpinnings of the marketplace in his lesser-known but important predecessor to *The Wealth of Nations* called *The Theory of Moral Sentiments* (1976/1759), which was published when he occupied the Chair of Moral Philosophy at the University of Glasgow. Indeed, Smith's famous characterization of the marketplace as "an invisible hand" (which actually first appeared in his earlier volume) was predicated on the assumption that the "rapacious" pursuit of self-interest, though reprehensible in light of his Stoic and Christian values, nevertheless produces beneficial results for the whole of society. (We will come back to this claim shortly.) Indeed, Smith (following Plato in the *Republic*) also stressed the vital importance to a civilized society of a division of labor, which depends upon close, sustained, and dependable cooperation. Recall his classic description of a pin factory.[3]

The key point is that cooperation produces synergies, but it also creates interdependence and a personal stake in those synergies. Again, it results in the "paradox of dependency." The more valuable the synergies produced by cooperation, the more likely we are to become dependent upon them. Thus, all of us have a vital stake in the viability of the collective survival enterprise, just as our remote australopithecine ancestors did. Moreover, we humans are hardly unique; the problem of harmonizing individual self-interest with group/collective interests is a central conundrum for superorganisms throughout the natural world, from leaf cutter ants to naked mole rats and savanna baboons.

Reclaiming the Middle Ground

I submit that this vision provides a new perspective on the vexed debate in political theory regarding the relationship between the individual and society (and the role of the state). Within the evolutionary/biological paradigm, both libertarian/individualist and communitarian/collectivist theories are partly valid and equally insufficient. There is a middle ground between them that might be called a "Darwinian community" (though the founding father of evolutionary ethics, Herbert Spencer, should also be mentioned—see chapter 15). It is somewhat similar to (but also different from) the "new communitarianism" of sociologist Amitai Etzioni (1993, 1995, 2000), which he also calls a "third way."[4]

On one side of the coin, personal self-interest is a major human motivator—a basic "module" of human nature, in the current jargon—and it is essential for the survival and reproduction of each individual. Accordingly, Adam Smith's enduring insight is time-tested. The genius of capitalism is that it harnesses self-interest and private wealth to innovative ideas and entrepreneurship (and sometimes gut-wrenching risks) to generate new wealth and material progress. It is a proven system, and it is currently transforming the global economy—though recent events have shown that it is subject to abuse and requires policing.

On the other side of the coin, the genius of cooperation is that it produces otherwise unattainable synergies. It harnesses individual resources, skills, and collective efforts to serve various aspects of the collective survival enterprise. This includes many nonmarket forms of cooperation, including the division/combination of labor known as government. At its best, government can play a vital role in the community. To cite one case in point, the U. S. government has historically subsidized (and protected) most of its major new industries; it has built and maintained its critically important infrastructure (from colonial-era canals and harbors to highways, reservoirs, power grids, and airways); it has regulated and policed the all-too-human ethical flaws in an imperfect marketplace; it has plugged major gaps in the ability of the markets to provide for the basic needs of the population; and it has been responsible for defending the country against major threats to its survival. The legendary national mobilization during World War II (see chapter 7) and the current war on terrorism are two striking examples.

However, the collective need for government in any given context is constantly changing, and the enduring political challenge is to recalibrate as necessary the precise relationship between individual rights and freedoms and the needs of society. One example among many is the intrusive and time-consuming security screening process now in place at U. S. airports, an unthinkable invasion of privacy in an earlier era. In other words, a moving balance must be maintained between the two competing claims to power (individual versus collective), and there is no all-purpose formula for how to do this or we would long ago have deployed it.

Beyond this broad generality, the concept of a Darwinian community embodies a number of other distinctive properties. Here I will mention just six of them:

1. **The "public interest" is nothing less than our shared stake in the continued viability and improvement of society as a superorganism.** Needless to say, this refers to meeting our basic survival needs and those of our posterity. Many of these needs may be satisfied in the marketplace. But sometimes it is only through nonmarket collective actions that we can provide for these needs. In modern societies, both the private sector and the public sector may serve the public interest. But sometimes, regrettably, neither one may do so very effectively.

2. **Social control in the name of the public interest is a two-way street.** In small, face-to-face tribal societies, the "social instincts," as Darwin called them—along with various informal social customs, practices, and sanctions—usually served well enough (and still do) to contain most antisocial behaviors, while reciprocity and sharing are ubiquitous phenomena. Anthropologists Christopher Boehm (1996, 1997, 1999) and Bruce Knauft (1991) have convincingly shown that so-called simple societies are, as a rule, egalitarian and that effective social restraints serve to keep a tight rein on aggressive individuals (cf. Erdal and Whiten 1994). However, large-scale human societies are at best "crude superorganisms," in the terminology of anthropologists Peter Richerson and Robert Boyd (1999); an array of artificial workarounds are essential for containing potentially destructive individual behaviors. But if markets cannot always be trusted to serve our needs, the same is equally true for the public sector. Governments can all too easily be corrupted. Institutional safeguards are important—checks and balances, free elections, secret ballots, a free press, etc. But so are legal constraints and reliable punishments for transgressions. We owe to the ancient Greeks the

legal principle, which we now take for granted, that the golden cord of law applies also to the rulers, and it is one of our most important safeguards.

3. **A Darwinian community must recognize and accommodate our inherent diversity.** Although there are many universals that we share (see D. E. Brown 1991), there are also significant differences among us in terms of personality, cognitive skills, values, age, sex, life experience, and so on. Basic personality differences have been well-documented by researchers in behavioral genetics and psychology (see Plomin et al. 1990; Lippa 1988), and there is a large body of research on personal interests and work objectives that is routinely utilized to help in selecting and training personnel for different occupations. Some of the statistics accumulated by one of the leading organizations in this field, Target Training International (1990), are illuminating. Only about thirty percent of the people who have taken TTI's assessments over the past twenty years have shown a dominant preference for economic and utilitarian objectives, and even fewer, only thirteen percent, are strongly motivated for political influence and power. On the other hand, fourteen percent are strongly motivated for social and humanitarian work, fifteen percent for learning and teaching, and seventeen percent for aesthetic and artistic ends. In short, human nature comes in many different colors. More than that, as Plato first pointed out, our differences can be a source of strength if society provides a diversity of niches that are able to utilize these gifts productively.

4. **Human nature is not fixed; it is labile and susceptible to a myriad of different social and cultural influences.** Many theorists—from Plato and Aristotle to Rousseau, Durkheim, and Marx (not to mention many modern-day social scientists)—have stressed the importance of "nurture" in shaping human behavior. Humans are, as a rule, greatly influenced by others, and by the "rules of the game" in their culture. To a significant degree, cultural influences can create self-fulfilling prophesies. If honesty, trust, mutual respect, courtesy, and the spirit of compromise are the prevailing norms while deviants are ostracized and penalized, a society and its institutions will likely reflect these values. Conversely, if the cultural climate encourages deception, demonization, vicious partisanship, and an uncompromising no-holds-barred attitude toward opponents, the social and political environment will more closely fit the paradigms of Machiavelli and Hobbes. By the same token, the well-known "contagion effects" to which we are so susceptible, from rock concerts to riots, can either have positive effects or be destructive. In other words, our cultures have the collective capacity to shape the ultimate expression of human nature, for better or worse. This

also has important implications for the issues of fairness and social justice, as we shall see.

5. **Though modern capitalist societies give priority to the private sector in meeting our basic needs and serving the public interest, its record has been decidedly mixed.** The private sector's claim to being assigned priority in serving our needs is based on the assertion that it can do the job better, cheaper, more efficiently, and with better outcomes. Unfortunately, this is not always true; there are major gaps in meeting the basic needs of our citizens, from health care to housing and adequate income. Accordingly, if the private sector fails to deliver on its promises to meet our basic needs, the community has a right—a collective self-interest—to undertake remedial "class action" through the legislative process, judicial system, shareholders, or other forms of cooperative action.

6. **Different modalities for satisfying our basic needs may lead to very different outcomes.** Free enterprise and markets are very often the most effective way to meet our basic needs. But this is not always the case. Health insurance, for instance, represents a vitally important need in our society. Yet some forty-four million Americans currently do not have any health insurance coverage at all, and an estimated forty million more have inadequate coverage. The reason, in a nutshell, is that we have entrusted the provision of this basic need to the private sector for those who are under retirement age (with some exceptions like the military). In the private sector, the criteria for coverage boil down to profitability, regardless of the need. Only those who can profitably be insured will be able to get coverage, which means high premiums and barriers to coverage for those who are at high risk. In insurance circles this is called "experience rating," and the effect is that the private sector screens out many of the most needy persons. The alternative approach is called "community rating," and it proceeds from the premise that everyone will be covered and that the risks and costs will be spread as widely as possible. The need will be fully satisfied, but the more affluent and lower-risk participants will pay relatively more, and the system may not be profitable. In other countries, this approach is called "national health insurance," and the United States is the only industrialized nation that does not have it. Medicare covers all Social Security retirees, and the administrative costs for Medicare run about three percent annually. The administrative costs alone for private health plans (including profits), run from fifteen to twenty-five percent (Brenden and Hamer 1998). If the criterion is meeting a basic need more completely and more efficiently, this is a case where the government can clearly do the job better.

A Return Engagement for Social Justice

In an era marked by unapologetic increases in the gap between the rich and
the poor (abetted by heavily weighted tax cuts) coupled with aggressive political attacks on the welfare state, the ancient, much debated concept of justice
has been deeply challenged.[5] Epicureans dismiss *justice* as a meaningless
term. Others paraphrase Thrasymachus in Plato's *Republic* (1985): justice
amounts to nothing more than "the interest of the stronger" (read the rich
and the powerful). The so-called realists invoke simplistic social Darwinist
stereotypes. Still others follow in the Lockean tradition and equate it with the
protection of property rights. Robert Nozick's famous libertarian tract,
Anarchy, State and Utopia (1974), is a classic statement of this position.[6]

To be sure, the concept of justice still has a secure place within our legal system. Indeed, the very concept of an independent judiciary represents, at heart,
an institutionalized instrument of justice. Thus, "procedural justice" refers to
such things as equality before the law, due process, and impartiality in the making of judicial decisions. For instance, most of us recoil from the idea of a
litigant who physically threatens or bribes a jury on his own behalf. Jury tam-
pering is illegal, and almost nobody thinks that a prohibition against this practice is unfair. On the other hand, "substantive justice" (letting the punishment
fit the crime and other legal principles), which can be traced all the way back
to the classical Greeks and the ancient Roman lawyers, is concerned with the
fairness and "equity" of the outcomes that are produced by a legal system.

However, social justice involves something more. *Social justice* is a term that
has invited many different definitions, but it generally refers to the distribution
of substantive rewards among the members of a society. Its origin traces back
at least to Periclean Athens. To Plato, justice is not primarily concerned with
some higher metaphysics or a tug-of-war over our rights as individuals. It is
concerned with equitable rewards for our conduct and the proper exercise of
our abilities in a network of interdependent economic relationships. Moreover,
and this point is crucial, Plato recognized that equity also has a floor—a minimum wage, so to speak. Here are Plato's words from the *Republic*:

> If we begin our inquiry by examining the beginning of a city, would that not
> aid us also in identifying the origins of justice and injustice? . . . A city—or a
> state—is a response to human needs. No human being is self-sufficient, and all
> of us have many wants . . . Since each person has many wants, many partners
> and purveyors will be required to furnish them. . . . Owing to this interchange
> of services, a multitude of persons will gather and dwell together in what we
> have come to call the city or the state. . . . [So] let us construct a city beginning

with its origins, keeping in mind that the origin of every real city is human necessity. . . . [However], we are not all alike. There is a diversity of talents among men; consequently, one man is best suited to one particular occupation and another to another. . . . We can conclude, then, that production in our city will be more abundant and the products more easily produced and of better quality if each does the work nature [and society] has equipped him to do, at the appropriate time, and is not required to spend time on other occupations. . . . Where, then, do we find justice and injustice? . . . Perhaps they have their origins in the mutual needs of the city's inhabitants. (book II, secs. 369a–369d, 370b, 370c, 372a)

Aristotle, in the *Politics,* supplemented his mentor's views in some very important ways. First, Aristotle emphasized that physical security—both external and internal—is also a fundamental function of the state. The *polis* is not exclusively an economic association. Aristotle also stressed that human nature is not an autonomous entity; it consists of a set of innate aptitudes that are uniquely fitted for society and that can only be developed in close social relationships. Thus, social life involves more than simply being a marketplace for economic transactions. It also involves a life in common; we are enriched by our membership in families and communities. A hermit is not only economically deprived; he/she is not fully human and, equally important, occupies an evolutionary dead end.

But more important for our purpose, Aristotle, in his *Nichomachean Ethics,* also elaborated on the concept of "just deserts," or "giving every man his due" as Plato's character Polymarchus put it. There have been countless debates through the centuries over what Plato and Aristotle meant by the word *due,* but a common sense interpretation is that the rewards provided by society should be proportionate to a person's contributions to society (and the same holds for crimes and punishments). It does not mean equality. Rather, it means an equitable portion in accordance with some measure of fairness— a "fair share." Aristotle himself used the term *proportionate equality.* There is, of course, a voluminous body of scholarly literature on the modern concept of "distributive equity," especially in welfare economics (see Sen 1992; Kolm 1996; Roemer 1996; Arrow et al. 1999 Zucker 2001; Kaplow and Shavell 2002). Here I will develop a variation on this theme.

Three Arguments for Social Justice

Plato and Aristotle were both acutely aware of the potential for destructive social conflict. Indeed, Aristotle and his students conducted a study of the political history of 158 different Greek cities (thus establishing his credentials as the first political scientist), and he knew full well what havoc could result when a state lost its legitimacy (the willing consent of the citizenry). Preserving the sometimes fragile stability of the community was a major concern for Plato and Aristotle.

Aristotle also devoted much attention to the fundamental political problem, also well appreciated by Plato, that the basic, seemingly inescapable cleavage between the few who are rich and the many who are poor is potentially the most dangerous social division of all and the underlying cause of much civil unrest. The key to preserving the community, therefore, is to strike a balance between these conflicting interests. To this end, the law must be "sovereign" and must serve as an impartial arbiter—"reason unaffected by desire." Moreover, there must be moral equality before the law. The law cannot be used as a tool to favor the rich and powerful but must be an instrument for achieving social justice. Otherwise it becomes a part of the problem. Aristotle was also mindful of the Greek playwright Euripides' admonition that the inherent conflict between the rich and the poor, if pushed to an extreme, can destroy a state. According to Euripides, it is the middle class that "saves states."

I submit that this insight still remains valid. There is a well-established empirical relationship between what contemporary political scientists call "relative deprivations" and the incidence of political turmoil (see Gurr 1970; Gurr and Bishop 1976). Therefore, one major argument for using an objective concept of social justice (beyond the workings of the hidden hand) is purely prudential—a matter of enlightened self-interest on the part of the "haves" in society. All of us depend upon the vast, interdependent (and always vulnerable) collective survival enterprise along with the willing cooperation of many others, as Adam Smith so eloquently expressed it. Recall that Darwin spoke of social "coherence." Sociologist Emile Durkheim (following Herbert Spencer) stressed the importance of "solidarity." And many others characterize it as "unity" or "patriotism"—the intangible spirit of cooperation, reciprocity, and fairness that undergirds a reasonably harmonious society.

A second argument for social justice is that it is rooted in biological imperatives that we all have in common—our basic survival and reproductive needs. Almost all of us are dependent upon the collective survival enterprise, and we are the beneficiaries of what our forebears collectively created for us over millions of years of evolution. But more to the point, there is no evolutionary future for any of us, or our legacy (whether it be our children or our personal accomplishments), apart from the collective survival enterprise. Moreover, our basic needs are not narrow, vague, or capriciously variable (recall chapter 11). They are concrete, measurable, and cut a very broad swath through our economy and society. Nor are they optional. Denying basic needs to any person, whether wittingly or not, unavoidably causes them harm.

Accordingly, the basic needs of the members of a society have a moral claim that is prior to, and ultimately limits, any claims to property rights. As many theorists have argued, property is ultimately a means to our biological ends, though other motives obviously come into play as well. To borrow an expression from the pro-life advocates, the "right to life" is prior to property rights, and it does not end at birth; it represents a lifelong claim on the resources of society. Indeed, both John Locke and the American founding fathers concurred with this rank ordering. Among our "inalienable rights," life comes before liberty, while property (or "the pursuit of happiness") comes last. And the proof of it is that most of us follow the same rank ordering of priorities when forced to choose.

A third argument for social justice, and the concept of fair shares, derives from the accumulating evidence that a sense of fairness is a deeply rooted aspect of human nature, as Darwin himself suggested. This was discussed briefly in earlier chapters; here I will elaborate.

In political scientist James Q. Wilson's (1993) characterization, most humans do have a "moral sense" (there are individual variations in this respect as always)—especially a sense of fairness, as well as concern for our own "rights." Wilson's important book discusses the subject at length. But see also Masters (1987) and Masters and Gruter (1992). Our sense of fairness appears to be a joint product of both nature and nurture. The "norm of fairness," as it has been called, first appears at a very early age. It involves, in essence, a recognition of entitlements that apply to others as well as to oneself. Simple decision-rules like equal shares or drawing straws work well enough at this age. But, as the child develops, the content of the sense of fairness changes and deepens as a rule (again, there are variations), and more complex criteria are used—age, merit, need, and even social relationships ("we versus they," or friends versus enemies). Also, needless to say, the content is influenced by the values, customs, rules, and practices of a given society (i.e, what others believe is fair). And, of course, we also have a propensity for rationalizing unfairness away when it suits our interests. Nevertheless, fairness also has a strong, if imperfect, pull on our conduct.

To reiterate, scientific evidence that a norm of fairness and reciprocity is a universal aspect of human nature can fairly be called robust and continues to grow. Indeed, it is found in virtually every society, and the few pathological exceptions seem to prove the rule. Among the many references on this topic, see especially Gouldner (1960); Westermarck (1971); Axelrod (1986); Alexander (1987); Mansbridge (1990); McShea (1990); Masters and Gruter (1992); Arnhart (1998); Crawford and Krebs (1997); Sober and Wilson (1998); Katz (2000); Leigh (2000); D.S. Wilson (2002); and Cela-Conde and Ayala (2004). Indeed, fairness is a daily issue in our own society. There is also a vast body of research on this in psychology, game theory, and experimental economics. Most noteworthy, perhaps, are the so-called ultimatum games, which have been used (repeatedly) to demonstrate that people are willing to share with others in ways that do not reflect their own rational self-interest but reflect instead a sense of fairness (see Gergen 1969; Greenberg and Shapiro 1971; Kahneman et al. 1986a, 1986b; Rabin 1993; Fehr and Schmidt 1999; Gintis 2000a, 2000b; Nowak et al. 2000; Henrich et al. 2001; Fehr and Fischbacher 2002, 2003; Sigmund et al. 2002).

Equally important, it appears that people are far more willing to invest in policing fairness and punishing deviants than classical economic theory predicts (see Sigmund et al. 2001; Bowles and Gintis 2002; Fehr and Gächter 2000a, 2000b, 2002; Gowdy 2004a, 2004b; M. E. Price et al. 2002; Gintis et al. 2003; for a more general discussion of evolutionary economics, see chapter 12). There are even some rudimentary examples of a sense of fairness in other species—the most conspicuous being sharing behaviors and reciprocity (see de Waal 1996, 1997, 2001; Wrangham and Peterson 1996; Dugatkin 1999). Finally, the accumulating psychological evidence has been given an evolutionary imprimatur by the resurgence of group selection theory in evolutionary biology, most notably the work by David Sloan Wilson and his colleague, Elliott Sober. As Wilson (1999) puts it in a recent article: "The idea that moral systems are designed to promote the common welfare of groups can be accepted at face value." See also Sober and D. S. Wilson (1998) and E.O. Wilson (1998).

To summarize the argument, then, the ideology of fair shares has three empirically grounded sources of justification:

1. It is an essential prerequisite for the stability and ultimate viability of the political order; in game theory terms, people are very likely to "defect" from the existing political order if their survival is threatened while others enjoy a huge surplus of resources.

2. A norm of fairness strikes a balance between the inescapable equalities in society (our basic needs) and the inevitable inequalities in inherited wealth, talent, risk taking, and hard work.

3. It meshes with a deep psychological sense of fairness, rooted in our evolved human nature, which goes to the very heart of the enduring political problem of how to secure legitimacy. In fact, fairness is our chief weapon in the age-old war with the centrifugal force of political alienation and social conflict.

The philosopher John Rawls, in his celebrated (and much-debated) theory of "justice as fairness," arrived at a somewhat similar place, albeit by a very different route. He too tried to wrestle with undeniable inequalities coupled with the existential needs of the least advantaged. Rawls did not succeed in convincing the utopians, who insist on radical equality by one means or another. Nor did he convince libertarians, who are in denial about our inextricable interdependency and do not recognize any social obligation to the disadvantaged. But Rawls did strike a chord, whatever the flaws in his fingering, with a broad spectrum of "fair-minded" individuals—those who seek a middle way between the revolutionary agenda of radical egalitarian socialists and the failed promise of "natural justice" purveyed by the *laissez faire* capitalists (or worse, the sanguinary "survival of the fittest" ethics of the social Darwinists). Rawls's instincts were correct. There is a middle way, and it has a biological foundation. (For more on Rawls's theory, see below.)

Fair Shares: A Synopsis

We need to begin by defining fairness. What is fairness? How do you know it when you see it? And why is it so hard to determine what is fair in any definitive, all-purpose way? The answer is that fairness is not some absolute principle, and there is no unambiguous measure of fairness. Fairness involves a value judgment that can only be made in the context of a specific set of social, economic, or political relationships. It refers to a certain quality, or property, of a particular relationship in a particular situation, and each instance must be judged in its own terms (or perhaps in terms of some precedent, norm, or decision rule for a class of similar situations). In other words, fairness is a relational concept.

But if fairness is context dependent and relative, does this mean that it is an illusion or a rationalization that serves to mask individual self-interest (as some cynics believe)? On the contrary, fairness, by definition, must take into account the interests and needs of all the parties to a relationship. Fairness is quintessentially about compromise and reconciling different, often conflicting, interests. It represents a middle ground where it is possible to replace coercion and zero-sum logic with the willing consent of the affected parties. Indeed, fairness is a basic prerequisite for a harmonious group or society, and it is (rightly) a core value of democratic societies.

Accordingly, two distinct, and frequently competing, fairness claims arise out of the imperatives of human nature and the nature of human society as a collective survival enterprise: (1) basic needs (or distributive equity) and (2) merit (giving every person his/her due). In the fair shares paradigm, our basic survival needs take precedence, but they do not nullify the claim to merit; they impose a constraint. The middle-ground position recognizes the validity of both capitalist and socialist moral claims. Accordingly, the fair shares framework rests on three major principles. Though these principles are not new, in combination they provide a synergistic third way. These principles are the following:

1. **Goods and services should be distributed to each according to his/her basic needs.** Though this may sound like an echo of Karl Marx, it is at once more specific and more limited. Here the term *basic needs* refers to the fourteen needs delineated in chapter 11. Our basic needs are not a vague, open-ended abstraction or a matter of personal preference. They constitute a concrete agenda, albeit subject to further refinement, with measurable indicators for assessing the outcomes. Also, this paradigm fully recognizes that there are individual and contextual differences, vitally important instrumental needs (which are subject to change throughout the life cycle), and that reproduction and the needs of dependent offspring must be included as well. It should go without saying that both the marketplace and other forms of collective action (including government actions) can play a vital role in meeting our basic needs.

2. **Surpluses beyond the provision for our basic needs should be distributed according to merit.** Like the term *fairness* itself, merit has an elusive quality; it does not denote some absolute standard or measuring rod. It too is relational, context-specific, and subject to all manner of culturally evolved norms and practices. In general, it implies that a person's rewards should be proportionate to his/her effort, investment, or contribution. For example, people would never claim that they won the lottery based on their merits. Merit has many facets, of course, but in the economic sphere, merit is not about what the recipient thinks is fair treatment but what is socially acceptable; like fairness, merit very often has to split the difference. In the fair shares paradigm, the ultimate criterion is that a person's rewards should reflect the contributions that are made to the collective survival enterprise and to our collective needs (and wants). Thus, the "merit principle" stakes a moral claim; it poses the right question. However, there is no formulaic way

of determining merit. Both the marketplace and a representative, mixed, democratic government can (and do) play a vital role in the imperfect art of determining what is fair compensation.

Does this paradigm imply a return to "welfare queens," or a culture of "freeloading" and an indolent class of economic "defectors" (in game theory terminology)? The answer, most emphatically, is no. Where's the equity in that? In fact, a crucial corollary of these two principles is that the collective survival enterprise has always been based on mutualism and reciprocity, with altruism being limited (typically) to special circumstances under a distinct moral claim—what could be called "no-fault needs." So, a third principle must be added to the fair shares paradigm.

3. **In return, each of us is obligated to contribute to the collective survival enterprise in accordance with his/her ability.** Needless to say, this principle—which is fundamental to Plato's definition of justice—applies equally to the rich and the poor, to wealthy matrons and welfare mothers. However, it also begs the question, How are abilities and contributions to be determined? Again, there are no formulaic answers, but societies have developed many different ways for permitting such collective judgments to be made, from markets to legislatures, election processes, lotteries, formal examinations, licenses, performance evaluations, progressive taxes, and many others.

An anonymous reviewer for this chapter characterized my ideological stance as "fuzzy Marxism." In fact, the opposite is true. Marxism is grounded in a fuzzy biology, along with a simplistic and one-sided model of human nature. Marxism actually violates the principles of fair shares. For one thing, Marx was quite diffident about specifying the content of our basic needs. He allowed the inference to be made that equality and equity are equivalent. In the fair shares paradigm, this is emphatically not the case. Also, Marx made no provision at all for merit, and he was quite hostile to capitalism. Remember, capitalists were viewed as the villains and were destined, in accordance with the imperatives of Marx's dialectical materialism, to end up in "the dustbin of history" (to use Bolshevik Leon Trotsky's epitaph for the Mensheviks). But most important, Marx's directive that everyone should contribute to society in accordance with his/her ability—in the absence of the other two principles articulated above, especially the claim for merit—could potentially be exploitative. Despite the similarities in phrasing, the Marxian "contribution principle" does not accord with the fair shares paradigm.

Many other issues and questions are begged by these principles, of course, and some very important qualifiers must also be added. I will very briefly outline a few of them.

- **First, there is the problem of the naturalistic fallacy and the "is-ought dichotomy."** A critic might ask, Why ought we to care about our survival and reproduction, much less that of anyone else in our society? More to the point, why should anyone—especially the "haves"—accept the fair shares ethic as a standard for guiding the policies and practices of a society? Even if we have been "programmed" by our evolutionary heritage to be concerned about fairness, how can anyone claim that this creates a normative imperative? Actually, these are the wrong questions. They amount to a sophist sand trap. The issue here is not whether we can justify some categorical imperative for morality. Rather, given the cardinal facts that (1) we do care—intensely—about satisfying our basic needs; (2) these needs must, by and large, be satisfied through cooperative activities associated with the collective survival enterprise; and (3) we do, after all, have a shared sense of fairness, then the fair shares ideology provides a compass for steering a society between the political shoals. It provides a set of prudential normative principles that direct us to navigate a middle course between free market capitalism and egalitarian socialism. Moreover, these principles represent existential imperatives in the sense that serious maladaptive consequences—both individually and collectively—will result from ignoring them and pursuing an alternative course.

- **How do we implement this ideology?** How do we go about ensuring that our basic needs are met? Does this imply an economic and social revolution of some sort? The answer is most certainly not. It implies a need to improve an evolved and well-tested economic system that has many virtues but also some serious deficiencies. There are currently many effective market-based instrumentalities for meeting our basic needs. These must be augmented, improved, and supplemented as needed; a better balance is required.

- **What about freedom and liberty—core values of Western democracies?** The response to this concern is that we need to move beyond the naive assumptions and self-serving rhetoric about freedom that is so prevalent today. As the social critic Charles Morgan put it long ago: "Liberty is the room created by the surrounding walls." In other words, freedom always has boundaries. What we are talking about here is an adjustment in the location of the walls. For some, the room will be expanded. For instance, more income for the poor would free them from some deep anxieties and severe, even life-threatening (or life-shortening), economic constraints. For others, there will be some shrinkage of freedom, but it would most likely be marginal. In any case, it is a trade-off we must accept for the viability of the collective survival enterprise. It is in our (enlightened) self-interest to do so.

- **How does the fair shares ideology affect our sacred property rights?** The fact is that property has no unqualified natural right beyond what the rest of us are willing to recognize—and defend. This observation goes back to Plato and was seconded by Bentham, Rousseau, Marx, and many others, including the Supreme Court of the United States. In the real world there are many political constraints on property rights. They are reflected in such things as eminent domain, restrictive zoning and building codes, fire codes, condominium covenants, property taxes, curfews, and other rules. Under the fair shares paradigm, property rights are further limited by what is compatible with meeting the basic needs of the rest of the population.

• **What about those who cannot contribute their fair share of productive capital and labor to society?** Is there some danger that unconstrained help for those who are truly needy would turn society into a vast charity ward, imposing an enormous economic burden on the rest of us? The fact is that we already willingly support our dependent children, the elderly, the disabled, and aging veterans (among others). We are more grudging, as a society, about aiding the poor and their children, and we know that many of them can contribute in various ways. Nevertheless, there is also a hard core of people in our society who, for one reason or another, will always be unable to contribute; they are victims, not wastrels. "Workfare", or welfare-to-work programs will never work for them. This is a political reality we seem reluctant to face. But if our evolved moral sensibilities can encompass the victims of highly visible disasters like floods, earthquakes, and terrorist attacks, there is no moral ground for excluding the less visible tragedies all around us, including those that are, sadly, biologically based. This is where the golden rule, and perhaps Rawls's "veil of ignorance," could be applied, especially since we ourselves, or people we love, could also end up in need. Indeed, one of every five families in the United States has a member who suffers from some form of mental illness. Thus, our welfare programs also represent a form of social insurance for the middle class, a principle that goes back to the Greek funeral societies.

• **What if the existing economic and political order fails to provide for our basic needs?** Both the historical record and the implicit terms of our biological contract warn us that all regimes are ultimately contingent. Indeed, the American Declaration of Independence contains an enduring justification for breaking the "political bands" of the existing order. Governments are "instituted among men" to secure our "inalienable rights," and derive their "just powers" from "the consent of the governed." Furthermore, whenever any form of government "becomes destructive of those ends," the people have the right "to alter or abolish it." Plato and Aristotle warned that no political order is immutable. And the modern game theorists, whose research has done so much to illuminate the foundations of social cooperation in evolution, have shown unequivocally that mutualism is essential. "Defection" (as the game theorists call it) is the likely response to an exploitative, asymmetrical interaction. To be sure, in human societies, coercive force is often used to prevent defections, but the costs and risks are always high, and the long-term outcome is always problematic. To repeat, extremes of wealth and poverty are the seedbeds of revolution.

- **Finally, there is the politically explosive issue of where we must draw the line or lines.** Is it realistic to have an open-ended commitment—an entitlement—to provide for the basic needs of all potential claimants? Should we accommodate an unrestricted number of babies born to welfare mothers or deadbeat fathers? And would this include the continuation in perpetuity of an open-door immigration policy or an unending flood of illegal immigrants? Finally, how do we draw lines in a global economy where more and more of our needs, and our wants, are satisfied by underpaid workers in other countries? Global poverty is a vast ocean of unmet needs. For example, in Mexico alone, some forty percent of its population of 97.5 million live in deep poverty. There are no easy answers to these questions, but I would reiterate a key point made earlier about the nature of the collective survival enterprise. It is based on mutualism and reciprocity, not altruism. So the general answer to the question above is that, in order to be consistent with the imperatives of the biological contract as articulated here, lines would eventually have to be drawn. This is an inescapable trade-off.

Conclusion

Fairness is the golden thread that binds a viable society together, and, when that thread breaks, the social fabric unravels. The response to the former British Prime Minister Margaret Thatcher's contemptuous claim that "there is no such thing as society" is that a society exists when people believe it does and act accordingly—and vice versa! But fairness is not an all-purpose formula or recipe. It is a general principle that recognizes the merit of competing interests and directs us to find equitable compromises—proportionate equality. In this paradigm, compromise is not necessarily a sellout of one's principles to political expediency. It may be, and often is, a superordinate principle with a higher moral claim. It recognizes and accommodates legitimate competing claims and serves the overarching goal of preserving a cooperative society. However, the evidence is all around us that fairness is often a matter of perspective; it can be very difficult to call. That is why we have a formal justice system, mediation, family counseling, contract negotiations, and markets. Indeed, every society has a toolkit of informal customs and practices for approximating fairness—from equal shares to first come, first served; drawing straws; and special accommodations for children, senior citizens, and the handicapped.

However, social justice can be defined, to a first approximation, within the framework of the fair shares principles. It is grounded in the bedrock imperatives of our basic needs, using the measuring rods provided by the survival indicators program. The fair shares ideology provides both a biological justification and, ultimately, a political imperative for striking a better balance between the provision for our basic needs and rewards for merit. More important, it provides specific tools for measuring where this balance can be found.

We conclude, then, by returning to where we began. Charles Darwin recognized that a human society is, quintessentially, an interdependent survival enterprise. The superorganism is the key to our survival and reproduction. However, this vision of our collective purpose does not negate or ignore our individual self-interests. Rather, it represents an aggregation of those interests into an immensely complex system of synergies based primarily on mutualism and reciprocity—as Adam Smith himself fully appreciated. The poet John Donne's famous line, "no man is an island," is also true in a very practical, bioeconomic sense.

Thus, we can affirm that the modern democratic state has evolved as an instrumentality for self-government and the pursuit of our common needs—the public interest—though its purpose is all too often subverted. Plato and Aristotle apprehended this basic purpose in their conception of the *polis,* and Aristotle prescribed a "mixed" government under law as our best hope for ensuring that the public interest would be served. Plato and Aristotle also recognized that a fair-minded form of justice is an essential part of the public interest; this is the only way to ensure long-term political stability and social harmony. Social justice consists of providing for both the basic needs of the population and rewards for merit. Indeed, merit provides important social incentives for ensuring that our needs are met.

Over the past two thousand years we have added very little to this vision that is fundamentally new, though we have made many important improvements to the machinery of self-government. The fair shares framework contributes to this effort by spelling out the principles for social justice in more specific detail. It enlists the growing power of modern evolutionary biology and the human sciences to shed light on this matter, and it articulates an explicit set of criteria for reconciling (if not harmonizing) the competing claims to social justice advocated by the New Left and the New Right. I believe that this ethic offers our best hope for achieving and maintaining that elusive state of willing consent that is the key to a harmonious society. It is an ideal worth striving for, because our own survival, and more certainly that of our descendants, may well depend upon it. Nothing less than our evolutionary future is at stake. To paraphrase Benjamin Franklin, in the long run either we will survive together or become extinct separately.

An Afterword

Three subjects that were touched upon in this chapter deserve more extended treatment. They were abbreviated sidebars for the original article on this subject. They include the basic tenets of capitalism and socialism, as well as John Rawls's effort to define a middle-ground. Although a full explication and analysis of these important topics is obviously not possible here (a vast scholarly body of literature is devoted to each one of them), an abbreviated discussion of some key points is provided below.

Capitalism: A Thumbnail Primer

Though there have been many variations over the years, the basic theme in conservative capitalist theory derives from Adam Smith's paradigm. Most fundamental, perhaps, is Smith's assumption that acquisitive self-interest is a primary human motivator and that a capitalist market economy harnesses private greed to serve the social good, thanks to the magic of the "invisible hand." The very essence of capitalism is that it gives full rein to individual entrepreneurship and provides commensurate rewards. Indeed, private wealth and free enterprise are touted as the engines of economic progress. Modern conservatives have also assured us—endlessly—that a rising tide lifts all boats; if the rich get richer, so will everyone else in due course. To quote the conservative economist Paul Rubin (2002, p. xiv): "In today's world . . . people mostly become wealthy by being productive and creating benefits for others, and, therefore, desires to punish or penalize the wealthy are misguided." Moreover, it is often claimed, an unfettered free market is vastly more efficient at satisfying human wants and preferences than is any centralized "command economy" (the former Soviet Union is usually cited as the poster child, rather than the contradictory example of the United States in World War II). Accordingly, the welfare state that was created by liberals in the New Deal era and then expanded further after the Second World War is often viewed as an impediment to free markets—or worse. Government services are often charged with preempting the supposedly more efficient private sector alternatives, and intrusive government regulations are resented as being a hindrance to the supposedly self-polic-ing, self-correcting mechanisms of the marketplace.

Like many overstated, often self-serving claims, this model has much to recommend it but also some serious shortcomings. To be specific:

1. Human nature is complex and diverse, and we are not all consumed either by the profit motive or by latent brotherly love.
2. Private enterprise has been only one of the engines of our progress as a civi-lization.
3. Sometimes, though not always, government *can* do the job better.
4. The wealthy do not always owe their wealth to their productivity; nor do they always use their wealth to enhance the productivity of society as a whole.
5. The private sector is obviously not reliably self-policing; indeed, Adam Smith (1759/1976) himself appreciated that the hidden hand sometimes morphs into a sleight of hand.
6. Finally, markets manifestly cannot be relied upon to meet the basic needs of the population as a whole; they respond mainly to supply and demand, which of course depends upon the ability to pay (see chapter 11).

Socialism-Liberalism: A Primer

It is often said that socialism traces its roots to Jean Jacques Rousseau's con-cept of the "noble savage." Be that as it may, the core assumption that has animated much of the socialist-liberal school over the years was recently restated by former senator George McGovern (2002), who relied on a quo-tation from Webster's Dictionary: "One cannot conceive of a nation dedi-cated to democracy that does not rest on faith in 'the essential goodness of man'" (p. 38). One corollary of this assumption is a commitment to egali-tarianism. This is a major theme in socialist and liberal theory. In practice, modern socialism-liberalism is often oriented toward the needs of the least advantaged in society, to use philosopher John Rawls's characterization, with government being utilized as the primary instrumentality for meeting these needs. As President Franklin D. Roosevelt expressed it in one of his famous radio *Fireside Chats*:

> One of the duties of the State is that of caring for those of its citizens who find themselves the victims of such adverse circumstances as makes them unable to obtain even the necessities for mere existence without the aid of others. That responsibility is recognized by every civilized nation. . . . To these

unfortunate citizens aid must be extended by Government—not as a matter of charity but as a matter of social duty. (quoted in Corning 1969, p. 29)

For moral support, Roosevelt invoked the words of Abraham Lincoln. Government, Lincoln had said, should "do for the people what they cannot do for themselves or cannot do so well for themselves" (quoted in Corning 1969, p. 29). Many modern, mainstream socialists and liberals also adhere to the view that the private sector and market mechanisms, while important, cannot be trusted to be self-policing or always to serve the public interest. Though President Reagan famously claimed that government is the problem, not the solution, many others believe that, as the old saying goes, the truth often runs well in reverse. Government action is sometimes the only effective way to defend the public against free market distortions and malfeasance—greed run amok.

Socialism-liberalism, like capitalist ideology, has its share of overclaims and warts, including a tendency toward oppressive overregulation, bureaucratic stagnation, gross inefficiencies, a stifling of innovation, and an all too common tendency to use governmental power for personal or narrowly partisan ends. But perhaps most serious is the charge that, despite good intentions, socialism-liberalism is sometimes the instigator of inequities and unwitting unfairness. This perception accounts for much of the recent animus against affirmative action and the public welfare program in the United States. However, the main battleground over fairness and equity in recent years has concerned taxes, where liberals and conservatives hold sharply different views.

A Critique of John Rawls's Theory of Justice

Rawls called his formulation a theory of justice, but it is not a causal/explanatory theory in the usual sense of the word. It is an effort to justify a normative stance—namely, that justice should be defined in terms of fairness. This aligned him with Plato and Aristotle, though his definition was significantly different. Rawls did not propose to do away with economic inequalities. Instead, he posited two broad principles. There should be (1) equality in the enjoyment of personal freedom and (2) a set of economic arrangements that allow for equal opportunity coupled with ways to allow the poor, or the "least advantaged," to benefit proportionately more when the rich get richer (to paraphrase his argument). Rawls's method for undergirding and supporting these principles was at once ingenious and frustrating. Like the social contract theorists, Rawls asked us to assume that we were in a hypothetical state of nature—an "original position"—in which we are behind a "veil of ignorance" about what our own station in life might end up being. In what amounted to an appeal to enlightened self-interest (not fairness, after all!), he argued that his principles are what we would rationally choose for organizing our society in a situation of uncertainty about our own circumstances. It is often termed hedging your risks. So Rawls's prescription really amounts to the golden rule in deep disguise: Do unto others as you would have them do unto you if you were the least advantaged.

Some critics have pointed out that it makes more sense (logically) to opt for economic equality under these circumstances. And, in fact, Rawls did just that in an earlier article, but he shifted to the "least favored" in his famous book. Others have questioned whether or not a hypothetical situation with no relationship to the real world—comparable to an unrealistic thought experiment in science—can legitimately be used to derive principles for real world application. Still others object that Rawls's two principles seem potentially to produce self-contradictions. On the one hand, if economic inequalities were allowed to persist, this would constrain the equality of freedom—the purchasing power and freedom from want—of the have-nots. On the other hand, imposing constraints on the rich that limit their ability to benefit from the fruits of their economic accomplishments represents a constraint on their freedom to hold property. Some critics also hold him accountable for not addressing the problem of implementation—how to move from a philosophical argument to the real world. Unfortunately, there is also a great deal of chaff in this critical literature—semantic hair splitting, willful misunderstanding or misinterpretation, and reveling in exceptions as if they were sufficient to disprove the general rule, among other things.

My own criticisms are more substantive and pragmatic. Rawls recognized what he called "primary goods"—that is, basic "rights and liberties, opportunities and powers, income and wealth" (1972, pp. 92–93)—which presumably include the wherewithal to satisfy basic needs. However, he did not explicitly give primacy in his theory to the satisfaction of basic needs per se; basic needs did not rise to a moral imperative for Rawls. Instead, his theory of justice would assure only that the poor get a piece of the action when the rich get richer. It amounts to a pledge that a rising tide should lift all boats. But what if the tide goes out? Or what if many of the boats develop leaks and are slowly sinking, as has been occurring in this country recently? In a later book, Rawls shifted the rank ordering of his normative principles and placed basic needs first, above liberty. But his rationale for doing so was curious. He argued that it was justified because meeting basic needs was a prerequisite for freedom. Another criticism is that Rawls tolerated inequalities, yet he did not make any explicit provision for merit— rewards for talent, effort, and achievement. As various critics have noted, Rawls was not much concerned about just deserts. Nor did he address the free-rider problem. The fair shares principles, on the other hand, do address these issues. For other critiques of Rawls's theory, see especially Wolff (1996), Alejandro (1998), and Raphael (2001).

Acknowledgments

I would especially like to acknowledge and thank my editors at the University of Chicago Press, beginning with senior editor Christie Henry, whose support, patience, and wise counsel were of great value in this undertaking. Also most helpful was the production assistance of Monica Holliday and the fine professional index prepared by Lys Ann Weiss. The many constructive and helpful suggestions offered by the anonymous reviewers of the manuscript were also of great value and contributed significantly to the final product. In addition, a number of the chapters in this book benefited from research assistance and reviews, including anonymous peer reviewers for various journals. I am grateful to all who provided these forms of assistance—both known and unknown. Below are the acknowledgments that accompanied these journal articles.

"Holistic Darwinism"

The author wishes to thank the following for encouragement and helpful suggestions on an earlier version of this paper: Howard Bloom, Michael Ghiselin, Lynn Margulis, Peter Meyer, John Maynard Smith, Ernst Mayr, Anatol Rapoport, John Paul Scott, David Smillie, Eörs Szathmáry, and David Sloan Wilson. I also benefited from an online discussion of some of these issues among the members of Howard Bloom's International Paleopsychology Project group. Particularly helpful were the comments of David Smillie, Timothy Perper, and Peter Frost. The diligent and resourceful research efforts of Patrick Tower are also greatly appreciated, along with the indispensable bibliographic assistance of Kitty Chiu.

"The Re-Emergence of Emergence"

The author would like to acknowledge the resourceful and diligent research assistance and insightful comments by Zachary Montz and the bibliographic assistance of Pamela Albert. Any errors are, of course, my responsibility. Also helpful was the online discussion of emergence hosted by the New England Complex Systems Institute (NECSI) during December 2000 and January 2001. The contributions of its various (unnamed) participants were valuable and much appreciated, even though I have disagreed with many of them. I also gratefully acknowledge Stanley Salthe's thoughtful reading and provocative

comments, even though we have agreed to di`sagree on some major issues. The comments
and helpful references provided by Geoffrey Hodgson are especially appreciated.

"Biological Adaptation in Human Societies"
      The author wishes to thank various commentators on earlier versions of this paper,
which were presented at the annual meetings of the Western Economic Association
International in 1996 and 1998, and at the annual meeting of the Public Choice Society in
1997. Particularly helpful were Gordon Tullock, Jack Hirshleifer, Michael Ghiselin, Janet
Landa, and Elihu Gerson. They do not necessarily endorse this project, or its conclusions.
The author also benefited from a challenging, and bruising, presentation-discussion sponsored by Helena Cronin and facilitated by Richard Webb at the London School of
Economics in the fall of 1996, with special thanks for the constructive criticisms and assistance of Max Steuer. Special thanks are also due to Patrick Tower and Connie Sutton for
their research assistance and to Kitty Chiu for preparing the graphics and the bibliography.

"Thermoeconomics"
      I am deeply indebted to the late Steve Kline, Woodard Professor of Science Technology
and Society and Mechanical Engineering, Emeritus, at Stanford University. I benefited
greatly from Professor Kline's widely acknowledged expertise in thermodynamics and his
patient mentoring. He was a good friend and is sorely missed. Needless to say, Prof. Kline
bears no responsibility for any errors that may be found in this paper. I am also grateful to
Ernst Mayr, Anatol Rapoport, and Howard Odum for their support of this work and to
Terrence Deacon for a lively and helpful discussion of these issues. I also sincerely thank
Stanley Salthe, a leader of the "infodynamics" school, for his willingness to read and comment upon a paper that was critical of his approach. I would also like to acknowledge the
insightful comments and helpful suggestions from two anonymous readers of the original
draft of this paper. Finally, I think Pamela Albert for her research and bibliographic
support. Her efforts contributed significantly to the final product.

"Control Information"
      This paper is the outgrowth of a close collaboration between this author and the late
Stephen Jay Kline, Woodard Professor of Science, Technology and Society, and of
Mechanical Engineeering, Emeritus, at Stanford University. Two jointly authored papers,
"Thermodynamics, Information and Life Revisited," (part I and part II) appeared in the
journal *Systems Research and Behavioral Science* (1998a, 1998b). However, the core concept of control information is one of the present author's contributions to this collaboration. This paper elaborates on the concept and relates it specifically to cybernetics and to
Norbert Wiener's theoretical framework. The author also wishes to thank the Collegium
Budapest (Institute for Advanced Study) in Hungary for a fellowship that was of great
assistance in completing this work, as well as Patrick Tower and Connie Sutton for their
diligent and capable research support, and Kitty Chiu for her varied contributions to the
production of the final result.

" 'Fair Shares': Holistic Darwinism and Social Justice"
      This paper was originally presented at the annual meeting of the Association for
Politics and the Life Sciences in 2002 with the title: " 'Fair Shares': Beyond Capitalism
and Socialism (The Biological Basis of Social Justice)." It has also been vetted by numer-

ous colleagues, some of whom were anonymous peer reviewers for *Politics and the Life Sciences*. I am also grateful for the favorable readings and helpful comments by Samuel Bowles, Ernst Fehr, Peter Richerson, Russ Genet, and especially Rob Sprinkle in his capacity as editor of the journal.

Citations for Publications and Presentations by Chapter

1. Modified from "Synergy: Another Idea Whose Time Has Come?" Commentary for the *Journal of Social and Evolutionary Systems* 21 (1):1–6 (1998). Courtesy of Elsevier Science.

2. Modified from "Holistic Darwinism: 'Synergistic Selection' and the Evolutionary Process." *Journal of Social and Evolutionary Systems* 20 (4): 363–400 (1997). Presented at the annual meeting, European Sociobiological Society, Cambridge, United Kingdom, 1996; also, the annual meeting of the Human Behavior and Evolution Society, Santa Barbara, CA, 1996; also, the annual meeting, International Society for Human Ethology, Vancouver, Canada, 1998; also, the Third International Congress, International Society for Endocytobiology, Freiburg, Germany, 1998. Courtesy of Elsevier Science.

3. Modified from "The 'Synergism Hypothesis': On the Concept of Synergy and Its Role in the Evolution of Complex Systems." *Journal of Social and Evolutionary Systems* 21 (2):133–72 (2000). Presented at the Annual Meeting, International Society for the Systems Sciences, Amsterdam, The Netherlands, 1995; also, at the Human Behavior and Evolution Society Annual Meeting, Davis, CA 1997. Courtesy of Elsevier Science.

4. Modified from "Synergy and Self-Organization in the Evolution of Complex Systems." *Systems Research* 12 (2):89–121 (1995). Presented at the annual meeting, International Society for the Systems Sciences, Asilomar CA 1994. Courtesy of John Wiley & Sons, Ltd.

5. Modified from "The Re-emergence of 'Emergence': A Venerable Concept in Search of a Theory." *Complexity* 7(6):18–30 (2002). Presented (with a slightly different title) at a Special Symposium on "The Role of Synergy in the Evolution of Complexity" at the annual meeting, Human Behavior and Evolution Society, London, 2001. Courtesy of John Wiley & Sons, Ltd.

6. Modified from "Synergy, Cybernetics and the Evolution of Politics." *International Political Science Review* 17 (1):91–119 (1996). Presented at the annual meeting, International Political Science Association, Munich, Germany, 1995. Courtesy of Sage Publications Ltd.

7. Modified from "'Devolution' as an Opportunity to Test the 'Synergism Hypothesis' and a Cybernetic Theory of Political Systems." *Systems Research and Behavioral Science* 19:3–26 (2001). Presented at the annual meeting, Association for Politics and the Life Sciences, Boston, MA, 1999. Courtesy of John Wiley & Sons, Ltd.

8. "Synergy and the Evolution of 'Superorganisms', Past, Present and Future." Presented at the annual meeting, Association for Politics and the Life Sciences, Montreal Canada, 2002.

9. Modified from "Evolutionary Economics: Metaphor or Unifying Paradigm?" *Journal of Social and Evolutionary Systems* 18 (4):421–35. Courtesy of Elsevier Science.

10. "Bioeconomics as a Subversive Science." Presented to a special workshop on "Evolutionary Concepts in Economics and Biology," Max Planck Institute, Evolutionary Economics Unit, Jena, Germany, December 2–4, 2004.

11. Modified from "Biological Adaptation in Human Societies: A 'Basic Needs' Approach." *Journal of Bioeconomics* 2:41–86 (2000). Presented at the annual meeting, Western Economic Association, Lake Tahoe, CA 1997. Courtesy of Kluwer Academic/Plenum Publisher.

12. Modified from "Thermodynamics, Information and Life Revisited, Part I: To Be or Entropy." (coauthored with Stephen J. Kline) *Systems Research and Behavioral Science* 15:273–95 (1998); "Part II: 'Thermoeconomics' and 'Control Information.'" *Systems Research and Behavioral Science* 15:453–82 (1998). Presented at the annual meeting, International Society for the Systems Sciences, Budapest, Hungary, 1996. Courtesy of John Wiley & Sons, Ltd.

13. Modified from "Thermoeconomics: Beyond the Second Law." *Journal of Bioeconomics* 4 (1): 57–88 (2002). Presented at the annual meeting, International Society for the Systems Sciences, Toronto, Canada, 2000. Courtesy of Kluwer Academic/Plenum Publisher.

14. Modified from "'Control Information': The Missing Element in Norbert Wiener's Cybernetic Paradigm?" *Kybernetes* 30 (9/10):1272–88 (2001). Presented at the annual meeting, International Society for the Systems Sciences, Atlanta, GA, 1998. Winner, UK Cybernetics Society 30th annual prize competition, 1999. Courtesy of the Emerald Group Publishing Ltd.

15. Modified from "Evolution and Ethics . . . An Idea Whose Time Has Come?" Part One, *Journal of Social and Evolutionary Systems*, 19 (3):277–85 (1996); Part Two, *Journal of Social and Evolutionary Systems*, 20 (3):323–31 (1997). Courtesy of Elsevier Science.

16. Modified from "The Sociobiology of Democracy: Is Authoritarianism in Our Genes?" *Politics and the Life Sciences* 19 (1):103–8 (2001). Courtesy of the Association for Politics and the Life Sciences.

17. Modified from "'Fair Shares': Beyond Capitalism and Socialism, Or the Biological Basis of Social Justice." *Politics and the Life Sciences* 22 (2):12–32 (2003). Presented (with a slightly different title) at the annual meeting, Association for Politics and the Life Sciences, Montreal Canada, August 11–14, 2002. Courtesy of the Association for Politics and the Life Sciences.

Notes

Chapter 2

1. Actually, it was W. D. Hamilton who started it. Hamilton had previously asserted that there were only three forms of social interaction—(1) altruism, (2) exploitative (zero-sum) selfishness, and (3) spite (E. O. Wilson 1975).

2. Douglas H. Boucher (1985), in an edited volume on mutualism, pointed out that there is a long-standing debate among ecologists over the relative importance of competition and cooperation in nature, which can be traced back at least to the 1920s. He noted the remarkable fact that, despite a general bias over the years in favor of competition as the basic organizing principle of nature and a concomitant preference among theoretical ecologists for using the famed Lotka-Volterra competition model in their analyses, in fact a cooperative version of the model (involving a simple sign change) has been reinvented (evidently independently) at least twenty-nine times since 1935. Boucher's volume reflected yet another of the periodic renewals of interest in the cooperative aspect of ecology. Similarly, in an overview and analysis of cooperative behaviors, Jerram Brown (1983, p. 29) noted: "Natural selection is an ecological process and cannot be understood solely from genetic considerations. Relatedness to non-descendants does not determine the direction or product of natural selection; it only supplies an additional cost or benefit." Also, Jon Seger (1991), echoing Darwin's proposed explanation for human evolution in *The Descent of Man,* points out that the various hypothesized explanations for social life are not mutually exclusive and in many cases might reinforce one another.

3. Another implication of this insight is that political theory can now be connected directly to evolutionary theory; insofar as politics and government are related to the problem of survival and reproduction, in any species, there is a common functional basis. Human political systems may thus be viewed as variations on an evolutionary theme. For an extended treatment of this paradigm, see Corning 1983, 1996b, 2004b. Also, see the offerings in Somit 1976; Willhoite 1976; Wiegele 1979; Somit et al. 1980; de Waal 1982; Schubert 1989; Masters 1989; Schubert and Masters 1991; Vanhanen 1992; Johnson 1995; Arnhart 1998. See also chapter 6.

4. Among the many recent publications that are relevant to this complex issue—including items in virtually every issue of the *American Anthropologist* and *Current*

Anthropology—see especially Johnson and Earle 1987; Byrne and Whiten 1988; Mellars 1989; Mithen 1990; Durham 1991; Jones et al. 1992; Maryanski and Turner 1992; Smith and Winterhalder 1992; Gibson and Ingold 1993; Quiatt and Reynolds 1993; Soltis et al. 1995; Holloway 1996; Boehm 1996; Feldman and Laland 1996; Flinn 1997. Particular note should be made of anthropologist William Durham's (1991) dualistic gene-culture coevolution paradigm, which directs our attention to the partially independent nature of human cultural evolution, even as it recognizes its interdependency with biological constraints and influences, and the interactions between the two evolutionary modes. See also Cavalli-Sforza and Feldman 1981; Corning 1983; Boyd and Richerson 1985. Ghiselin (1993), on the other hand, sharply criticizes Durham for using information as the basis for his definition of culture, an obvious analogy with the informational properties of genes. Ghiselin rightly observes that this confuses a means, or an instrumentality of transmission and replication, with the functional product: "The recipe is not the cake" (p. 124). It should also be noted that Flinn (1997), an evolutionary psychologist, stresses the influence of evolved, biologically-based psychological mechanisms in cultural selection and evolution, but this is not incompatible with the notion that there are multiple, interacting levels of causation involved in cultural processes.

5. In support of this scenario, anthropologist Christopher Boehm (1996) has proposed that a suite of behavioral/cultural "factors" that are widely observed in contemporary egalitarian foraging bands might also have enhanced the influence of group selection among prehistoric human groups (whose lifestyles are presumed to have been similar). These factors are (1) internal social leveling pressures, (2) moralistic policing of cheats and shirkers, (3) consensual decision-making and shared within-group adaptive strategies, and (4) marked differences between groups in adaptive strategies. Together, these behaviors could have had the effect of dampening within-group variation and selection pressures while augmenting between-group variation and selection. Once again, this implies a behavioral pacemaker for natural selection. Also consistent is the novel proposal of Wilson and Dugatkin (1997) that "assortative interactions" among individual organisms at the cognitive/ behavioral level may also play a role in determining differential reproductive success. (It could serve as an alternative mechanism of group selection.) As Wilson and Dugatkin note, this scenario is particularly relevant to human evolution. On the other hand, Soltis et al. (1995), have questioned what they call "cultural group selection" (the differential survival of different cultural traits) as a significant factor in human evolution. Utilizing data from New Guinea, they conclude that it would take five hundred to one thousand years for a cultural trait to spread by a process of differential group extinction. However, as various commentators on their paper pointed out, much depends upon the assumptions used, how groups are defined, which data set is used, and, indeed, which trait is involved. A food taboo is one thing, but a weapon—say thrown spears or horse cavalry, or wheeled chariots, or Greek Fire, or the phalanx, or the cross-bow, or siege cannons, or tanks—can be used by one group to gain a military advantage that results in a rapid process of differential group selection. History is replete with examples of such military breakthroughs. Conversely, many cultural traits diffuse between groups without discernable (biological) selective consequences.

6. This is not to deny either the partial autonomy of cultural processes/systems or our unique (evolved) biological needs and psychological capabilities—human nature.

The mechanism of cultural evolution, which I also refer to as "teleonomic selection," involves an interaction between these two sets of causal factors (and much more besides). Teleonomic selection obviously plays a unique role in shaping the course of humankind. But the rudiments of this mechanism can be found elsewhere in nature as well. (For a full discussion of this issue, see Corning 1983, 2003 and chapter 4; also see Barkow et al. 1992; Flinn 1997.)

Chapter 4

1. Indeed, there is no agreed-upon definition of complexity, much less a theoretically-rigorous formalization, despite the fact that complexity is currently a "hot" research topic. Many books and innumerable scholarly papers have been published on the subject in the past few years, and there is even a new journal, *Complexity*, devoted to this nascent science. The problem with algorithmic complexity, as Chaitin concedes, is that random sequences are invariably *more* complex because in each case the recipe is as long as the whole thing being specified; it cannot be compressed. More recently, Charles Bennett has focused on the concept of logical depth—the computational requirements for converting a recipe into a finished product. Though useful, it seems to be limited to processes in which there is a logical structure of some sort. It would seem to exclude the "booming, buzzing confusion" of the real world, where the internal logic may be problematical or only partially knowable—say the immense number of context-specific chaotic interactions that are responsible for producing global weather patterns, or the imponderable forces that will determine the future course of the evolutionary process itself.

A number of researchers, especially those who are associated with the Santa Fe Institute, believe that the key lies in the so-called phase transitions between highly ordered and highly disordered physical systems. An often-cited analogy is water, whose complex physical properties lie between the highly ordered state of ice crystals and the highly disordered movements of steam molecules. While the "Santa Fe Paradigm" may be useful, it also sets strict limits on what can be termed "complex." For instance, it seems to exclude the extremes associated with highly ordered or strictly random phenomena, even though there can be more or less complex patterns of order and more or less complex forms of disorder—degrees of complexity that are not associated with phase transitions. (Indeed, random phenomena seem to be excluded by fiat from some definitions of complexity.)

To confuse matters further, a distinction must be made between what could be labeled "objective complexity"—the embedded properties of a physical phenomenon and "subjective complexity"—its meaning to a human observer. As Timothy Perper has observed (personal communication), the equation $w = f(z)$ is structurally simple, but it might have a universe of meaning depending upon how its terms are defined. Indeed, information theory is notorious for its reliance on quantitative, statistical measures and its blindness to meaning—where much can be made of very few words. The telephone directory for a large metropolitan area contains many more words than a Shakespeare play, but is it more complex? Furthermore, as Elisabet Sahtouris has pointed out (personal communication), the degree of complexity that we might impute to a phenomenon can depend upon our frame of reference for viewing it. If we adopt a broad, ecological perspective we will see many more factors, and relationships, at work than if we adopt a physiological perspective. When Howard Bloom (personal communication) quotes the line "to see the World in a Grain of Sand . . ." from William Blake's famous poem, "Auguries of Innocence," it reminds us that even a simple object can denote a vast pattern of relationships, if we choose to see it that way. Accordingly, subjective complexity is a highly variable property of the phenomenal world.

Perhaps we need to go back to the drawing board. Complexity is, after all, a word— a verbal construct, a mental image. Like the words "electron" or "snow" or "blue" or "tree," complexity is a shorthand tool for thinking and communicating about various aspects of the phenomenal world. Some words may be very narrow in scope. (Presumably all electrons are alike in their basic properties, although their behavior can vary greatly.) However, many other words may hold a potful of meaning. We often use the word "snow" in conversation without taking the trouble to differentiate among the many different kinds of snow, as serious skiers (and Inuit Eskimos) routinely do. Similarly, the English word "blue" refers to a broad band of hues in the color spectrum, and we must drape the word with various qualifiers, from navy blue to royal blue to robin's egg blue (and many more), to denote the subtle differences among them. So it is also, I believe, with the word "complexity"; it is used in many different ways and encompasses a great variety of phenomena. (Indeed, it seems that many theorists, to suit their own purposes, prefer *not* to define complexity too precisely.) The utility of any word, whether broad or narrow in scope, is always a function of how much information it imparts to the user(s). Take the word "tree", for example. It tells you about certain fundamental properties that all trees have in common. But it does not tell you whether or not a given tree is deciduous, whether it is tall or short, or even whether it is living or dead. The same shortcoming applies also to the concept of "complexity." Although there may be some commonalities between a complex personality, a complex wine, a complex piece of music and a complex machine, the similarities are not obvious. Each is complex in a different way, and their complexities cannot be reduced to an all-purpose algorithm. Moreover, the differences among them are at least as important as any common properties.

What in fact does the word "complexity" connote? One of the leaders in the complexity field, Seth Lloyd of MIT, took the trouble to compile a list of some three dozen different ways in which the term is used in scientific discourse. However, this exercise produced no blinding insight. When asked to define complexity, Lloyd told a reporter: "I can't define it for you, but I know it when I see it" (Johnson 1997). Rather than trying to define what complexity is, perhaps it would be more useful to identify the properties that are commonly associated with the term. We would suggest that complexity often (not always) implies the following attributes: (1) a complex phenomenon consists of many parts (or items, or units, or individuals); (2) there are many relationships/interactions among the parts; and (3) the parts produce combined effects (synergies) that are not easily predicted and may often be novel, unexpected, even surprising. At the risk of inviting the wrath of the researchers in this field, we would argue that complexity per se is one of the less interesting properties of complex phenomena. We believe the differences among them are more important than the commonalities. (For further discussion of this issue see chapter 13, note 12.)

Chapter 6

1. A few points made in chapter 4 deserve to be repeated. Cybernetic mechanisms are not limited only to one level of organization. We have come to appreciate the fact

that they exist at many levels of living systems. They can be observed in, among other things, morphogenesis, cellular activity and neuronal network operation, as well as in the orchestration of animal behavior. Also, the cybernetic model encompasses processes that conform to physicist Herman Haken's (1988b) paradigm of "distributed control." It should also be noted that cybernetic control processes may produce results that resemble Boolean dynamical attractors, but they are achieved in a very different way. By the same token, the cybernetic model, properly applied, calls into question the hypothesis (e.g., Lovelock 1990) that the biosphere is controlled by "automatic" non-teleological feedback relationships. Without some internal "reference signal" (teleonomy), there can be no feedback control, although there can certainly be self-ordered processes of reciprocal causation at work, or perhaps Darwinian processes of coevolution and stabilizing selection. Indeed, the existence of systemic purposiveness (teleonomy) is what distinguishes organisms (and superorganisms) from ecosystems (see D. S. Wilson and Sober 1989). The mere fact of functional interdependence is insufficient to justify the use of an organismic/cybernetic analogy.

2. In his introduction to a special issue of the journal *Human Ecology* devoted to climate and human affairs, Gunn (1994) noted the fact that many suggestive linkages have been shown to exist between major climate changes and significant economic and cultural changes over the past 19,000 years, a period which encompasses the emergence of agriculture and the rise of urban civilization. His assertion is backed by many specific anthropological and archeological studies. In addition to the sudden collapse of the Akkadian empire in about 2200 BC (Weiss et al. 1993), climate changes have also been implicated in the fall of Mayan civilization (Hodell 1995; Sabloff 1995) and of Teotihuacán. Many less traumatic economic changes may also have been affected by climatic shifts. Guillet (1987) studied the historical cycles of terrace construction, maintenance and abandonment, along with the use of water conservation measures, among the aboriginal inhabitants of the Colca Valley of southern Peru and showed that strategic shifts over the course of time have been closely correlated with ecological changes that produced variations in the availability of fresh water. Likewise, Stanley and Warne (1993) have linked climate changes to the deposition of cultivatable silts in the Nile delta during the period from 6500-5500 BC, a development which preceded the emergence of agriculture in the Nile valley by only a few centuries. The earliest settlement was probably occupied around 4900–5000 BC (For more on this subject, see chapter 7.)

3. Some students of the *origins* of warfare tend to focus on the role of various underlying or predisposing causal factors. Peter Meyer (1987, 1990) identifies the motivating effect of "fear itself" (fear of other potentially hostile human groups) and the closely linked psychology of ethnocentrism and xenophobia (see also Reynolds et al.1987; van der Dennen and Falger 1990). John Tooby and Leda Cosmides (1988) argue the case for various "cognitive preadaptations" that they believe were necessary preconditions for the orchestration of social behaviors. R. B. Ferguson (1984) stresses the material bases: conflicts over land, protein, women, etc., (see also Durham 1976), while Paul Shaw and Yuwa Wong (1989) have proposed a multifactorial theory that encompasses competition over scarce resources, psychological predispositions, and weapons development. But see also the masterful study *A History of Warfare* by John Keegan (1993). See also Corning and van der Dennen (2005).

Chapter 7

1. One reviewer charged that this theory of political systems is circular. Basic survival needs "by definition" determine the survival of a society's individual members, and if they don't survive neither does the macrolevel political regime, it was argued. Unfortunately, it is not quite that simple. First, it should be noted that many other, non–survival-related explanations for political devolution have been advanced in the past, from the overthrow of capitalist economies or political elites by the oppressed masses (the Marxists) to moral declines (Gibbon) and the marginal value of complexity itself (Tainter). Nor is basic-needs satisfaction an either-or thing; there can be more or less (look at North Korea and Cuba) and it is always subject to empirical verification. But more important, the relationship between basic-needs satisfaction at the individual level and the survival of the political system per se (or even a specific regime) is by no means deterministic and axiomatic. People die in a complex society every day without threatening the viability of the system, and millions of citizens may die defending their country without major consequence for the political system. Finally, and most crucial of all, political/cybernetic systems as defined here have all manner of purposes. The subset of systems that is concerned with securing and advancing the macrolevel "collective survival enterprise" is unique in having a special purpose that is closely tied to the basic survival needs of the population as a whole. And, as the case-study regarding the U.S. in World War II (below) clearly shows, this relationship is direct and nontrivial.

2. Sources used for the following discussion include Snyder (1960); Blum (1976); R. H. Bailey (1978); M. J. Harris, Mitchell, and Schechter (1984); Sidey (1994); *Historical Statistics of the United States* (U. S. Department of Commerce 1975); and *Statistical Abstract of the United States* (U. S. Department of Commerce 1953, 1997).

3. One other alternative approach to the explanation of political evolution should be mentioned in this regard. Jong Heon Byeon (1999) has proposed that political change is a self-organizing process, with a prevailing tendency (along with all other fundamental processes) toward greater complexity. Over time, Byeon claims, entropy (defined as "disorder") decreases and "order" (i.e., a patterning or thermodynamic order) increases. As noted elsewhere (Corning and Kline 1998a, 1998b), this popular formulation (Byeon follows the lead of many other contemporary theorists) involves a serious and unwarranted conflation of energetic and physical order; a concept of complexity (order) that cannot be operationalized; the use of statistical information concepts from information theory that cannot be applied to cybernetic, feedback-controlled systems; and, most serious, a core premise that can readily be falsified. As noted earlier, modern evolutionary biologists find the postulate of an inherent, orthogenetic trend in evolution to be unsupportable and in fundamental conflict with Darwin's theory. Indeed, if there is an inherent tendency toward political complexity, how can the many examples of political devolution be accounted for? The cybernetic, functional theory of political complexity and the Synergism Hypothesis predict what will happen to a complex society that suffers a prolonged, severe drought. A thermodynamic theory cannot.

Chapter 8

1. In his 1911 essay, "The Ant Colony as an Organism," Wheeler utilized only the traditional organismic analogy. He shifted to using the Spencerian term in his 1928

book, *The Social Insects: Their Origin and Evolution.* "We have seen that the insect colony or society may be regarded as a super-organism and hence as a living whole bent on preserving its moving equilibrium and integrity" (quoted in Hölldobler and Wilson 1990, p. 358). (It is noteworthy that Spencer also hyphenated the term. Modern theorists on the whole do not.)

2. Allee explicitly acknowledged the Spencerian origin of Wheeler's analogy in his classic book, *Animal Aggregations: A Study in General Sociology* (1931, p. 353), but Allee was more interested in exploring the general phenomenon of cooperation in nature.

3. That same year, Thomas Seeley, a specialist in honeybees, published an article in the *American Scientist* (1989) on "The Honeybee Colony as a Superorganism." In Seeley's definition, a group of organisms may be classified as a superorganism when they "form a cooperative unit to propagate their genes." Many social groups are imperfect superorganisms due to intragroup reproductive conflicts, Seeley argued, but honeybees come much closer to this ideal.

4. A problem with more expansive definitions, like those of Lovelock and Goodwin, is that they are so broad as to be of little value; they don't draw a useful distinction between what is a superorganism and what is simply an "interaction" of some sort. A functional, cybernetic definition identifies a specific class of evolved "purposive" phenomena, and it retains the important boundary between organisms and their environments, or more generic systems. In the Lovelock/Goodwin definitions, a bird and its perch constitute a superorganism; so do goldfish and their ponds, a tree and its soil, a baboon and its water hole — in short, almost everything in nature. A more circumscribed, Spencerian definition provides a more useful analytical framework.

5. To some extent, this distinction may involve no more than a different perspective on the same phenomenon. Thus, it could be said that the individuals, or participants, in a division of labor are contributing their specialized skills and efforts to the production of a combined result. However, the terminological distinction may also be useful in drawing attention to some important functional differences. For instance, many collective efforts, in nature and human societies alike, do not entail a task specialization. The joint efforts of various honey bee workers that serve to warm or cool the interior of the hive, or the combined efforts of the players in a classic tug-of-war game, do not entail a breakdown of the task at hand into specialized roles. Nevertheless, individual efforts are combined to produce collective results. A categorical distinction between a division and a combination of labor may also be useful in drawing attention to how a particular collaborative activity arose, or evolved. The eukaryotic cell, it is now generally agreed, provides an example of both. Some of its major organelles are most likely the product of a symbiotic union—a joint venture among specialists—whereas other organelles are believed to have resulted from an internal differentiation process over the course of time (Margulis 1993). In human societies, various combinations of labor go under such headings as partnerships, strategic alliances, joint ventures, even cooperatives.

6. In many parts of the globe, in fact, human populations have manifestly not grown larger over the past few millennia but have remained relatively stable, or declined. For instance, very low population densities are found, and for good reason, in the open grasslands areas with low water resources that constitute about 30 percent of the earth's surface. There are also many cases where human populations have expanded at one point in their history and then later on declined. It has not always been upward and onward if

you look at both sides of the growth curve. A reasonable alternative hypothesis is that population growth has been both a cause and an effect of cultural evolution, along with many other influences; the causal dynamic over time has been reciprocal.

7. Depending upon the terminology that is adopted, all superorganisms—or none —could be called "crude." However, it is useful to make a distinction between species that can sustain superorganisms without the need for elaborate external, cultural supports and those that cannot. The manifestly precarious superorganisms in complex human societies, which are regularly created, modified, supplanted, disbanded and overthrown, often are sustained only through coercive force and/or other social "incentives". However, these crude superorganisms are not a recent invention, I argue. Their roots can be traced back perhaps five million years.

8. The stunning, unexpected demise of the Soviet empire is especially instructive. The Soviet Union arose out of the centuries-long dynamic of economic, military and political competition (and wars) that had forged the nation-states of Europe in the first place. The culmination of this dynamic was World War Two, and the Soviet empire was a product of its military success in that titanic struggle. Economic objectives (notably the so-called COMECON) were secondary. After many years of Cold War, however, a suite of factors "worked together" to undermine the Soviet regime. A gradual reduction in Russia's sense of vulnerability to external attack was an important factor; it diminished the perceived need for its precarious empire. A further blow was Russia's disastrous, bitterly disillusioning war in Afghanistan. The challenge from the emerging global economy and the growing need to reform a sclerotic and uncompetitive internal economy provided another powerful impetus. Also important were the rising demands for autonomy from Soviet "republics" that had been culturally and politically distinct nations before World War Two. In other words, the functional underpinnings of the empire had dissolved; the calculus of perceived benefits and costs had shifted for those who had the power to defend it.

Chapter 10

1. The noted biologist George Williams, in a rare lapse, asserted otherwise. In his legendary book, *Adaptation and Natural Selection* (1966, p. 159), Williams wrote: "The central biological problem is not survival as such, but design for survival." On the contrary, survival *is* the central problem for organisms, and designs for survival are contingent solutions. Perhaps he meant that designs for survival are the central problem for evolutionary biologists.

2. Harsanyi later adds two qualifiers that would appear to contradict this bald assertion. One, following the example of Adam Smith, imposes the precondition of a "moral community" (There can be no anti-social preferences.) The other addresses the problem of having imperfect information. Harsanyi's principle applies only if a person's "true preferences" are involved—i.e., if the actor has access to all the relevant information and has reasoned with care about all of the costs and benefits involved. As Scanlon (1991) points out, these qualifiers effectively nullify the claim that "autonomous" individual choice is the ultimate criterion of what is good for a person.

3. Given the economists' traditional disinterest in biology, and especially the concept of basic needs, the notion of "necessities," or "demand inelasticity," has generally been treated in a wooden and constricted way. Indeed, there are no economic substitutes for

such essentials as sleep, defecation, thermoregulation, a balanced diet, fresh water, fresh air, or physical safety, among other needs (see chapter 11). Markets have no solution for a drought; they can only allocate the starvation.

Chapter 11

1. There are many other discussions of the concept of adaptation in the biological literature, including Williams (1966), Dobzhansky (1970), Lewontin (1979, 1984), Burian (1992), West-Eberhard (1992) inter alia, not to mention the abbreviated definitions that can be found in the glossaries for most standard biology textbooks (e.g., Wilson 1975; Curtis and Barnes 1989; Ricklefs 1996). Some of these definitions emphasize traits that "enhance" the survival chances of an organism, or increase its fitness, which limits the term to relative advantages (positive selection) rather than the totality of an organism's survival needs and functional capabilities. Likewise, Burian (1992) insists that a distinction must be made between an adaptation as an artifact (as it were) of an organism's past evolutionary history and an adaptation in relation to the organism's current fitness, since these two foci (history and current function) may be at odds with each other. To further confuse matters, the term "adaptation" can be used either as a noun or a verb. For our purposes, we favor Huxley's broader, functional (nominative) definition of the term adaptation, which Mayr (1988) suggested should be called "adaptedness".

2. A contemporary example of a maladapted society is Hungary, which the World Health Organization has cited as (currently) the least healthy nation in Europe. Hungary has a very high incidence of such risk factors as alcoholism, heavy smoking and high fat diets. As a result, that country has alarmingly poor health statistics and deteriorating life-expectancies (McKinsey 1997).

3. To reiterate a crucial point made earlier: An important parallel in biological evolution involves "phenotypic plasticity"—which encompasses a range of phenomena, from "physiological adjustments" (say calluses or high altitude changes in blood chemistry) to behavioral flexibility, found even in bacteria and single-celled protists. Phenotypic plasticity is not, therefore, something recent or associated only with complex organisms or human cultures. It is a primitive and very basic "survival strategy" in the living world.

4. There has been a sharp debate in recent years among evolutionary anthropologists, evolutionary psychologists, and others regarding the proper relationship between Darwinian theory and the causal dynamics of human behavior and adaptation. Andrew Vayda (1995), in a sharp critique of what he terms "Darwinian Ecological Anthropology" (or DEA), attacks what he characterizes as their "naive functionalism" (a term borrowed from both Elster and Schwartz). Although DEA theorists claim to be "unleashing the power of Darwinism," Vayda charges that they "cavalierly" posit explanations for every seemingly beneficial trait in terms of natural selection and fitness maximization, without regard for its evolutionary origins or, more important, the proximate causes involved. Vayda contrasts what he calls the "cardboard Darwinism" of DEA theorists with the work of the evolutionary psychologists, whose primary focus is the specific mechanisms of causation. While evolutionary psychologists account for the evolution of various mechanisms/traits in terms of our "ancestral" conditions as a species (i.e., the adaptive context within which human evolution took place), they do not necessarily see these mechanisms as being currently adaptive. (See also Caro and

Borgerhoff Mulder 1987 and Symons 1989.) A more ecumenical position was adopted by Blurton Jones (1990), who argued for the complementary value of both adaptationist analyses and the identification of psychological mechanisms. Alexander (1990) went even further. The adaptive significance and underlying mechanisms associated with any trait must be studied together. Behavioral adaptations are usually manifested in the phenotype—the behavior itself, he points out. Moreover, many behavioral adaptations may not be linked to specific "alleles" but to general abilities (supported by many genes) to assess costs and benefits and engage in conditional behaviors using "mental scenario-building." We agree with Alexander. In our view, both ecological and psychological approaches (and more) are relevant to the explanation of human behavior. We espouse a multi-leveled "interactional paradigm" (see Corning 1983).

5. Of course, the normative basis of welfare economics is axiomatic. To quote one of the classic economics textbooks (Samuelson and Nordhaus 1992, p. 295): "Economics cannot have the final word on these controversial [ethical and political] problems. For underlying all these issues are normative assumptions and value judgments about what is good and right and just. What an economist does, therefore, is to try very hard to keep positive science cleanly separated from normative judgments." Even the traditional emphasis in economics on maximizing "efficiency" is, after all, normatively-driven (see especially the discussion of this issue in Hanley and Spash 1993).

6. Actually the United Nations efforts trace back to the so-called Bariloche Model, developed by workers at the Fundacion Bariloche in the early 1970s. The Bariloche Model in turn inspired the landmark "Declaration of Principles and Programme of Action for a Basic Needs Strategy of Development" that emerged from the World Employment Conference of the International Labour Office in 1976 (see Chichilnisky 1982). Also notable is the work of the World Health Organization (WHO 1980), and the United Nations Development Program (UNDP 1990). Other important theoretical works on basic needs include Lederer (1980), Braybrooke (1987) and Thomson (1987).

7. Doyal and Gough claim to have developed a "theory" of human needs, but it is not strictly speaking an empirical theory with associated testable hypotheses. Their framework is grounded in a set of propositions regarding the existence of two overarching human needs—physical health and autonomy. As they put it: "the target standard of satisfaction of each characteristic is the minimum necessary to secure the *optimum* individual health and autonomy, in turn defined as the highest standard achieved in any nation state [italics added]" (p. 169). Although their case is carefully presented and well-argued, in point of fact their core propositions must be accepted as a moral imperative, or an act of faith, not a decision based on the weight of the evidence. Once accepted, the rest of the enterprise flows logically from their premise, but the premise ultimately amounts to an "ought", not an "is". So, their framework begs the question: Is "optimal autonomy" a basic need which, if denied, will cause "serious harm?" Harm in what sense? In the final analysis, the Doyal and Gough framework represents a melding of a strong moral agenda with the pragmatic measurement tools associated with the social indicators movement. However appealing (indeed useful) it may be, their framework does not in the end establish a rigorous theoretical foundation for the concept of basic needs, in our view.

8. The precise relationship between our basic needs, our motivational states, prior learning (including cultural influences) and our manifest behavioral patterns is very

complex and still imperfectly understood. This subject was discussed at length in Corning (1983). A more recent treatment can be found in Flinn 1997.

9. Actually, this approach most nearly accords with the version of Darwinism that is often referred to as "normalizing" or "stabilizing" selection—that is, selection that maintains a gene, or an adaptation, or a population, in a more or less stable relationship with its environment (as distinct from "positive" or "negative" selection).

10. Unfortunately, there are many gaps in the Poverty Indicators tables. For instance, poverty line income statistics are available for only about 25 countries. Unskilled and non-cash wages in Third World countries are especially hard to estimate, and household surveys of needs-satisfaction are non existent in many countries. The lack of an economic safety net in the form of social insurance is also a conspicuous problem in many of these countries.

11. One critic of this paradigm asked: "How does this help explain and predict economic behavior? Human behavior is complex. By attempting to unite all these complexities and relating them to biological needs may be a useful exercise. But will it result in a theory of behavior? I'm skeptical." This and similar criticisms totally miss the point of the "survival indicators" project. It does not claim to be a theory of human behavior but only (obliquely) a theory about the consequences of our behavior for an objective, inescapable and multi-faceted human problem. It involves an analytical framework that attempts to specify, to a first approximation, the dimensionality of the survival/adaptation problem for humankind and the degree to which our biological/survival and reproductive needs are satisfied in a given situation. A full discussion of the many-sided complexities of human behavior can be found in a chapter of Corning (1983) entitled "The Interactional Paradigm."

Chapter 12

1. Clausius's original formalization was $S'' - S = dQ_r/T$, where $S'' - S$ refers to the entropy change when a system changes from a state x to another state x'', which is calculated by dividing each increment of reversible heat addition (dQ_r) by the absolute temperature (T) of transfer and adding the quotients over the relevant change of states. To differentiate, we propose that Clausius's entropy be designated S_C.

2. Boltzmann's entropy (which we will designate S_B) is expressed as $S_B = -k \ln P$, where k refers to Boltzmann's famous constant (1.38×10^{-16} erg/°C) and P is the reciprocal of the number of equiprobable "complexions" (W) that may exist in any bounded system. Gibbs's equation for entropy (we will designate it S_G) was $S_G = -kP_i \ln P_i$, where P_i refers to the energy-dependent probabilities of various alternative microstates.

3. Schrödinger's formalization is $N = k \log (1/D)$, where k refers to Boltzmann's constant and D summarizes a separately derived expression for atomic disorder.

4. A variation on this "paradigm" can be found in Morowitz 1968 (see part 2); also, see Fivaz (1991). An alternative approach has been developed by Eric D. Schneider and James J. Kay (1994, 1995) (also see Schneider and Sagan 2005). These theorists view living systems as, quintessentially, a means for dissipating solar energy; their function is "only" to provide a means for resisting the tendency of the solar energy gradient to perturb the equilibrium state of the "system" that encompasses the earth. Thus, energy flows "determine the direction" of evolution and the development of living systems over

time. Our objections to all such monolithic thermodynamic visions, and our alternative paradigm, will be elaborated upon below and in chapter 13.

5. More formally, $I_x = \log_2 (1/P_x)$, where the information content (I) of an event (x) in bits is the base-2 logarithm of the reciprocal of its probability. Shannon's expression for entropy is $H = -K\, P_i \log_2 P_i$ (which we will designate S_s). Despite many claims to the contrary, Shannon's K cannot be equated to Boltzmann's constant, because they involve different dimensions. However, the relationship among the variables (aside from scaling functions) is the same as that in Gibbs's statistical analogues for entropy.

6. Biophysicist Rupert Riedl (1978), for instance, proposed a rigorous formulation. Order, he declared, is "law times the number of instances when the law applies" (p. 16). The problem with this definition is that it is rooted in a nineteenth century model of science, and of the phenomenal world. "All events which can be studied by scientific method," he tells us, "can be regarded as either accidental or necessary. This world of accident and necessity seems to contain no third alternative" (p. 4). (It should be noted that Riedl is here explicitly paraphrasing molecular biologist Jacques Monod in his famous volume on *Chance and Necessity*, 1971.) Of course, Riedl wrote before the emergence of chaos theory, a relatively new branch of mathematical physics, which posits a third alternative, namely fully determinate systems whose dynamics are nonetheless unpredictable. (We will describe below a fourth, cybernetic alternative, which is unique to living systems.)

7. Jeffrey Wicken (1987), whose work was noted above, is another theorist who pursued a thermodynamic *weltanschauung*. In our view, he correctly identified functional information as a significant factor in the evolutionary process. (Recall his characterization of an organism as an "informed thermodynamic system.") He was well aware of the distinction between the structural order found in a crystal and the functional properties of organisms. And yet, information did not play a direct role in his theory. Instead, Wicken kept information in the background because he could not "operationalize" it: "All these considerations," he noted, "make quantification of 'information content' extremely problematic, and pursuing that theme would only serve to reduce focus on the primary issue" (p. 50). Wicken did suggest the use of informational "compressibility"—certain statistical properties (algorithmic or probabilistic) associated with various informational "units"—as a measure of ordered complexity, but this still did not solve the problem of defining information in functional terms. In the end, Wicken adopted the kind of thermodynamic determinism that characterizes most of the other theoretical efforts in this vein. Although he spoke of "informed dissipation" and "informed transformations," he claimed that it is the second law that ultimately propels the evolutionary process. How? His answer is entropy—not the informational entropy employed by Brooks and Wiley but thermodynamic entropy. It is the relentless tendency toward the degradation of energetic order in irreversible thermodynamic processes, as formalized in the second law, which generates the variations that lead to directional changes in evolution (actually there are many sources of variability), subject to the (subsidiary) screening effect of Darwinian differential selection processes. Wicken described this model of evolutionary complexification as a "teleomatic" process, meaning that the functional organization of living systems is ultimately an autocatalytic product of a "teleomatic drive" that arises from, or is "governed by," the "randomizing forces" of the entropy principle. (Compare this to the energy-driven theories of Prigogine and Schneider and Kay cited above.)

8. There are, in fact, several different formalizations for thermodynamic entropy in the literature. The classical equation developed by Clausius is shown in note 1 above. The statistical equations developed by Boltzmann and Gibbs are shown in note 2. Subsequently, other more detailed entropy functions were added to model different ensembles of interacting particles. Three examples are known as the Maxwell-Boltzmann, Fermi-Dirac, and Bose-Einstein equations, and we have designated them S_{MB}, S_{FD}, and S_{BE}, respectively.

All of these statistical expressions for entropy share one property in common. They reach maximum values for any assembly of interacting particles at an equilibrium condition. However, they also differ in important ways. It is analogous to saying that apples and oranges are both fruit and grow on trees, but their differences also matter. The relationship between the various forms of entropy can be represented by a classical Venn diagram, drawn in the form of two intersecting circles that portray an area of overlap (or isomorphy) between two "sets" of items, as well as areas where the sets are distinct from one another. Examples of systems where classical entropy analyses can be used and statistical analyses cannot are jet engines, pumps, turbines, and heat exchangers. Examples of systems where both classical and statistical entropies can be derived are homogenous "chunks" of matter at equilibrium (perfect gases, crystalline solids), where the collisions or lattices of atoms/molecules can be modeled. And examples of a system where only statistical approaches can be used are ensembles of interacting particles in non-equilibrium states (say a chemical mixture irradiated by photons). In sum: despite many claims to the contrary, neither macroscopic nor statistical forms of thermodynamic entropy encompass one another.

In light of the great variety of thermodynamic processes that occur in complex living systems, we believe that it is essential for biologists to use both macroscopic (Clausius) and various statistical analyses of entropy, in their proper places. Since each formalism has its own distinct properties, we believe that it is also incumbent upon an analyst to spell out which form is being used in a given analysis. By the same token, we believe that Shannon's informational entropy should be differentiated from various thermodynamic forms of entropy. This is why we have suggested the notational distinctions utilized above, namely: S_C, S_B, S_G, S_S, S_{MB}, S_{FD}, and S_{BE}.

Chapter 13

1. Not all of these theorists deny the relevance of natural selection, needless to say, but in various ways they downgrade its importance. For instance, Stuart Kauffman (1995) acknowledges that natural selection is not irrelevant to the trajectory of evolution, but he pushes it into the background as an agency that provides "fine tuning" and "modest improvements" to the order that arises spontaneously in nature (see also Salthe 1998, who claims that adaptation is "not essential to life"). John Collier (1986) asserts that natural selection does not determine the "intrinsic dynamics" of evolution; it is merely "a rate-determining extrinsic factor." Vilmos Csányi (1998) likewise acknowledges a subsidiary role for natural selection but gives primacy to an "autogenetic model" of evolution in which the main source of creativity involves "hidden properties" that emerge from an inherent "drive to be." Biologist Jeffrey Wicken (1987, 1988, 1989), who acknowledges that there has been an "over extension of the entropy concept" among the members of the thermodynamics school, nevertheless argues that thermodynamic "forces" underlie the principles of variation and selection in nature (1988, p. 141). Even

Depew and Weber (1988, 1995), in the course of presenting perhaps the most balanced view of the relationship between thermodynamics and selection (they speak of a dualistic process involving both autocatalysis and natural selection), circumscribe its role by excluding what they call "physical selection," "chemical selection," and even "thermodynamic selection." In their view, only gene-based organic selection processes count as natural selection. We disagree. Natural selection applies to differential survival and "replication" at any biological level, whenever varying functional properties are responsible for the outcome (more on this in note 10 below).

2. Brooks and Wiley are also representative of recent efforts to incorporate information theory into the thermodynamics paradigm. Stanley Salthe (1993) calls it "infodynamics." While this is certainly a salutary development, it suffers from the long-standing problem that physics cannot provide a functional definition of information, which is essential to understanding its role in living systems (but see note 7 below regarding the concept of "control information"; also Corning and Kline 1998b, and chapter 14). The root of this problem traces to the pioneering work of physicist Claude Shannon (1948; also Shannon and Weaver 1949) on what he initially called "communications theory" but is now (perhaps inappropriately) called "information theory." Shannon, who worked at the Bell Laboratories, was concerned with the problem of measuring uncertainty in the communication of messages between a sender and a receiver. At the suggestion of mathematician John von Neumann, Shannon adopted the term "entropy" to describe his measure. However, his form of entropy referred only to the degree of statistical uncertainty (disorder) in a given communications context before the fact, while "information" in his terms referred only to the capacity to reduce statistical uncertainty. If one uses the binary bit as a basic unit of measure, the degree of informational uncertainty (entropy) can therefore be defined empirically as a function of the number of bits required for its elimination. Attracted by the mathematical isomorphism between Shannon's entropy and the Boltzmann/Gibbs formalizations for statistical entropy in thermodynamics, many other theorists since the 1940s have tried to apply information theory directly to thermodynamics, an enterprise Shannon himself is said to have discouraged. In general, these efforts share a tendency to lose sight of the original (energy-related) purpose of statistical entropy measures. For instance, physicist David Layzer (1988, p. 29) defines information as the difference between the observed entropy state of any system and the maximum possible entropy. Wicken (1987) makes a convincing case against the notion that Shannon's information/entropy concepts can be treated as generalized measures of order/disorder in nature. As Wicken notes, Shannon's entropy bears no relationship to the state of the phenomenal world; it relates to the efficiency or effectiveness with which a "message" is communicated from a sender to a receiver and the degree of "uncertainty" reduction that occurs. Equally important, in the biological realm information is a functional phenomenon; it controls the work that is done via cybernetic control processes. Thus, it is an ontological (or at least semantic) error to use the same concept both as a measuring rod for uncertainty/predictability and as a causal agency in the production of order/organization in the real world. Brooks and Wiley (1988), like other theorists of this school, have sought to circumvent this problem by differentiating between "structural information," which is derived from what they claim are the "inherent" self-organizing capabilities of living systems and "instructional information," which they assert (after Collier 1986) is a "physical array." The latter form of information provides a description of the state of the system, they say, and its "flow" is

subject to informational entropy. A comment by biophysicist Harold Morowitz (1992, pp. 73, 77) may be relevant here. "It is possibly the success of thermodynamics that has led to excesses by biological theorists looking for global extremum principles of biology in terms of parameters and variables that have little meaning in the domains in which they operate; to think in terms of predictive grand, unified theories based on thermodynamics is simply dreaming."

3. In his important book *Evolution, Thermodynamics and Information* (1987), Wicken initially adopts Shannon's concept of information, a formulation that refers to certain statistical and quantitative properties associated with the "messages" that are transmitted in formal communications systems. Then, in an acknowledged theoretical segue, Wicken proceeds to deploy the concept of information as a *causal agency* in biological evolution. In order to do so, however, Wicken must shift to using a *functional* definition of information as an evolved, purposive artifact, a definition that more nearly accords with our common sense understanding of the term. Wicken advances the notion that organisms are "informed thermodynamic systems," although he demurs from addressing the unresolved challenge of how to measure functional information empirically. He characterizes it as "a very perilous enterprise. We aren't even close to knowing how to quantify it" (pp. 27-28). Wicken is well aware of the distinction between physical order and biological organization (see below), and he was among the first members of this school to recognize that biological organization depends upon functional (cybernetic) information. But he also acknowledged that he could not operationalize it: "All these considerations," he noted, "make quantification of "information content" extremely problematic, and pursuing that theme would only serve to reduce focus on the primary issue" (p. 50). Wicken did suggest the use of informational "compressibility"—certain statistical properties (algorithmic or probabilistic) associated with various informational "units"—as a measure of ordered complexity, but this still did not solve the problem of defining information in functional terms. (Again, see the discussion of "control information" in chapter 14.) It should also be noted that a number of these theorists have recently established linkages with the field of semiotics, which has developed a much more compatible approach to biological information than the Shannon-Weaver paradigm (see especially Sebeok 1986; Nöth 1990; Brier 1992; Qvortrup 1993; Hoffmeyer 1997; Van de Vijver et al., 1998).

4. Schneider and Kay (1994, 1995) view living systems as being, quintessentially, a means for dissipating solar energy. The purpose of life, they assert, is "only" to provide a means for resisting the tendency of the solar energy gradient to perturb the equilibrium state of the "system" that encompasses the Earth. They view the evolutionary process as self-organizing because they posit an inherent tendency of any "system" to resist being "removed" from an equilibrium state. They describe evolution as "a march away from disorder." Thus, energy flows "determine the direction" of evolution and the development of living systems over time. Below we will detail why we believe that any such monolithic thermodynamic determinism is inadequate as an explanation of the evolutionary process; we view biological evolution as a vastly more complex, multifaceted "survival enterprise." The devil is in the details that Schneider and Kay allude to as "environmental conditions". (See also Schneider and Sagan 2005.)

5. "Available energy" is a precisely defined technical term in thermodynamics that we much prefer to the more commonly used Helmholz or Gibbs "free energy" functions. The distinctions between them, and reasons behind our preference, are detailed in Corning and

Kline (1998a, appendix B) and in chapter 12. To reiterate, the availability function allows one to calculate the work potential in any given environment, net of entropy, for both control mass and control volume situations. Though use of the control mass paradigm is more common in biology, we maintain that this category of systems is in fact inappropriate for the analysis of whole organisms, ecosystems and macro-evolutionary processes, because living systems at these levels are not systems of fixed mass; the flow of matter and energy through these systems more nearly resembles a jet engine than a bottle containing a fixed quantity of gas molecules. In any case, the availability function enjoys the advantage that it properly accounts for entropy without making entropy the analytical focus.

6. To put this issue into perspective, the available energy associated with the part of the total solar flux that actually impinges on the Earth has been estimated to be about 13×10^{23} calories of radiant energy per year (Curtis and Barnes 1989). Of this total flux, less than 1 percent is "captured" (a number of variables affect the quantity of incident sunlight) and is put to use to support life (Hubbert 1971, Harold 1986). The majority of the energy in the solar flux (about 80 percent) is reflected or entropically returned to space. The remaining 20 percent drives hydrological cycles, geological processes, the dynamics of the atmosphere, and so on, in addition to sustaining life (Davis 1990). But, in any case, the Earth itself is a far greater source of "wasted" entropic energy (more than 99 percent) than is all of the Earth's biological activity put together. Living systems contribute a trivial amount of entropy to the universe.

7. Actually, Szilard's influential paper was preceded by a similar line of argument in a thermodynamics textbook by Lewis and Randall in 1923 and by Szilard himself in his 1925 doctoral dissertation at the University of Berlin (see Leff and Rex 1990).

8. Kline (1997) has shown that Maxwell's demon is "wildly unfeasible" for any one of several reasons. (He defines "wildly" as meaning that it is currently beyond our technical capabilities by a factor of more than one million.) The demon would require capabilities for perception/detection, data collection, mechanical operation, and feedback control that appear to be totally impracticable, not to mention being totally uneconomic. Kline points out that it is bad science to base theories and thought experiments on events that have no reasonable likelihood of occurring.

9. Charles H. Bennett is well known as a theorist on the thermodynamics of information. His work on the reversibility of (Shannon) information was inspired by earlier work in this area by a colleague at the IBM Thomas J. Watson Research Center, Rolf Landauer. Bennett showed that information might (theoretically) be reversible, both logically and in thermodynamic (entropy) terms. However, Bennett also supported Landauer's conclusion that there is an inescapable thermodynamic cost for "erasing" information to start a new measurement, and he applied this to Maxwell's demon. Thus, Bennett concluded, it was not the cost of acquiring information (as Szilard supposed) but the cost of destroying it that makes the demon infeasible. The problem with this line of reasoning is that the calculations are all "internal"; they include only the thermodynamic costs of the information process itself. Landauer and Bennett both overlooked the real-world "economic" costs—the work associated with building and operating the demon, and in particular the work associated with "acquiring" and using (control) information. Indeed, Bennett (1988, pp. 70–71) approvingly quotes at length from Maxwell's original passage in the *Theory of Heat* (1871), including the author's claim that the demon could operate "without expenditure of work."

10. We prefer to define natural selection as the differential survival and "replication" among functional variants at all levels of living systems and at all stages of evolution. This point was emphasized by Morowitz (1992, pp. 49, 53) in his book on biogenesis. He pointed out that the conversion of photon energy to chemical energy in a biologically useful way was no simple matter; severe restrictions had to be overcome. Likewise, the biological information that is stored in DNA molecules is costly to maintain; DNA is constantly undergoing thermal degradation and requires energy inputs for its maintenance. This is not an entropic process, however, because the instabilities are energy-related; they are induced by the temperature of their surroundings.

11. An alternative scenario for eukaryote evolution was recently proposed by William Martin and Miklós Müller (1998). It is called the hydrogen hypothesis, and it is supported by a variety of genetic and biochemical data. Martin and Müller believe that the process of symbiogenesis was cooperative from the start. In their view, a mutually beneficial association developed between ancient hydrogen-producing bacteria and a methanogen—a microbe that can utilize hydrogen to extract energy and make sugars, leaving methane as a waste product. The idea came to Martin one day when he was viewing a modern analogue, a one-celled eukaryote called *Plagiopyla*.

12. In a recent commentary entitled "Complexity Is Just a Word!" (Corning 1998b), it was argued that there is no agreed-upon definition of complexity, and for very good reason. There are, in fact, many different kinds of complexity. It is a qualitative property that we apply to both apples and oranges—to borrow a cliché— that are both fruits and grow in trees but also differ from each other in important ways. Despite the many fruitless attempts (pardon the pun) to develop a general definition for the term, there are a number of commonly associated properties. Often (not always) these include the following attributes: (1) a complex phenomenon consists of many parts (or items, or units, or individuals); (2) there are many relationships/interactions among the parts; and (3) the parts produce combined effects (synergies) that are not always predictable and may often be novel, or unexpected. Kline (1995) has also provided a useful index for measuring the complexity of a cybernetic control system. His "complexity index" (denoted C) contains three quantities: V for the number of independent variables needed to describe the state of the system, P for the number of independent parameters needed to distinguish the system from like systems, and L for the number of feedback loops. A highly imaginative and practicable new approach to measuring complexity in specifically in biological systems has recently been proposed by Szathmáry et al. (2001). Their indices are focused on the number of interactions that occur in various networks. (See also chapter 4, note 1.)

Chapter 14

1. Actually, the use of feedback mechanisms in technological systems dates back to antiquity (see O. Mayr 1970). However, Wiener provided a broader framework for understanding feedback processes in relation to goal-directed behaviors of all kinds. It should be stressed again that there is a fundamental difference between evolved, internal purposiveness (teleonomy) and an externally imposed "teleology."

2. The other leading figure among the pioneers in cybernetics, H. Ross Ashby, was even less helpful. In his much-cited classic, *Design for a Brain* (1960), Ashby barely mentioned communications, and the term "information" was not even referenced in

his index. Even the all-important concept of feedback merited only two index references. There are occasional allusions to information, however. Thus, in one place Ashby describes trial-and-error learning as a valuable part of information gathering for an animal, which he notes is essential to adaptation (p. 83). However, there is no explicit treatment of information in Ashby's book, much less the problem of measuring it.

3. A simple thought experiment can be used to illustrate. Imagine two alternative experimental designs. In one case, there is a delicately structured, heated crystal inside an isolated system with Gibbsian constraints (no gravity or other extraneous influences). It is in a highly ordered state and also has a certain heat content and available energy. Now imagine a second isolated system containing an identical crystal with the same available energy but in the form of a pile of disordered shards. Is there any difference in the ability of the two crystals to do work?

4. Another problem with defining information as equivalent to physical order is that it entails the same kind of semantic pettifoggery that is associated with the concept of negative entropy. In fact, the term *negative entropy* is really a convoluted synonym for thermodynamic order. To repeat, it means, literally, an absence of an absence of order. If information is equivalent to order/negentropy, then it is inextricably tied to available energy, or physical order of all kinds (or both), depending upon how the term negentropy is defined. If so, information is highly inflammable; it is consumed every time irreversible work is performed and every time entropy increases, for whatever reason.

Chapter 16

1. The research literature in these two closely related fields is vast. Still relevant is R. W. White's seminal article "Motivation Reconsidered: The Concept of Competence" in the *Psychological Review* (1959); also Stanley Coopersmith's classic *The Antecedents of Self-Esteem* (1967). Among the many more recent publications in this area, the research of psychologists Edward Deci and Richard Ryan is especially notable; also the writings of Donald Vickery, Kenneth Pelletier and Kate Lorig.

2. Among the many references in this area, some standouts include Peter Senge's *The Fifth Discipline* (1990); Warren Bennis's *Why Leaders Can't Lead* (1990); Robert Greenleaf's much acclaimed and reprinted *Servant Leadership: A Journey into the Nature of Legitimate Power and Greatness* (1991); Russel Ackoff's *The Democratic Corporation* (1994); and, of course, the prolific outpouring of volumes by management icon Peter Drucker. A good overview of this subject can be found in the edited volume by Michael Ray and Alan Rinzler, *The New Paradigm in Business* (1993).

Chapter 17

1. Many neo-Darwinians, and their followers in evolutionary psychology, argue that the focus of all our efforts, and those of all other species, is maximizing our reproductive success. "The point of life is the proliferation of life," according to one of the leading students of human nature, Laura Betzig (1997, p. 1). However, survival is prior to reproduction—a prerequisite—and this fundamental objective often constrains, and may even thwart, reproduction. Thus, the large number of forty-to-fifty-year-old women in our society who have sacrificed reproduction for the sake of their careers are not a total aberration. As biologist Paul Ehrlich points out, "People deliberately choose to limit their

reproduction at levels that do not maximize an individual's genetic contribution to future generations—and we've done it for thousands, maybe tens of thousands of years." (2000, p. 71). Indeed, the differential reproductive success of different groups has been a key to human evolution.

2. Although there are many facets to human nature, perhaps the most significant for political theory is the inherent, often conflicting, duality between our intense egoism (self-interest) and an equally intense concern for the larger community (sociality), as Plato and Aristotle recognized. However, there are also many complications associated with this dichotomy. As Paul Ehrlich (2000) points out, we should really call it human natures (plural), because we are not all biologically alike any more than we share exactly the same culture and life experiences (more on this below). While we can properly speak of norms and averages with respect to human personalities (and there are universals like smiling, language, the experience of grief, etc.), there are also wide individual variations in many human traits, from personality characteristics to cognitive abilities. Some sex differences also appear to exist on average (but not necessarily for any given individual). Human nature also changes during the course of the life cycle, a point that is often overlooked by those who would reduce the concept to a simple stereotype. Finally, nurture is also an ineluctable part of our fully developed nature; we are shaped by the interactions that occur during ontogeny. A full-length treatment of this interaction can be found in Lippa (1988).

3. Smith (like Darwin) often gets a bum rap for the misuses that are made of his ideas. However, Smith's moral foundation was the Stoic philosophy of world citizenship, the good of the community as a whole, and the Christian teaching of the golden rule. Smith even quotes the biblical injunction to "love our neighbour as we love ourselves" (Smith 1976, I.i.5.5). Moreover, according to Smith, virtue consists of exercising "self-command" over our baser impulses and having sympathy toward others (1976, II.3.34). Indeed, self-command is essential to a civilized society (1976, VI.iii.II). Moreover, Smith's justification for the invisible hand was that it would actually benefit society because the rich could not actually consume a much greater proportion of the necessities of life; their share would only be of better quality (1976, IV.I.10). In other words, Smith was not endorsing a zero-sum game in which the rich get richer at the expense of the poor.

4. Etzioni has published numerous works on this theme. In a nutshell, Etzioni advances an overarching ethical agenda. His fundamental premise is that our deepest aspiration is for the achievement of a "good society"—one in which people are treated as "ends, not means." Markets certainly play a role in this enterprise, but a good society requires a balance between the marketplace, the state, and communities. Indeed, Etzioni argues that it is our relationships with others—loved ones, families, and communities— that give life meaning and purpose. Etzioni does acknowledge the practical need to provide for "a rich basic minimum for all," as he puts it. But he views this only as a means for achieving larger social and ethical ends. He also shows little concern for achieving economic fairness and social justice, and he has been sharply criticized for being one-sided and overly optimistic about human nature, among other things. See especially Etzioni (1993, 1995, 2000). For an alternative approach, see Peter Singer (2000).

5. There is a vast body of literature on the concept of justice, going back to the ancient Greeks. An excellent set of readings can be found in Solomon and Murphy (2000). Broad overviews of the subject were written by D. Miller (1976) and Pettit

(1980). A more recent treatment, emphasizing theories of distributive justice, can be found by Kolm (1996). An especially perceptive treatment can be found by Raphael (2001). See also Masters and Gruter (1992).

6. A word is in order regarding the libertarian position. A desire for personal freedom and the pursuit of self-interest are perfectly consistent with a Darwinian, evolutionary perspective, and there is good evidence in the literature of experimental psychology that a need for personal autonomy is an important (if variable) facet of human nature. Competitiveness and striving for influence and power are also important facets of human nature. However, we are also deeply social beings, and, most important, we are compelled to satisfy our needs within a complex economic system. Freedom and social responsibility are the two sides of the social contract. However, some extreme libertarians take a one-sided view; their claims for individual freedom have no regard for social obligations. Indeed, in the lexicon of modern-day *laissez faire* capitalists, freedom is the highest social good. In the words of the eccentric conservative novelist Ayn Rand (1943), who remains the soul mate of many libertarians and free-market romantics, "civilization is the process of setting man free from men" (p. 685). Rand's protagonists are always defiant individualists. "Just as life is an end in itself, so every living human being is an end in himself, not the means to the ends or welfare of others—and, therefore, man must live for his own sake, neither sacrificing himself to others nor sacrificing others to himself" (Rand 1962, vol. 1, p. 35). The problem is that this position is ultimately exploitative. In game theory, it's called defection or cheating, and it is unsustainable. Why should the rest of us accede to this view? As the old saying goes, he who takes from society without giving back is a thief.

References

Ackoff, R. L. 1958. "Towards a Behavioral Theory of Communications." *Management Science* 4:218–34.

———. 1994. *The Democratic Corporation.* New York: Oxford University Press.

Adams, R. 1975. *Energy and Structure.* Austin: University of Texas Press.

———. 1988. *The Eighth Day: Social Evolution as the Self-Organization of Energy.* Austin: University of Texas Press.

Adler-Karlsson, G. 1977. "Bioeconomics: A Coming Subject." In *Economics in Institutional Perspective,* eds. R. Steppacher, B. Zogg-Walz, and H. Hatzfeldt, pp. 85–92. Lexington, MA: Lexington Books.

Ahmadjian, V., and S. Paracer. 1966. *Symbiosis: An Introduction to Biological Associations.* Hanover, NH: University Press of New England (for Clark University).

Akhavan, P., and R. Howse. 1995. *Yugoslavia: The Former and Future: Perspectives of Scholars from the Region.* Washington, DC: The Brookings Institution and Geneva: The United Nations Research Institute for Social Development.

Alchian, A. A. 1950. "Uncertainty, Evolution and Economic Theory." *Journal of Political Economy* 58:211–21.

Aldrich, H. 1999. *Organizations Evolving.* London: Sage Publications.

Alejandro, R. 1998. *The Limits of Rawlsian Justice.* Baltimore: Johns Hopkins University Press.

Alexander, R. D. 1979. *Darwinism and Human Affairs.* Seattle: University of Washington Press.

———. 1987. *The Biology of Moral Systems.* New York: Aldine de Gruyter.

———. 1990. "Epigenetic Rules and Darwinian Algorithms: The Adaptive Study of Learning and Development." *Ethology and Sociobiology* 11:241–303.

Allardt, E. 1973. "Individual Needs, Social Structures, and Indicators of National Development." In *Building States and Nations: Models and Data Resources,* eds. S. N. Eisenstadt and S. Rokkan, pp. 259–73. Beverly Hills, CA: Sage Publications.

Allee, W. C. 1931. *Animal Aggregations: A Study in General Sociology.* Chicago: University of Chicago Press.

———. 1938. *The Social Life of Animals.* New York: W. W. Norton.

———. 1951. *Cooperation among Animals: With Human Implications.* New York: Henry Schuman. (Orig. pub. 1938.)

Allen, J. R. M., U. Brandt, A. Brauer, H.-W. Hubbertena, B. Huntley, J. Kellerk, M. Kraml et al. 1999. "Rapid Environmental Changes in Southern Europe during the Last Glacial Period." *Nature* 400:740–43.

Allen, T. F. H., J. A. Tainter, and T. W. Hoekstra. 1999. "Supply-Side Sustainability." *Systems Research and Behavioral Science* 16 (5):403–27.

Alley, R. B. 2000. *The Two-Mile Time Machine: Ice Cores, Abrupt Climate Change, and Our Future.* Princeton, NJ: Princeton University Press.

Amato, I. 1992. "A New Blueprint for Water's Architecture." *Science* 256:1764.

Amsden, A. H. 2002. "Why Are Globalizers so Provincial?" [op. ed.] *New York Times,* January 31, p. A27.

Anderson, P. W. 1972. "'More Is Different': Broken Symmetry and the Nature of the Hierarchical Structure of Science." *Science* 177:393–96.

Ardrey, R. 1966. *The Territorial Imperative.* New York: Atheneum.

———. 1976. *The Hunting Hypothesis.* New York: Atheneum.

Aristotle. 1946. *The Politics,* trans. E. Barker. Oxford: Oxford University Press.

———. 1961. *The Metaphysics,* trans. H. Tredennick. Cambridge, MA: Harvard University Press.

———. 1985. *Nichomachean Ethics,* trans. T. Irwin. Indianapolis, IN: Hackett.

Arnhart, L. 1998. *Darwinian Natural Right.* Albany: State University of New York Press.

Arrow, K., S. Bowles, and S. Durlauf, eds. 1999. *Meritocracy and Economic Inequality.* Princeton, NJ: Princeton University Press.

Arthur, W. 2002. "The Emerging Conceptual Framework of Evolutionary Developmental Biology." *Nature* 415:757–64.

Arthur, W. B. 1988. "Self-Reinforcing Mechanisms in Economics." In *The Economy as an Evolving Complex System,* eds. P. W. Anderson, K. J. Arrow, and D. Pines, pp. 9–10. Reading, MA: Addison-Wesley.

———. 1990. "Positive Feedbacks in the Economy." *Scientific American* 266 (2):92–99.

Ashby, H. R. 1958. "General Systems Theory as a New Discipline." *General Systems (Yearbook of the Society for the Advancement of General Systems Theory)* 3:1–6.

Ashby, W. R. 1956. *An Introduction to Cybernetics.* New York: John Wiley.

———. 1960. *Design for a Brain.* 2nd ed. New York: John Wiley. (Orig. pub. 1952.)

Ashcroft, F. 2002. *Life at the Extremes: The Science of Survival.* Berkeley: University of California Press.

Atsatt, P. R. 1988. "Are Vascular Plants Inside-Out Lichens?" *Ecology* 69:17–23.

———. 1991. "Fungi and the Origin of Land Plants." In *Symbiosis as a Source of Evolutionary Innovation,* eds. L. Margulis and R. Fester, pp. 301–5. Cambridge, MA: MIT Press.

Avital, E., and E. Jablonka. 1994. "Social Learning and the Evolution of Behaviour." *Animal Behaviour* 48:1195–99.

———. 2000. *Animal Traditions: Behavioural Inheritance in Evolution.* Cambridge: Cambridge University Press.

Axelrod, R. 1984. *The Evolution of Cooperation.* New York: Basic Books.

————. 1986. "An Evolutionary Approach to Norms." *American Political Science Review* 80:1095–1111.

————. 2001. *The Complexity of Cooperation: Agent-Based Models of Competition and Collaboration.* Princeton, NJ: Princeton University Press.

Axelrod, R., and D. Dion. 1988. "The Further Evolution of Cooperation." *Science* 242:1385–89.

Axelrod, R., and W. Hamilton. 1981. "The Evolution of Cooperation." *Science* 211:1390.

Ayala, F. J. 1970. "Teleological Explanations in Evolutionary Biology." *Philosophy of Science* 37:1–15.

Ayres, R., and I. Nair. 1984. "Thermodynamics and Economics." *Physics Today* 37:62–71.

Bailey, K. D. 1990. *Social Entropy Theory.* Albany: State University of New York Press.

Bailey, R. H. 1978. *The Home Front: U.S.A.* Alexandria, VA: Time-Life Books.

Bak, P., and K. Chen. 1991. "Self-Organized Criticality." *Scientific American* 261 (1):46–53.

Banerjee, S., P. R. Sibbald, and J. Maze. 1990. "Quantifying the Dynamics of Order and Organization in Biological Systems." *Journal of Theoretical Biology* 143:91–111.

Barabási, A. L. 2002. *Linked: The New Science of Networks.* Cambridge, MA: Perseus Books.

Barash, D. 1977. *Sociobiology and Behavior.* New York: Elsevier North-Holland.

————. 1986. *The Hare and the Tortoise: Culture, Biology and Human Nature.* New York: Viking.

Barclay, P. 2004. "Trustworthiness and Competitive Altruism Can Also Solve the 'Tragedy of the Commons'" *Evolution and Human Behavior* 25:209–20.

Barkow, J. H. 1989. *Darwin, Sex and Status: Biological Approaches to Mind and Culture.* Toronto: University of Toronto Press.

Barkow, J. H., L. Cosmides, and J. Tooby. 1992. *The Adapted Mind: Evolutionary Psychology and the Generation of Culture.* Oxford: Oxford University Press.

Bateson, G. 1972. *Steps to an Ecology of the Mind.* New York: Ballantine.

————. 1979. *Mind and Nature: A Necessary Unity.* New York: E. P. Dutton.

Bateson, P. P. G. 1988. "The Active Role of Behavior in Evolution." In *Evolutionary Processes and Metaphors,* eds. M. W. Ho and S. W. Fox, pp. 191–207. New York: John Wiley.

Bateson, P. P. G., P. Klopfer, and N. Thompson, eds. 1993. *Perspectives in Ethology, Vol. 10. Behaviour and Evolution.* New York: Plenum Press.

Bauer, R. A., ed. 1966. *Social Indicators.* Cambridge, MA: MIT Press.

Baum, J. A. C., and J. V. Singh 1994. *The Evolutionary Dynamics of Organizations.* New York: Oxford University Press.

Baumol, W. J., and A. S. Blinder. 1991. *Economics: Principles and Policy.* 5th ed. San Diego, CA: Harcourt Brace Jovanovich.

Bechtel, W., ed. 1986. *Science and Philosophy: Integrating Scientific Disciplines.* Dordrecht: Martinus Nijhoff Publishers.

Beck, A. 1967. *Depression: Clinical, Experimental and Theoretical Aspects.* New York: Harper & Row.

Becker, G. S. 1974. "A Theory of Social Interactions." *Journal of Political Economy* 82:1063–93.

———. 1976. *The Economic Approach to Human Behavior.* Chicago: University of Chicago Press.

Becker, G. S., and N. Tomes. 1979. "An Equilibrium Theory of the Distribution of Income and Intergenerational Mobility." *Journal of Political Economy* 1153–89.

Beiser, M. 1985. "A Study of Depression among Traditional Africans, Urban North Americans and Southeast Asian Refugees." In *Culture and Depression: Studies in the Anthropology and Cross-Cultural Psychiatry of Affect and Disorder,* eds. A. Kleinman and B. Good, pp. 272–98. Berkeley: University of California Press.

Bell, G. 1985. "Origin and Early Evolution of Germ Cells as Illustrated by the Volvocales." In *Origin and Evolution of Sex,* eds. H. O. Halvorson and A. Monroy, pp. 221–56. New York: Alan R. Liss.

Bennett, C. 1995. *Yugoslavia's Bloody Collapse: Causes, Course and Consequences.* London: Hurst & Company.

Bennett, C. H. 1988. "Logical Depth and Physical Complexity." In *The Universal Turing Machine: A Half Century Survey,* ed. R. Herken. Oxford: Oxford University Press.

Bennett, J. W. 1969. *Northern Plainsmen: Adaptive Strategy and Agrarian Life.* Chicago: Aldine-Atherton.

———. 1976. *The Ecological Transition: Cultural Anthropology and Human Adaptation.* New York: Pergamon.

Bennis, W. 1990. *Why Leaders Can't Lead.* San Francisco: Jossey-Bass.

Bercovitch, F. B. 1991. "Social Stratification, Social Strategies, and Reproductive Success in Primates." *Ethology and Sociobiology* 12:315–33.

Bergner, M., R. A. Bobbitt, S. Kressel, W. E. Pollard, B. S. Gilson, and J. R. Morris. 1976. "The Sickness Impact Profile: Conceptual Formulation and Methodology for the Development of a Health Status Measure." *International Journal of Health Sciences* 6:393–415.

Bermudes, D., and L. Margulis. 1987. "Symbiont Acquisition as Neoseme: Origin of Species and Higher Taxa." *Symbiosis* 4:185–98.

Berndt, E. 1978. "Aggregate Energy, Efficiency, and Productivity Measurement." *Annual Review of Energy* 9:409–26.

Berry, R. S., P. Salamon, and G. Heal. 1978. "On a Relation between Economic and Thermodynamic Optima." *Resources and Energy* 1:125–37.

Bethlenfalvay, G. J., H. G. Bayne, and R. S. Pacovsky. 1983. "Parasitic and Mutualistic Associations between a Mycorrhizal Fungus and Soybean: The Effect of Phosphorus on Host Plant-Endophyte Interactions." *Journal of Plant Physiology (Stuttgart)* 57:543–48.

Betzig, L., ed. 1997. *Human Nature: A Critical Reader.* New York: Oxford University Press.

Bierce, A. 1967. *The Devil's Dictionary.* New York: Doubleday. (Orig. pub. 1906.)

Bigelow, R. 1969. *The Dawn Warriors: Man's Evolution Towards Peace.* Boston: Little Brown.

Binmore, K. 1994a. *Game Theory and the Social Contract. Volume I: Playing Fair.* Cambridge, MA: MIT Press.

———. 1994b. *Game Theory and the Social Contract. Volume II: Just Playing.* Cambridge, MA: MIT Press.

———. 2004. "Reciprocity and the Social Contract." *Politics, Philosophy & Economics* 3 (1): 5–35.

Blake, R. W., ed. 1991. *Efficiency and Economy in Animal Physiology.* New York: Cambridge University Press.

Blitz, D. 1992. *Emergent Evolution: Qualitative Novelty and the Levels of Reality.* Dordrecht, Netherlands: Kluwer Academic Publishers.

Bloom, H. 1997. "A History of the Global Brain: Creative Nets in the Pre-Cambrian Age." *ASCAP (Across Species Comparisons and Psychopathology Society)* 10 (3):7–11.

———. 2000. *Global Brain: The Evolution of Mass Mind from the Big Bang to the 21st Century.* New York: John Wiley and Sons.

Blum, J. M. 1976. *V Was for Victory: Politics and American Culture during World War II.* New York: Harcourt Brace Jovanovich.

Blurton Jones, N. G. 1990. "Three Sensible Paradigms for Research on Evolution and Human Behavior?" *Ethology and Sociobiology* 11:353–59.

Boehm, C. 1993. "Egalitarian Behavior and Reverse Dominance Hierarchy." *Current Anthropology* 34:227–54.

———. 1996. "Emergency Decisions, Cultural-Selection Mechanics, and Group Selection." *Current Anthropology* 37:763–93.

———. 1997. "Impact of the Human Egalitarian Syndrome on Darwinian Selection Mechanics." *The American Naturalist* 150:5100–5121.

———. 1999. *Hierarchy in the Forest: The Evolution of Egalitarian Behavior.* Cambridge, MA: Harvard University Press.

Boesch, C., and M. Tomasello. 1998. "Chimpanzee and Human Cultures." *Current Anthropology* 39 (5):591–614.

Bolt, R. 1960. *A Man for All Seasons.* New York: Vintage.

Boltzmann, L. 1909. *Wissenschafliche Abhandlungen,* ed. F. Hasenöhrl. 3 vols. Leipzig: J. A. Barth.

Bongaarts, J. 1994. "Can the Growing Human Population Feed Itself?" *Scientific American* 270 (3):36–42.

Bonner, J. T. 1980. *The Evolution of Culture in Animals.* Princeton, NJ: Princeton University Press.

———. 1988. *The Evolution of Complexity by Means of Natural Selection.* Princeton, NJ: Princeton University Press.

Boserup, E. 1965. *The Conditions of Agricultural Growth: The Economies of Agrarian Change under Population Pressure.* Chicago: Aldine.

Boucher, D. H., ed. 1985. *The Biology of Mutualism: Ecology and Evolution.* New York: Oxford University Press.

Boulding, K. E. 1956. "General System Theory—The Skeleton of Science." *General Systems (Yearbook of the Society for the Advancement of General Systems Theory),* eds.

L. von Bertalanffy and A. Rapoport, pp.11–17. Ann Arbor, MI: Society for General Systems Research.

———. 1977. "The Universe as a General System. Fourth Annual Ludwig von Bertalanffy Memorial Lecture." *Behavioral Science* 22:229–306.

———. 1978. *Ecodynamics: A New Theory of Societal Evolution.* Beverly Hills, CA: Sage Publications.

———. 1981. *Evolutionary Economics.* Beverly Hills, CA: Sage Publications.

Bowles, S., and H. Gintis. 2002. "Homo Reciprocans." *Nature* 415:125–28.

Bowles, S., and H. Gintis. 2004. "The Evolution of Strong Reciprocity: Cooperation in Heterogeneous Populations." *Theoretical Population Biology* 65:17–28.

Bowles, S., J.-K. Choi, and A. Hopfensitz. 2003. "The Co-evolution of Individual Behaviors and Social Institutions." *Journal of Theoretical Biology* 223:135–47.

Boyd, R., and P. J. Richerson. 1985. *Culture and the Evolutionary Process.* Chicago: University of Chicago Press.

———. 1989. "The Evolution of Indirect Reciprocity." *Social Networks* 11:213–36.

———. 1992. "Punishment Allows the Evolution of Cooperation (or Anything Else) in Sizable Groups." *Ethology and Sociobiology* 13:171–95.

———. 2002. "Group Beneficial Norms Can Spread Rapidly in a Structured Population." *Journal of Theoretical Biology* 215:287–96.

Boyd, R., H. Gintis, S. Bowles, and P. J. Richerson. 2003. "The Evolution of Altruistic Punishment." *Proceedings of the National Academy of Sciences (USA)* 100 (6):3531–35.

Brain, C. K. 1981. *The Hunters or the Hunted?* Chicago: University of Chicago Press.

———. 1985. "Interpreting Early Hominid Death Assemblies: The Use of Taphonomy Since 1925." In *Hominid Evolution: Past, Present and Future. Proceedings of the Taung Diamond Jubilee International Symposium,* ed. P. V. Tobias, pp. 41–46. New York: Alan R. Liss.

Brandon, R. N., and R. M. Burian, eds. 1984. *Genes, Organisms, Populations: Controversies over the Units of Selection.* Cambridge, MA: MIT Press.

Braybrooke, D. 1987. *Meeting Needs.* Princeton, NJ: Princeton University Press.

Breed, M. D. 1988. "Genetics and Labour in Bees." *Nature* 333:299.

Brembs, B. 1996. "Chaos, Cheating and Cooperation: Potential Solutions to the Prisoner's Dilemma." *Oikos* 76:14–24.

Brenden, J. A., and R. L. Hamer. 1998. *The Interstudy HMO Trend Report.* St. Paul, MN: Interstudy Publications. Also available at http://www.statehealthfacts.kff.org

Bridgman, P. 1941. *The Nature of Thermodynamics.* Cambridge, MA: Harvard University Press.

Brier, S. 1992. "Information and Consciousness: A Critique of the Mechanistic Concept of Information." *Cybernetics and Human Knowing* 1 (2/3): 71–94.

Brillouin, L. 1949. "Life, Thermodynamics and Cybernetics." *American Scientist* 37:554–68.

———. 1962. *Science and Information Theory.* New York: Academic Press.

———. 1968. "Life, Thermodynamics, and Cybernetics." In *Modern Systems Research for the Behavioral Scientist,* ed. W. Buckley, pp. 147–56. Chicago: Aldine Publishing Company.

Brock, D. 1993. "Quality of Life Measures in Health Care and Medical Ethics." In *The Quality of Life,* eds. M. Nussbaum and A. Sen, pp. 95–139. Oxford: Clarendon Press.

Broda, E. 1975. *The Evolution of Bioenergetic Processes.* New York: Pergamon Press.

Brooks, D. R., and E. O. Wiley. 1988. *Evolution as Entropy: Toward a Unified Theory of Biology.* 2nd ed. Chicago: University of Chicago Press.

Brooks, D. R., J. Collier, B. A. Maurer, J. D. H. Smith, and E. O. Wiley. 1989. "Entropy and Information in Evolving Biological Systems." *Biology and Philosophy* 4:407–32.

Brostow, W. 1972. "Between Laws of Thermodynamics and Coding of Information." *Science* 178 (4057):123–126.

Brown, D. E. 1991. *Human Universals.* Philadelphia: Temple University Press.

Brown, J. L. 1983. "Cooperation—A Biologists' Dilemma." *Advances in the Study of Behavior* 13:1–37.

———. 1987. *Helping and Communal Breeding in Birds.* Princeton, NJ: Princeton University Press.

Bryant, J. 1982. "A Thermodynamic Approach to Economics." *Energy Economics* 36–50.

Buchanan, M. 2000. *Ubiquity: Why the World Is Simpler than We Think.* London: Weidenfeld & Nicolson.

———. 2002. *Nexus: Small Worlds and the Groundbreaking Science of Networks.* New York, London: W. W. Norton.

Buckley, W., ed. 1968. *Modern Systems Research for the Behavioral Scientist.* Chicago: Aldine Publishing Co.

Bueno de Mesquita, B. 1981. *The War Trap.* New Haven, CT: Yale University Press.

Bunge, M. 1986. "Review of C. Truesdell *Rational Thermodynamics* 1984." *Philosophy of Science* 53:305–6.

Burian, R. M. 1992. "Adaptation: Historical Perspectives." In *Keywords in Evolutionary Biology,* eds. E. F. Keller and E. A. Lloyd. Cambridge, MA: Harvard University Press.

Burke, J. A., and R. L. Hamer, eds. 1999. *Works of Edmund Burke,* vol. 2. New York: Harper & Brothers.

Busfield, J. 1986. *Managing Madness: Changing Ideas and Practice.* London: Hutchinson.

Buss, L. W. 1987. *The Evolution of Individuality.* Princeton, NJ: Princeton University Press.

Byeon, J. H. 1999. "Non-Equilibrium Thermodynamic Approach to the Change in Political Systems." *Systems Research and Behavioral Science* 16:283–91.

Byrne, R. W., and A. Whiten, eds. 1988. *Machiavellian Intelligence: Social Expertise and the Evolution of Intellect in Monkeys, Apes and Humans.* Oxford: Clarendon Press.

Cairns, J., J. Overbaugh, and S. Miller. 1988. "The Origin of Mutants." *Nature* 335:142–45.

Calderone, N. W., and R. E. Page. 1992. "Effects of Interactions among Genotypically Diverse Nestmates on Task Specialization by Foraging Honey Bees (*Apis mellifera*)." *Behavioral Ecology and Sociobiology* 30:219–26.

Camazine, S., J.-L. Deneubourg, N. R. Franks, J. Sneyd, G. Theraulaz, and E. Bonabeau. 2001. *Self-Organization in Biological Systems.* Princeton, NJ: Princeton University Press.

Camerer, C. F., G. Loewenstein, and M. Rabin, eds. 2003. *Advances in Behavioral Economics.* Princeton, NJ: Princeton University Press.

Campagna, C., C. Bisioli, F. Quintana, F. Perez, and A Vila. 1992. "Group Breeding in Sea Lions: Pups Survive Better in Colonies." *Animal Behaviour* 43:541–48.

Campbell, A., P. E. Converse, and W. L. Rodgers. 1976. *The Quality of American Life: Perceptions, Evaluations, and Satisfactions.* New York: Russell Sage Foundation.

Campbell, D. T. 1974. "Downward Causation in Hierarchically Organized Biological Systems." In *Studies in the Philosophy of Biology,* eds. T. Dobzhansky and F. J. Ayala, pp. 85–90. Berkeley: University of California Press.

Campbell, J. H. 1994. "Organisms Create Evolution?!" In *Creative Evolution?!,* eds. J. H. Campbell and J. W. Schopf, pp. 85–100. Boston: Jones and Bartlett Publishers.

Capitanio, J. P. 1993. "More on the Relation of Inheritance to Dominance." *Animal Behaviour* 46:600–602.

Caplan, A. L., H. T. Engelhardt, and J. M. McCartney, eds. 1981. *Concepts of Health and Disease: Interdisciplinary Perspectives.* Reading, MA: Addison-Wesley.

Caraco, T., S. Martindale, and H. R. Pulliam. 1980. "Avian Flocking in the Presence of a Predator." *Nature* 285:400–401.

Caraco, T., and L. Wolf. 1975. "Ecological Determinants of Group Sizes of Foraging Lions." *The American Naturalist* 109:343–52.

Carnegie, A. 1992. "Wealth," reprinted in *The Andrew Carnegie Reader,* ed. J. F. Wall, pp. 129–54. Pittsburgh, PA: University of Pittsburgh Press. (Orig. pub. 1889.)

Carneiro, R. L. 1967. "On the Relationship between Size of Population and Complexity of Social Organization." *Southwestern Journal of Anthropology* 23:234–43.

———. 1970. "A Theory of the Origin of the State." *Science* 169:733–38.

———. 1972. "The Devolution of Evolution." *Social Biology* 19 (3):248–58.

———. 1973. "The Four Faces of Evolution." In *Handbook of Social and Cultural Anthropology,* ed. J. H. Honigmann, pp. 89–110. Chicago: Rand McNally.

———. 1978. "Political Expansion as an Expression of the Principle of Competitive Exclusion." In *Origins of the State,* eds. R. N. Cohen and E. R. Service, pp. 205–23. Philadelphia: Institute for the Study of Human Issues.

———. 1987. "The Evolution of Complexity in Human Societies and Its Mathematical Expression." *International Journal of Sociology* 28 (3–4):111–28.

Caro, T. M., and M. Borgerhoff Mulder. 1987. "The Problem of Adaptation in the Study of Human Behavior." *Ethology and Sociobiology* 8:61–72.

Carrier, D. R. 1984. "The Energetic Paradox of Human Running and Hominid Evolution." *Current Anthropology* 25:483–89.

Casti, J. L. 1979. *Connectivity, Complexity, and Catastrophe in Large-Scale Systems.* New York: John Wiley.

———. 1995. *Complexification: Explaining a Paradoxical World through the Science of Surprise.* New York: Harper Perennial.

———. 1997. *Would-Be Worlds: How Simulation Is Changing the Frontiers of Science.* New York: John Wiley.

Cavalli-Sforza, L. L., and M. W. Feldman. 1981. *Cultural Transmission and Evolution: A Quantitative Approach.* Princeton, NJ: Princeton University Press.

Cela-Conde, C. J., and F. J. Ayala. 2004. The Evolution of Morality. In *Handbook of Evolution,* Vol. I, eds. F. M. Wuketits and C. Antweiler, pp. 171–90. Weinheim, Germany: Wiley-VCH Verlag.

Chase, R. X. 1985. "A Theory of Socioeconomic Change: Entropic Processes, Technology, and Evolutionary Development." *Journal of Economic Issues* 19 (4):797–823.

Chauvet, G. A. 1993. "Hierarchical Functional Organization of Formal Biological Systems: A Dynamical Approach." *Philosophical Transactions, Royal Society of London B* 339:425–81.

Chen, M. M., J. W. Bush, and D. L. Patrick. 1975. "Social Indicators for Health Planning and Policy Analysis." *Policy Sciences* 6:71–89.

Cherry, C. 1978. *On Human Communication.* 3rd ed. Cambridge, MA: MIT Press.

Chichilnisky, G. 1982. *Basic Needs and the North/South Debate.* World Order Models Project Working Paper No. 21. New York: Institute for World Order.

Childe, V. G. 1951. *Man Makes Himself.* New York: New American Library. (Orig. pub. 1936.)

Chong, L., and L. B. Ray. 2002. "Whole-istic Biology." *Science* 295:1661.

Choucri, N., and R. North. 1975. *Nations in Conflict: National Growth and International Violence.* San Francisco: W. H. Freeman.

Clare, A. 1980. *Psychiatry in Dissent.* London: Tavistock.

Clark, C. W. 1990. *Mathematical Bioeconomics: The Optimal Management of Renewable Resources.* 2nd ed. New York: Wiley-Interscience.

Clarke, B. 1975. "The Causes of Biological Diversity." *Scientific American* 232 (2): 50–60.

Clarke, M. F. 1989. "The Pattern of Helping in the Bell Miner." *Ethology* 80:292–306.

Clausius, R. 1864–1867. *Abhandlugen uber die Mechanische Warmetheorie.* Braunschweig, Germany: F. Vieweg und Sohn.

Clutton-Brock, T. H. 2002. "Breeding Together: Kin Selection and Mutualism in Cooperative Vertebrates." *Science* 296:69–72.

Clutton-Brock, T. H., P. N. M. Brotherton, M. J. O'Riain, A. S. Griffin, D. Gaynor, R. Kansky, L. Sharpe, and G. M. McIlrath. 2001. "Contributions to Cooperative Rearing in Meerkats, *Suricata suricatta.*" *Animal Behaviour* 61 (4):705–10.

Clutton-Brock, T. H., and G. A. Parker. 1995. "Punishment in Animal Societies." *Nature* 373:209–16.

Coale, A. J., and S. C. Watkins, eds. 1986. *The Decline of Fertility in Europe.* Princeton, NJ: Princeton University Press.

Coelho, G. V., D. A. Hamburg, and J. E. Adams, eds. 1974. *Coping and Adaptation.* New York: Basic Books.

Coghlan, A. 1996. "Slime City." *New Scientist* 151 (2045):31–36.

Cohen, D. 1988. *Forgotten Millions: The Treatment of the Mentally Ill: A Global Perspective.* London: Paladin Grafton Books.

Cohen, M. N. 1977. *The Food Crisis in Prehistory: Overpopulation and the Origins of Agriculture.* New Haven, CT: Yale University Press.

Colby, B. N. 1987. "Well-Being: A Theoretical Paradigm." *American Anthropologist* 89:879–95.

Colby, B. N., C. M. Aldwin, L. Price, C. Stegemann, and S. Mishra. 1985. "Adaptive Potential, Stress and Illness in the Elderly." *Medical Anthropology* 9 (4):283–96.

Collier, J. 1986. "Entropy in Evolution." *Biology and Philosophy* 1:5–24.

Colson, S. D., and T. H. Dunning, Jr. 1994. "The Structure of Nature's Solvent: Water." *Science* 265:43–44.

Combs, G. F., Jr., R. M. Welch, J. M. Duxbury, N. T. Uphoff, and M. C. Nesheim. 1996. *Food-Based Approaches to Preventing Micronutrient Malnutrition: An International Research Agenda.* Ithaca, NY: Cornell Institute for Food, Agriculture and Development.

Comfort, L. K. 1994a. "Risk and Resilience: Inter-organizational Learning Following the Northridge Earthquake of 17 January 1994." *Journal of Contingencies and Crisis Management* 2 (3):157–70.

———. 1994b. "Self-Organization in Complex Systems." *Journal of Public Administration Research and Theory* 4 (3):393–410.

———. 1998. "Shared Risk: A Dynamic Model of Organizational Learning and Action." In *Handbook of Administrative Communication,* eds. J. L. Garnett and A. Kouzmin, pp. 395–411. New York: Marcel Dekker.

Conradt, L., and T. J. Roper. 2003. "Group Decision-Making in Animals." *Nature* 421:155–58.

Coopersmith, S. 1967. *The Antecedents of Self-Esteem.* San Francisco: W. H. Freeman.

Corning, P. A. 1969. "The Evolution of Medicare . . . from Idea to Law." Research Report No. 29, Office of Research and Statistics, Social Security Administration, Washington, DC.

———. 1970. "Evolutionary Indicators: Applying the Theory of Evolution to Political Science." Prepared for the Biennial Meeting of the International Political Science Association, Munich, Germany.

———. 1971a. "The Biological Bases of Behavior and Some Implications for Political Science." *World Politics* 23:321–70.

———. 1971b. "The Theory of Evolution as a Paradigm for the Analysis of Social and Political Phenomena" Doctoral Dissertation, New York University.

———. 1974. "Politics and the Evolutionary Process." In *Evolutionary Biology,* Vol. VIII, ed. T. Dobzhansky, pp. 253–94. New York: Plenum.

———. 1976. "Toward a Survival-Oriented Policy Science." In *Biology and Politics,* ed. A. Somit, pp.127–54. Paris: Mouton.

———. 1977. "Human Nature *Redivivus.*" In *Human Nature in Politics* (Nomos XVII), eds. J. R. Pennock and J. W. Chapman, pp.19–68. New York: New York University Press.

———. 1978. *A Basic Needs Approach to Measuring the Quality of Life* (*Final Report*). Sacramento: State of California, Department of Benefit Payments.

———. 1979. *A Basic Needs Approach to Measuring the Quality of Life: Data from a Survey of Public Assistance Recipients.* Sacramento: State of California, Department of Benefit Payments.

———. 1982. "Durkheim and Spencer." *The British Journal of Sociology* 33 (3):359–82.

———. 1983. *The Synergism Hypothesis: A Theory of Progressive Evolution.* New York: McGraw-Hill.

———. 1987. "Evolution and Political Control: A Synopsis of a General Theory of Politics." In *Evolutionary Theory in Social Science,* eds. M. Schmid and F. M. Wuketis, pp. 127–70. Dordrecht: D. Reidel.

———. 1992. "The Power of Information" Prepared for the 36th Annual Meeting, International Society for the Systems Sciences, Denver, CO.

———. 1995. "Synergy and Self-Organization in the Evolution of Complex Systems." *Systems Research* 12:89–121.

———. 1996a. "The Co-operative Gene: On the Role of Synergy in Evolution." *Evolutionary Theory* 11:183–207.

———. 1996b. "Synergy, Cybernetics and the Evolution of Politics." *International Political Science Review* 17 (1):91–119.

———. 1997a. "Holistic Darwinism: 'Synergistic Selection' and the Evolutionary Process." *Journal of Social and Evolutionary Systems* 20:363–400.

———. 1997b. "Biopolitical Economy: A Trail-Guide for an Inevitable Discipline." In *Research in Biopolitics,* Vol. 5, eds. A. Somit and S. Peterson, pp. 247–77. Greenwich, CT: JAI Press.

———. 1998. "Complexity Is Just a Word!" *Technological Forecasting and Social Change* 58:1–4.

———. 2000. "Biological Adaptation in Human Societies: A 'Basic Needs' Approach." *Journal of Bioeconomics* 2:41–86.

———. 2001a. "Synergy Goes to War: An Evolutionary Theory of Collective Violence." Prepared for the Annual Meeting, Association for Politics and the Life Sciences, Charleston, SC, October 18–21, 2001.

———. 2001b. "The Sociobiology of Democracy Revisited: A Reply and a Reiteration." *Politics and the Life Sciences* 20 (1):213–16

———. 2002a. "Fair Shares: Beyond Capitalism and Socialism, or the Biological Basis of Social Justice." *Politics and the Life Sciences* 22(2):12–32.

———. 2002b. "Synergy and the Evolution of Superorganisms: Past, Present, and Future." Prepared for the Annual Meeting, Association for Politics and the Life Sciences, Montreal Canada, August 11–14, 2002.

———. 2002c. "Thermoeconomics: Beyond the Second Law." *Journal of Bioeconomics* 4:57–88.

———. 2003. *Nature's Magic: Synergy in Evolution and the Fate of Humankind.* New York: Cambridge University Press.

———. 2004. "The Evolution of Politics" In *Handbook of Evolution*, Vol. I, eds. F. W. Wuketits and C. Antweiler, pp. 191–252. Weinheim, Germany: Wiley-VCH Verlag.

———. 2005. "Why We Need a Strategic Plan for 'Spaceship Earth.'" *Technological Forecasting and Social Change*, In Press.

Corning, P. A., and S. M. Hines, Jr. 1988. "Political Development and Political Evolution." *Politics and the Life Sciences* 6 (2):140–72.

Corning, P. A., and S. J. Kline. 1998a. "Thermodynamics, Information and Life Revisited, Part I: To Be or Entropy." *Systems Research and Behavioral Science* 15:273–95.

Corning, P. A., and S. J. Kline. 1998b. "Thermodynamics, Information and Life Revisited, Part II: Thermoeconomics and Control Information." *Systems Research and Behavioral Science* 15:453–82.

Corning, P. A., and J. M. G. van der Dennen, eds. 2005. *The Evolution of War.* Unpublished manuscript.

Costanza, R. 1980. "Embodied Energy and Economic Valuation." *Science* 210:1219–24.

———. 1989. "What is Ecological Economics?" *Ecological Economics* 1:1–7.

Cottrell, F. 1953. *Energy and Society.* New York: McGraw Hill.

———. 1972. *Technology, Man and Progress.* Columbus, OH: Merrill.

Crawford, C., and D. L. Krebs, eds.1997. *Handbook of Evolutionary Psychology: Ideas, Issues and Applications.* Mahwah, NJ: Lawrence Erlbaum Associates.

Crick, F. 1994. *The Astonishing Hypothesis: The Scientific Search for the Soul.* New York: Charles Scribner's Sons.

Cronin, H. 1993. *The Ant and the Peacock.* New York: Cambridge University Press. (Orig. pub. 1991.)

Crutchfield, J. P., J. Doyne Farmer, N. H. Packard, and R. S. Shaw. 1986. "Chaos." *Scientific American* 255 (6):46–57.

Csányi, V. 1989. *Evolutionary Systems and Society: A General Theory.* Durham, NC: Duke University Press.

———. 1998. "Evolution: Model or Metaphor?" In *Evolutionary Systems: Biological and Epistemological Perspectives in Selection and Self-Organization,* eds. G. Van de Vijver, S. N. Salthe, and M. Delpos. pp. 1–12. Dordrecht, Netherlands: Kluwer Academic Publishers.

Csete, M. E., and J. C. Doyle. 2002. "Reverse Engineering of Biological Complexity." *Science* 295:1664–69.

Culyer, A. J., R. J. Lavers, and A. Williams. 1972. "Health Indicators." In *Social Indicators and Social Policy,* eds. A. Shonfield and S. Shaw, pp. 94–118. London: Heinemann.

Curl, R. F., and R. E. Smalley. 1988. "Probing C_{60}." *Science* 242:1017–22.

Currie, C. R. 2001. "A Community of Ants, Fungi, and Bacteria: A Multilateral Approach to Studying Symbiosis." *Annual Review of Microbiology* 55:357–80.

Curtis, H., and N. S. Barnes. 1989. *Biology.* 5th ed. New York: Worth Publishers.

Daft, M. J., and A. A. El-Giahmi. 1978. "Effect of Arbuscular Mycorrhiza on Plant Growth, VIII: Effects of Defoliation and Light on Selected Hosts." *New Phytology* 80:365–72.

Dahl, R. A. 1970. *Modern Political Analysis.* Englewood Cliffs, NJ: Prentice Hall.

Darnell, J. E.,H. F. Lodish, and D. Baltimor. 1990. *Molecular Cell Biology.* 2nd ed. New York: Scientific American Books.

Darwin, C. R. 1874. *The Descent of Man, and Selection in Relation to Sex.* New York: A. L. Burt. (Orig. pub. 1871.)

———. 1965. *The Expression of the Emotions in Man and Animals.* London: John Murray. (Orig. pub. 1873.)

———. 1968. *On the Origin of Species by Means of Natural Selection, or the Preservation of Favoured Races in the Struggle for Life.* Baltimore: Penguin. (Orig. pub. 1851.)

Davis, G. R. 1990. "Energy for Planet Earth." *Scientific American* 263 (3):55–62.

Dawkins, R. 1976. *The Selfish Gene*. New York: Oxford University Press.

———. 1982. *The Extended Phenotype*. New York: W. H. Freeman.

———. 1983. "Universal Darwinism." In *Evolution from Molecules to Man*, ed. D. S. Bendall, pp. 403–25. Cambridge: Cambridge University Press.

———. 1987. *The Blind Watchmaker: Why the Evidence of Evolution Reveals a Universe without Design*. New York: W. W. Norton. (Orig. pub. 1986.)

———. 1989. *The Selfish Gene*. 2nd ed. Oxford: Oxford University Press.

Deacon, T. W. 1997. *The Symbolic Species: The Co-Evolution of Language and the Brain*. New York: W. W. Norton.

Deamer, D. W., ed. 1978. *Light Transcending Membranes: Structure, Function and Evolution*. New York: Academic Press.

Deamer, D. W., and J. Oro. 1980. "Role of Lipids in Prebiotic Structures." *Biosystems* 12:167–75.

Deamer, D. W., and R. M. Pashley. 1989. "Amphiphilic Components of the Murchison Carbonaceous Chondrite: Surface Properties and Membrane Formation." *Origins of Life* 74:319–27.

de Bary, H. A. 1879. *Die Erscheinung der Symbiose*. Strasburg, Germany: Verlag von K. J. Trubner.

Deci, E. L., and R. M. Ryan. 1985. *Intrinsic Motivation and Self-Determination in Human Behavior*. New York: Plenum.

de Groot, P. 1980. "Information Transfer in a Socially Roosting Weaver Bird *(Quelea quelea; Ploceinae)*: An Experimental Study." *Animal Behavior* 28:1249–54.

Depew, D. J., and B. H. Weber. 1988. "Consequences of Nonequilibrium Thermodynamics for the Darwinian Tradition." In *Entropy, Information, and Evolution: New Perspectives on Physical and Biological Evolution*, eds. B. H. Weber, D. J. Depew, and J. D. Smith, pp. 317–54. Cambridge, MA: MIT Press.

———. 1995. *Darwinism Evolving: Systems Dynamics and the Genealogy of Natural Selection*. Cambridge, MA: MIT Press.

de Quervain, D. J., U. Fischbacher, V. Treyer, M. Schellhammer, U. Schnyder, A. Buck, and E. Fehr. 2004. "The Neural Basis of Altruistic Punishment." *Science* 305:1254–58.

De Robertis, E. M., G. Oliver, and C. V. Wright. 1990. "Homeobox Genes and the Vertebrate Body Plan." *Scientific American* 263 (1):46–52

Detrain, C., J. L. Deneubourg, and J. M. Pasteels. 1999. *Information Processing in Social Insects*. Boston: Birkhäuser Verlag.

Deutsch, K. W. 1963. *The Nerves of Government: Models of Political Communication and Control*. New York: Free Press.

———. 1979. *Tides among Nations*. New York: Free Press.

Devezas, T., and G. Modelski. 2003. "Power Law Behavior and World System Evolution: A Millenial Learning Process." *Technological Forecasting and Social Change* 70:819–59.

de Waal, F. B. M. 1982. *Chimpanzee Politics: Power and Sex among Apes*. New York: Harper & Row.

———. 1989. *Peacemaking among Primates.* Cambridge, MA: Harvard University Press.

———. 1996. *Good Natured: The Origin of Right and Wrong in Humans and Other Animals.* Cambridge, MA: Harvard University Press.

———. 1997. *Bonobo: The Forgotten Ape.* Berkeley: University of California Press

———. 1999. "Cultural Primatology Comes of Age." *Nature* 399:635–36.

———, ed. 2001. *Tree of Origin: What Primate Behavior Can Tell Us about Human Social Evolution.* Cambridge, MA: Harvard University Press.

Diamond, J. M. 1997. *Guns, Germs, and Steel: The Fates of Human Societies.* New York: W. W. Norton.

———. 2005. *Collapse: How Societies Choose to Fail or Succeed.* New York: Viking Penguin.

Diener, E. E. N. Suh, R. E. Lucas, and H. L. Smith. 1999. "Subjective Well-Being: Three Decades of Progress." *Psychological Bulletin* 125 (2): 276–302.

Ditto, W. L., and L. M. Pecora. 1993. "Mastering Chaos." *Scientific American* 269 (2):78–84.

Dobzhansky, T. 1937. "Further Data on *Drosophila miranda* and Its Hybrid with *Drosophila pseudoobscura.*" *Journal of Genetics* 34:135–51.

———. 1962. *Mankind Evolving: The Evolution of the Human Species.* New Haven, CT: Yale University Press.

———. 1967. *The Biology of Ultimate Concern.* New York: New American Library.

———. 1970. *Genetics of the Evolutionary Process.* New York: Columbia University Press.

———. 1974. "Introductory Remarks." In *Studies in the Philosophy of Biology,* eds. F. J. Ayala and T. Dobzhansky, pp. 1–2. New York: Macmillan.

———. 1975. "Darwinian or 'Oriented' Evolution?" *Evolution* 29:376–78.

Dobzhansky, T., F. J. Ayala, G. L. Stebbins, and J. W. Valentine, eds. 1977. *Evolution.* San Francisco: W. H. Freeman.

Doolan, S. P., and D. W. Macdonald. 1996a. "Diet and Foraging Behaviour of Group-Living Meerkats, *Suricata suricatta,* in the Southern Kalahari." *The Zoological Society of London* 239:697–716.

———. 1996b. "Dispersal and Extra-Territorial Prospecting by Slender-Tailed Meerkats (*Suricata suricatta*) in the South-Western Kalahari." *The Zoological Society of London* 240:59–73.

———. 1997. "Breeding and Juvenile Survival among Slender-Tailed Meerkats (*Suricata suricatta*) in the South-Western Kalahari: Ecological and Social Influences." *The Zoological Society of London* 242:309–27.

Dosi, G. 2000. *Innovation, Organization and Economics Dynamics.* Cheltenham, UK: Edward Elgar.

Dosi, G., R. R. Nelson, and S. G. Winter, eds. 2000. *The Nature and Dynamics of Organizational Capabilities.* New York: Oxford University Press.

Dosi, G., D. J. Teece, and J. Chytry, eds. 1998. *Technology, Organization and Competitiveness.* Oxford: Oxford University Press.

Doyal, L., and I. Gough. 1991. *A Theory of Human Need.* London: MacMillan Education Ltd.

Dragan, J. C., and M. C. Demetrescu. 1986. *Entropy and Bioeconomics*. Pelham, NY: Nagard Publishers.

Drews, C. 1993. "The Concept and Definition of Dominance in Animal Behaviour." *Behaviour* 125 (3–4):283–313.

Drèze, J., and A. Sen. 1989. *Hunger and Public Action*. Oxford: Clarendon Press.

Drèze, J., A. Sen, and A. Hussain, eds. 1995. *The Political Economy of Hunger: Selected Essays* (*Wider Studies in Development Economics*). Oxford: Clarendon Press.

Drucker, B. 1978. "The Price of Progress in the Philippines." *Sierra* 63:22–26.

Dugatkin, L. A. 1997. *Cooperation among Animals: An Evolutionary Perspective*. New York: Oxford University Press.

———. 1999. *Cheating Monkeys and Citizen Bees*. New York: Free Press.

Dugatkin, L. A., and M. Mesterton-Gibbons. 1996. "Cooperation among Unrelated Individuals: Reciprocal Altruism, By-Product Mutualism and Group Selection in Fishes." *BioSystems* 37:19–30.

Dugatkin, L. A., M. Mesterton-Gibbons, and A. I. Houston. 1992. "Beyond the Prisoner's Dilemma: Towards Models to Discriminate among Mechanisms of Cooperation in Nature." *Trends in Ecology and Evolution* 7:202–5.

Dugatkin, L. A., and H. K. Reeve. 1994. "Behavioral Ecology and Levels of Selection: Dissolving the Group Selection Controversy." *Advances in the Study of Behavior* 23:101–33.

Dumond, D. E. 1965. "Population Growth and Cultural Change." *Southwestern Journal of Anthropology* 21:302–24.

Dunbar, R. I. M. 1988. *Primate Social Systems*. London: Croom Helm.

———. 2001. "Brain on Two Legs: Group Size and the Evolution of Intelligence." In *Tree of Origin: What Primate Behavior Can Tell Us about Human Social Evolution*, ed. F. B. M. de Waal, pp. 173–91. Cambridge, MA: Harvard University Press.

Dunbar, R. I. M., C. Knight, and C. Power, eds. 1999. *The Evolution of Culture*. Edinburgh, UK: Edinburgh University Press.

Du Plessis, M. A. 1993. "Do Group-Territorial Green Woodhoopoes Choose Roosting Partners on the Basis of Relatedness?" *Animal Behaviour* 46:612–15.

Durham, W. H. 1976. "Resource Competition and Human Aggression. Part I: A Review of Primitive War." *Quarterly Review of Biology* 51:385–415.

———. 1991. *Coevolution: Genes, Culture and Human Diversity*. Stanford, CA: Stanford University Press.

Durkheim, E. 1938. *The Rules of Sociological Method*. Chicago, University of Chicago. (Orig. pub. 1895.)

Dyson, F. J. 1971. "Energy in the Universe." In *Energy and Power* (A Scientific American Book), pp. 19–27. San Francisco: W. H. Freeman.

Easterlin, R. A., ed. 2002. *Happiness in Economics*. Cheltenham: Edward Elgar.

Easton, D. 1965. *A Systems Analysis to Political Life*. New York: John Wiley.

———. 1993. *An Approach to the Analysis of Political Systems*. New York: Irvington Publishers.

Eco, U. 1986. *Semiotics and the Philosophy of Meaning*. Bloomington: Indiana University Press.

Edgerton, R. B. 1992. *Sick Societies: Challenging the Myth of Primitive Harmony.* New York: Free Press.

Ehrlich, A. H. 1998. "The Human Predicament: Where Do We Stand Now?" [Presentation.] Center for the Evolution of Culture, Palo Alto, CA.

Ehrlich, P. R. 1989. "The Limits to Substitution: Meta-Resource Depletion and a New Economic-Ecological Paradigm." *Ecological Economics* 1:9–16.

———. 2000. *Human Natures: Genes Cultures and the Human Prospect.* Washington, DC: Island Press/Shearwater Books.

Ehrlich, P. R., and P. H. Raven. 1964. "Butterflies and Plants: A Study in Coevolution." *Evolution* 18:586–608.

Eibl-Eibesfeldt, I., and F. K. Salter, eds. 1998. *Indoctrinability, Ideology, and Warfare: Evolutionary Perspectives.* New York: Berghahn Books.

Eigen, M., and P. Schuster. 1977. "The Hypercycle: A Principle of Natural Self-Organization." *Die Naturwissenschaften* 64 (11):541–65.

———. 1979. *The Hypercycle: A Principle of Natural Self-Organization.* Berlin: Springer-Verlag.

Eigen, M., W. Gardiner, P. Schuster, and R. Winkler-Oswatitsch. 1981. "The Origin of Genetic Information." *Scientific American* 244 (4):88–118.

Ekman, P., ed. 1973. *Darwin and Facial Expression: A Century of Research in Review.* New York: Academic Press.

———, ed. 1982. *Emotion in the Human Face.* 2nd ed. New York: Cambridge University Press.

Eldredge, N. 1985. *Unfinished Synthesis: Biological Hierarchies and Modern Evolutionary Thought.* New York: Oxford University Press.

———. 1995. *Reinventing Darwin.* New York: John Wiley.

Eldredge, N., and S. N. Salthe. 1984. "Hierarchy and Evolution." In *Oxford Surveys in Evolutionary Biology,* Vol. 1, eds. R. Dawkins and M. Ridley, pp. 184–208. Oxford: Oxford University Press.

Elster, J., and A. Hylland, eds. 1986. *Foundations of Social Choice Theory.* Cambridge: Cambridge University Press.

Elster, J., and J. E. Roemer, eds. 1991. *Interpersonal Comparisons of Well-Being.* Cambridge: Cambridge University Press.

Emlen, S. T. 1996. "Living with Relatives: Lessons from Avian Family Systems." *The International Journal of Avian Science* 138:87–100.

Endler, J. A. 1992. "Natural Selection: Current Usages." In *Key Words in Evolutionary Biology,* eds. E. F. Keller and E. A. Lloyd, pp. 220–24. Cambridge, MA: Harvard University Press.

Erdal, D., and A. Whiten. 1994. "On Human Egalitarianism: An Evolutionary Product of Machiavellian Status Escalation?" *Current Anthropology* 35 (2):175–83.

Erikson, R. 1993. "Descriptions of Inequality: The Swedish Approach to Welfare Research." In *The Quality of Life,* eds. M. Nussbaum and A. Sen. Oxford: Clarendon Press.

Erikson, R., E. J. Hansen, S. Ringen, and H. Uusitalo, eds. 1987. *The Scandinavian Model: Welfare States and Welfare Research.* London: M. E. Sharpe.

Estes, R. D., and J. Goddard. 1967. "Prey Selection and Hunting Behavior of the African Wild Dog." *Journal of Wildlife Management* 31:52–70.

Etzioni, A. 1993. *The Spirit of Community.* New York: Crown Publishers.

———, ed. 1995. *New Communitarian Thinking.* Charlottesville: University of Virginia Press.

———. 2000. *The Third Way to a Good Society.* London: Demos.

Faber, M. 1985. "A Biophysical Approach to the Economy Entropy, Environment and Resources." In *Energy and Time in the Economic and Physical Sciences,* eds. W. van Gool and J. J. C. Bruggink, pp. 315–35. New York: Elsevier Science Publishers.

Faber, M., and J. L. R. Proops. 1990. *Evolution, Time, Production and the Environment.* Berlin: Springer-Verlag.

Falk, A., E. Fehr, and U. Fischbacher. 2001. Driving Forces of Informal Sanctions. Working Paper No. 59. Working Paper Series ISSN 1424–0459. Zurich, Switzerland: University of Zurich.

Famintsyn, A. S. 1907a. "Concerning the Role of Symbiosis in the Evolution of Organisms." *Mémoirs Acad. Sci., Ser. 8, Physical-Mathematical Division* 20 (3):1–14.

———. 1907b. "Concerning the Role of Symbiosis in the Evolution of Organisms." *Transactions of the St. Petersburg Society of Natural Science, v. 38, Issue 1, Minutes of Session* 4:141–43.

———. 1918. "What Is Going On with Lichens?" *Nature* (April–May): 266–82.

Faux, J. 2002. "A Deal Built on Sand: At the Doha Meetings, the World Trade Organization Got Back on the Fast Track. But a Train Wreck Lies Just Ahead." *American Prospect* 13 (1):A22–25.

Fehr, E., and U. Fischbacher. 2002. "Why Social Preferences Matter—The Impact of Non-Selfish Motives on Competition, Cooperation and Incentives." *The Economics Journal* 112:C1–C33.

———. 2003. "The Nature of Human Altruism." *Nature* 425:785–91.

Fehr, E., U. Fischbacher, and S. Gächter. 2002. "Strong Reciprocity, Human Cooperation, and the Enforcement of Social Norms." *Human Nature* 13:1–25.

Fehr, E., and S. Gächter. 2000a. "Cooperation and Punishment in Public Goods Experiment." *American Economic Review* 90:980–94.

———. 2000b. "Fairness and Retaliation: The Economics of Reciprocity." *Journal of Economic Perspectives* 14:159-81.

———. 2002. "Altruistic Punishment in Humans." *Nature* 415:137–40.

Fehr, E., and K. M. Schmidt. 1999. "A Theory of Fairness, Competition, and Co-operation." *Quarterly Journal of Economics* 114:817–68.

Feldman, M. W., and K. N. Laland. 1996. "Gene-Culture Coevolutionary Theory." *Trends in Ecology and Evolution* 11:453–57.

Fenchel, T., and B. J. Finlay. 1994. "The Evolution of Life without Oxygen." *American Scientist* 82:22–29.

Ferguson, R. B., ed. 1984. *Warfare, Culture and Environment.* New York: Academic Press.

Ferguson, Y. H., and R. W. Mansbach. 1999. "History's Revenge and Future Shock." In *Approaches to Global Governance Theory,* eds. M. Hewson and T. J. Sinclair, pp. 197–238. Albany: State University of New York Press.

Fewell, J. H. 2003. "Social Insect Networks." *Science* 301:1867–70.

Fewell, J. H., and M. L. Winston. 1992. "Colony State and Regulation of Pollen Foraging in the Honey Bee, *Apis mellifera* L." *Behavioral Ecology and Sociobiology* 30:387–93.

Fitzgerald, R., ed. 1977. *Human Needs and Politics.* Oxford: Pergamon.

Fivaz, R. 1991. "Thermodynamics of Complexity." *Systems Research* 8 (1):19–32.

Flinn, M. V. 1997. "Culture and the Evolution of Social Learning." *Evolution and Human Behavior* 18:23–67.

Foley, R. 1995. *Humans before Humanity: An Evolutionary Perspective.* Oxford: Blackwell Publishers.

Foster, G., and B. Anderson. 1978. *Medical Anthropology.* New York: John Wiley.

François, C. 2004. *International Encyclopedia of Systems and Cybernetics.* London: Thomson Learning.

Frank, S. A. 1995. "Mutual Policing and Repression of Competition in the Evolution of Cooperative Groups." *Nature* (London) 377:520–22.

———. 1996. "Policing and Group Cohesion When Resources Vary." *Animal Behaviour* 52:1163–69.

———. 1998. *Foundations of Social Evolution.* Princeton, NJ: Princeton University Press.

———. 2003. "Perspective: Repression of Competition and the Evolution of Cooperation." *Evolution* 57 (4):693–705.

Franks, N. R. 1989. "Army Ants: A Collective Intelligence." *American Scientist* 77(2):139–45.

Franks, N. R., N. Gomez, S. Goss, and J. L. Deneubourg. 1991. "The Blind Leading the Blind in Army Ant Raid Patterns: Testing a Model of Self-Organization (*Hymenoptera: Formicidae*)." *Journal of Insect Behavior* 4(5):583–607.

Freud, S. 1961. *Civilisation and Its Discontents. The Standard Edition of the Complete Works of Freud, vol. 21,* trans. J. Strachey. New York: W. W. Norton. (Orig. pub. 1930.)

Frey, B. S., and A. Stutzer. 2002. *Happiness and Economics: How the Economy and Institutions Affect Well-Being.* Princeton, NJ and Oxford: Princeton University Press.

———. 2004. "Happiness Research: State and Prospects." Working Paper No. 190, Institute for Empirical Research in Economics, University of Zurich.

Frölich, H. 1970. "Long Range Coherence and the Actions of Enzymes." *Nature* 228:1093.

———. 1975. "The Extraordinary Dialectric Properties of Biological Materials and the Action of the Enzyme." *Proceedings of the National Academy of Sciences (USA)* 72 (11):4211–15.

Fukuyama, F. 1992. *The End of History and the Last Man.* New York: Free Press.

Futuyma, D. J., and M. Slatkin, eds. 1983. *Coevolution.* Sunderland, MA: Sinauer.

Gadagkar, R. 2001. *The Social Biology of Ropalidia marginata: Toward Understanding the Evolution of Eusociality.* Cambridge, MA: Harvard University Press.

Gage, D. H.,M. Schiffer, S. J. Kline, and W. C. Reynolds. 1966. "The Non-Existence of a General Thermokinetic Variational Principle." In *Non-Equilibrium*

Thermodynamics: Variational Techniques and Stability, eds. R.J. Donnelly, R. Herman, and I. Prigogine, pp. 283–86. Chicago: University of Chicago Press.

Galtung, J. 1980. "The Basic Needs Approach." In *Human Needs: A Contribution to the Current Debate,* ed. K. Lederer, pp. 55–125. Cambridge, MA: Oelgeschlager, Gunn and Hain.

Gatlin, L. 1972. *Information Theory and Living Systems.* New York: Columbia University Press.

Gehring, W. J. 1985. "The Molecular Basis of Development." *Scientific American* 253 (4):153–62.

Geiger, G. 1988. "On the Evolutionary Origins and Function of Political Power." *Journal of Social and Biological Structures* 11:235–50.

Geist, V. 1978. *Life Strategies, Human Evolution, Environmental Design: Toward a Biological Theory of Health.* New York: Springer-Verlag.

Georgescu-Roegen, N. 1971. *The Entropy Law and Economic Process.* Cambridge, MA: Harvard University Press.

———. 1976a. *Energy and Economic Myths.* New York: Pergamon.

———. 1976b. "Bioeconomics: A New Look at the Nature of Economic Activity." In *The Political Economy of Food and Energy,* ed. L. Junker, pp. 105–34. Ann Arbor: The University of Michigan Press.

———. 1977a. "A Bioeconomic Viewpoint." *Review of Social Economy* 35:361–75.

———. 1977b. "The Steady State and Ecological Salvation: A Thermodynamic Analysis." *BioScience* 27:266–70.

———. 1977c. "Inequality, Limits and Growth from a Bioeconomic Viewpoint." *Review of Social Economy* 35:361–75.

———. 1979. "Energy Analysis and Economic Valuation." *Southern Economic Journal* 45:1023–58.

Gergen, K. J. 1969. *The Psychology of Behavior Exchange.* Reading, MA: Addison-Wesley.

Ghiselin, M. T. 1969. *The Triumph of the Darwinian Method.* Berkeley: University of California Press.

———. 1974. *The Economy of Nature and the Evolution of Sex.* Berkeley: University of California Press.

———. 1978. "The Economy of the Body." *American Economic Review* 68:233–37.

———. 1981. "Categories, Life and Thinking." *Behavioral and Brain Sciences* 4:269–313.

———. 1986. "Principles and Prospects for General Economy." In *Economic Imperialism: The Economic Approach Applied outside the Field of Economics,* eds. G. Radnitzky and P. Bernholz, pp. 21–31. New York: Paragon House.

———. 1987. "Bioeconomics and the Metaphysics of Selection." *Journal of Social and Biological Structures* 10:361–69.

———. 1992. "Biology, Economics, and Bioeconomics." In *Universal Economics: Assessing the Achievements of the Economic Approach,* ed. G. Radnitzky, pp. 71–118. New York: Paragon House.

———. 1993. "Book Review: *Coevolution: Genes, Culture, and Human Diversity.*" *Politics and the Life Sciences* 12:123–24.

————. 1997. *Metaphysics and the Origin of Species.* Albany: State University of New York Press.

Giampietro, M., S. G. F. Bukkens, and D. Pimentel. 1993. "Labor Productivity: A Biophysical Definition and Assessment." *Human Ecology* 21:229–59.

Gibbs, J. W. 1906. *The Scientific Papers of J. Willard Gibbs,* eds. H. A. Bumstead and R. G. Van Name. 2 vols. New York: Longmans, Green.

Gibson, R., and T. Ingold, eds. 1993. *Tools, Language, and Cognition in Human Evolution.* Cambridge: Cambridge University Press.

Gilbert, P. 1984. *Depression, from Psychology to Brain State.* London: Lawrence Erlbaum.

Gilliland, M. 1975. "Energy Analysis and Public Policy." *Science* 189:1051–56.

Gilpin, R. 1981. *War and Change in World Politics.* Cambridge: Cambridge University Press.

Gilroy, S., and A. Trewavas. 2001. "Signal Processing and Transduction in Plant Cells: The End and the Beginning." *Nature Reviews (Molecular Cell Biology)* 2:307–14.

Gintis, H. 2000a. *Game Theory Evolving: A Problem-Centered Introduction to Modeling Strategic Behavior.* Princeton, NJ: Princeton University Press

————. 2000b. "Strong Reciprocity and Human Sociality." *Journal of Theoretical Biology* 206:169–79.

————. 2003. "The Hitchhiker's Guide to Altruism: Gene-Culture Coevolution, and the Internationalization of Norms." *Journal of Theoretical Biology* 220:407–18.

Gintis, H. S. Bowles, R. Boyd, and E. Fehr. 2003. "Explaining Altruistic Behavior in Humans." *Evolution and Human Behavior* 24:153–72.

Gleick, P. H. 1993. *Water in Crisis.* New York: Oxford University Press.

Gluckman, M. 1940. "The Kingdom of the Zulu of South Africa." In *African Political Systems,* eds. M. Fortes and E. E. Evans-Pritchard, pp. 25–55. London: Oxford University Press.

————. 1969. "The Rise of a Zulu Empire." *Scientific American* 202:157–68.

Goldberg, L. P. 1994. *Interconnectedness in Nature and Cooperation in Science: The Case of Climate and Climate Modeling.* Doctoral Dissertation, University of Colorado.

Goldstein, J. 1999. "Emergence as a Construct: History and Issues." *Emergence* 11:49–72.

————. 2002. "The Singular Nature of Emergent Levels: Suggestions for a Theory of Emergence." *Nonlinear Dynamics, Psychology and Life Sciences* 6 (4):293–309.

Goldstein, J. S. 1985. "Basic Human Needs: The Plateau Curve." *World Development* 13:595–609.

Goodall, J. 1986. *The Chimpanzees of Gombe: Patterns of Behavior.* Cambridge, MA: Harvard University Press.

Goodfellow, P. 1995. "The Genome Directory: Complementary Endeavours." *Nature* 377:285.

Goodnight, C., and L. Stevens. 1997. "Experimental Studies of Group Selection: What They Tell Us about Group Selection in Nature." *Bulletin of the Ecological Society of America* 77 (3 Suppl. Part 2): 168.

Goodwin, B. C. 1994. *How the Leopard Changed Its Spots: The Evolution of Complexity.* London: Weidenfeld & Nicolson.

Gordon, D. M. 1987. "Group-Level Dynamics in Harvester Ants: Young Colonies and the Role of Patrolling." *Animal Behaviour* 35:833–43.

———. 1999. *Ants at Work: How an Insect Society is Organized.* New York: Free Press.

Gould, J. L., and C. G. Gould. 1995. *The Honey Bee.* New York: Scientific American Library.

Gould, S. J. 1996. *Full House: The Spread of Excellence from Plato to Darwin.* New York: Harmony Books.

———. 2002. *The Structure of Evolutionary Theory.* Cambridge, MA: Belknap (Harvard University Press).

Gould, S. J., and R. C. Lewontin. 1979. "The Spandrels of San Marco and the Panglosssian Paradigm: A Critique of the Adaptationist Programme." *Proceedings of the Royal Society* 205:581–98.

Gouldner, A. 1960. "The Norm of Reciprocity." *American Sociological Review* 25:161–78.

Gowdy, J. M. 1985. "Evolutionary Theory and Economic Theory: Some Methodological Issues." *Review of Social Economy* 43 (3):316–24.

———. 1987. "Bio-Economics: Social Economy versus the Chicago School." *International Journal of Social Economics* 14(1):32–42.

———. 1991. "Bioeconomics and Post Keynesian Economics: A Search for Common Ground." *Ecological Economics* 3:77–87.

———. 1994. *Coevolutionary Economics: The Economy, Society and the Environment.* Boston/Dordrecht/London: Kluwer Academic Publishers.

———. 2004a. "Evolution of Economics." In *Handbook of Evolution,* vol. I, eds. F. M. Wuketits and C. Antweiler, pp. 253–98. Weinheim, Germany: Wiley-VCH Verlag.

———. 2004b. "Economic Man and Selfish Genes: The Implications of Group Selection for Economic Valuation and Policy." *The Journal of Socioeconomics* 33 (3):343–58.

Gowdy, J. M., and S. Mesner 1998. "The Evolution of Georgescu-Roegen's Bioeconomics." *Review of Social Economy* 56 (2): 136–56.

Grady, M. F., and M. T. McGuire. 1999. "The Nature of Constitutions." *Journal of Bioeconomics* 1:227–40.

Grand, S. 2001. *Creation: Life and How to Make It.* Cambridge, MA: Harvard University Press.

Granger, H. 1985. "The *Scala Naturae* and the Continuity of Kinds." *Phonesis* 30 (2):181–200.

Grant, B. R., and P. R. Grant. 1979. "Darwin's Finches: Population Variation and Sympatric Speciation." *Proceedings of the National Academy of Sciences* (USA) 76:2359–63.

———. 1989. "Natural Selection in a Population of Darwin's Finches." *American Naturalist* 133 (3):377–93.

———. 1993. "Evolution of Darwin's Finches Caused by a Rare Climatic Event." *Proceedings of the Royal Society of London B,* 251:111–17.

———. 2002. "Adaptive Radiation of Darwin's Finches." *American Scientist* 90 (2):130–39.

Green, D. 2002. "The Rough Guide to the WTO." CAFOD Policy Paper, www.wtowtach.org

Greenberg, M. S., and S. P. Shapiro. 1971. "Indebtedness: An Adverse Aspect of Asking For and Receiving Help." *Sociometry* 34:290–301.

Greene, S. 1998. "The Most Tainted Place on Earth." *The New York Times Magazine,* February 8, pp. 48–52.

Greenleaf, R. 1991. *Servant Leadership: A Journey into the Nature of Legitimate Power and Greatness.* Mahwah, NJ: Paulist Press. (Orig. pub. 1983.)

Gregerson, N. H., ed. 2002. *From Complexity to Life: Explaining the Emergence of Life and Meaning.* New York: Oxford University Press.

Grene, M. 1987. "Hierarchies in Biology." *American Scientist* 75:504–610.

Guilford, T., and I. Cuthill. 1991. "The Evolution of Aposematism in Marine Gastropods." *Evolution* 45:449–51.

Guillet, D. 1987. "Terracing and Irrigation in the Peruvian Highlands." *Current Anthropology* 28:409–30.

Gunn, J. D. 1994. "Introduction: A Perspective from the Humanities-Science Boundary." *Human Ecology* 22:1–22.

Gurr, T. R. 1970. *Why Men Rebel.* Princeton, NJ: Princeton University Press.

———. 1988. "War, Revolution and the Growth of the Coercive State." *Comparative Political Studies* 21:45–65.

Gurr, T. R., and V. F. Bishop. 1976. "Violent Nations, and Others." *The Journal of Conflict Resolution* 20 (1):79–110.

Guzmán-Novoa, E., R. E. Page Jr., and N. E. Gary. 1994. "Behavioral and Life-History Components of Division of Labor in Honey Bees, (*Apis mellifera* L.)." *Behavioral Ecology and Sociobiology* 34:409–17.

Gygax, L. 2002. "Evolution of Group Size in the Dolphins and Porpoises: Interspecific Consistency of Intraspecific Patterns." *Behavioral Ecology* 13 (5):583–90.

Haig, S. M., J. R. Walters, and J. H. Plissner. 1994. "Genetic Evidence for Monogamy in the Cooperatively Breeding Red-Cockaded Woodpecker." *Behavioral Ecology and Sociobiology* 34:295–303.

Haisch, B. A. Rueda, and H. E. Puthoff. 1994. "Beyond E=mc^2." *The Sciences* 34 (6):26–31.

Haken, H. 1973. *Cooperative Phenomena.* New York: Springer-Verlag.

———. 1974. *Cooperative Effects.* New York: American Elsevier.

———. 1977. *Synergetics.* Berlin: Springer-Verlag.

———. 1983. *Advanced Synergetics.* Berlin: Springer-Verlag.

———. 1988a. *Dynamic Patterns in Complex Systems.* Singapore: World Scientific.

———. 1988b. *Information and Self-Organization.* Berlin: Springer-Verlag.

———. 1990. *Synergetics of Cognition.* Berlin: Springer-Verlag.

Hallpike, C. R. 1986. *The Principles of Social Evolution.* Oxford: Clarendon Press.

Hamilton, W. D. 1964a. "The Genetical Evolution of Social Behavior, I." *Journal of Theoretical Biology* 7:1–16.

———. 1964b. "The Genetical Evolution of Social Behavior, II." *Journal of Theoretical Biology* 7:17–52.

Hammerstein, P., ed. 2003. *Genetic and Cultural Evolution of Cooperation.* Cambridge, MA: MIT Press.

Hancock, T. 1993. "The Evolution, Impact and Significance of the Healthy Cities/Healthy Communities Movement." *Journal of Public Health Policy* 14:5–18.

Hanley, N., and C. L. Spash. 1993. *Cost-Benefit Analysis and the Environment.* Aldershot: Edward Elgar Publishing.

Hannon, B. 1973. "An Energy Standard of Value." *Annals of the American Academy of Political Science* 410:139–53,

Harcourt, A. H., and F. B. M. de Waal, eds. 1992. *Coalitions and Alliances in Humans and Other Animals.* Oxford: Oxford University Press.

Hardesty, D. L. 1977. *Ecological Anthropology.* New York: John Wiley.

Hardin, G. 1968. "The Tragedy of the Commons." *Science* 162:1243–48.

Hardin, G. 1972. "Genetic Consequences of Cultural Decisions in the Realm of Population." *Social Biology* 19 (4):350–61.

Harold, F. M. 1986. *The Vital Force: A Study of Bioenergetics.* New York: W. H. Freeman and Co.

Harris, M. 1968. *The Rise of Anthropological Theory: A History of Theories of Culture.* New York: Crowell.

Harris, M. J., F. Mitchell, and S. Schechter. 1984. *The Homefront: America during World War II.* New York: G.P. Putnam's Sons.

Harrison, D. M. 1995. *The Organisation of Europe: Developing a Continental Market Order.* London: Routledge.

Harsanyi, J. D. 1982. "Morality and the Theory of Rational Behavior." In *Utilitarianism and Beyond*, eds. A. K. Sen and B. Williams, pp. 39–62. Cambridge: Cambridge University Press.

Harvey, P. H. 1986. "Energetic Costs of Reproduction." *Nature* 321:648–49.

Hawking, S. W. 1988. *A Brief History of Time: From the Big Bang to Black Holes.* New York: Bantam Books.

Hawksworth, P. L. 1988. "Coevolution of Fungi with Algae and Cyanobacteria in Lichen Symbiosis." In *Coevolution of Fungi with Plants and Animals,* eds. K. A. Pirozynski and D. L. Hawksworth, pp. 125–48. New York: Academic Press.

Hayek, F. A. 1988. *Collected Works of F. A. Hayek.* London: Routledge.

Haynes, R. H. 1991. "Modes of Mutation and Repair in Evolutionary Rhythms." In *Symbiosis as a Source of Evolutionary Innovation,* eds. L. Margulis and R. Fester, pp. 40–56. Cambridge, MA: MIT Press.

Heidemann, S. R. 1993. "A New Twist on Integrins and the Cytoskeleton." *Science* 260:1080–81.

Helman, C. 1990. *Culture, Health and Illness.* London: Wright.

Henrich, J., and R. Boyd. 2001. "Why People Punish Defectors." *Journal of Theoretical Biology* 208:79–89.

Henrich, J., R. Boyd, S. Bowles, C. Camerer, H. Gintis, R. McElreath, and E. Fehr. 2001. "In Search of Homo Economicus: Behavioral Experiments Conducted in 15 Small-Scale Societies." *American Economic Review* 91 (2):73–78.

Hess, B., and A. Mikhailov. 1994. "Self-Organization in Living Cells." *Science* 264:223–24.

Hewson, M., and T. J. Sinclair. 1999. *Approaches to Global Governance Theory.* Albany: State University of New York Press.

Heylighen, F., J. Bollen, and A. Riegler, eds. 1999. *The Evolution of Complexity: The Violet Book of "Einstein Meets Magritte."* Dordrecht, Netherlands: Kluwer Academic Publishers.

Heywood, A. 1999. *Political Theory: An Introduction.* 2nd ed. London: Macmillan.

Hicks, N., and P. Streeten. 1979. "Indicators of Development: The Search for a Basic Needs Yardstick." *World Development* 7:567–80.

Hilbert, D., and W. Ackerman. 1950. *Principles of Mathematical Logic.* New York: Chelsea Publishing.

Hill, T. L. 1985. *Cooperativity Theory in Biochemistry: Steady-State and Equilibrium Systems.* New York: Springer-Verlag.

Hirshleifer, J. 1977. "Economics from a Biological Viewpoint." *Journal of Law and Economics* 20:1–52.

———. 1978a. "Competition, Cooperation, and Conflict in Economics and Biology." *American Economic Review* 68:238–43.

———. 1978b. "Natural Economy versus Political Economy." *Journal of Social and Biological Structures* 1:319–37.

———. 1982. "Evolutionary Models in Economics and Law: Cooperation versus Conflict Strategies." *Research in Law and Economics* 4:1–60.

———. 1985. "The Expanding Domain of Economics." *American Economic Review* 75:53–68.

———. 1986. "The Economic Approach to Conflict." In *Economic Imperialism: The Economic Approach Applied outside the Field of Economics,* eds. G. Radnitzky and P. Bernholz, pp. 335–64. New York: Paragon House.

———. 1999. "There Are Many Evolutionary Pathways to Cooperation." *Journal of Bioeconomics* 1:73–93.

Hirst, P., and P. Woolley. 1982. *Social Relations and Human Attributes.* London: Tavistock.

Hodell, D. A., J. H. Curtis, and M. Brenner. 1995. "Possible Role of Climate in the Collapse of Classic Maya Civilization." *Nature* 375:391–94.

Hodgson, G. M. 1993. *Economics and Evolution: Bringing Life Back into Economics.* Cambridge: Polity Press (Blackwell).

———. 1998a. "On the Evolution of Thorstein Veblen's Evolutionary Economics." *Cambridge Journal of Economics* 22 (4):415–31.

———. 1998b. "Veblen's Theory of the Leisure Class and the Genesis of Evolutionary Economics." In *The Founding of Evolutionary Economics,* ed. W. J. Samuels, pp. 170–200. London: Routledge.

———. 1999. *Evolution and Institutions: On Evolutionary Economics and the Evolution of Economics.* Cheltenham, UK: Edward Elgar.

———. 2001a. "Darwin, Veblen and the Problem of Causality and Economics." *History and Philosophy of the Life Sciences* 23:385–423.

———. 2001b. *How Economics Forgot History: The Problem of Historical Specificity in Social Science.* London and New York: Routledge.

———. 2001c. "Is Social Revolution Lamarckian or Darwinian?" In *Darwinism and Evolutionary Economics,* eds. J. Laurent and J. Nightingale, pp. 87–118. Cheltenham, UK: Edward Elgar.

———. 2003a. "The Hidden Persuaders: Institutions and Individuals in Economic Theory." *Cambridge Journal of Economics* 27 (2):159–75.

———. 2003b. "Darwinism and Institutional Economics." *Journal of Economic Issues* 37:85–97.

———. 2004a. "Reclaiming Habit for Institutional Economics." *Journal of Economic Psychology* 25 (5):651–60.

———. 2004b. *The Evolution of Institutional Economics: Agency, Structure and Darwinism in American Institutionalism.* London: Routledge.

Hodgson, G. M., and T. Knudsen. 2004a. "The Firm as an Interactor: Firms as Vehicles for Habits and Routines." *Journal of Evolutionary Economics* 14 (3):281–307.

———. 2004b. "The Complex Evolution of a Simple Traffic Convention: The Functions and Implications of Habit." *Economic Behavior and Organization* 54:19–47.

Hoffmeyer, J. 1997. "Biosemiotics: Towards a New Synthesis in Biology." *European Journal for Semiotic Studies* 9:355–76.

Holland, J. H. 1992. "Complex Adaptive Systems." *Daedalus* 121 (1):17–30.

———. 1995. *Hidden Order: How Adaptation Builds Complexity.* Reading, MA: Addison-Wesley (Helix Books).

———. 1998. *Emergence: From Chaos to Order.* Reading, MA: Addison-Wesley (Helix Books).

Hölldobler, B., and E. O. Wilson. 1990. *The Ants.* Cambridge, MA: Harvard University Press.

———. 1994. *Journey to the Ants.* Cambridge, MA: Belknap Press.

Holloway, R. L. 1996. "Evolution of the Human Brain." In *Handbook of Human Symbolic Evolution,* eds. A. Lock and C. R. Peters, pp. 74–108. Oxford: Oxford Science Publications.

Hoogland, J. L. 1981. "The Evolution of Coloniality in White-Tailed and Black-Tailed Prairie Dogs (Sciuridae: *Cynomys leucurus* and *C. ludovicianus*)." *Ecology* 62:252–72.

Hoogland, J. L., and P. W. Sherman. 1976. "Advantages and Disadvantages of Bank Swallow (*Riparia riparia*) Coloniality." *Ecological Monographs* 46:33–58.

Hopf, F. A. 1988. "Entropy and Evolution: Sorting through the Confusion." In *Entropy, Information and Evolution: New Perspective on Physical and Biological Evolution,* eds. W. H. Weber, D. J. Depew, and J. D. Smith, pp.163–274. Cambridge, MA: MIT Press.

Howells, W. 1993. *Getting Here: The Story of Human Evolution.* Washington, DC: Compass.

Hubbert, M. K. 1971. "The Energy Resources of the Earth." In *Energy and Power* (A Scientific American book), pp. 31–40. San Francisco: W. H. Freeman.

Huettner, D. 1976. "Net Energy Analysis: An Economic Assessment." *Science* 192:101–4.

Hull, D. L. 1980. "Individuality and Selection." *Annual Review of Ecology and Systematics* 11:311–32.

Hunter, A. F., and L. W. Aarssen. 1988. "Plants Helping Plants." *BioScience* 38:34–40.

Huntington, S. 1996. *The Clash of Civilizations and the Remaking of World Order.* New York: Simon & Schuster.

Hurst, L. D. 1990. "Parasite Diversity and the Evolution of Diploidy, Multicellularity and Anisogamy." *Journal of Theoretical Biology* 144:429–43.

Huxley, J. S. 1942. *Evolution: The Modern Synthesis.* New York: Harper & Row.

Huxley, J. S., and T. H. Huxley. 1947. *Evolution and Ethics: 1893–1943.* London: The Pilot Press.

Huxley, T. H. 1993. "Evolution and Ethics." Reprinted in *Evolutionary Ethics*, eds. M. H. Nitecki and D. V. Nitecki, pp.29–80. Albany: State University of New York Press.

International Labour Organisation (ILO). 1976. *Employment, Growth and Basic Needs: A One-World Problem.* International Labour Organisation: Geneva.

Isack, H. A., and H. U. Reyer. 1989. "Honeyguides and Honey Gatherers: Interspecific Communication in a Symbiotic Relationship." *Science* 243:1343–46.

Isbell, L. A. 1995. "Predation on Primates: Ecological Patterns and Evolutionary Consequences." *Evolutionary Anthropology* 3 (2):61–71.

James, S. R. 1989. "Hominid Use of Fire in the Lower and Middle Pleistocene." *Current Anthropology* 30:1–26.

Jantsch, E. 1980. *The Self-Organizing Universe: Scientific and Human Implications of the Emerging Paradigm of Evolution.* New York: Pergamon Press.

Jaynes, E. T. 1983. *Papers on Probability, Statistics and Statistical Physics,* ed. R. D. Rosenkrantz. Boston: D. Reidel.

Jeanne, R. L. 1986. "The Organization of Work in *Polybia occidentalis*: Costs and Benefits of Specialization in a Social Wasp." *Behavioral Ecology and Sociobiology* 19:333–41.

Jeon, K. W. 1972. "Development of Cellular Dependence in Infective Organisms: Micurgical Studies in Amoebas." *Science* 176:1122–23.

———. 1983. "Integration of Bacterial Endosymbionts in Amoebae." *International Review of Cytology* (Suppl. 14): 29–47.

———. 1992. "Macromolecules Involved in the Amoeba-Bacteria Symbiosis." *Journal of Protozoolgy* 39:199–204.

Johnson, A. W., and T. Earle. 1987. *The Evolution of Human Societies: From Foraging Group to Agrarian State.* Stanford, CA: Stanford University Press.

Johnson, G. 1997. "Researchers on Complexity Ponder What It's All About." *The New York Times,* May 5, pp. B9, B13.

Johnson, G. R. 1995. "The Evolutionary Origins Government and Politics." In *Human Nature and Politics,* eds. A. Somit and J. Losco, pp. 243–305. Greenwich, CT: JAI Press.

Johnson, S. 2001. *Emergence: The Connected Lives of Ants, Brains, Cities and Software.* New York: Charles Scribner's Sons.

Jones, S., , R. D. Martin, and D. R. Pilbeam, eds. 1992. *The Cambridge Encyclopedia of Human Evolution.* Cambridge: Cambridge University Press.

Kahneman, D., E. Diener, and N. Schwarz, eds. 1999. *Well-Being: The Foundations of Hedonic Psychology.* New York: Russell Sage Foundation.

Kahneman, D., J. L. Knetsch, and R. Thaler. 1986a. "Fairness as a Constraint on Profit Seeking: Entitlements in the Market." *The American Economic Review* 76:728–41.

———. 1986b. "Fairness and the Assumptions of Economics." *The Journal of Business* 59:S285–S300.

Kaplow, L., and S. Shavell. 2002. *Fairness and Welfare.* Cambridge, MA: Harvard University Press.

Karasov, W. H., and J. M. Diamond. 1985. "Digestive Adaptations for Fueling the Cost of Endothermy." *Science* 228:202–4.

Katz, L. D. 2000. *Evolutionary Origins of Morality.* Thorverton, UK: Imprint Academic.

Kauffman, S. A. 1993. *The Origins of Order: Self-Organization and Selection in Evolution.* New York: Oxford University Press.

———. 1995. *At Home in the Universe: The Search for the Laws of Self-Organization and Complexity.* New York: Oxford University Press.

———. 2000. *Investigations.* New York: Oxford University Press.

Keegan, J. 1993. *A History of Warfare.* New York: Alfred A. Knopf.

Keenan, J. H. 1941. *Thermodynamics.* New York: John Wiley.

———. 1951. *Availability and Irreversibility in Thermodynamics.* Proceedings, Institution of Mechanical Engineers (Great Britain).

Keith, A. 1949. *A New Theory of Human Evolution.* New York: Philosophical Library.

Keller, A. 1931. *Societal Evolution.* New Haven, CT: Yale University Press. (Orig. pub. 1915.)

Keller, E. F. 1983. *A Feeling for the Organism.* New York: W. H. Freeman.

Kellmer-Pringle, M. 1980. *The Needs of Children.* 2nd ed. London: Hutchinson.

Kendrick, B. 1991. "Fungi Symbioses and Evolutionary Innovation." In *Symbiosis as a Source of Evolutionary Innovation,* eds. L. Margulis and R. Fester, pp. 249–61. Cambridge, MA: MIT Press.

Kennedy, P. M. 1993. *Preparing for the Twenty-first Century.* New York: Random House.

Kettlewell, H. B. D. 1973. *The Evolution of Melanism: The Study of a Recurring Necessity.* Oxford: Clarendon Press.

Khakhina, L. N. 1979. *Concepts of Symbiogensis.* [In Russian.] Leningrad, USSR: Akademie NAUK.

———. 1992. "Evolutionary Significance of Symbiosis: Development of the Symbiogenesis Concept." *Symbiosis* 14:217–28.

King, J. A. 1955. "Social Behavior, Social Organization, and Population Dynamics in a Black-Tailed Prairiedog Town in the Black Hills of South Dakota." *Contributions from the Laboratory of Vertebrate Biology* (University of Michigan, Ann Arbor) 67:123.

Kingdon, J. 1993. *Self-Made Man: Human Evolution from Eden to Extinction?* New York: John Wiley.

Kitano, H. 2001. *Foundations of Systems Biology.* Cambridge, MA: MIT Press.

———. 2002. "Systems Biology: A Brief Overview." *Science* 295:1662–64.

Klein, R. G. 1999. *The Human Career: Human Biological and Cultural Origins.* 2nd ed. Chicago: University of Chicago Press.

———. 2000. "Archeology and the Evolution of Human Behavior." *Evolutionary Anthropology* 9 (1):17–36.

Kline, S. J. 1995. *Conceptual Foundations for Multidisciplinary Thinking.* Stanford, CA: Stanford University Press.

———. 1996. "The Powers and Limitations of Reductionism and Synoptism." Program in Science, Technology and Society, Report CF, Stanford University, Stanford, CA.

———. 1997. *The Semantics and Meaning of the Entropies.* Report CB-1, Department of Mechanical Engineering, Stanford University, Stanford, CA.

Knauft, B. M. 1987. "Divergence between Cultural Success and Reproductive Fitness in Preindustrial Cities." *Cultural Anthropology* 2:94–114.

———. 1991. "Violence and Sociality in Human Evolution." *Current Anthropology* 32 (4):391–409.

Knudsen, T. 2001. "Nesting Lamarckism within Darwinian Explanations: Necessity in Economics and Possibility in Biology." In *Darwinism and Evolutionary Economics,* eds. J. Laurent and J. Nightingale, pp. 121–59. Cheltenham, UK: Edward Elgar.

Koestler, A. 1967. *The Ghost in the Machine.* New York: Macmillan.

———. 1978. *Janus: A Summing Up.* New York: Random House.

Koestler, A., and J. R. Smythies, eds. 1969. *Beyond Reductionism: New Perspectives in the Life Sciences.* London: Hutchinson.

Kolm, S.-C. 1996. *Modern Theories of Justice.* Cambridge, MA: MIT Press.

———. 1997. *Justice and Equity,* trans. H. F. See. Cambridge, MA: MIT Press.

Kondrashov, A. S. 1982. "Selection against Harmful Mutations in Large Sexual and Asexual Populations." *Genetical Research* 40:325–32.

———. 1988. "Deleterious Mutations and the Evolution of Sexual Reproduction." *Nature* 336:435–40.

Konner, M. 2002. "Seeking Universals." *Nature* 415:121.

Korzybski, A. 1933. *Science and Sanity: An Introduction to Non-Aristotelian Systems and General Semantics.* Lancaster, PA: The International Non-Aristotelian Library Publishing Company (distributed by The Science Press Printing Company).

Kozo-Polyansky, B. M. 1924. *A New Principle of Biology. Essay on the Theory of Symbiogenesis.* [In Russian.] Voronezh.

———. 1932. *Introduction to Darwinism.* [In Russian.] Voronezh.

Krippendorff, K., ed. 1979. *Communication and Control in Society.* New York: Gordon and Breach Science Publishers.

Kropotkin, P. 1902. *Mutual Aid, A Factor of Evolution.* New York: McClure Phillips & Co.

Kuhn, T. 1962. *The Structure of Scientific Revolutions.* Chicago: University of Chicago Press.

Kummer, H. 1968. *Social Organization of Hamadryas Baboons: A Field Study.* Chicago: University of Chicago Press.

———. 1971. *Primate Societies: Group Techniques of Ecological Adaptation.* Chicago: Aldine-Atherton.

Küppers, B. 1990. *Information and the Origin of Life*. Cambridge, MA: MIT Press.

Kushmerick, M. J., R. E. Larson, and R. D. Davies. 1969. "The Chemical Energies of Muscle Contraction (II)." *Proceedings of the Royal Society (London)* 174:315–53.

Lal, R., and B. A. Stewart. 1990. *Soil Degradation*. New York: Springer-Verlag.

Laland, K. N., F. J. Odling-Smee, and M. W. Feldman. 2000. "Niche Construction, Biological Evolution, and Cultural Change." *Behavioral and Brain Sciences* 23:131–75.

———. 2001. "Cultural Niche Construction and Human Evolution." *Journal of Evolutionary Biology* 14:22–33.

Lamarck, J. B. de 1963. *Zoological Philosophy*, trans. H. Elliot. New York: Hafner. (Orig. pub. 1809.)

Landauer, R. 1996. "Minimal Energy Requirements in Communications." *Science* 272:1914–18.

Lansing, J. S., and J. N. Kremer. 1993. "Emergent Properties of Balinese Water Temple Networks: Coadaptation on a Rugged Fitness Landscape." *American Anthropologist* 95 (1):97–114.

Laszlo, E., 1972. *The Systems View of the World*. New York: George Braziller.

Laurent, J., and J. Nightingale, eds. 2001. *Darwinism and Evolutionary Economics*. Cheltenham, UK: Edward Elgar.

Layzer, D. 1988. "Growth of Order in the Universe." In *Entropy, Information, and Evolution: New Perspectives on Physical and Biological Evolution*, eds. B. H. Weber, D. J. Depew, and J. D. Smith, pp. 23–40. Cambridge, MA: MIT Press.

Leakey, R. 1994. *The Origin of Humankind*. New York: Basic Books.

Le Boeuf, B. J. 1985. *Elephant Seals*. Pacific Grove, CA: Boxwood Press.

Le Boeuf, B. J., and R. M. Laws. 1994. *Elephant Seals: Population Ecology, Behavior and Physiology*. Berkeley: University of California Press.

Lederer, K., ed. 1980. *Human Needs: A Contribution to the Current Debate*. Cambridge, MA: Oelgeschlager, Gunn and Hain, Publishers.

Lee, R. B. 1968. "What Hunters Do for a Living, or, How to Make Out on Scarce Resources." In *Man the Hunter*, eds. R. B. Lee and I. DeVore, pp. 30–48. Chicago: Aldine.

Leff, H. S., and A. F. Rex. 1990. *Maxwell's Demon, Entropy, Information, Computing*. Princeton, NJ: Princeton University Press.

Lehn, J. M. 1993. "Supramolecular Chemistry." *Science* 260:1762–63.

Lehninger, A. L. 1971. *Bioenergetics: The Molecular Basis of Biological Energy Transformations*. Menlo Park, CA: Benjamin/Cummings.

Leigh, E. 1971. *Adaptation and Diversity*. San Francisco: Freeman, Cooper and Co.

———. 1977. "How Does Selection Reconcile Individual Advantage with the Good of the Group?" *Proceedings of the National Academy of Sciences (USA)* 74:4542–46.

———. 1983. "When Does the Good of the Group Override the Advantage of the Individual?" *Proceedings of the National Academy of Sciences (USA)* 80:2985–89.

———. 1991. "Genes, Bees and Ecosystems: The Evolution of a Common Interest among Individuals." *Trends in Ecology and Evolution* 6:257–62.

————. 2000. "Evolutionary Biology and Human Altruism: A Review of Three Books." *Evolution* 54 (1):328–32.

Le Maho, Y. 1977. "The Emperor Penguin: A Strategy to Live and Breed in the Cold." *American Scientist* 65:680–93.

Leopold, C. A., and R. Ardrey. 1972. "Toxic Substances in Plants and the Food Habits of Early Man." *Science* 176:512–14.

Leslie, P. W., J. R. Bindon, and P. T. Baker. 1984. "Caloric Requirements of Human Populations: A Model." *Human Ecology* 12:137–62.

Levine, R. D., and M. Tribus, eds. 1979. *The Maximum Entropy Formalism.* Cambridge, MA: MIT Press.

Lewes, G. H. 1874–1879. *Problems of Life and Mind.* London: Truebner.

Lewin, R. 1992. *Complexity: Life at the Edge of Chaos.* New York: Macmillan.

————. 1993. *Human Evolution: An Illustrated Introduction.* 3rd ed. Boston: Blackwell Scientific.

————. 1996. "All for One." *New Scientist* 152 (2060):28–33.

Lewis, D. H. 1991. "Mutualistic Symbioses in the Origin and Evolution of Land Plants." In *Symbiosis as a Source of Evolutionary Innovation,* eds. L. Margulis and R. Fester, pp. 288–300. Cambridge, MA: MIT Press.

Lewontin, R. C. 1978. "Adaptation." *Scientific American* 239 (3):213–30.

————. 1979. "Sociobiology as an Adaptationist Program." *Behavioral Science* 24:5–14.

————. 1984. "Adaptation." In *Conceptual Issues in Evolutionary Biology,* ed. E. Sober. Cambridge, MA: Harvard University Press.

Ligon, J. D., and S. H. Ligon. 1978. "Communal Breeding in Green Woodhoopoes as a Case for Reciprocity." *Nature* 276:496–98.

————. 1982. "The Cooperative Breeding Behavior of the Green Woodhoopoe." *Scientific American* 247 (1):126–34.

Lima, S. L. 1989. "Iterated Prisoner's Dilemma: An Approach to Evolutionarily Stable Cooperation." *American Naturalist* 134:828–34.

Lippa, R. A. 1988. *Gender, Nature, and Nurture.* Mahwah, NJ: Lawrence Erlbaum Associates.

Lissack, M. R. 1999. "Complexity: The Science, Its Vocabulary, and Its Relation to Organizations." *Emergence* 1:110–25.

Little, P. 1995. "The Genome Directory: Navigational Progress." *Nature* 377:288.

Lloyd Morgan, C. 1926. *Life, Mind and Spirit.* London: Williams and Norgate.

————. 1933. *The Emergence of Novelty.* New York: Henry Holt.

Lopez, B. 1978. *Of Wolves and Men.* New York: Scribner.

Lorenz, K. 1966. *On Aggression,* trans. M. K. Wilson. New York: Harcourt, Brace, World.

Lotka, A. J. 1922. "Contribution to the Energetics of Evolution." *Proceedings of the National Academy of Sciences (USA)* 8:147–55.

————. 1945. "The Law of Evolution as a Maximal Principle." *Human Biology* 17:167–94.

Lovejoy, A. O. 1936. *The Great Chain of Being: A Study of the History of an Idea.* Cambridge, MA: Harvard University Press.

Lovelock, J. E. 1990. "Hands up for the Gaia Hypothesis." *Nature* 344:100–102.

———. 1993. "Gaia: Science or Myth?" *New Statesman and Society* 6 (235):37–38.

———. 2003. "The Living Earth." *Nature* 426:769–70.

Luna, F., and A. Perrone. 2001. *Agent-Based Methods in Economics and Finance.* Dordrecht, Netherlands: Kluwer Academic Publishers.

Lux, K. 1990. *Adam Smith's Mistake.* Boston: Shambhala.

Lwoff, A. 1962. *Biological Order.* Cambridge, MA: MIT Press.

MacDonald, D. W. 1986. "A Meerkat Volunteers for Guard Duty So Its Comrades Can Live in Peace." *Smithsonian* 17 (11):55–64.

MacDonald, K. 1997. "Life History Theory and Human Reproductive Behavior." *Human Nature* 8:327–59.

MacKay, D. M. 1968. "The Informational Analysis of Questions and Commands." In *Modern Systems Research for the Behavioral Scientist,* ed. W. Buckley, pp. 204–9. Chicago: Aldine. (Orig. pub. 1961.)

Mackie, G. O. 1990. "The Elementary Nervous System Revisited." *American Zoologist* 30:907–20.

MacPherson, S. 1987. *Five Hundred Million Children: Poverty and Child Welfare in the Third World.* Brighton: Wheatsheaf.

Malinowski, B. 1944. *A Scientific Theory of Culture and Other Essays.* Chapel Hill: University of North Carolina Press.

Malthus, T. R. 1989. *Principles of Political Economy.* Cambridge: Cambridge University Press. (Orig. pub. 1820.)

Mann, J., R. C. Connor, P. L. Tyack, and H. Whitehead, eds. 2000. *Cetacean Societies: Field Studies of Dolphins and Whales.* Chicago: University of Chicago Press.

Mansbridge, J. J., ed. 1990. *Beyond Self Interest.* Chicago: University of Chicago Press.

Margulis, L. 1970. *Origin of Eukaryotic Cells.* New Haven, CT: Yale University Press.

———. 1981. *Symbiosis in Cell Evolution.* San Francisco: W. H. Freeman.

———. 1993. *Symbiosis in Cell Evolution.* 2nd ed. New York: W. H. Freeman.

Margulis, L., and R. Fester, eds. 1991. *Symbiosis as a Source of Evolutionary Innovation: Speciation and Morphogenesis.* Cambridge, MA: MIT Press.

Margulis, L., and M. McMenamin, eds. 1993. *Concepts of Symbiogenesis: A Historical and Critical Study of the Research of Russian Botanists.* New Haven, CT: Yale University Press.

Margulis, L., and D. Sagan. 1995. *What is Life?* New York: Simon & Schuster.

Margulis, L., and K. V. Schwartz. 1982. *Five Kingdoms: An Illustrated Guide to the Phyla of Life on Earth.* San Francisco: W. H. Freeman.

Marshall, A. 1890. *Principles of Economics.* London: Macmillan.

Martin, W., and M. Müller. 1998. "The Hydrogen Hypothesis for the First Eukaryote." *Nature* 392:37–41.

Maryanski, A., and J. H. Turner. 1992. *The Social Cage: Human Nature and the Evolution of Society.* Stanford, CA: Stanford University Press.

Marzluff, J. M., and B. Heinrich. 1991. "Foraging by Common Ravens in the Presence and Absence of Territory Holders: An Experimental Analysis of Social Foraging." *Animal Behaviour* 42:755–70.

Maslow, A. H. 1954. *Motivation and Personality.* New York: Harper & Row.

———. 1962. *Toward a Psychology of Being.* Princeton, NJ: Van Nostrand.

———. 1967. "A Theory of Metamotivation: The Biological Rooting of the Value-Life." *Journal of Humanistic Psychology* 7:38–39, 58–61.

Masters, R. D. 1987. "Evolutionary Biology and Natural Right." In *The Crisis of Liberal Democracy,* eds. W. Soffer and K. Deutsch, pp. 48–66. Albany: State University New York Press.

———. 1989. *The Nature of Politics.* New Haven: Yale University Press.

Masters, R. D., and D. M. Gruter, eds. 1992. *The Sense of Justice: Biological Foundations of Law.* Newbury Park, CA: Sage Publications.

Matessi, C., and S. D. Jayakar. 1976. "Conditions for the Evolution of Altruism under Darwinian Selection." *Theoretical Population Biology* 9:360–87.

Mathieu, L. G., and S. Sonea. 1995. "A Powerful Bacterial World." *Endeavour* 19 (3): 112–17.

Maturana, H. R., and F. Varela. 1980. *Autopoiesis and Cognition: The Realization of Living.* Dordrecht, Netherlands: D. Reidel.

Maughan, W. S. 1941. *The Moon and Sixpence.* New York: Heritage Press.

Max-Neef, M. A. 1991. *Human Scale Development: Conception, Application and Further Reflections.* New York: Apex Press.

Max-Neef, M. A.A. Elizalde, and M. Hopenhayn. 1989. *Human Scale Development: An Option for the Future.* Development Dialogue. CEPAUR, Dag Hamarskjöld Foundation. Santiago de Chile.

Maxwell, J. C. 1871. *Theory of Heat.* London: Longman's, Green and Co.

May, R. M. 1992. "How Many Species Inhabit the Earth?" *Scientific American* 267 (4):42–48.

Maynard Smith, J. 1975. *The Theory of Evolution.* 3rd. ed. New York: Penguin.

———. 1978. "Optimization Theory in Evolution." *Annual Review of Ecology and Systematics* 9:31–56.

———. 1982a. "The Evolution of Social Behavior—A Classification of Models." In *Current Problems in Sociobiology,* ed. King's College Sociobiology Group, pp. 28–44. Cambridge: Cambridge University Press.

———. 1982b. *Evolution and the Theory of Games.* Cambridge: Cambridge University Press.

———. 1983. "Models of Evolution." *Proceedings of the Royal Society of London B* 219:315–25.

———. 1984. "Game Theory and the Evolution of Behaviour." *The Behavioral and Brain Sciences* 7:95–125.

———. 1989. *Evolutionary Genetics.* Oxford: Oxford University Press.

———. 1998. *Shaping Life: Genes, Embryos and Evolution.* New Haven, CT: Yale University Press.

Maynard Smith, J., and E. Szathmáry. 1993. "The Origin of Chromosomes I. Selection for Linkage." *Journal of Theoretical Biology* 164:437–46.

———. 1995. *The Major Transitions in Evolution.* Oxford: Freeman Press.

———. 1999. *The Origins of Life: From the Birth of Life to the Origin of Language.* Oxford: Oxford University Press.

Mayr, E. 1960. "The Emergence of Evolutionary Novelties." In *Evolution after Darwin,* vol I., ed. S. Tax, pp. 349–80. Chicago: University of Chicago Press.

———. 1963. *Animal Species and Evolution.* Cambridge, MA: Harvard University Press.

———. 1965. "Cause and Effect in Biology." In *Cause and Effect,* ed. D. Lerner, pp. 33–49. New York: Free Press.

———. 1974a. Behavior Programs and Evolutionary Strategies. *American Scientist* 62:650–59.

———. 1974b. "Teleological and Teleonomic: A New Analysis." In *Boston Studies in the Philosophy of Science,* vol. XIV, eds. R. S. Cohen and M. W. Wartofsky, pp. 91–117. Boston: D. Reidel.

———. 1976. *Evolution and the Diversity of Life: Selected Essays.* Cambridge, MA: Harvard University Press.

———. 1982. *The Growth of Biological Thought: Diversity, Evolution, and Inheritance.* Cambridge, MA: Harvard University Press.

———. 1988. *Toward a New Philosophy of Biology: Observations of an Evolutionist.* Cambridge, MA: Harvard University Press.

———. 2001. *What Evolution Is.* New York: Basic Books.

Mayr, O. 1970. *The Origins of Feedback Control.* Cambridge, MA: MIT Press.

Mazess, R. B. 1975. "Adaptation: A Conceptual Framework." In *Evolutionary Models and Studies in Human Diversity,* eds. R. J. Meier, C. M. Otten, and F. Abdel-Hameed, pp. 9–15. The Hague: Mouton.

McClare, C. W. F. 1971. "Chemical Machines, Maxwell's Demon and Living Organisms." *Journal of Theoretical Biology* 30:1–34.

———. 1972. "A 'Molecular Energy' Muscle Model." *Journal of Theoretical Biology* 35:569–95.

McFarlane, B. 1988. *Yugoslavia: Politics, Economics and Society.* London: Pinter Publishers.

McGovern, G. 2002. "The Case for Liberalism: A Defense of the Future against the Past." *Harpers Magazine* 305 (1831):37–42.

McHale, J., and M. C. McHale. 1978. *Basic Human Needs: A Framework for Action.* New Brunswick, NJ: Transaction.

McKinney, H. L. 1972. *Wallace and Natural Selection.* New Haven, CT: Yale University Press.

McKinsey, K. 1997. "Hungarians' Lifestyle Is Killing Them." *San Francisco Chronicle,* 6 October: A1.

McShea, R. J. 1990. *Morality and Human Nature: A New Route to Ethical Theory.* Philadelphia: Temple University Press.

Mech, L. D. 1981. *The Wolf: The Ecology and Behavior of an Endangered Species.* Minneapolis: University of Minnesota Press. (Orig. pub. 1970.)

———. 1988. *The Arctic Wolf: Ten Years with the Pack.* Stillwater, MN: Voyageur Press.

Mellars, P. 1989. "Major Issues in the Emergence of Modern Humans." *Current Anthropology* 30:349–85.

Mereschkovsky, K. C. 1909. "The Theory of Two Plasms as the Foundation of Symbiogenesis: A New Doctrine about the Origins of Organisms." [In Russian.] *Proceedings of the Imperial Kazan University* 12:1–102.

———. 1920. "La Plante Considérée Comme un Complexe Symbiotique." *Societé des Sciences Naturelles de l'Quest de la France, Bulletin* 6:17–98.

Mesterton-Gibbons, M., and L. A. Dugatkin. 1992. "Cooperation among Unrelated Individuals: Evolutionary Factors." *The Quarterly Review of Biology* 67 (3):267–81.

Metcalfe, S. 1998. *Evolutionary Economics and Creative Destruction.* London and New York: Routledge.

Meyer, P. 1987. "Ethnocentrism in Human Social Behavior." In *The Sociobiology of Ethnocentrism,* eds. V. Reynolds, V. S. E. Falger, and I. Vine, pp. 81–93. London: Croom Helm.

———. 1990. "Human Nature and the Function of War in Social Evolution." In *Sociobiology and Conflict,* eds. J. M. G. van der Dennen and V. Falger, pp. 227–40. London: Chapman Hall.

Michod, R. E. 1996. "Cooperation and Conflict in the Evolution of Individuality. II. Conflict Mediation." *Proceedings of the Royal Society of London B* 263:813–22.

———. 1997. "Cooperation and Conflict in the Evolution of Individuality. I. Multilevel Selection of the Organism." *American Naturalist* 149:607–45.

———. 1999. *Darwinian Dynamics, Evolutionary Transitions in Fitness and Individuality.* Princeton, NJ: Princeton University Press.

Miele, F. 1998. "The Ionian Instauration." *Skeptic* 6 (1):76–85.

Miles, I. 1985. *Social Indicators for Human Development.* New York: St. Martin's Press.

Miller, D. 1976. *Social Justice.* Oxford: Oxford University Press.

Miller, J. G. 1995. *Living Systems.* Niwot: University Press of Colorado. (Orig. pub. 1978.)

Mirowski, P. 1988a. *More Heat Than Light: Economics as Social Physics: Physics as Nature's Economics.* New York: Cambridge University Press.

———. 1988b. "Energy and Energetics in Economic Theory: A Review Essay." *Journal of Economic Issues* 22:811–30.

Mithen, S. J. 1990. *Thoughtful Foragers: A Study of Prehistoric Decision Making.* Cambridge: Cambridge University Press.

Mittenthal, J. E., and A. B. Baskin, eds. 1992. *Workshop on Principles of Organization in Organisms.* Reading, MA: Addison-Wesley.

Modelski, G. 1987. *Long Cycles in World Politics.* London: Macmillan.

———. 1994a. "Evolutionary Paradigms in the Social Sciences." Report of a Workshop, Seattle Battelle Research Center, May 13–14, 1994. Seattle: University of Washington.

———. 1994b. "An Evolutionary Paradigm for World Politics." Prepared for Delivery at the Workshop on Evolutionary Paradigms in the Social Sciences, Batelle Seattle Conference Center, May 13–14, 1994.

———. 1999. "From Leadership to Organization: The Evolution of Global Politics." In *The Future of Global Conflict,* eds. V. Bornschier and C. Chase-Dunn, pp.11–39. London: Sage Studies in International Sociology.

Modelski, G., and W. R. Thompson. 1999. "The Long and the Short of Global Politics in the Twenty-first Century: An Evolutionary Approach." *The International Studies Review* 1 (2): 110–40.

Mokyr, J. 1999. "Science, Technology, and Knowledge: What Historians Can Learn from an Evolutionary Approach." Working Paper, Department of Economics, Northwestern University, Evanston, IL.

Møller, A. P. 1987. "Advantages and Disadvantages of Coloniality in the Swallow, *Hirundo rustica.*" *Animal Behaviour* 35:819–32.

Monod, J. 1971. *Chance and Necessity,* trans. A. Wainhouse. New York: Knopf.

Moore, A. J. 1993. "Towards an Evolutionary View of Social Dominance." *Animal Behaviour* 46:594–96.

Moore, J. 1984. "The Evolution of Reciprocal Sharing." *Ethology and Sociobiology* 5:5–14.

Morgan, C. L. 1923. *Emergent Evolution.* New York: Henry Holt.

Moritz, R. F. A., and S. Fuchs. 1998. "Organization of Honeybee Colonies: Characteristics and Consequences of a Superorganism Concept." *Apidologie* 29 (1–2):7–21.

Moritz, R. F. A., and E. E. Southwick. 1992. *Bees as Superorganisms: An Evolutionary Reality.* New York: Springer-Verlag.

Morowitz, H. J. 1968. *Energy Flow in Biology.* New York: Academic Press.

———. 1978a. *Foundations of Bioenergetics.* New York: Academic Press.

———. 1978b. "Proton Semiconductors and Energy Transduction in Biological Systems." *American Journal of Physiology* 235:R99–114.

———. 1981. "Phase Separation, Charge Separation and Biogenesis." *Biosystems* 14:41–47.

———. 1992. *Beginnings of Cellular Life: Metabolism Recapitulates Biogenesis.* New Haven, CT: Yale University Press.

———. 2002. *The Emergence of Everything.* Oxford: Oxford University Press.

Morowitz, H. J., B. Heinz, and D. W. Deamer. 1988. "The Chemical Logic of a Minimum Protocell." *Origins of Life* 18:281–87.

Morris, D. R. 1965. *The Washing of Spears.* New York: Simon & Schuster.

Morris, M. D. 1979. *Measuring the Condition of the World's Poor.* Oxford: Pergamon Press.

Mosekilde, E., and L. Mosekilde, eds. 1991. *Complexity, Chaos and Biological Evolution.* New York: Plenum Press.

Mumme, R. L. 1992. "Do Helpers Increase Reproductive Success?" *Behavioral Ecology and Sociobiology* 31:319–28.

Mumme, R. L., W. D. Koenig, and F. A. Pitelka. 1988. "Costs and Benefits of Joint Nesting in the Acorn Woodpecker." *The American Naturalist* 131:654–77.

Mumme, R. L., W. D. Koenig, and F. L. W. Ratnieks. 1989. "Helping Behaviour, Reproductive Value, and the Future Component of Indirect Fitness." *Animal Behaviour* 38:331–43.

Murmann, J. P. 2003. *Knowledge and Competitive Advantage.* Cambridge: Cambridge University Press.

Murphy, H. B. M. 1982. *Comparative Psychiatry: The International and Intercultural Distribution of Mental Illness*. Berlin: Springer-Verlag.

Naroll, R. 1983. *The Moral Order: An Introduction to the Human Situation*. Beverly Hills, CA: Sage Publications.

Needham, J. 1937. *Integrative Levels: A Reevaluation of the Idea of Progress*. Oxford: Clarendon Press.

Nelson, R. R. 1996. *The Sources of Economic Growth*. Cambridge, MA: Harvard University Press.

Nelson, R. R., and S. G. Winter. 1982. *An Evolutionary Theory of Economic Change*. Cambridge, MA: Harvard University Press.

Nicholls, D. G., and S. J. Ferguson. 1992. *Bioenergetics 2*. San Diego: Academic Press.

Nickles, T., ed. 1980. *Scientific Discovery, Logic and Rationality*. Dordrecht, Netherlands: D. Reidel.

Nicolis, G., and I. Prigogine. 1977. *Self-Organization in Nonequilibrium Systems*. New York: John Wiley.

———. 1989. *Exploring Complexity*. New York: W. H. Freeman.

Nitecki, M. H., ed. 1983. *Coevolution*. Chicago: University of Chicago Press.

———, ed. 1988. *Evolutionary Progress*. Chicago: University of Chicago Press.

Nitecki, M. H., and D. Nitecki. 1993. *Evolutionary Ethics*. New York: State University of New York Press.

Nossal, G. J. V. 1993. "Life, Death and the Immune System." *Scientific American* 269 (3):53–62.

Nöth, W. 1990. *Handbook of Semiotics*. Bloomington: University of Indiana Press.

Novikoff, A. 1945. "The Concept of Integrative Levels in Biology." *Science* 101:209–15.

Nowak, M. A., and K. Sigmund. 1993. "A Strategy of Win-Stay, Lose-Shift That Outperforms Tit-for-Tat in the Prisoner's Dilemma Game." *Nature* 364:56–58.

———. 1998a. "Evolution of Indirect Reciprocity by Image Scoring." *Nature* 394:573–78.

———. 1998b. "The Dynamics of Indirect Reciprocity." *Journal of Theoretical Biology* 194:561–74.

Nowak, M. A., K. M. Page, and K. Sigmund. 2000. "Fairness versus Reason in the Ultimatum Game." *Science* 289:1773–75.

Nozick, R. 1974. *Anarchy, State and Utopia*. Oxford: Blackwell.

Nussbaum, M. 1988. "Nature, Function and Capability: Aristotle on Political Distribution." *Oxford Studies in Ancient Philosophy* (Suppl.): 145–84.

———. 1993. "Non-Relative Virtues: An Aristotelian Approach." In *The Quality of Life*, eds. M. Nussbaum and A. Sen, pp. 1–6. Oxford: Clarendon Press.

Nussbaum, M., and A. Sen, eds. 1993. *The Quality of Life*. Oxford: Clarendon Press.

Nye, J. 2002. *The Paradox of American Power: Why the World's Only Superpower Can't Go It Alone*. Oxford: Oxford University Press.

Odling-Smee, F. J., K. N. Laland, and M. W. Feldman. 2003. *Niche Construction: The Neglected Process in Evolution*. [Monographs in Popular Biology series no. 37.] Princeton, NJ: Princeton University Press.

Odum, E. P. 1971. *Fundamentals of Ecology*. 3rd ed. Philadelphia: W. B. Saunders.

———. 1983. *Basic Ecology.* Philadelphia: Saunders College Publishers.

Odum, H. T. 1971. *Environment, Power and Society.* London: John Wiley.

———. 1983. *Systems Ecology: An Introduction.* New York: John Wiley

———. 1986. "Emergy in Ecosystems" in *Environmental Monographs and Symposia,* ed. N. Polunin, pp. 337–69. New York: John Wiley.

———. 1988. "Self-Organization, Transformity and Information." *Science* 242:1132–39.

———. 1996. *Environmental Accounting, Emergy and Decision Making.* New York: John Wiley.

Odum, H. T., and E. P. Odum. 1982. *Energy Basis for Man and Nature.* 2nd ed. New York: McGraw-Hill.

Ogburn, W. F., ed. 1929. *Social Changes in 1928.* Chicago: University of Chicago Press.

Oldroyd, B. P., T. E. Rinderer, and S. M. Buco. 1992a. "Intra-Colonial Foraging Specialism by Honey Bees *(Apis mellifera)* (Hymenoptera: Apidae)." *Behavioral Ecology and Sociobiology* 30:291–95.

Oldroyd, B. P., T. E. Rinderer, J. R. Harbo, and S. M. Buco. 1992b. "Effects of Intracolonial Genetic Diversity on Honey Bee (Hymenoptera: Apidae) Colony Performance." *Annals of the Entomological Society of America* 85 (3):335–43.

Olendorf, R., and T. Getty. 2004. "Cooperative Nest Defence in Red-Winged Blackbirds: Reciprocal Altruism, Kinship or By-product Mutualism?" *Proceedings Royal Society of London B* 271:177–82.

Olson, M., Jr. 1969. "The Plan and Purpose of a Social Report." *The Public Interest* Spring (15):85–97.

Organisation for Economic Co-operation and Development. 1976. *Measuring Social Well-Being.* Paris: OECD.

———. 1985. *Measuring Health Care 1960–83: Expenditure, Costs and Performance.* Paris: OECD.

———. 1995. *Regional Integration and the Multilateral Trading System: Synergy and Divergence.* Paris: OECD.

Orwell, G. 1984. *1984.* San Diego: Harcourt Brace Jovanovich. (Orig. pub. 1948.)

Ostrom, E. 1990. *Governing the Commons.* Cambridge: Cambridge University Press.

———. 1998. "A Behavioral Approach to the Rational Choice Theory of Collective Action" *American Political Science Review* 92:1–22.

———. 2000. "Collective Action and the Evolution of Social Norms." *Journal of Economic Perspectives* 14:137–58.

Oswald, A. J. 1997. "Happiness and Economic Performance." *Economic Journal* 107:1815–31.

Overbye, D. 2002. "The End of Everything." *The New York Times* (*Science Times*), January 1, pp. D1, D7.

Oyama, S. 2000. *Evolution's Eye: A Systems View of the Biology-Culture Divide.* Durham, NC: Duke University Press.

Packer, C., L. Herbst, A. E. Pusey, J. D. Bygott, J. P. Hanby, S. J. Cairns, and M. Borgerhoff-Mulder. 1990. "Reproductive Success of Lions." In *Reproductive Success,* ed. T. H. Clutton-Brock. Chicago: University of Chicago Press.

Packer, C., and A. E. Pusey. 1982. "Cooperation and Competition within Coalitions of Male Lions: Kin Selection or Game Theory?" *Nature* 296:740–42.

Packer, C., A. E. Pusey, and L. E. Eberly. 2001. "Egalitarianism in Female African Lions." *Science* 293:690–93.

Packer, C., and L. Ruttan. 1988. "The Evolution of Cooperative Hunting." *The American Naturalist* 132 (2):159–98.

Page, R. E., and S. D. Mitchell. 1998. "Self-Organization and the Evolution of Division of Labor." *Apidologie* 29:171–90.

Page, R. E., and G. E. Robinson. 1991. "The Genetics of Division of Labour in Honey Bee Colonies." *Advances in Insect Physiology* 23:118–69.

Page, R. E., G. E. Robinson, and M. K. Fondrk. 1989. "Genetic Specialists, Kin Recognition and Nepotism in Honey Bee Colonies." *Nature* 338:576–81.

Pahl-Wostl, C. 1993. "The Hierarchical Organization of the Aquatic Ecosystem: An Outline How Reductionism and Holism May Be Reconciled. *Ecological Modelling* 66:81–100.

———. 1995. *The Dynamic Nature of Ecosystems: Chaos and Order Entwined.* London: John Wiley.

Paine, R. T. 1966. "Food Web Complexity and Species Diversity." *The American Naturalist* 100:65–75.

Pankiw, P. 1967. "Studies of Honey Bees on Alfalfa Flowers." *Journal of Apicultural Research* 6:105–12.

Papaseit, C., N. Pochon, and J. Tabony. 2000. "Microtubule Self-Organization Is Gravity Dependent." *Proceedings of the National Academy of Sciences (USA)* 97:8364–68.

Parker, P., T. Waite, B. Heinrich, and J. M. Marzluff. 1994. "Do Common Ravens Share Ephemeral Food Resources with Kin? DNA Fingerprinting Evidence." *Animal Behaviour* 48:1085–93.

Parsons, T., and B. Harrison. 1981. "Energy Utilization and Evaluation." *Journal of Social and Biological Structures* 4:1–15.

Partridge, B. L. 1982. "The Structure and Function of Fish Schools." *Scientific American* 246 (6):114–23.

Pattee, H. H., ed. 1973. *Hierarchy Theory.* New York: Braziller.

Peck, J. R. 1993. "Friendship and the Evolution of Co-operation in Honey Bee Colonies." *Journal of Theoretical Biology* 162:195–228.

Penrose, R. 1989. *The Emperor's New Mind: Concerning Computers, Minds, and the Laws of Physics.* New York: Oxford University Press.

Perry, N. 1983. *Symbiosis: Close Encounters of the Natural Kind.* Poole, Dorset: Blandford Press.

Perutz, M. F. 1987. "Physics and the Riddle of Life." *Nature* 326:555–58.

Pettersson, M. 1996. *Complexity and Evolution.* New York: Cambridge University Press.

Pettit, P. 1980. *Judging Justice: An Introduction to Contemporary Political Philosophy.* London: Routledge.

Pigliucci, M. 2001. *Phenotypic Plasticity: Beyond Nature and Nurture.* Baltimore: Johns Hopkins University Press.

Pimentel, D., and M. Pimentel, eds. 1996. *Food, Energy, and Society.* Rev. ed. Niwot: University Press of Colorado.

Pimentel, D., C. Harvey, P. Resosudarmo, K. Sinclair, D. Kurz, M. McNair, S. Crist et al. 1995. "Environmental and Economic Costs of Soil Erosion and Conservation Benefits." *Science* 267:1117–23.

Pimentel, D., J. Houser, E. Preiss, O. White, H. Fang, L. Mesnick, T. Barsky, S. Tariche, J. Schreck, and S. Alpert. 1997. "Water Resources: Agriculture, the Environment and Society." *BioScience* 47:97–106.

Pirenne, J. 1962. *The Tides of History.* New York: Dutton

Pirozynski, K. A., and D. W. Malloch. 1975. "The Origins of Land Plants: A Matter of Mycotrophism." *Biosystems* 6:153–64.

Pittendrigh, C. S. 1958. "Adaptation, Natural Selection and Behavior." In *Behavior and Evolution,* eds. A. Roe and G. G. Simpson, pp. 390–416. New Haven, CT: Yale University Press.

Plato. 1985. *The Republic,* trans. R. W. Sterling and W. C. Scott. New York: W. W. Norton.

Plomin, R., G. E. McLearn, J. C. DeFries, and M. Rutter. 1990. *Behavioral Genetics: A Primer.* 2nd ed. New York: W. H. Freeman.

Plotkin, H. C., ed. 1988. *The Role of Behavior in Evolution.* Cambridge, MA: MIT Press.

———. 1994. *Darwin Machines and the Nature of Knowledge.* Cambridge, MA: Harvard University Press.

Polanyi, M. 1968. "Life's Irreducible Structure." *Science* 160:1308–12.

Popper, K. R. 1965. "Time's Arrow and Entropy." *Nature* 207:233–34.

Postel, S. 1992. *The Last Oasis: Facing Water Scarcity.* New York: W. W. Norton.

Potts, J. D. 2000. *The New Evolutionary Microeconomics: Complexity, Competence, and Adaptive Behavior.* Northampton, MA: Edward Elgar.

Powers, W. T. 1973. *Behavior: The Control of Perception.* Chicago: Aldine.

Price, M. E., L. Cosmides, and J. Tooby. 2002. "Punitive Sentiment as an Anti-Free Rider Psychological Device." *Evolution and Human Behavior* 23:203–31.

Price, P. W. 1991. "The Web of Life: Development over 3.8 Billion Years of Trophic Relationship." In *Symbiosis as a Source of Evolutionary Innovation,* eds. L. Margulis and R. Fester, pp. 262–72. Cambridge, MA: MIT Press.

Prigogine, I. 1978. "Time, Structure and Fluctuation." *Science* 201:777–84.

———. 1980. *From Being to Becoming: Time and Complexity in the Physical Sciences.* San Francisco: W. H. Freeman.

Prigogine, I., P. Allen, and R. Herman. 1977. "The Evolution of Complexity and the Laws of Nature." In *Goals in a Global Society,* eds. E. Laszlo and J. Bierman, pp. 1–63. New York: Pergamon.

Prigogine, I., and G. Nicolis. 1971. "Biological Order, Structure and Instabilities." *Quarterly Review of Biophysics* 4:107–48.

Prigogine, I., G. Nicolis, and A. Babloyants. 1972a. "Thermodynamics of Evolution (I)." *Physics Today* 25:23–28.

———. 1972b. "Thermodynamics of Evolution (II)." *Physics Today* 25:38–44.

Proops, J. L. R. 1983. "Organization and Dissipation in Economic Systems." *Journal of Social and Biological Structures* 6:353–66.

———. 1985. "Thermodynamics and Economics: From Analogy to Physical Functioning." In *Energy and Time in the Economic and Physical Sciences,* eds. W. van Gool and J. J. C. Bruggink, pp. 155–74. Amsterdam: Elsevier Science Publishers B.V.

———. 1987. "Entropy, Information and Confusion in the Social Sciences." *The Journal of Interdisciplinary Economics* 1:225–42.

———. 1989. "Ecological Economics: Rationale and Problem Areas." *Ecological Economics* 1:59–76.

Pusey, A., J. Williams, and J. Goodall. 1997. "The Influence of Dominance Rank on the Reproductive Success of Female Chimpanzees." *Science* 277:828–31.

Queller, D. C. 1985. "Kinship, Reciprocity and Synergism in the Evolution of Social Behavior." *Nature* 318:366–67.

Queller, D. C., C. R. Hughes, and J. E. Strassmann. 1988. "Genetic Relatedness in Colonies of Tropical Wasps with Multiple Queens." *Science* 242:1155–57.

Quiatt, D. D., and V. Reynolds. 1993. *Primate Behaviour: Information, Social Knowledge, and the Evolution of Culture.* Cambridge: Cambridge University Press.

Qvortrup, L. 1993. "The Controversy over the Concept of Information." *Cybernetics and Human Knowing* 1 (4):3–24.

Rabin, M. 1993. "Incorporating Fairness into Game Theory and Economics." *American Economic Review* 83:1281-1302.

Rainey, P. B., and K. Rainey. 2003. "Evolution of Cooperation and Conflict in Experimental Bacterial Populations." *Nature* 425:72–74.

Ram, R. 1985. "The Role of Real Income Level and Income Distribution in the Fulfillment of Basic Needs." *World Development* 13:589–94.

Rand, A. 1943. *The Fountainhead.* New York: Bobbs-Merrill Company.

———. 1962. "The Objectivist Newsletter." vol. 1. New York: Times-Mirror.

Raphael, D. D. 2001. *Concepts of Justice.* Oxford: Clarendon Press.

Rapoport, A. 1956. "The Promise and Pitfalls of Information Theory." *Behavioral Science* 1:303–9.

———. 1968. "Foreword." In *Modern Systems Research for the Behavioral Scientist,* ed. W. Buckley, pp. xiii–xxii. Chicago: Aldine Publishing Co.

Ratnieks, F. L. W., and P. K. Visscher. 1989. "Worker Policing in the Honey Bee." *Nature* 342:796–97.

Raven, J. A. 1992. "Energy and Nutrient Acquisition by Autotrophic Symbioses and Their Asymbiotic Ancestors." *Symbiosis* 14:33–60.

Rawls, J. 1972. *A Theory of Justice.* Cambridge, MA: Harvard University Press.

Ray, M., and A. Rinzler. 1993. *The New Paradigm in Business.* New York: J. P. Taracher/Perigee.

Raymond, R. C. 1950. "Communication, Entropy, and Life." *American Scientist* 38:273–78.

Read, A. F., and P. H. Harvey. 1993. "Evolving in a Dynamic World." *Science* 260:1760–62.

Reader, S. M., and K. N. Laland. 2002. "Social Intelligence, Innovation, and Enhanced Brain Size in Primates." *Proceedings of the National Academy of Sciences (USA)* 99 (7):4436–41.

Reinhardt, J. F. 1952. "Responses of Honey Bees to Alfalfa Flowers." *American Naturalist* 86:257–75.

Reinheimer, H. 1913. *Evolution by Co-operation: A Study in Bioeconomics.* London: Kegan, Paul, Trench, Trubner.

Reynolds, V., V. Falger, and I. Vine. 1987. *The Sociobiology of Ethnocentrism: Evolutionary Dimensions of Xenophobia, Discrimination, Racism and Nationalism.* London: Croom Helm.

Reynolds, W. C. 1965. *Thermodynamics.* New York: McGraw-Hill Book Co.

Richardson, R. C. 2001. "Complexity, Self-Organization and Selection." *Biology and Philosophy* 16:655–83.

Richerson, P. J., and R. Boyd. 1992. "Cultural Inheritance and Evolutionary Ecology." In *Evolutionary Ecology and Human Behavior,* eds. E. A. Smith and B. Winterhalder, pp. 62–92. New York: Aldine de Gruyter.

———. 1999. "Complex Societies: The Evolutionary Origins of a Crude Superorganism." *Human Nature* 10:253–89.

———. 2004. *The Nature of Cultures.* Chicago: University of Chicago Press.

Riches, G. 1997. *First World Hunger: Food Security and Welfare Politics.* New York: St. Martins Press.

Ridley, M. 1997. *The Origins of Virtue: Human Justice and the Evolution of Cooperation.* New York: Viking.

Ridley, M. 2001. *The Cooperative Gene: How Mendel's Demon Explains the Evolution of Complex Beings.* New York: Free Press.

Riedl, R. 1978. *Order in Living Organisms: A Systems Analysis of Evolution,* trans. R. P. S. Jefferies. New York: John Wiley.

Rifkin, J. (with T. Howard). 1980. *Entropy: A New World View.* New York: Viking.

Rissing, S. W., G. B. Pollock, M. R. Higgins, R. H. Hagen, and D. R. Smith. 1989. "Foraging Specialization without Relatedness or Dominance among Co-founding Ant Queens." *Nature* 338:420–22.

Rissing, S. W., and G. B. Pollack. 1991. "An Experimental Analysis of Pleometrotic Advantage in the Desert Seed-Harvester Ant *Messor pergandei* (Hymenoptera: Formicidae)." *Insect Society* 38:205–11.

Rist, G. 1980. "Basic Questions about Basic Human Needs." In *Human Needs: A Contribution to the Current Debate,* ed. K. Lederer, pp. 233–53. Cambridge, MA: Oelgeschlager, Gunn and Hain.

Roberts, P. 1982. "Energy and Value." *Energy Policy* 10:171–80.

Robinson, C. A., Jr. 1951. *Ancient History: From Prehistoric Times to the Death of Justinian.* New York: Macmillan.

Robinson, G. E., and R. E. Page. 1988. "Genetic Determination of Guarding and Undertaking in Honey Bee Colonies." *Nature* 333:356–61.

Rock, I., and S. Palmer. 1990. "The Legacy of Gestalt Psychology." *Scientific American* 263 (6):84–90.

Roe, A., and G. G. Simpson, eds. 1958. *Behavior and Evolution.* New Haven, CT: Yale University Press.

Roemer, J. E. 1996. *Theories of Distributed Justice.* Cambridge, MA: Harvard University Press.

Rogel, C. 1998. *The Breakup of Yugoslavia and the War in Bosnia.* Westport, CT: Greenwood Press.

Roitt, I. M. 1988. *Essential Immunology.* 6th ed. Oxford: Blackwell Scientific.

Rollo, C. D. 1995. *Phenotypes: Their Epigenetics, Ecology and Evolution.* New York: Chapman & Hall.

Rosen, D. E. 1978. "Darwin's Demon." *Systematic Zoology* 27:370–73.

Rosen, R. 1985. "Organisms as Causal Systems Which Are Not Mechanisms: An Essay into the Nature of Complexity." In *Theoretical Biology and Complexity,* ed. R. Rosen, pp. 165–203. New York: Academic Press.

Rosenau, J. N. 1999. "Toward an Ontology for Global Governance." In *Approaches to Global Governance Theory,* eds. M. Hewson and T. J. Sinclair, pp. 287–301. Albany: State University of New York Press.

Rosenberg, A. 1992. *Economics: Mathematical Politics or Science of Diminishing Returns?* Chicago: University of Chicago Press.

Rosenberg, G. 1991. "Aposematism and Synergistic Selection in Marine Gastropods." *Evolution* 45:451–54.

Rosenbleuth, A., N. Wiener, and J. Bigelow. 1943. "Behavior, Purpose and Teleology." *Philosophy of Science* 10:18–24.

Rothschild, M. 1990. *Bionomics: Economy as Ecosystem.* New York: Henry Holt and Company.

Rubin, P. H. 2002. *Darwinian Politics: The Evolutionary Origins of Freedom.* New Brunswick, NJ: Rutgers University Press.

Ruby, E. G., and J. G. Morin. 1979. "Luminousenteric Bacteria of Marine Fishes in a Study of Their Distribution, Density and Dispersion." *Applied and Environmental Microbiology* 38:406–11.

Russell, B. 1927. *The Analysis of Matter.* London: Allen and Unwin.

Russett, B. M., ed. 1972. *Peace, War and Numbers.* Beverly Hills, CA: Sage Publications.

Ruyle, E. E. 1973. "Genetic and Cultural Pools: Some Suggestions for a Unified Theory of Biocultural Evolution." *Human Ecology* 1:201–15.

Sabloff, J. A. 1995. "Drought and Decline." *Nature* 375:357.

Sahlins, M. 1972. *Stone Age Economics.* Chicago and New York: Aldine Atherton.

Salisbury, R. F. 1973. "Economic Anthropology." *Annual Review of Anthropology* 2:85–94.

Salthe, S. N. 1985. *Evolving Hierarchical Systems.* Columbia University Press.

———. 1993. *Development and Evolution: Complexity and Change in Biology.* Cambridge, MA: MIT Press (Bradford Books).

Samuelson, P. A., and W. D. Nordhaus. 1992. *Economics.* 14th ed. New York: McGraw-Hill.

Santillán-Doherty, A. M., J. L. Diaz, R. Mondragon-Ceballos. 1991. "Synergistic Effects of Kinship, Sex and Rank in the Behavioural Interactions of Captive Stump-Tailed Macques." *Folia Primatologica* 56:177–89

Scanlon, T. M. 1991. "The Moral Basis of Interpersonal Comparisons." In *Interpersonal Comparisons of Well-Being*, eds. J. Elster and J. E. Roemer, pp.17–44. Cambridge: Cambridge University Press.

———. 1993. "Value, Desire and Quality of Life." In *The Quality of Life*, eds. M. Nussbaum and A. Sen, pp. 185–205. Oxford: Clarendon Press.

Schaller, G. B. 1972. *The Serengeti Lion: A Study of Predator Prey Relations* Chicago: University of Chicago Press.

Scheel, D., and C. Packer. 1991. "Group Hunting Behaviour of Lions: A Search for Cooperation." *Animal Behaviour* 41 (4):697–710.

Schjelderup-Ebbe, T. 1922. "Beiträge zur Sozialpsychologie des Haushuhns." *Zeitschrift für Psychologie* 88:226–52.

Schlesinger, A. M. 1939. "Tides of American Politics." *Yale Review* 29:217–30.

Schmidt-Nielsen, K. S. 1972. *How Animals Work*. Cambridge: Cambridge University Press.

Schneider, E. D., and J. J. Kay. 1994. "Life as a Manifestation of the Second Law of Thermodynamics." *Mathematical Computing Modeling* 19:25–48.

———. 1995. "Order from Disorder: The Thermodynamics of Complexity in Biology." In *What Is Life? The Next Fifty Years*, eds. M. P. Murphy and L. A. J. O'Neill, pp. 161–73. New York: Cambridge University Press.

Schneider, E. D., and D. Sagan. 2005. *Into the Cool: Energy Flow, Thermodynamics and Life*. Chicago: University of Chicago Press.

Schrödinger, E. 1945. *What Is Life? The Physical Aspect of the Living Cell*. New York: The Macmillan Co.

Schubert, G. 1989. *Evolutionary Politics*. Carbondale: Southern Illinois University Press.

Schubert, G., and R. Masters, eds. 1991. *Primate Politics*. Carbondale: Southern Illinois University Press.

Schumpter, J. A. 1934. *The Theory of Economic Development*. Cambridge, MA: Harvard University Press.

Schwemmler, W. 1989. *Symbiogenesis: A Macro-Mechanism of Evolution*. New York: Walter de Gruyter.

Schwemmler, W., and H. E. A. Schenck, eds. 1980. *Endocytobiology—Endosymbiosis and Cell Biology: A Synthesis of Recent Research*. New York: Walter de Gruyter.

Scott, J. P. 1989. *The Evolution of Social Systems*. New York: Gordon and Breach Science Publishers.

Scott, M. P. 1994. "Competition with Flies Promotes Communal Breeding in the Burying Beetle (*Nicrophorus tomentosus*)." *Behavioral Ecology and Sociobiology* 34:367–73.

Sebeok, T. A. 1986. "The Doctrine of Signs." *Journal of Social and Biological Structures* 9:345–52.

Seeley, T. D. 1989. "The Honey Bee Colony as a Super-organism." *American Scientist* 77:546–53.

————. 1995. *The Wisdom of the Hive: The Social Physiology of Honey Bee Colonies.* Cambridge, MA: Harvard University Press.

Seeley, T. D., and R. A. Levien. 1987. "A Colony of Mind: The Beehive as Thinking Machine." *The Sciences* 27 (4):38–43.

Seger, J. 1991. "Cooperation and Conflict in Social Insects." In *Behavioural Ecology: An Evolutionary Approach.* 2nd ed., eds. J. R. Krebs and N. B. Davies, pp. 338–73. Oxford: Blackwell Scientific.

Selowsky, M. 1981. "Income Distribution, Basic Needs and Trade-Offs with Growth." *World Development* 9 (1):73–92.

Sen, A. K. 1982. *Choice, Welfare and Measurement.* Cambridge, MA: MIT Press.

————. 1985. *Commodities and Capabilities.* Amsterdam: Elsevier.

————. 1992. *Inequality Reexamined.* Oxford: Clarendon Press.

————. 1993. "Capability and Well-Being." In *The Quality of Life,* eds. M. Nussbaum and A. Sen. pp. 46–47.Oxford: Clarendon Press.

Senge, P. 1990. *The Fifth Discipline.* New York: Doubleday.

Service, E. R. 1971. *Cultural Evolutionism: Theory in Practice.* New York: Holt, Rinehart, and Winston.

Sethi, R., and E. Somanathan.1996. "The Evolution of Social Norms in Common Property Resource Use." *The American Economic Review* 86 (4):766–88.

————. 2001. "Preference Evolution and Reciprocity." *Journal of Economic Theory* 97:273–97.

Shannon, C. E. 1948. "A Mathematical Theory of Communication." *Bell System Technical Journal* 27:379–423, 623–56.

Shannon, C. E., and W. Weaver. 1949. *The Mathematical Theory of Communication.* Urbana: University of Illinois Press.

Shapiro, A. H. 1953. *The Dynamics and Thermodynamics of Compressible Fluid Flow.* New York: Ronald Press.

Shapiro, J. A. 1988. "Bacteria as Multicellular Organisms." *Scientific American* 258 (6):82–89.

————. 1991. "Genomes as Smart Systems." *Genetica* 84:3–4.

————. 1992. "Natural Genetic Engineering in Evolution." *Genetica* 86:99–111.

Shapiro, J. A., and M. Dworkin, eds. 1997. *Bacteria as Multicellular Organisms.* New York: Oxford University Press.

Shaw, P., and Y. Wong. 1989. *Genetic Seeds of Warfare: Evolution, Nationalism and Patriotism.* London: Unwin Hyman.

Sherman, P. W., J. U. M. Jarvis, and R. D. Alexander, eds. 1991. *The Biology of the Naked Mole-Rat.* Princeton, NJ: Princeton University Press.

Sherman, P. W., J. U. M. Jarvis, and S. Braude. 1992. "Naked Mole Rats." *Scientific American* 267 (2):72–78.

Sherman, P. W., T. D. Seeley, and H. K. Reeve. 1988. "Parasites, Pathogens, and Polyandry in Social Hymenoptera." *The American Naturalist* 131 (4):602–10.

Shettleworth, S. J. 1998. *Cognition, Evolution, and Behavior.* New York: Oxford University Press.

Shykoff, J. A., and P. Schmid-Hempel. 1991. "Genetic Relatedness and Eusociality: Parasite-Mediated Selection on the Genetic Composition of Groups." *Behavioral Ecology and Sociobiology* 38:371–76.

Sidey, H. 1994. "The Home Front." *Time* 143 (24):48–49.

Sigmund, K. 1993. *Games of Life: Explorations in Ecology, Evolution, and Behaviour.* Oxford: Oxford University Press.

Sigmund, K., E. Fehr, and M. A. Nowak. 2002. "The Economics of Fair Play." *Scientific American* 286 (1):83–87.

Sigmund, K., C. Hauert, and M. A. Nowak. 2001. "Reward and Punishment." *Proceedings of the National Academy of Sciences (USA)* 98:10757–62.

Simms, J. R. 1999. *Principles of Quantitative Living Systems Science.* New York: Kluwer Academic Publishers.

Simon, H. A. 1957. *Models of Man: Social and Rational.* New York: John Wiley.

———. 1962. "The Architecture of Complexity." *Proceedings of the American Philosophical Society* 106:467–82.

Simpson, G. G. 1953. "The Baldwin Effect." *Evolution* 2:110–17.

———. 1967. *The Meaning of Evolution.* Rev. ed. New Haven, CT: Yale University Press.

Singer, P. 2000. *A Darwinian Left: Politics, Evolution and Cooperation.* London: Weidenfeld & Nicolson.

Skinner, B. F. 1981. "Selection by Consequences." *Science* 213:501–4.

Slesser, M. 1975. "Accounting for Energy." *Nature* 254:170–72.

Smail, J. K. 1997. "Beyond Population Stabilization: The Case for Dramatically Reducing Global Human Numbers." *Politics and the Life Sciences* 16:183–92.

Smarr, L. L. 1985. "An Approach to Complexity: Numerical Computations." *Science* 228:403–8.

Smillie, D. 1993. "Darwin's Tangled Bank: The Role of Social Environments." In *Perspectives in Ethology, Volume 10: Behavior and Evolution*, eds. P. P. G. Bateson, P. H. Klopfer, and N. S. Thompson, pp. 119–41. New York: Plenum Press.

Smith, A. 1964. *The Wealth of Nations.* 2 vols. London: Dent. (Orig. pub. 1776.)

———. 1976. *The Theory of Moral Sentiments.* Oxford: Clarendon Press. (Orig. pub. 1759.)

Smith, D.C. 1992. "The Symbiotic Condition." *Symbiosis* 14:3–15.

Smith, D. C., and A. E. Douglas. 1987. *The Biology of Symbiosis.* Baltimore: Edward Arnold.

Smith, E. A., and B. Winterhalder, eds. 1992. *Evolutionary Ecology and Human Behavior.* New York: Aldine de Gruyter.

Smuts, J. C. 1926. *Holism and Evolution.* New York: Macmillan Co.

Snyder, L. L. 1960. *The War: A Concise History, 1939–1945.* New York: Jillian Messner.

Sober, E., and D. S. Wilson. 1998. *Unto Others: The Evolution and Psychology of Unselfish Behavior.* Cambridge, MA: Harvard University Press.

Soddy, F. 1933. *Wealth, Virtual Wealth and Debt: The Solution of the Economic Paradox.* New York: Dutton.

Solomon, R. C., and M. C. Murphy, eds. 2000. *What Is Justice: Classic and Contemporary Readings*. New York: Oxford University Press.

Soltis, J., R. Boyd, and P. J. Richerson. 1995. "Can Group-Functional Behaviors Evolve by Cultural Group Selection? An Empirical Test." *Current Anthropology* 36 (3):473–84.

Somit, A., ed. 1976. *Biology and Politics: Recent Explorations*. The Hague: Mouton.

Somit, A., and S. A. Peterson. 1997. *Darwinism, Dominance, and Democracy: The Biological Bases of Authoritarianism*. Westport, CT: Praeger.

———. 2001. "Darwinism, Dominance and Democracy: A Reaffirmation." *Politics and the Life Sciences* 20 (1):213–16.

Somit, A., S. A. Peterson, and W. Richardson. 1980. *The Literature of Biopolitics*. Rev. ed. Dekalb, IL: The Center for Biopolitical Research.

Sonea, S. 1991. "Bacterial Evolution without Speciation." In *Symbiosis as a Source of Evolutionary Innovation,* eds. L. Margulis and R. Fester, pp. 95–105. Cambridge, MA: MIT Press.

Sonea, S., and M. Panisset. 1983. *A New Bacteriology.* Boston: Jones and Bartlett.

Spencer, H. 1850. *Social Statics*. New York: D. Appleton.

———. 1852. "A Theory of Population, Deduced from the General Law of Animal Fertility." *Westminster Review* 57:468–501.

———. 1892. "The Development Hypothesis." In *Essays: Scientific, Political and Speculative, pp.* 1–7. New York: Appleton. (Orig. pub. 1852.)

———. 1897. *The Principles of Sociology*. 3rd ed. 3 vols. New York: Appleton. (Orig. pub. 1874–82.)

———. 1898. *The Principles of Ethics*. 2 vols. London: Williams and Norgate.

Sperry, R. W. 1964. "Problems Outstanding in the Evolution of Brain Function." James Arthur Lecture Series, American Museum of Natural History.

———. 1969. "A Modified Concept of Consciousness." *Psychological Review* 76:532–36.

———. 1991. "In Defense of Mentalism and Emergent Interaction." *The Journal of Mind and Behavior* 12 (2):221–46.

Spitzer, W. O., A. J. Dobson, J. Hall, E. Chesterman, J. Levi, R. Shepherd, R. N. Battista, and B. R. Catchlove. 1981. "Measuring the Quality of Life of Cancer Patients: A Concise QL-Index for Use by Physicians." *Journal of Chronic Diseases* 34:585–97.

Stahl, A. B. 1984. "Hominid Dietary Selection before Fire." *Current Anthropology* 25:151–68.

Stander, P. E. 1992. "Cooperative Hunting in Lions: The Role of the Individual." *Behavioral Ecology and Sociobiology* 29:445–54.

Stanley, D. J., and A. G. Warne. 1993. "Sea Level and Initiation of Predynastic Culture in the Nile Delta." *Nature* 363:435–38.

Stanford, C. B. 1992. "Costs and Benefits of Allomothering in Wild Capped Langurs (*Presbytis pileata*)." *Behavioral Ecology and Sociobiology* 30:29–34.

Stanford Research Institute. 1975. *Minimum Standards for Quality of Life*. Menlo Park, CA: Stanford Research Institute.

Stebbins, G. L. 1969. *The Basis of Progressive Evolution*. Chapel Hill: University of North Carolina Press.

Stebbins, G. L., and F. Ayala. 1985. "The Evolution of Darwinism." *Scientific American* 253 (1):72–82.

Steinbruner, J. D. 1974. *The Cybernetic Theory of Decision: New Dimensions of Political Analysis.* Princeton, NJ: Princeton University Press.

Stephens, D. W., C. M. McLinn, and J. R. Stevens. 2002. "Discounting and Reciprocity in an Iterated Prisoner's Dilemma." *Science* 298:2216–18.

Steward, J. H. 1938. *Basin-Plateau Aboriginal Sociopolitical Groups.* Washington, DC: U.S. Government Printing Office.

Stewart, F. 1985. *Planning to Meet Basic Needs.* London: Macmillan.

Stewart, J. 1997. "Evolutionary Progress." *Journal of Social and Evolutionary Systems* 20 (4):335–62.

Stipp, D. 2004. "Climate Collapse: The Pentagon's Weather Nightmare." *Fortune* 149 (3):100–108.

Stock, G. 1993. *Metaman: The Merging of Humans and Machines into a Global Superorganism.* New York: Simon & Schuster.

Stonier, T. 1990. *Information and the Internal Structure of the Universe.* London: Springer-Verlag.

Stoto, M. A. 1992. "Public Health Assessment in the 1990s." *Annual Review of Public Health* 13:59–78.

Strassman, J.E., C. R. Hughes, S. Turillazzi, C. R. Solís, and D. C. Queller. 1994. "Genetic Relatedness and Incipient Eusociality in Stenogastrine Wasps." *Animal Behaviour* 48:813–21.

Streeten, P. 1977. "Distinctive Features of a Basic Needs Approach to Development." *International Development Review* 19:8–16.

———. 1979. "Basic Needs: Premises and Promises." *Journal of Policy Modelling* 1:136–46.

———. 1981. *Development Perspectives.* London: MacMillan.

———. 1984. "Basic Needs: Some Unsettled Questions." *World Development* 12:973–78.

Streeten, P., and S. J. Burki. 1978. "Basic Needs: Some Issues." *World Development* 6 (3):411–21.

Streeten, P. (with S. J. Burki, M. ul Haq, N. Hicks, and F. Stewart). 1982. *First Things First: Meeting Basic Human Needs in Developing Countries.* Oxford: Oxford University Press.

Strogatz, S. 2003. *Sync: The Emerging Science of Spontaneous Order.* New York: Hyperion.

Strohman, R. 2002. "Maneuvering in the Complex Path from Genotype to Phenotype." *Science* 296:701–3.

Strum, S. C. 1987. *Almost Human: A Journey into the World of Baboons.* New York: Random House.

Sullivan, R. 1998. "Sleeping Well." *Life* 21 (2):56–66.

Swenson, R. 1989. "Emergent Attractors and the Law of Maximum Entropy Production: Foundations to a Theory of General Evolution." *Systems Research* 6 (3):187–97.

Swenson, R., and M. T. Turvey. 1991. "Thermodynamic Reasons for Perception-Action Cycles." *Ecological Psychology* 3 (4):317–48.

Symons, D. 1989. "A Critique of Darwinian Anthropology." *Ethology and Sociobiology* 10:131–44.

Szasz, T. 1961. *The Myth of Mental Illness.* New York: Harper & Row.

Szathmáry, E. 1993. "Do Deleterious Mutations Act Synergistically? Metabolic Control Theory Provides a Partial Answer." *Genetics* 133:127–32.

Szathmáry, E., and L. Demeter. 1987. "Group Selection of Early Replicators and the Origin of Life." *Journal of Theoretical Biology* 128:463–86.

Szathmáry, E., F. Jordán, and C. Pál. 2001. "Can Genes Explain Biological Complexity?" *Science* 292:1315–16.

Szilard, L. 1929. "On the Increase of Entropy in a Thermodynamic System by the Intervention of Intelligent Beings" [trans. A. Rapoport and M. Knoller]. *Behavioral Science* 9:302–10.

Tainter, J. A. 1988. *The Collapse of Complex Societies.* Cambridge: Cambridge University Press.

Tansley, A. G. 1935. "The Use and Abuse of Vegetational Concepts and Terms." *Ecology* 16:284–307.

Target Training International, Ltd. 1990. *Personal Interests, Attitudes and Values Assessment.* Scottsdale, AZ: Target Training International.

Tattersall, I. 1995. *The Fossil Trail: How We Know What We Think We Know about Human Evolution.* Oxford: Oxford University Press.

Taylor, J. B. 1998. *Economics.* 2nd ed. Boston: Houghton Mifflin.

Thaler, D. S. 1994. "The Evolution of Genetic Intelligence." *Science* 264:224–25.

Thayer, B. A. 2004. *Darwin and International Relations.* Lexington: The University Press of Kentucky.

Thompson, J. N. 1982. *Interaction and Coevolution.* New York: John Wiley.

———. 1994. *The Coevolutionary Process.* Chicago: University of Chicago Press

Thomson, G. 1987. *Needs.* London: Routledge.

Thorndike, E. L. 1965. *Animal Intelligence: Experimental Studies.* New York: Hafner. (Orig. pub. 1911.)

Thornhill, N. W., ed. 1993. *The Natural History of Inbreeding and Outbreeding: Theoretical and Empirical Perspectives.* Chicago: University of Chicago Press.

Thucydides. 1962. *History of the Peloponnesian War* [with an English translation by C. F. Smith]. London: W. Heinemann.

Tiger, L., and R. Fox. 1971. *The Imperial Animal.* New York: Holt, Rinehart & Winston.

Tooby, J., and L. Cosmides. 1988. "The Evolution of War and Its Cognitive Foundations." *Proceedings of The Institute of Evolutionary Studies* 88:1–5.

———. 1990. "The Past Explains the Present: Emotional Adaptations and the Structure of Ancestral Environments." *Ethology and Sociobiology* 11:375–424.

Tribus, M. 1961. "Information Theory as a Basis for Thermostatics and Thermodynamics." *Journal of Applied Mechanics* 28:1–8.

Tribus, M., and E. C. McIrvine. 1971. "Energy and Information." *Scientific American* 225 (3):179–88.

Trivers, R. L. 1971. "The Evolution of Reciprocal Altruism." *Quarterly Review of Biology* 46: 35–57.

———. 1985. *Social Evolution*. Menlo Park, CA: Benjamin/Cummings.

Tullock, G. 1970. "Switching in General Predators: A Comment." *Bulletin of the Ecological Society of America* 51:21–24.

———. 1971a. "Biological Externalities." *Journal of Theoretical Biology* 33:565–76.

———. 1971b. "The Coal Tit as a Careful Shopper." *The American Naturalist* 105:77–80.

———. 1976. "Economics and Sociobiology: A Comment." *Journal of Economic Literature* 502–5.

———. 1979. "Sociobiology and Economics." *Atlantic Economic Journal* 7:1–10.

———. 1990. "The Economics of (Very) Primitive Societies." *Journal of Social and Biological Structures* 13:151–62.

Tuomi, J., and M. Augner. 1993. "Synergistic Selection of Unpalatability in Plants." *Evolution* 47:668–72.

Turnbull, C. 1972. *The Mountain People*. New York: Simon & Schuster.

Tylor, E. B. 1889. *Primitive Culture: Researches into the Development of Mythology, Philosophy, Religion, Art and Customs*. New York: Henry Holt. (Orig. pub. 1871.)

Ulanowicz, R. E. 1980. "An Hypothesis on the Development of Natural Communities." *Journal of Theoretical Biology* 85:223–45.

———. 1983. "Identifying the Structure of Cycling in Ecosystems." *Mathematical Bioscience* 65:219–37.

———. 1986. *Growth and Development: Ecosystems Phenomenology*. New York: Springer-Verlag.

UNICEF. 1987. *The State of the World's Children*. New York: Oxford University Press.

United Nations Development Programme. 1990. *Human Development Report 1990*. New York: Oxford University Press. Also available at http://hdr.undp.org/reports/global/1990/en/

———. 2002. *Human Development Report 2002: Deepening Democracy in a Fragmented World*. New York: Oxford University Press. Also available at http://hdr.undp.org/reports/global/2002/en/

Ursprung, H. W. 1988. "Evolution and the Economic Approach to Human Behavior." *Journal of Social Biological Structures* 11:257–79.

U.S. Department of Commerce. 1953. *Statistical Abstract of the United States*. Washington, DC: U.S. Government Printing Office.

———. 1975. *Historical Statistics of the United States: Colonial Times to 1970*. Washington, DC: U.S. Government Printing Office.

———. 1997. *Statistical Abstract of the United States*. Washington, DC: U.S. Government Printing Office.

U.S. Department of Health Education and Welfare. 1970. *Toward a Social Report* [with introductory commentary by W. J. Cohen]. Ann Arbor: University of Michigan Press.

U.S. Environmental Protection Agency. 1973. *The Quality of Life Concept: A Potential New Tool for Decision-Makers*. Washington, DC: U.S. Environmental Protection Agency.

van der Dennen, J. M. G. 1991. "Studies of Conflict." In *The Sociobiological Imagination,* ed. M. Maxwell, pp. 223–41. Albany: State University of New York Press.

———. 1995. *The Origin of War: The Evolution of a Male-Coalitional Reproductive Strategy.* Groningen: Origin Press.

———. 1999. "Human Evolution and the Origin of War: A Darwinian Heritage." In *The Darwinian Heritage and Sociobiology,* eds. J. M. G. van der Dennen, D. Smillie, and D. R. Wilson, pp.163–86. Westport, CT: Praeger.

van der Dennen, J. M. G., and V. S. E. Falger, eds. 1990. *Sociobiology and Conflict: Evolutionary Perspectives on Competition, Cooperation, Violence and Warfare.* London: Chapman and Hall.

Van de Vijver, G., S. N. Salthe, and M. Delpos. 1998. *Evolutionary Systems: Biological and Epistemological Perspectives in Selection and Self-Organization.* Dordrecht, Netherlands: Kluwer Academic Publishers.

Van Gool, W., and J. Bruggink, eds. 1985. *Energy and Time in the Economic and Physical Sciences.* Amsterdam: North Holland.

Van Horn, R. C., A. L. Engh, K. T. Scribner, S. M. Funk, and K. E. Holekamp. 2004. "Behavioral Structuring of Relatedness in the Spotted Hyena (*Crocuta crocuta*) Suggests Direct Fitness Benefits of Clan-Level Cooperation." *Molecular Ecology* 13:449–58.

van Schaik, C. P., M. Ancrenaz, G. Borgen, B. Galdikas, C. D. Knott, I. Singleton, A. Suzuki, S. S. Utami, and M. Merrill. 2003. "Orangutan Cultures and the Evolution of Material Culture." *Science* 299:102–5.

Van Valen, L. M. 1976. "Energy and Evolution." *Evolutionary Theory* 1:179–229.

Vanhanen, T. 1992. *On the Evolutionary Roots of Politics.* New Delhi: Sterling.

Vayda, A. P. 1995. "Failures of Explanation in Darwinian Ecological Anthropology: Part I." *Philosophy of Social Science* 25:219–49.

Vayda, A. P., and B. J. McCay. 1975. "New Directions in Ecology and Ecological Anthropology." *Annual Review of Anthropology* 4:293–306.

Veblen, T. B. 1899. *The Theory of the Leisure Class: An Economic Study of Institutions.* New York: Macmillan.

———. 1919. *The Place of Science in Modern Civilization and Other Essays.* New York: Huelosch. Reprinted with a new introduction by W. J. Samuels (New Brunswick, NJ: Transaction.)

Vedeld, P. O. 1994. "The Environment and Interdisciplinarity: Ecological and Neoclassical Economical Approaches to the Use of Natural Resources." *Ecological Economics* 10:1–13.

Vermeij, G. J. 1987. *Evolution and Escalation: An Ecological History of Life.* Princeton, NJ: Princeton University Press.

Vetter, R. D. 1991. "Symbiosis and the Evolution of Novel Trophic Strategies: Thiotrophic Organisms at Hydrothermal Vents." In *Symbiosis as a Source of Evolutionary Innovation,* eds. L. Margulis and R. Fester, pp. 219–45. Cambridge, MA: MIT Press.

von Bertalanffy, L. 1950. "The Theory of Open Systems in Physics and Biology." *Science* 111:23–29.

———. 1952. *Problems of Life: An Evaluation of Modern Biological Thought.* New York: John Wiley. (Orig. pub. 1949.)

———. 1956. "General System Theory." In *General Systems (Yearbook of the Society for the Advancement of General Systems Theory)*, ed. L. von Bertalanffy and A. Rapoport, pp. 1–10. Ann Arbor, MI: Society for General Systems Research.

———. 1967. *Robots, Men and Minds.* New York: George Braziller.

———. 1968. *General System Theory: Foundations, Development, Applications.* New York: George Braziller.

von Clausewitz, C. 1968. *On War.* [Edited and with introduction by A. Rapoport.] Baltimore: Penguin Books. (Orig. pub. 1832.)

Von Foerster, H. 1966. "From Stimulus to Symbol: The Economy of Biological Computation." In *Sign, Image, Symbol,* ed. G. Kepes, pp. 42–61. New York: George Braziller.

———. 1980. "Epistemology of Communication." In *The Myths of Information,* ed. K. Woodward, pp. 18–27. Madison: University of Wisconsin Press.

von Wagner, H. O. 1954. "Massenansammlungen von Weberknechten in Mexiko." *Zeitschrift für Tierpsychologie* 11:349–52.

Vromen, J. 1995. *Economic Evolution: An Enquiry into the Foundations of New Institutional Economics.* London: Routledge.

———. 2001. "The Human Agent in Evolutionary Economics," In *Darwinism and Evolutionary Economics,* eds. J. Laurent and J. Nightingale, pp. 184–208. Cheltenham, UK: Edward Elgar.

Wächtershäuser, G. 1988. "Before Templates: Theory of Surface Metabolism." *Microbiological Reviews* 52:452–84.

———. 1990. "Evolution of the First Metabolic Cycles." *Proceedings of the National Academy of Sciences (USA)* 87:200–204.

Waddington, C. H. 1962. *New Patterns in Genetics and Development.* New York: Columbia University Press.

———, ed. 1968. *Towards a Theoretical Biology, vol. 1: Prolegomena.* Chicago: Aldine Press.

———. 1975. *The Evolution of An Evolutionist.* Ithaca, NY: Cornell University Press.

Wade, M. J. 1977. "An Experimental Study of Group Selection." *Evolution* 31:134–53.

———. 1985. "Soft Selection, Hard Selection, Kin Selection, and Group Selection." *The American Naturalist* 125 (1):61–73.

Waldrop, M. M. 1992. *Complexity: The Emerging Science at the Edge of Order and Chaos.* New York: Touchstone (Simon & Schuster).

Wallerstein, I. 1996. "The Global Picture, 1945–90, and The Global Possibilities, 1990–2025." In *The Age of Transition: Trajectory of the World System, 1945–2025,* eds. T. K. Hopkins and I. Wallerstein, pp. 209–25. London: Zed Books.

Wallin, I. E. 1927. *Symbionticism and the Origin of Species.* Baltimore: Williams and Wilkins.

Waltz, K. N. 1975. "Theory of International Politics." In *Handbook of Political Science, vol. 8,* eds. F. I. Greenstein and N. W. Polsby, pp. 161–93. Reading, MA: Addison-Wesley.

———. 1979. *Theory of International Politics.* Reading, MA: Addison-Wesley.

———. 1993. "The Emerging Structure of International Politics." *International Security* 18:44–79.

Wang, N., J. P. Butler, and D. E. Ingber. 1993. "Mechanotransduction across the Cell Surface and through the Cytoskeleton." *Science* 260:1124–27.

Ware, J. E., Jr. 1993. *SF-36 Health Survey: Manual and Interpretation Guide.* Boston: The Health Institute, New England Medical Center.

Warr, P. 1987. *Work, Unemployment and Mental Health.* Oxford: Clarendon Press.

Weber, B. H., and D. J. Depew, eds. 2003. *Evolution and Learning: The Baldwin Effect Reconsidered.* Cambridge, MA: MIT Press.

Weber, B. H., D. J. Depew, and J. D. Smith, eds. 1988. *Entropy, Information, and Evolution: New Perspectives on Physical and Biological Evolution.* Cambridge, MA: MIT Press.

Weinberg, R. A. 1985. "The Molecules of Life." *Scientific American* 253 (4):48–57.

Weiner, J. 1994. *The Beak of the Finch.* New York: Vintage Books.

Weingart, P., S. D. Mitchell, P. J. Richerson, and S. Maasen, eds. 1997. *Human by Nature: Between Biology and the Social Sciences.* Mahwah, NJ: Lawrence Erlbaum.

Weiss, H. 1996. "Desert Storm: Weather Brought Destruction to the First Ancient Civilization." *Science* 36 (3):30–37.

Weiss, H., and R. S. Bradley. 2001. "What Drives Societal Collapse?" *Science* 291:995–1005.

Weiss, H., M.-A. Courty, W. Wetterstrom, F. Guichard, R. Meadow, L. Senior, and A. Curnow. 1993. "The Genesis and Collapse of Third Millennium North Mesopotamian Civilization." *Science* 261:995–1004.

Weiss, P. A. 1969. "The Living System: Determinism Stratified." In *Beyond Reductionism,* eds. A. Koestler and J. R. Smythies, pp. 3–35. London: Hutchinson.

———, ed. 1971. *Hierarchically Organized Systems in Theory and Practice.* New York: Hafner.

Wesley, J. P. 1989. "Life and Thermodynamic Ordering of the Earth's Surface." *Evolutionary Theory* 9:45–56.

West-Eberhard, M. J. 1975. "The Evolution of Social Behavior by Kin Selection." *Quarterly Review of Biology* 50:1–33.

———. 1992. "Adaptation: Current Usages." In *Keywords in Evolutionary Biology,* eds. E. F. Keller and E. A. Lloyd, pp.13–18. Cambridge, MA: Harvard University Press.

———. 2003. *Developmental Plasticity and Evolution.* Oxford: Oxford University Press.

Westermarck, E. 1971. *The Origin and Development of Moral Ideas.* Freeport, NY: Books for Libraries Press. (Orig. pub. 1906.)

Wheeler, W. M. 1927. *Emergent Evolution and the Social.* London: Kegan, Paul, Trench, Trubner.

———. 1928. *The Social Insects: Their Origin and Evolution.* London: Kegan, Paul, Trench, Trubner.

White, L. A. 1943. "Energy and the Evolution of Culture." *American Anthropologist* 45:335–56.

———. 1949. *The Science of Culture: A Study of Man and Civilization.* New York: Grove Press.

———. 1959. *The Evolution of Culture.* New York: McGraw-Hill.

White, R. W. 1959. "Motivation Reconsidered: The Concept of Competence." *Psychological Review* 66:297–333.

———. 1974. "Strategies of Adaptation: An Attempt at Systematic Description." In *Coping and Adaptation,* eds. G. V. Coelho, D. A. Hamburg, and J. E. Adams, pp. 47–68. New York: Basic Books.

Whitehead, H., and L. Rendell. 2004. "Movements, Habitat Use and Feeding Success of Cultural Clans of South Pacific Sperm Whales." *Journal of Animal Ecology* 73:190–96.

Whiten, A., J. Goodall, W. C. McGrew, T. Nishida, V. Reynolds, Y. Sugiyama, C. E. G. Tutin, R. W. Wrangham, and C. Boesch. 1999. "Cultures in Chimpanzees." *Nature* 399:682–85.

Whitesides, G. M., and B. Grzybowski. 2002. "Self-Assembly at All Scales." *Science* 295:2418–21.

Wicken, J. S. 1987. *Evolution, Thermodynamics, and Information: Extending the Darwinian Program.* New York: Oxford University Press.

———. 1988. "Thermodynamics, Evolution, and Emergence: Ingredients for a New Synthesis." In *Entropy, Information, and Evolution: New Perspectives on Physical and Biological Evolution,* eds. B. H. Weber, D. J. Depew, and J. D. Smith, pp. 139–69. Cambridge, MA: MIT Press.

———. 1989. "Evolution and Thermodynamics: The New Paradigm." *Systems Research* 6 (3):181–86.

Wiegele, T. C. 1979. *Biopolitics: Search for a More Human Political Science.* Boulder, CO: Westview Press.

Wiener, N. 1948. *Cybernetics: Or Control and Communication in the Animal and the Machine.* Cambridge, MA: MIT Press.

———. 1950. "Entropy and Information." *Proceedings of the Symposia in Applied Mathematics (American Mathematical Society)* 2:89.

Wilkins, J. S. 2001. "The Appearance of Lamarckism in the Evolution of Culture," In *Darwinism and Evolutionary Economics,* eds. J. Laurent and J. Nightingale, pp. 160–83. Cheltenham, UK: Edward Elgar.

Wilkinson, G. S. 1984. "Reciprocal Food Sharing in the Vampire Bat." *Nature* 308:181–84.

———. 1988. "Reciprocal Altruism in Bats and Other Mammals." *Ethology and Sociobiology* 9:85–100.

———. 1990. "Food Sharing in Vampire Bats." *Scientific American* 262 (2):76–82.

———. 1992. "Communal Nursing in the Evening Bat, *Nycticeius humeralis.*" *Behavioral Ecology and Sociobiology* 31:225–35.

Wilkinson, R. G. 1996. *Unhealthy Societies: The Afflictions of Inequality.* London: Routledge.

———. 2001. *Mind the Gap: Hierarchies, Health and Human Evolution.* New Haven, CT: Yale University Press.

Willhoite, F. H., Sr. 1976. "Primates and Political Authority: A Biobehavioral Perspective." *The American Political Science Review* 70:1110–26.

Williams, G. C. 1966. *Adaptation and Natural Selection: A Critique of Some Current Evolutionary Thought.* Princeton, NJ: Princeton University Press.

———. 1992. *Natural Selection: Domains, Levels, and Challenges.* New York: Oxford University Press.

Wilson, D. S. 1975. "A General Theory of Group Selection." *Proceedings of the National Academy of Sciences (USA)* 72:143–46.

———. 1980. *The Natural Selection of Populations and Communities.* Menlo Park, CA: Benjamin/Cummings.

———. 1997a. "Introduction: Multilevel Selection Theory Comes of Age." *The American Naturalist* 150 (Suppl.): S1–S4.

———. 1997b. "Altruism and Organism: Disentangling the Themes of Multilevel Selection Theory." *The American Naturalist* 150 (Suppl.): S122–S124.

———. 1999. "A Critique of R. D. Alexander's Views on Group Selection." *Biology and Philosophy* 14:431–49.

———. 2002. *Darwin's Cathedral.* Chicago: University of Chicago Press.

Wilson, D. S., and L. A. Dugatkin. 1997. "Group Selection and Assortative Interactions." *The American Naturalist* 149:336–51.

Wilson, D. S., and E. Sober. 1989. "Reviving the Superorganism." *Journal of Theoretical Biology* 136:337–56.

———. 1994. "Reintroducing Group Selection to the Human Behavioral Sciences." *Behavioral and Brain Sciences* 17:585–608.

Wilson, E. O. 1971. *The Insect Societies.* Cambridge, MA: Belknap Press.

———. 1975. *Sociobiology: The New Synthesis.* Cambridge, MA: Harvard University Press.

———. 1985. "The Sociogenesis of Insect Colonies." *Science* 228:1489–95.

———. 1987. "Causes of Ecological Success: The Case of the Ants." *Journal of Animal Ecology* 56:1–9.

———. 1998. *Consilience: The Unity of Knowledge.* New York: Alfred A. Knopf.

Wilson, J. A. 1968. "Entropy, Not Negentropy." *Nature* 219:535–36.

Wilson, J. Q. 1993. *The Moral Sense.* New York: Free Press.

Wimsatt, W. C. 1974. "Complexity and Organization." In *Boston Studies in the Philosophy of Science* vol. 20, eds. K. F. Schaffner and R. S. Cohen, pp. 67–86. Boston: D. Reidel.

———. 1980. "Reductionistic Research Strategies and Their Biases in the Units of Selection Controversy." In *Scientific Discovery: Case Studies,* ed. T. Nickles, pp. 213–59. Boston: D. Reidel.

Wisner, B. 1988. *Power and Need in Africa: Basic Human Needs and Development.* Trenton, NJ: Africa World Press.

Witt, U. 1991. *Individualistic Foundations of Evolutionary Economics.* Cambridge: Cambridge University Press.

———. 1992. *Evolutionary Economics.* London: Edward Elgar.

———. 2003. *The Evolving Economy.* Cheltenham, UK: Edward Elgar.

Wolff, J. 1996. *An Introduction to Political Philosophy.* Oxford: Oxford University Press.

Wolpoff, M. H. 1999a. *Paleoanthropology*. 2nd ed. New York: McGraw-Hill.

———. 1999b. "The Systematics of *Homo*." *Science* 248:1774–75.

World Bank. 1996. *Social Indicators of Development, 1996*. Baltimore: The Johns Hopkins University Press.

World Health Organization. 1980. *International Classification of Impairments, Disabilities, and Handicaps*. Geneva: World Health Organization.

———. 1983. *Depressive Disorders in Different Cultures*. Geneva: World Health Organization.

———. 1992. *Our Planet, Our Health: Report of the WHO Commission on Health and the Environment*. Geneva: World Health Organization.

———. 1995. *Bridging the Gaps*. Geneva: World Health Organization.

Worster, D. 1977. *Nature's Economy: A History of Ecological Ideas*. Cambridge: Cambridge University Press.

Woyciechowski, M. 1990. "Do Honey Bee, *Apis mellifera* L., Workers Favour Sibling Eggs and Larvae in Queen Rearing?" *Animal Behaviour* 39:1220–22.

Wrangham, R. W. 1997. "Subtle, Secret Female Chimpanzees." *Science* 277:774–75.

———. 2001. "Out of the *Pan,* into the Fire: How Our Ancestors' Evolution Depended on What They Ate." In *Tree of Origin: What Primate Behavior Can Tell Us about Human Social Evolution,* ed. F. B. M. de Waal, pp.121–43. Cambridge, MA: Harvard University Press.

Wrangham, R. W., J. H. Jones, G. Laden, D. Pilbeam, and N. Conklin-Brittain. 1999. "The Raw and the Stolen: Cooking and the Ecology of Human Origins." *Current Anthropology* 40 (5):567–94.

Wrangham, R. W., W. C. McGrew, F. B. M. de Wall, and P. G. Heltne, eds. 1994. *Chimpanzee Cultures*. Cambridge, MA: Harvard University Press.

Wrangham, R. W., and D. Peterson. 1996. *Demonic Males: Apes and the Origins of Human Violence*. Boston: Houghton Mifflin.

Wright, R. 1994. *The Moral Animal: Evolutionary Psychology and Everyday Life*. New York: Pantheon Books.

———. 2000. *Non Zero: The Logic of Human Destiny*. New York: Pantheon Books.

Wright, S. 1968–78. *Evolution and the Genetics of Populations: A Treatise*. 4 vols. Chicago: University of Chicago Press.

———. 1980. "Genic and Organismic Selection." *Evolution* 34:825–43.

Würsig, B. 1988. "The Behavior of Baleen Whales." *Scientific American* 258 (4):102–7.

———. 1989. "Cetaceans." *Science* 244:1550–57.

Wynne-Edwards, V. C. 1962. *Animal Dispersion in Relation to Social Behaviour*. New York: Hafner.

———. 1963. "Intergroup Selection in the Evolution of Social Systems." *Nature* 200:623–26.

Yates, F. E., ed. 1987. *Self-Organizing Systems: The Emergence of Order*. New York: Plenum.

Yockey, H. P. 1977. "A Calculation of the Probability of Spontaneous Biogenesis by Information Theory." *Journal of Theoretical Biology* 67:377–98.

Yoffee, N., and G. L. Cowgill, eds. 1988. *The Collapse of Ancient States and Civilizations.* Tucson: University of Arizona Press.

Young, H. P. 2003. "The Power of Norms." In *Genetic and Cultural Evolution of Cooperation,* ed. P. Hammerstein, pp. 389–400. Cambridge, MA: MIT Press.

Yurk, H., L. Barrett-Lennard, J. K. B. Ford, and C. O. Matkin. 2002. "Cultural Transmission within Maternal Lineages: Vocal Clans in Resident Killer Whales in Southern Alaska." *Animal Behaviour* 63:1103–19.

Zinnes, D. A. 1980. "Why War? Evidence on the Outbreak of International Conflict." In *Handbook of Political Conflict: Theory and Research,* ed. T. Gurr, pp. 331–60. New York: Free Press.

Zucker, R. 2001. *Democratic Distributive Justice.* Cambridge: Cambridge University Press.

Index